Plant Structure: Function and Development

A Treatise on Anatomy and
Vegetative Development,
with Special Reference to Woody Plants

THE BLACKBURN PRESS

J. A. Romberger · Z. Hejnowicz · J. F. Hill

Plant Structure: Function and Development

A Treatise on Anatomy and
Vegetative Development,
with Special Reference to Woody Plants

With 57 Figures

DR. JOHN A. ROMBERGER (Emeritus Collaborator)
US Department of Agriculture
Beltsville Agricultural Research Center – West
Climate Stress Laboratory, Bldg. 046 A
10300 Baltimore Avenue, Beltsville, Maryland 20705, USA

And: 2005 Forest Hill Drive, Silver Spring, Maryland 20903, USA

PROF. ZYGMUNT HEJNOWICZ
Department of Biophysics and Cell Biology
Silesian University
ul. Jagiellońska 28
40-032 Katowice, Poland

And: Botanical Institute, University of Bonn,
        Venusbergweg 22, D-5300 Bonn 1, FRG

DR. JANE F. HILL (Volunteer Collaborator)
US Department of Agriculture
Beltsville Agricultural Research Center – West
Climate Stress Laboratory, Bldg. 046 A
10300 Baltimore Avenue, Beltsville, Maryland 20705, USA

And: 8211 Hawthorne Road, Bethesda, Maryland 20817, USA

Reprint of First Edition, Copyright © 1993

Plant Structure: Function and Development
A Treatise on Anatomy and Vegetative Development,
with Special Reference to Woody Plants

ISBN-10: 1-930665-95-4
ISBN-13: 978-1-930665-95-8

Library of Congress Control Number: 2004100206

THE BLACKBURN PRESS
P. O. Box 287
Caldwell, New Jersey   07006 U.S.A.
973-228-7077
www.BlackburnPress.com

Preface

This book is about the developmental anatomy of large, complex plants, particularly of the woody plants that grow and survive for decades or centuries. It is focused on the meaning of that anatomy, that integrated structure, as a determinant of effective function. A pervading theme is that the plant structures that have survived "selection" processes during the eons of organismal evolution, within the larger context of geologic and climatic evolution, are well attuned to biochemical and biophysical principles that determine and define efficient function.

The sets of structure-and-function couples existing in the various plant taxa differ so widely that generalities are often difficult to discern. This diversity is due partly to the broad range of ecological conditions to which higher plant organisms have become adapted under stresses imposed by competition and continual climatic change. It is also due to the tendency of different taxa, with their different complements of inherited information, to respond to similar situations in different ways. Cognizant of this reality, we have tried throughout the book to avoid generalizing too broadly on the basis of data from the relatively small fraction of plant species that have as yet been studied.

This book is intended for those who have already studied the anatomy and development of plants. It is addressed to advanced students, teachers, and researchers in the interrelated fields of botany, forestry, horticulture, and agronomy, and to others having professional interests in the culture of woody plants and the stewardship of ecosystems. The book is also addressed to those, of whatever broader discipline, who invoke theoretical biology in efforts to understand how the higher plant, as a complex organism, copes with its biotic and abiotic environment, grows, develops, and reproduces its kind — while also supporting the animal world (including us). Especially it is addressed to those who, by study and research, seek to narrow the wide gap between the cellular and molecular biology approaches to understanding the format and content of inherited information, and the actual morphogenesis and integrated functioning of higher plant organisms. We feel that presuming to study or practice the various botanical subdisciplines without first gaining a full appreciation of the meanings inherent in developmental plant anatomy is somewhat like attempting to practice medicine without first having acquired an understanding of the anatomy, development, and functioning of the normal human organism.

By citation of a range of significant references, including some less recent ones, we have sought to point out access routes to the origins of ideas, to the rich research traditions, and the stores of basic information in the world literature — lest we become too strongly influenced by recent thinking and work still to be thoroughly evaluated by broad scientific opinion as it evolves over time.

We hope that the ideas and interpretations embodied in this volume will serve as stimulants to those planning and conducting research on plant growth and development, and also that the book will be useful as a guide to the thinking and literature pertinent to advanced courses on plant structure and development; for some it may, perhaps, serve as a textbook.

Though we initially had broader aspirations, this book is focused on vegetative growth and development. Limitations of space precluded a treatment of reproductive development, and of morphogenesis in fruits and seeds. We have, however, included a chapter on embryogeny as the beginning of development of the individual higher plant organism.

October, 1992   The Authors

Acknowledgements

The authors express their appreciation of the encouragement and moral support given by their various present and former colleagues in Poland, Germany, and the United States during preparation of this book. The senior author (JAR) gratefully acknowledges support from the Interacademy Exchange Program between the counterpart organizations in the USA and Poland, which enabled him to work at Silesian University in Katowice with the second author (ZH) in 1981 and 1983. He also acknowledges the gracious hospitality extended to him by that institution. This book was planned and partially drafted during those visits. The senior author further gratefully acknowledges the support and hospitality received from The Institute of Agricultural Sciences, College of Agriculture (A. R. Lublin) in Zamość, Poland, during a visit in 1985. Work on this book was continued both jointly (by ZH and JAR) during such visits, and separately as conditions allowed.

Two of us (JAR and JFH), both having been affiliated with the former Forest Physiology Laboratory of the US Department of Agriculture, Forest Service, in Beltsville, Maryland, are pleased to acknowledge the hospitality and professional recognition afforded by appointments as Volunteer Collaborators at the Climate Stress Laboratory of the US Department of Agriculture, Agricultural Research Service, in Beltsville, Maryland. We also acknowledge, with sincere appreciation, the splendid cooperation and assistance provided to us over the years by the staff members of The National Agricultural Library in Beltsville. Their help was truly indispensable.

The second author (ZH) expresses his appreciation to the administration of Silesian University for enabling him to accept a Visiting Professorship at the University of Bonn, and to the latter institution for allowing him to continue working part-time on this book while in residence there. The senior author (JAR) also expresses his appreciation of hospitality and working space provided to him by the University of Bonn during visits in 1983 and 1988.

The authors collectively express their indebtedness to Dorota Hejnowicz and to Karen L. Parker for their dedicated work in skillfully drawing, or redrawing, the technically accurate illustrations that the book required. We also acknowledge the invaluable help of many of our associates who consented to read critically, review, or proofread the entire manuscript or parts of it. We want especially to mention the contributions of Robert D. Warmbrodt, William Van Der Woode, Claude L. Brown, Robert McGaughy, George C. Martin, and Kitren G. Weis. The senior author also acknowledges the genuinely valuable proofreading done by his wife, Margery D. Romberger.

The authors gratefully acknowledge the permissions to redraw or reproduce illustrations, from copyrighted publications, granted by the holders of those copyrights. We are pleased specifically to mention: Państwowe Wydawnictwo Naukowe, Warsaw (publisher of Hejnowicz 1973, 1980); Wiley and Sons, New York (publisher of Esau 1965a); Wissenschaftliche Verlagsgesellschaft mBH, Stuttgart (publisher of Braun 1963); SIR Publishing, Wellington (publisher of New Zealand Journal of Botany); Blackwell Scientific Publications, Oxford (publisher of Clowes 1961); McGraw-Hill Publishing Co., New York (publisher of Eames and MacDaniels 1947); Springer-Verlag (Wien), Vienna (publisher of Protoplasma); Botanical Society of America, Ames (publisher of American Journal of Botany); Cambridge University Press, New York (publisher of Williams 1975); Cambridge University Press, Sheffield (publisher of New Phytologist); University of Chicago Press (publisher of Botanical Gazette); Carl Winter Universitätsverlag, GmbH, Heidelberg (publisher of Mägdefrau 1951); Forest Products Research Society, Madison (publisher of Wood Science); Society of Wood Science and Technology, Madison (publisher of Wood and Fiber); and The National Research Council of Canada, Ottawa (publisher of Canadian Journal of Botany). Details pertaining to sources of specific illustrations are given in the figure captions and in the References.

Contents

Part II Developmental Anatomy

Part I Functional Anatomy

1 Introduction

1.1 The Subject

This book is about plant anatomy and about the relation between anatomy and function. It is also about how, during ontogeny, the small, simple, and immature becomes large, complex, and mature. The term "plant anatomy" often has had connotations of static description. We stress its dynamic aspects: changes in structure, and how those changes may be responses both to changing mechanical forces operating on and within the plant body and to physiological factors. That is, we focus on organismal growth and development. While we see the species as an idealized representation of a dynamic population of individuals having a common fund of inherited information, possibilities and limitations, we see the individual plant as a dynamic organism that is always responding and becoming, rather than merely being.

Every individual plant, through the details of its structure of cells, tissues, and organs, with their spatial relations and physiological states, brings to the present a great store of information about its past life and environment. This organismal information, accreted in the past, can determine or constrain a plant's responses to conditions of the present. It is from this perspective that we write about anatomy, morphogenesis, and development.

Plant anatomy as a subject focuses on structural differences between cells that are grouped into tissues and organs having a range of functions, positions, and developmental pathways within the plant body. In contrast, attributes that all cells have in common are the proper concern of cytology. We here discuss cytological aspects of cells only as necessary to the exposition of our broader, organism-centered subject.

The great diversity of anatomical detail of plants can be classified into subject fields, each based on one of the following themes: (1) a common plan of structure (comparative anatomy); (2) origin and development of structure on the time scale of organic evolution (phylogeny or phylogenesis); (3) the competence of a structure to perform or satisfy a function (structural physiology, physiological anatomy); and (4) the development of a class of elements (ontogenesis, morphogenesis, embryogeny, cytogenesis, histogenesis, organogenesis, etc.). We concentrate on themes (3) and (4). But, recognizing that in science all subject fields, and even the most carefully delineated questions, are only artificially separable, we also use concepts and terminology derived from fields focused on themes (1) and (2).

We have chosen to address several questions: what are the structural bases for functional processes in plants, how do these structures develop, and what types of control and integration systems impose organ- or organism-centered attributes and constraints on differentiating cells? What can be learned about function by studying the details of structure and its genesis? What can be deduced about conditions in the past by interpreting anatomical details recorded and preserved in wood, in herbarium specimens, or in fossils? In short, what meaning in terms of physiology, function, or past conditions can we find in anatomy?

Although there is no linear relation, the number and difficulty of physiological, mechanical, and anatomical problems increase with plant size, complexity, and longevity, all of which approach maxima in trees. In this sense, trees dominate the plant world. Though we can surely learn much about plants from the more modest and inconspicuous herbs, and even from algae, we should recall that modern herbaceous angiosperms may have evolved from ancestral woody forms (but see Taylor and Hickey 1990). Though the earliest plants to colonize the land were herbaceous, lacking vascular-cambial activity and stems and roots with secondary mechanical and conductive tissue, these ancient herbs may not have been the immediate ancestors of modern herbaceous angiosperms.

Confronted by the dominance of arborescent plants over much of the Earth's land surface, we, as biologists, cannot ignore the challenge to understanding posed by the living presence of trees. And, as we study the anatomy, function, and ontogeny of trees, we learn anew that growth and development are dynamic processes, meaningful only in relation to time. In this sense, plant anatomy is dynamic also. Some of its aspects are commonly illustrated in two or three dimensions, but in reality it has a fourth dimension — time. This temporality is obvious in the structure of tree stems. When its dynamic aspects are considered, plant anatomy becomes inseparable from morphogenesis.

During recent decades, the popularity of biochemical and molecular-biology approaches has led to an emphasis on cellular and subcellular components, processes, and dynamics, and to virtual fragmentation of the organism. Meanwhile, organism-level attributes have been partially eclipsed and neglected, though it is obvious that an organism is not merely a population of autonomous cells. Questions of hierarchy have been neglected. Although reductionism is a valid approach to discovering more about some life processes, it can hardly be applied effectively without appreciation of the hierarchical nature of complex systems, their dynamics, and their controls.

This chapter deals with spatial systems, structures, and hierarchical relations (Sect. 1.2) common to the entire plant and not just to any one functional system, tissue, or organ. The properties of cell walls are discussed in Section 1.4. The major spatial systems of the plant — the apoplasm and symplasm — are described in Section 1.5. General aspects of cell dimensions and geometry, and the concept of hierarchical relations controlling the growth and development of cells, tissues, organs, and ultimately the entire organism, are also treated at some length.

The following chapters are organized into two sets. In the first set, consisting of mostly short chapters, we view the functional systems of the vascular-plant body from a physiological-anatomical perspective, somewhat in the spirit of Haberlandt's (1924) *Physiologische Pflanzenanatomie*. Such functional tissue systems, of course, do not exist with any discreteness in the plant. Instead, most tissues have several complementary functions. Tissues that are primarily "supportive" may also be "protective", and so on. Nevertheless, functional "systems" serve well as a didactic device in organizing information and ideas. Such "systems" also illustrate the existence of a functional hierarchy that is not obvious on the basis of structural considerations alone.

Chapters in the second set are longer. In them, we emphasize cell and tissue differentiation, organ and organismal development, and the genesis of patterns in time and space (morphogenesis). We also encourage awareness of integrating and control systems that, operating above the cellular level, allow complex plants to function, grow, compete, and reproduce themselves.

1.2 Hierarchy: A Concept of Ordered Relations

The principle of hierarchy, as used in describing and analyzing complex structures such as living organisms, is easier to understand on the basis of common sense and intuition than it is to define rigorously. Examples of hierarchies, depending on how we choose to use the term, could be series such as the following: division, class, order, family, genus, species, individual (a taxonomic hierarchy); or organism, organ, tissue, cell, organelle, macromolecule, monomer, atom (a structural hierarchy); or chapter, paragraph, sentence, word, letter (a structural and informational hierarchy).

These examples may be criticized and questioned: do such hierarchies exist primarily in the minds of those who become aware of, arrange, and label them? Perhaps yes, but in the same sense that species and genera exist as idealized concepts in the minds of those who have accepted and applied them. Are such hierarchical series irrelevant to understanding organismal development and function? Probably not. Biological science is now so diffuse and fragmented that the fundamental principles of organismal life tend to become obscured by accumulations of data calling for interpretation. During the study of any complex system, the problem of handling myriads of information fragments is made more manageable by giving some attention to the probable hierarchical organization of the system. Then, one can ask: to what level in what type of hierarchy does this bit of information pertain, and to what others is it irrelevant?

From our perspective, if a system is so complex that as it functions there are, simultaneously, distinguishable activities having such a wide range of attributes that qualitatively different system specifications and techniques are necessary for the study of each level, then the system can be analyzed as a **functional hierarchy**. The same is true for hierarchies based on structure. Each level of a hierarchy consists of an arrangement of constituent parts having certain properties and degrees of freedom. But the integrated properties of all the parts of a level *are not* simply the sum of properties that the same parts have when isolated. Consequently, a level consisting of assemblies of parts from the next lower level can have new properties. Thus, increasing complexity can be accompanied by synthesis of new forms (Whyte 1969). It can be argued that evolutionary processes tend to produce systems that are hierarchically organized, both functionally and structurally (Lumsden 1985; Salthe 1985).

Living organisms, human societies, and even governments and multinational corporations, can be understood as functioning through the interaction of hierarchies of different kinds. As the many activities within systems as complex as living organisms must be integrated in time, and responses must be dynamic rather than static in their capabilities, the elements constituting the different hierarchies must be subject to some authority. There must be a hierarchy of authority, different from the hierarchies of physical structure and of function. Further, there must be a hierarchy of information, with data and information bits at different levels having qualitatively different attributes.

It is evident that in the organism that we call "tree" there are hierarchies of structure, function, information, and seemingly also of authority. The latter, of course, is not seated in a central brain or computer having decision-making capabilities. Even in taxa in which they are present, brains do not direct the embryogeny and organogenesis that produce them. Developing organisms, through their hierarchies of information, operate as automata, and their hierarchies of authority may be regarded as operating through a diffuse biological cybernetic system (Calow 1976). Accordingly, those who study plant development primarily as biologists need also to be aware of research by those who are more oriented towards cybernetics, synergetics, information theory, and computer science (Gatlin 1972; Herman and Rozenberg 1975; Meinhardt 1982; Harte and Lindenmayer 1983; Silk 1984).

In a broad sense, the elements of the different levels of a structural hierarchy, such as a plant body, are related to the levels of a hierarchy of dimensional scale: (1) macroscopic — organs and some tissues; (2) microscopic — most tissues, cells, and organelles; (3) ultramicroscopic and macromolecular — suborganellar structural components of cells; (4) molecular — groupings and alignments of atoms and molecules. At each hierarchical level, the dimensions and properties of the elements of the next lower level are very relevant. As noted above, though, the properties of one level are not reducible to those of the elements of the next lower level, nor do they accurately predict the properties of the elements of the next higher level. This is why studies of ultrastructural and cellular features, though of value and interest, cannot in themselves answer the greater questions of control and integration of development and function at the organ and organismal levels.

In this book, we discuss topics in a convenient, linear order. No ranking or arrangement of topics into a hierarchical series is intended. Though all these aspects coexist and simultaneously contribute to organismal dynamism, in a book such as this they must be discussed linearly. We have attempted to alleviate this limitation somewhat by liberal use of cross references.

1.3 Geometry and Dimensions of Cells

Plant and animal cells have many features in common — but also some fundamental differences. The most striking difference is that the cells of higher animals, rather than having peripheral coats or walls, are generally bounded only by thin, nonrigid membranes, and their naked protoplasts need have no fixed shape. In contrast, plant-cell protoplasts in most situations surround themselves with relatively rigid walls. These walls determine the shapes of plant cells. Enzymatic digestion of the walls releases the protoplasts, which then typically assume a spheroidal shape because of surface tensions in their plasmalemmae. Some plant cells, though, are naturally naked, or nearly so, and are not spheroidal. In these, a well-developed cytoskeleton within the protoplast gives it a discrete shape in spite of surface tensions (Marchant 1982; Parthasarathy et al. 1985).

Plant cells within tissues are polyhedroids. Geometrically, internal cells are characterized by their faces, edges, and vertices. Two cells are in contact along a face. Three cells (with rare exceptions) are in contact along an edge. Four cells (with rare exceptions) are in contact at a vertex. If cells physically separate, leaving intercellular gas spaces at vertices and along edges, they are no longer true polyhedra, because the regions where the walls confront gas spaces are curved. Cells nevertheless can be classified geometrically as though they were strict polyhedra with faces varying in number from 4 to 50 or more. Cells internally located within tissues, if they are of equal size, shape, and orientation, tend to have 14 faces, as was rigorously proved by Lord Kelvin (published as Thomson 1887; see also Dormer 1980). Because cells in real tissues are not of uniform size, shape, and orientation, the theoretical mean number of 14 is attained only approximately.

The gross shape of a cell can be characterized by the relative dimensions of the cell in three orthogonal directions. If the three dimensions are similar, the cell is **isodiametric,** or **cuboidal.** If one dimension is much greater than the others, the cell is **prosenchymatous** (a rarely used term). If one dimension is much smaller than the others, the cell is **tabular.** A cell may become quite irregular in shape if it has multiple regions of localized intrusive growth between neighboring cells.

Cell diameters are limited because wall stress at a given turgor pressure increases in proportion to the diameter (Sect. 4.4.1). The length of cylindrical cells is not so limited and, in the specialized tissues where these cells commonly occur, is highly variable. In homogeneous tissue, the diameters of individual cells vary within a narrow range, as if in accord with the traditional mason's rule that, in a well-built wall, one should never use a stone more than twice as large as the smallest one (Korn 1984).

1.4 Cell Walls

1.4.1 General Aspects

A cell wall is a specialized structure that accrued just outside the plasmalemma. Walls bind cells together to form tissues. Many morphological and physiological characteristics of plant tissues arise from the properties of the walls of their cells. Cells having different major functions tend to have walls of different composition and structure. For example, cell walls of aerial organs exposed to the external atmosphere are specialized so as to minimize movement of water across them (Sect. 2.2.1.2), whereas most other cell walls contain ultra-microspaces that allow slow diffusional transport of water.

Much more rapid flow of water, from roots to leaves, occurs through the lumina of specialized, dead cells, which constitute microcapillary tubes. While living, these cells develop the wall characteristics requisite to their function in transport after death.

Walls of living cells are barriers to many pathogens. Some of the proteins in plant walls can specifically inhibit the wall-degrading enzymes secreted by microbes. Other wall proteins can degrade the microbial wall and still others can release from microbial walls molecules that activate a defense response in the plant (Darvill et al. 1985). Interaction of wall-associated substances also underlies cell-to-cell recognition during pollination and fertilization in plants (Clarke et al. 1985). Wall-bound enzymes may modify incoming extracellular chemical signal substances and thus may act as primary signal receptors able to generate secondary signals.

The wall of an expanding cell is called a primary wall. Typically, it is only about 0.1 μm thick, but it may become several μm thick, as, for example, in collenchyma. Layers of cell wall deposited after cell expansion ceases constitute the secondary wall. Usually, formation of these layers precludes further expansion of the cell surface.

1.4.2 Pits

The primary walls of neighboring cell faces are cemented together by a middle lamella and constitute a double wall between cells (Fig. 1.1A). While the cells are living, this double wall typically is traversed by numerous **plasmodesmata**, which are extremely slender, intercellular, plasmatic channels (Sect. 1.5.1). Small, localized areas of primary wall having numerous plasmodesmata are called **primary pit fields**. Usually, no secondary wall material is deposited over a pit field, and the discontinuity in the secondary wall becomes a **pit**. The field becomes the "bottom", and the margins of the surrounding secondary-wall deposits become the "sides", of the pit cavity (Fig. 1.1B).

Pits are important to the overall functioning of a plant because they have a high frequency of plasmodesmata, which link most of the protoplasts of the plant into a single symplasm (Sect. 1.5.1). In addition, pits between functional tracheary elements are important in the flow of water within the plant (Sect. 7.4.1). Tracheary elements, being dead, have no protoplasts and are not part of the symplasm, and pits between these elements lack plasmodesmata.

Generally, the pits in walls on opposite sides of a middle lamella are arranged in complementary pairs (Fig. 1.1). The double primary wall between the cavities of a pit pair is not simply an open passageway, but is equipped with a closing membrane, the **pit membrane**, which is composed of cellulose. If the secondary wall layer tends to overarch the pit, the pit is **bordered**. If there is no such tendency, the pit is **simple**. Bordered pits occur generally in the walls of tracheary elements, which collectively function in water transport after their protoplasts have died. Living cells with secondary walls usually have simple pits (Fig. 1.1B).

In a bordered pit, that part of the pit cavity beneath the overarching secondary wall, or pit "ceiling", is the **pit chamber**. The opening in the pit "ceiling" is the **pit aperture**. If the secondary wall deposits are very thick, the pit aperture becomes a pit canal between the pit chamber and the cell lumen. Even simple (unbordered) pits in thick-walled cells have such canals. If the canal in a bordered pit is very long, an inner and an outer aperture can be distinguished. In extremely thick walls, the canals of neighboring pits may coalesce near the lumen, giving the appearance of canals branching outwardly.

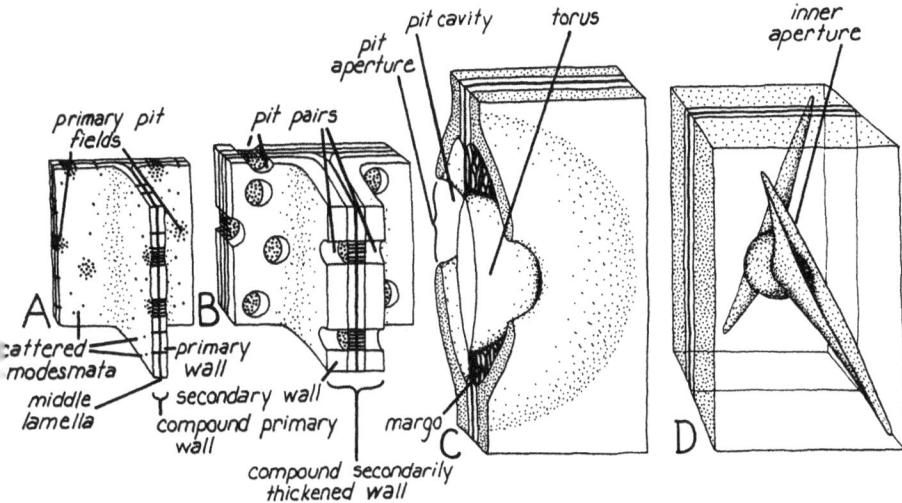

Fig. 1.1 A-D. Cell walls with complementary pit pairs. **A** Double primary wall with primary pit fields. **B** Similar wall areas after deposition of secondary wall layers. Primary pit fields have become simple pits. **C** Section of secondarily thickened walls of tracheary elements with bordered pit pair showing torus and margo. **D** Pair of bordered pits with slit-like inner apertures

Typically, the outer (first-formed) apertures of pit canals are round, whereas the inner apertures, formed after extensive secondary wall thickening, are elliptical or slit-like. The pit canals thus tend toward the form of somewhat flattened funnels (Fig. 1.1D). The transition from round to oblong or slit-like apertures, as secondary walls increase in thickness, is due to the predominantly helical orientation of microfibrils in the secondary-wall lamellae (Sect. 19.2.1). The prevailing orientation of the microfibrils in a secondary wall thus determines the orientation of the inner pit aperture. The direction of the S_2 helix is the same in all tracheids within a tree, and possibly the same in all trees of a species. Accordingly the microfibrils in two contacting cell walls tend to be inclined in opposite directions. Usually, if a pit pair is viewed tangentially to the cell wall in microtome sections, one pit will be seen as from the cell lumen looking out, and the other from the outside looking in, because the section includes the thickness of the two cell walls. The long axes of the two opposite inner apertures of a pit pair thus seem to form a three-dimensional figure X (Fig. 1.1D).

In tracheids of gymnosperms, the marginal part of the pit membrane of a bordered-pit pair is porous, whereas the central part is solid, and in most species is thickened. In these tracheids, the porous, marginal part is the **margo** and the thickened, central part is the **torus** (Fig. 1.1C). Some authors also use the term "torus" to refer to a central, solid — but unthickened — area (Bauch et al. 1972).

Because pits having a torus-and-margo structure occur mainly in conifer tracheids, where they are involved in translocating water from one tracheid to another, we defer our description of the process by which these specialized pits are formed to our discussion of secondary xylem (Sect. 19.2.1.2). Tori also occur in vascular elements of some angiosperms (Dute and Rushing 1987, 1990). The torus-like structures of some angiosperm fiber-tracheid pit membranes may be eccentrically located and possibly have a function different from that of the thicker tori of gymnosperms (Barnett 1987).

Torus diameter is larger than the effective outer aperture of a pit canal. Because the margo is a flexible mesh, the torus can be pushed against one of the pit-aperture rims and act as a self-closing valve, if there is a large pressure differential between the two sides. The valve action of the torus is important in conifers because the openings in the meshwork of the margo are too large to prevent spreading of air embolisms simply through the surface tension of water if negative pressure in the tracheid is great (Sect. 7.4.3.4).

Sometimes a torus may become irreversibly "glued" to a pit aperture rim, thereby closing the pit, or the pit membrane may be coated by secretion products of xylem parenchyma cells such that its permeability is reduced. This may occur during the formation of heartwood (Sect. 19.5.2.3). The tori in the bordered pits of angiosperm tracheary elements may have a different function. They may help prevent the rather fragile margo membrane from tearing during closure of the valve in vivo, and, incidentally, during aspiration of samples in the laboratory (Dute and Rushing 1990).

1.4.3 Cell-Wall Polymers

Cell walls are often considered to consist of long microfibrils embedded in an amorphous matrix of proteinaceous and carbohydrate materials. A more accurate view probably is that the molecules of microfibrils and matrix materials are bound together either covalently or by hydrogen bonds, and that at least the primary wall is analogous to one giant macromolecular complex (Albersheim 1975; Läuchli 1976).

The microfibrils are long bundles of polymers, the cores of which are cellulose chains consisting of partly crystalline arrangements of ß 1-4-linked glucose residues (Preston 1974, 1979). They are oval or round in transection, and mostly have diameters of 5 to 10 nm. Their length is indefinite because of the merging of one microfibril into another. The surface of an elementary microfibril has numerous accessible -OH groups. These promote binding of other compatible polymers by covalent or hydrogen bonds. Each microfibril may thus be thought of as a central core of partly crystalline cellulose surrounded by an amorphus cortex of chains of matrix polymers.

The carbohydrates of this embedding matrix vary in composition among taxa, but may be highly specific to a particular taxon. They usually consist of mixed polymers of glucose, mannose, galactose, arabinose, xylose, rhamnose, glucuronic acid, galacturonic acid, and other, less common residues. Testimony to the extent of specificity of cell-wall materials is the fact that various pathogens and parasites identify their hosts by finding "recognition sites", characterized by specific spatial arrangements of residues in their cell walls (Darvill and Albersheim 1984).

The various carbohydrate **matrix polymers** are customarily grouped into hemicelluloses and pectins (Selvendran 1985). Both are heterogeneous and branched. **Hemicelluloses** have a long, ß 1-4-linked linear backbone of residues of one kind of saccharide (glucan, mannan, xylan), from which short side chains of other residues protrude. **Pectins**, in contrast, have a backbone of segments of α 1-4-linked galacturonic acid residues connected by residues of other types to which side chains may be attached. Because of their negatively charged galacturonic-acid residues, pectins avidly bind cations, especially Ca^{++}.

Carbohydrate macromolecules of the cell-wall matrix can also be classified according to their physical configuration as: (1) **ribbon-like chains**, which include α or ß 1-4-linked linear backbones (or segments of such) consisting of residues of a single kind of saccharide; (2) **helical chains** based on 1-3-linked polysaccharides; or (3) **extensively branched chains** such as arabinogalactan, with a backbone of 1-3-linked galactose residues, all bearing galactosyl or arabinosyl side branches.

The primary cell wall also contains glycoproteins, which, though small in amount, are important components. These include an insoluble structural wall component, **extensin**. Along its chains, extensin has sequences of four hydroxyproline residues accompanied by a serine residue at one end. Because hydroxyproline is only a minor constituent of the cytoplasm and the plasmalemma, extensin must be a structural component of the wall, not a contaminant. It is noteworthy that hydroxyproline is also a component of collagen, one of the major macromolecules of the extracellular matrix of animal cells.

Extensin putatively functions in controlling cell extension. It is cross-linked and appears to form an independent network in the cell wall. Enhanced synthesis, secretion, and cross-linking of extensin are associated with regulation of cell expansion during normal development and with defense against pathogens (Roberts et al. 1985). Studies with antibody labelling indicate that newly synthesized extensin is added to the wall by intussusception (Stafstrom and Staehelin 1988).

Another group of glycoproteins, the lectins, often occur in cell walls in many types of vegetative tissues as well as in seeds (Nsimba-Lubaki et al. 1986). Lectin molecules have binding sites that "recognize" specific saccharide sequences either in complex sugars or in sugar-containing macromolecules. They function in cell-to-cell recognition within an organism and in recognition of "foreign" macromolecules. Their probable roles include protection against pathogenic and predatory organisms and regulation of pollen compatibility, pollen germination, and related processes (Rüdiger 1984; Etzler 1985).

1.4.4 Water in Cell Walls

Water accounts for 90% or more of the fresh weight of primary cell walls. Of the dry matter, typically only about one-third is cellulose (Frey-Wyssling 1976). Nevertheless, cellulose is a major determinant of water status in the wall. Water cannot enter the crystalline regions of the cellulose microfibrils (the crystallites, or micelles), but most other regions, though permeated by matrix substances, are accessible to water via intermicrofibrillar, interfibrillar, and intercrystallite spaces. Water in all but the smallest of these spaces is mobile and its volume is the "water free-space" discussed in the older literature. Free carboxyl groups (with their fixed negative charges) of wall constituents affect the behavior of both cations and anions in the spaces.

These water-filled spaces are the physical basis of apoplasmic transport within cell walls (Sect. 1.5.2). Resistance to water transport across cell walls is low compared with that imposed by the plasmalemma. Thus, water and small solute molecules diffuse through the walls rapidly. In contrast, macromolecules with molecular weights above 15 000 daltons diffuse through typical plant cell walls only slowly.

1.4.5 Genesis and Ultrastructure of Cell Walls

As the cell wall is a major determinant of cell shape and cell growth, the control of cell-wall assembly is basic to control of morphogenesis. To provide a conceptual framework for understanding the spatial relations and control of wall assembly, we accept several assumptions as being almost self-evident: (1) cellulose microfibrils reinforce and constrain the cell wall; (2) the orientation of microfibrils imposes a directionality on the constraint, and thus also on wall expansion; and (3) the driving force of cell expansion is turgor.

Cell-wall formation is a secretion process in the sense that precursors are prepared inside, and products appear outside, of the plasmalemma. Cellulose is synthesized within **terminal complexes** embedded in the plasmalemma. In many plant cells in both angiosperms and gymnosperms, a terminal complex appears to consist of a **globular** and a **rosette** component, as inferred from electron micrographs of freeze-fractured plasmalemma membranes from cell regions in which cellulose was being deposited (Brown 1985). The globular components are located in the outer "leaflet" of the split plasmalemma, whereas the rosette components are embedded in the inner "leaflet". There is evidence that a globular unit and a rosette (with its six subparticles) are apposed in the intact plasmalemma (Fig. 1.2). The concept of the terminal complex may not apply to all plants, however (Schneider and Herth 1986).

In contrast to *Acetobacter xylinum,* in which cellulose is synthesized at fixed sites in the cell membrane and released into the medium, in eukaryotic cells the synthesizing complexes seem to move in the plasmalemma, which may itself be fluid. Great numbers of these mobile complexes can move over, or with, the surface of the protoplast and systematically encase it in a wall of cellulose microfibrils. If the synthesizing complexes are embedded in the plasmalemma, then the orientation of cellulose microfibrils is related to membrane fluidity and systematically directed membrane flow (Mueller and Brown 1982a,b). A plausible concept is that a microfibril, while being assembled, is anchored in the existing wall at its outer end (formed earlier in time) while its inner end (currently being formed) is continuous with a synthesizing complex embedded in the plasmalemma. The complex moves in or with the plasmalemma, leaving the microfibril trailing behind. If the membrane flow is at least as fast as synthesis, the microfibril will be oriented parallel to the flow. Though partly conjectural, the concept of directed membrane flow as an orienting system is consistent with the "flow patterns" of secondary-wall microfibrils observed around pit fields.

The dynamics of microfibril synthesis are poorly understood. Active synthesizing complexes, at least in their rosette aspect, are rather ephemeral. In *Funaria* protonema tips, each rosette probably persists for only about ten minutes, during which time it produces a unit of 36 glucan chains, each about 10 μm long and consisting of about

20 000 residues (Schnepf et al. 1985). Whether a rosette disintegrates after producing a single such unit, or whether it is able to reorganize and produce another, is uncertain. The relation between the globular complexes in the outer plasmalemma leaflet and the rosettes in the inner one is also unclear (Schneider and Herth 1986).

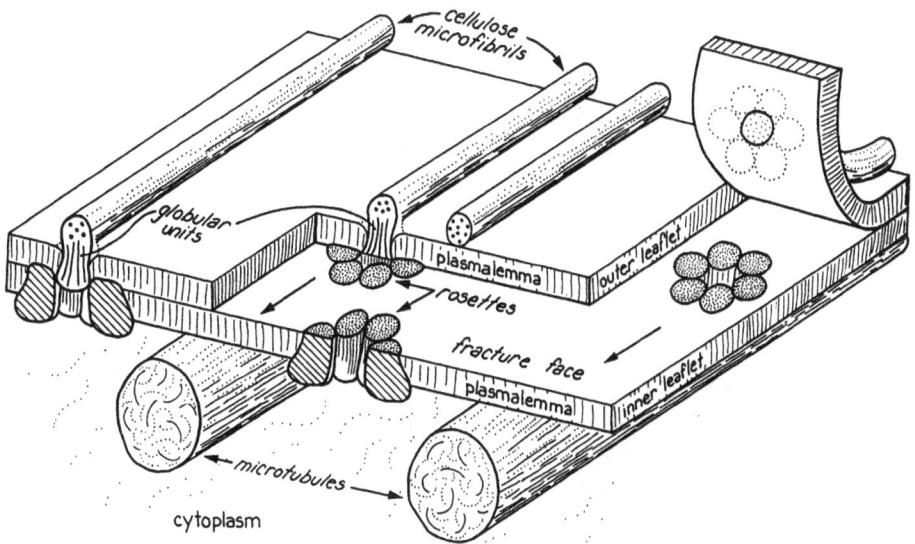

Fig. 1.2. Conjectural scheme representing possible spatial relations between microtubules, plasmalemma leaflets, rosettes, globular units, and cellulose microfibrils, during assembly of the microfibrils and their deposition into the cell wall. *Long arrows* indicate direction of movement of rosettes (and globular units) in or with the plasma membrane, which may be flowing. (Based on information published by Lloyd 1984; Schneider and Herth 1986; Seagull 1989

The system guiding the movement of cellulose-synthesizing complexes in a fluid plasmamembrane, such that the microfibrils are systematically oriented, may involve microtubules (Mueller and Brown 1982b; Lloyd 1984; Giddings and Staehelin 1991). The microtubules may, at times, lie so close to the plasmalemma that "adhesion" is suggested and they might then delineate boundaries of membrane channels within which the rosettes are constrained to travel (Fig. 1.2). It has been demonstrated that, during the deposition of secondary walls in primary-xylem cells, the rosettes are concentrated over the thickenings and are arranged irregularly in tracks between microtubule arrays (Schneider and Herth 1986). However, microtubules seem not always to be necessary for cellulose microfibril formation, nor is there always parallelism between newly formed microfibrils and microtubules.

The orientation of microfibrils may possibly be guided in another way: existing cell-wall lamellae could serve as templates for self-assembly of subsequent lamellae. The efficacy of self-assembly probably depends on the spacing between microfibrils and between lamellae. That spacing in turn depends on the extent of binding of other polymers, especially hemicelluloses, to the microfibrils (Neville 1988).

The importance of hemicelluloses and pectins as structural components was long underestimated. Once thought to serve only as a filling matrix, they are now known to be important structural and functional components of the wall. The synthesis of hemicelluloses and pectins, unlike that of cellulose, is not confined to the outer surface of the plasmalemma. These substances are partially synthesized in dictyosomes (Golgi bodies), are carried by secretory (Golgi) vesicles to the plasmalemma, and are then released to the cell wall. The possibility that matrix materials released into the "periplast" (the space just outside the plasmalemma) undergo self-assembly processes related to the morphogenesis of liquid crystals, and that the new cellulose microfibrils follow the pattern thus established as they are assembled, is well supported (Vian 1982; Neville 1988). In this process, the chains of pectins and hemicelluloses become bound to newly formed cellulose microfibrils by covalent or hydrogen bonds, so that the whole primary wall, or each of its lamellae, approximates an enormous, complex macromolecule.

The multinet hypothesis of wall growth, widely accepted as plausible, is based on the fact that the primary wall is polylamellate. This hypothesis proposes that recently deposited microfibrils are mostly parallel to each other and are transversely or obliquely oriented relative to the major axis of growth. During wall growth, these microfibrils are thought to be passively reoriented and to become nearly parallel to the major axis (Preston 1974, 1982). However, the structure of many polylamellate walls is somewhat different from that proposed by the multinet growth hypothesis, in that there is a so-called **helicoidal** pattern similar to that in many animal exoskeletal systems (Neville and Levy 1985).

A helicoidal cell wall consists of a stack of lamellae with microfibrils parallel within each lamella but with a small angle between microfibrils of successive layers. Indeed, helicoidal structure can be represented by a stack of planks or rods, each of which is rotated, in the same direction, through a small angle relative to its neighbor beneath (Fig. 1.3A). A reliable indicator of helicoidal structure is that a cut made obliquely with respect to the layers results in an "arced" (or "herringbone") pattern (Fig. 1.3B). In a helicoidal cell wall, the arcs are made up of short segments of parallel microfibrils from the successively formed lamellae. Each set of arcs evidences a rotation of 180°, which brings the microfibrils into an antiparallel orientation. Two sets of arcs evidence a complete 360° rotation, from parallel to parallel orientation. The chiral sense of the arcs reverses with reversal of the obliquity of a section (compare the upper and lower oblique views in Fig. 1.3C). This is a critical test for verifying helicoidal structure from two-dimensional sections.

The helicoidal wall concept, which is not necessarily irreconcilable with the multinet growth hypothesis, allows the primary wall to be an active rather than a merely passive, mechanical participant in determining the direction and extent of growth. An essential feature of the ordered-subunit hypothesis proposed by Roland et al. (1987) is controlled, differential loosening of local regions of lamellae (ordered subunits). The orientational attributes of the various lamellae would appear to control the direction and extent of growth rather than being, as implied by the multinet hypothesis, a consequence of growth (Roland and Vian 1979; Roland et al. 1987).

During expansion of a primary wall with helicoidal structure, the helicoidal ordering of lamellae is progressively distorted. Apparently, the more helicoidally ordered the

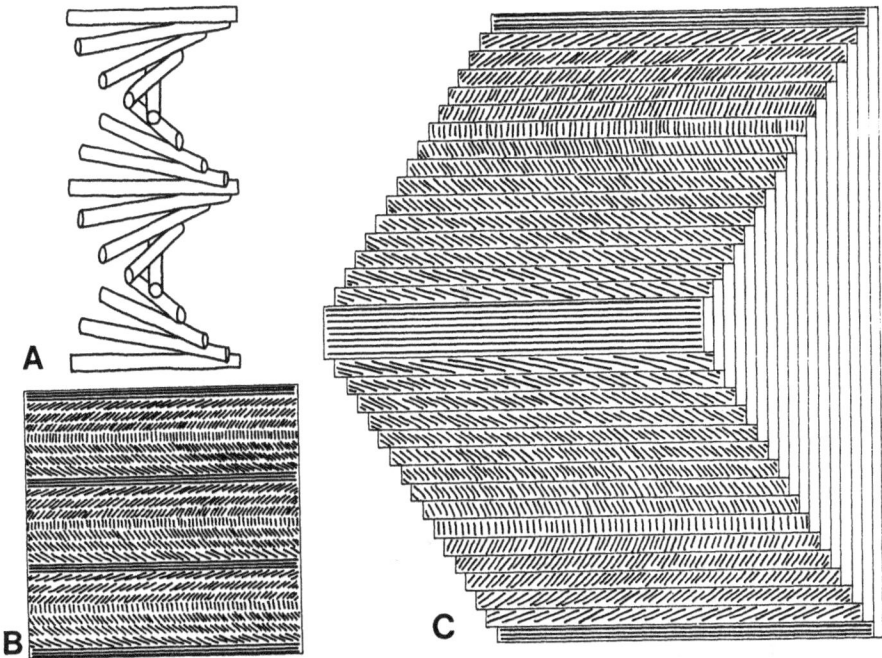

Fig. 1.3 A-C. Basic features of helicoidal wall structure. **A** Rod segments represented in helicoidal arrangement. **B** An oblique section of a polylamellate, helicoidal wall exhibits illusory "arced" or "herringbone" patterns. Each set of arcs evidences a 180° rotation of major orientation of microfibrils in successively deposited lamellae. **C** Reversing the sense of the obliquity of a wall section reverses the chiral sense of the illusory arcs. See text for further interpretation and references

primary wall is, the more potentially extensible it is. In *Phaseolus aureus* hypocotyls, growth ceases when the helicoidal order has been completely dissipated (Vian et al. 1982).

In secondary walls, which are deposited in nongrowing cells, the helicoidal structure is stable. With cell maturation, the molecular network becomes locked by permanent bonds, but may retain some of its original pliability, elasticity, and flexibility.

Helicoidal patterns are not to be confused with the helical courses taken by single microfibrils through space. Helicoidal structures are based on systematically defined relations between lamellae successively deposited in the same locality. In the analogy of stacked rods, "helicoidal" refers to the manner of stacking, whereas "helical" would pertain to the long-distance course of a single rod (Roland et al. 1987). Structures with helical components are not necessarily helicoidal. For example, in xylem fibers and tracheids, the secondary-wall layers may be grossly helical, but are not helicoidal. Nevertheless, polylamellate helicoidal transition zones may occur between the various secondary-wall layers (Roland and Mosiniak 1983; Vian et al. 1986).

Helicoidal structure has been observed in representatives of plant taxa from algae to angiosperms. It has been found in a wide variety of cell types, including parenchyma, collenchyma, epidermal cells, tracheids, and sclereids, though it does not occur in all such cells. A functional aspect of primary cell walls with helicoidal structure is their mechanical versatility in providing strength. Like plywood, helicoidal structures provide strength in all directions, whereas structures based on a narrow range of orientations may allow adaptation to stresses in only a few directions.

1.4.6 Incrustation of Cell Walls

Incrustation of walls results from migration of precursor molecules into the cell wall proper and their subsequent polymerization or condensation therein. Lignin is the most common wall-incrusting substance, contributing greatly to rigidity and strength. It is a mixed polymer of high molecular weight, consisting mostly of units of coniferyl, sinapyl, and coumaryl alcohols. Its many free hydroxyl groups make it readily wettable by water. It is, nonetheless, insoluble. Lignin is a major constituent of cell walls of vascular plants and is conspicuously absent from the nonvascular groups, such as fungi and algae.

The rigidity of walls impregnated with lignin allows them to persist and function in support and transport long after the protoplasts that produced them have disintegrated, leaving the cell lumina as microcapillaries. This persistence of function beyond cell death provides a plant with a conductive and supportive skeleton that need not be discarded and reconstructed with each cycle of the seasons, but that can instead be extended and strengthened in seasonal or other increments.

Incrustation of cell walls with lignin makes possible the efficient function of a super-apoplasm. At the same time, it greatly inhibits water movement in the cell-wall apoplasm. The basis of this will become obvious as we discuss the two apoplasmic spatial systems (Sect. 1.5). Although lignification of cell walls can be general, it often is confined to specifically patterned regions or bands. Patterned lignification can provide barriers to the movement of water within the apoplasm. The lignified (and suberized) Casparian bands of the root endodermal cells are a prime example of such a barrier.

Lignification commonly begins as cell expansion ceases. It generally occurs in the middle lamella, the primary wall, and the various lamellae of the secondary wall. The middle lamella generally accrues the highest lignin content of all. Sometimes, the wall on only one side of the middle lamella is lignified. Incrustation with lignin can occur in precisely delineated local areas. For example, the primary-wall areas between secondary thickenings in protoxylem cells, and the primary pit fields of the prospective perforation plates between vessel members, remain unlignified. When the protoplast of a maturing tracheary element undergoes autolysis, the lignified wall areas are protected while the unlignified areas undergo partial or complete hydrolysis (O'Brien 1970).

1.4.7 Adcrustation of Cell Walls

Substances different from incrusting substances may be deposited by adcrustation on the outer surface of the wall (as is cuticle) or on the inner wall surface, just outside the plasmalemma (as is suberin). Adcrustations initially deposited on the inside of the wall may be covered by further cellulosic lamellae and thus may eventually be located within the thickness of the wall, though they remain physically separate and do not impregnate neighboring lamellae.

The most common adcrusting substances are cutin, suberin, and waxes. Cutin and suberin are polymers of cross-esterified fatty and hydroxy-fatty acids. The waxes that adcrust cell walls are complex mixtures of long-chain alkanes, alcohols, ketones, and fatty acids having chain lengths of 15 to 35 carbon atoms (Northcote 1972). Waxes and cutin are the major constituents of cuticle (Sect. 2.2.1.1), whereas waxes and suberin typically adcrust cell walls in the phellem of the periderm. Suberin is also adcrusted in exodermis, in endodermis in its secondary stage of development (Sect. 17.2.4), and in root apices during dormancy (Sect. 17.4.2).

The deposition of callose, a ß 1-3-linked glucan, only sometimes qualifies as adcrustation. Callose deposits can seal plasmodesmata and block the connecting strands between sieve elements, thus isolating a group of cells or region of tissues from the greater symplasm (Sect. 1.5). This can be a response to wounding or to various stress stimuli (Fincher and Stone 1981). Callose also appears at the surfaces of certain cell types at defined stages of development (Waterkeyn 1981). If cells are wounded, callose can be polymerized locally from glucosyl precursors in a few seconds (Kauss 1987). The polymer quickly becomes highly hydrated, and in this swollen and viscous state it restricts diffusion of solutes.

1.5 The Duality of Spatial Systems Within Plants

Two interlaced and continuous spatial systems coexist within the plant body. One system consists of all protoplasts interconnected by plasmodesmata, and is bounded by continuous plasmalemmae. The other system is extraprotoplasmic, encompassing microspaces within cell walls, some intercellular spaces, and the lumina of dead cells. The continuity of the plasmalemma barrier is a basic principle of the dual-spatial-system concept. Large particles are known to be transferred across the membrane by exo- and endocytosis, so no pores need be invoked to explain passage between the two spatial systems.

The terminology that has been applied to the two spatial systems is inconsistent. Haberlandt (1904) introduced the term "symplast" for the single protoplasmic system that is continuous throughout the living parts of the plant body via plasmodesmata. He considered the "symplast" to be the physical basis for transmission of stimuli in plants. Münch (1930) later called this system the "symplasm" rather than "symplast", while accepting the term "apoplast" (not "apoplasm") for the extraprotoplasmic system. In contrast, Tyree (1970), Lüttge and Higinbotham (1979), and many others have used the fully parallel terms "symplast" and "apoplast", along with the adjectival forms, "symplastic" and "apoplastic".

The term "symplastic", however, has another well-established meaning, referring to growth of cells within tissues and organs in such a way that expansion rates are fully integrated and there is no tearing or sliding along the middle lamellae of adjacent cells

(Priestley 1930). We accept this meaning, and use the term "symplastic" only in this sense. When referring to space within the protoplasmic system, or within or between cell walls, we use the terms **"symplasmic"** and **"apoplasmic"** (and appropriate derivative terms), respectively, as proposed by Erickson (1986). An appreciation of the meaning and physical basis of the concepts of symplasmic and apoplasmic space is basic to understanding material transport and the propagation of disturbances of electrical potential in plants.

1.5.1 The Symplasm

The existence of an organism-level symplasm obviously depends on the myriad plasmodesmata that are characteristic of cells of plants but not of animals. The gap junctions between some closely associated animal cells are only partly analogous (Juniper 1976; Olesen and Robards 1990).

Although plasmodesmata, strictly speaking, are not resolvable by light microscopy, their existence was deduced (Tangl 1879) long before electron microscopes were developed. Certain structures detectable by light microscopy, and long called "plasmodesmata", are apparently tubes or sleeves about 0.2 μm in diameter, which can be interpreted as "macrodesmata". They surround the ultramicroscopic channels that are the actual plasmodesmata (previously called "microplasmodesmata"). The actual plasmodesmata have inside diameters of only about 25 nm (Taiz and Jones 1973). The structure and probable functions of plasmodesmata are further discussed in Section 7.3.

Though most plasmodesmata are formed when cell plates are organized during cytokinesis (Sec.12.2.1), they may also appear de novo in existing walls (Kollmann et al. 1985; see also Sects. 12.2.3.1; 14.1.5.1). Plasmodesmata may sometimes be lost during cell maturation.

A small fraction of higher-plant cells do not have plasmodesmatal connections with their neighbors. Usually this is not due to random variation, but is developmentally determined. Within a complex plant body, there may be small communities of cells constituting sub-symplasms that are isolated from the major organism-level symplasm by discrete apoplasmic boundaries. A common example is mature stomatal guard cells, which are not connected with adjacent, non-guard epidermal cells, although the two guard cells of a pair may be interconnected by protoplasmic strands larger than ordinary plasmodesmata. Another example is the symplasmic separation of one organismal generation from the next (Sect. 13.6).

In addition, transfer cells, which are especially adapted to extensive symplasm-apoplasm exchange (Sect. 1.5.3), often have no plasmodesmata on the cell faces that have the characteristic wall ingrowths. For example, companion cells in the phloem in minor veins of leaves may have transfer zones, instead of plasmodesmata, on the faces confronting the mesophyll.

In angiosperms, protoplasmic strands larger than plasmodesmata typically pass through pores in the sieve plates of adjoining sieve elements in the phloem (Sects. 7.2.2; 20.3.1.1). These connecting strands may have diameters as large as 14 μm. However, in the sieve areas of some primitive vascular plants, these strands may be so slender that they are not readily distinguishable from plasmodesmata.

We consider functioning sieve elements to be part of the symplasm, though not all botanists do so. Sieve elements are undeniably a part of the internal spatial system bounded by membranes — which includes the protoplasts of most other living cells as well. Because long-distance transport in the sieve elements is much more rapid than the diffusional symplasmic transport characteristic of protoplasts of other cells, the sieve elements greatly augment the communication system based on plasmodesmata (Esau and Thorsch 1985). To be rigorous, one must differentiate between the very rapid, "phloic symplasmic transport", and the much slower, "extraphloic symplasmic transport".

The symplasmic continuity of most living cells of a plant, except, for example, between stomatal guard cells and adjacent epidermal cells and between one organismal generation and the next, has been demonstrated by several techniques (Spanswick 1976; Erwee et al. 1985). This continuity may sometimes be limited or completely blocked, either as a normal part of development or as a result of wounding followed by localized deposition of callose (Sect. 1.4.7).

1.5.2 The Apoplasm

Within a plant body, space that is not membrane-limited — and thus not part of the organism-wide symplasm or some smaller local symplasm — is apoplasmic space. The apoplasm has three major classes of substituents: ultramicrospaces within the thickness of cell walls, lumina of dead cells, and intercellular spaces. The ultra-microspaces between and within the cellulose microfibrils of the wall are essentially always water filled. Such spaces make up the "water free-space" of the older literature. The lumina of tracheary elements also are usually filled with water, which is continuous with the apoplasmic water in the neighboring walls. But, another fraction of apoplasmic space typically is gas-filled. This usually includes an anastomosing system of intercellular spaces, lumina of some dead cells, and, sometimes, larger cavities or channels. The aqueous apoplasm is usually referred to simply as "apoplasm", while the gas-filled fraction is the "inner atmosphere" or "inner gas space". We will conform to this usage, applying the term "apoplasm" only to the water-filled fraction.

In absorbing organs, such as young roots, the apoplasmic water is continuous with external water, as has been demonstrated by using tracers followed by electron micros-copy (Peterson et al. 1986). In contrast, the apoplasmic water of the shoot typically is discontinuous with external liquid water, even if the shoot is temporarily submerged, because cuticle or other protective and hydrophobic layers provide a barrier (Sect. 2.2.1).

Barriers to water movement within the apoplasm may functionally divide the apo-plasm into two domains (Epel and Bandurski 1990). Common examples of barriers are Casparian bands (Sect. 17.2.3) in endodermis of most roots and in exodermis of some (Peterson and Perumalla 1984). The existence of this apoplasmic duality compli-cates interpretation of apoplasmic volume measurements in plant organs (Beeson et al. 1986).

The lumina of tracheary elements are microcapillaries and thus are vastly greater in diameter than are the water-filled, ultramicrospaces within the thickness of the cell wall. Tracheary lumina, in fact, comprise most of the apoplasmic space of vascular plants and provide most of the apoplasmic water-transporting capacity. Throughout the remainder of this book, we apply the term **super apoplasm** to the lumen component of the apoplasm, to distinguish it from the wall component. In all large vascular plants, the leaves are functionally connected with the roots mainly by the many files of tracheary elements constituting the super apoplasm and accounting for much of the bulk of the stem.

1.5.3 Transfer Between Apoplasm and Symplasm

The plasmalemma controls transport between symplasm and apoplasm. Here "control" implies discrimination as to molecular species as well as nonspecific inhibition or stimulation. Cells specialized for apoplasm-symplasm exchange have a structural adaptation that increases the plasmalemma surface available for this exchange. These cells, called **transfer cells**, have extensive ingrowths of wall material and consequently have infoldings of the plasmalemma that greatly increase the effective area of membrane (Gunning and Pate 1969; Pate and Gunning 1972).

Transfer cells occur in epidermis, phloem and xylem parenchyma, pericycle, some glands, some reproductive organs, and elsewhere. They are quite evident at interfaces between sporophytic and gametophytic generations. The wall ingrowths occur in the secondary wall only and thus are formed after cell expansion is complete. Aside from greatly increasing the plasmalemma surface, the secondary-wall ingrowths, which may extend deeply into the cytoplasm, may also greatly increase the local volume of apoplasmic space, thus increasing the amount of water and solutes that can be held near the region of plasmalemma immediately available for apoplasm-symplasm transfer. The structure and function of transfer cells are discussed further in Section 3.3.1.

Transport between apoplasm and symplasm can also be controlled or redirected by layers of living cells having walls locally impregnated by lignin and suberin, which block apoplasmic transport through the wall microspaces. The prime example of such a layer is the endodermis in roots (Sect. 17.2.3). The lignified and suberized Casparian bands of the endodermis prevent apoplasmic "leakage" from or to the vascular tissue of the root across the cortex. Transport that would otherwise be apoplasmic is then diverted to a symplasmic route and brought under control of the protoplasts of the endodermal cells.

1.6 The Morphological Classification of Tissues

We here present a brief morphological survey of tissues, primarily to introduce some terminology. More detailed descriptions will follow when appropriate and relevant to the understanding of development and active functioning of specific tissues.

A tissue is an assemblage of cells distinguishable by structure, position, origin, or developmental phase; it collectively performs a function and usually is physically continuous. If the cells constituting a tissue are contiguous, they may form layers, sheaths, strands (or bundles), groups, or simply a "ground" mass. However, if cells that are similar based on structural or functional criteria are not physically together, then they cannot be considered an assemblage. Spatial distribution of structurally and functionally similar cells in a plant body thus may be either **contagious**, in which similar cells occur in sets or groups, or **repulsive**, in which a cell (or sometimes a pair of cells) is notably different from its immediate neighbors. Logically, then, a tissue might consist of contiguous and functionally similar cells in one locality, or of all cells in the plant body that collectively perform a specific function. For example, "sclerenchyma" can denote either a local complex of cells, as in some seed coats, or all the sclereids and fibers in the plant. In this book, we use the term "tissue" in its contagious sense only, reserving the term "idioblasts" for repulsively differentiated cells — that is, scattered cells different from their neighbors.

If developmental phase, or extent of differentiation, is the leading criterion for defining a tissue, then tissues can be classified as meristematic, differentiating, or mature. Meristematic tissues consist of cells that retain their ability (sometimes latently) to divide. They can be further classified, according to their position within the plant body, as apical (at the tips of roots, shoots, or appendages), lateral (located periclinally, beneath an organ surface), or intercalary (between nonmeristematic segments, as in some leaf bases). It is customary to consider the apical meristems as primary meristems. These meristems typically consist of lineages of cells that have maintained continuous meristematic potential since embryonic development began, though their activity may have been episodic (Sects. 14.1.3.2; 17.1). In contrast, lateral meristems are considered to be secondary because their cells typically are not members of such direct and continuously meristematic lineages. Primary meristems produce the cells of the primary plant body, whereas secondary meristems produce the cells of the secondary body.

Three basic tissue systems can be recognized in all vascular plants: epidermal (or rhizodermal), vascular, and ground. The differentiating, but still meristematic, tissues leading to these are the protoderm, the provascular strands, and the ground meristems, respectively. The lateral meristem that continues production of vascular tissues, and typically is developmentally continuous with the provascular tissues, is the vascular cambium.

Epidermal (and rhizodermal) tissues are distinguished by their position on the surface of the primary plant body. Vascular tissues are internal and in the mature state consist of xylem and phloem. Xylem is specialized for rapid long-distance apoplasmic transport (Sect. 7.4), mostly through the lumina of files of dead tracheary elements, which constitute a super apoplasm. Phloem tissue, with its files of sieve elements, is specialized for rapid (usually long-distance) symplasmic transport (Sect. 7.2), as distinguished from the slow symplasmic transport in other living cells.

Older literature on the arrangement of primary vascular tissues in stems and roots includes a concept of "stelar" organization. The term "stele" refers to the usually central, vascular part of a plant axis. The concept is not quite consonant with contemporary knowledge, but many terms adopted when it was accepted doctrine are still used and are defined with reference to a stele. Thus, the endodermis is a sheath of parenchymatous cells surrounding a stele; the pericycle is a layer of parenchymatous cells on the periphery of a stele and in turn surrounded by an endodermis; a leaf gap (in stelae that have a medulla or pith inside) is a parenchyma-filled gap in a stele just at the point of insertion of a leaf trace; and a leaf trace is the extension of a branch of a stele into a leaf. As knowledge of vascular development has accumulated, the stelar concept has gradually been replaced by or modified into the idea that the unit of vascular tissue organization is a continuous "strand" or "bundle" running between root and leaf.

Ground tissue is that part of the primary plant body that is neither dermal nor vascular. In stems and roots, the ground tissue between the vascular tissue and the epidermis (or rhizodermis) is the cortex. Most primary stems and some primary roots have a second, central zone of ground tissue, the pith, that is nearly surrounded by vascular tissue.

On the basis of structural criteria, there are four major tissue types: parenchyma, collenchyma, sclerenchyma, and tracheary elements, many of which also function as sclerenchyma. Parenchyma is the fundamental ground tissue. The walls of parenchyma cells are typically thin, but are sometimes secondarily thickened. Collenchyma cells characteristically have thickened *primary* walls. Sclerenchyma and conducting cells have secondary walls that are both thickened and lignified. Two morphological types of sclerenchyma cells may be distinguished: fibers, which are extremely elongated and have tapered ends; and sclereids, which are of various shapes and may be moderately elongated.

2 Protective Systems — Boundaries

2.1 Introduction

In higher plants, protective systems defend or guard against pathogens and other harmful agents originating outside the boundaries of the organism, and also have other functions. For example, the various dermal tissues — mostly the rhizodermis and epidermis — of the primary plant body are boundary layers, often serving as barriers to free exchange of a variety of substances. They function as multicellular "membranes" separating the ordered, low-entropy world of the society of cells that is the organism from the less ordered, high-entropy world outside. "Protective", in this broader sense of controlling passage across the organismal boundary, often includes an inhibition of water loss. Typical land plants minimize water loss while permitting the exchange of essential gases between water-saturated cell surfaces and the external environment. "Protective" thus implies not only passive exclusion and defense against invaders, but also a selective control of cross-boundary traffic. In this sense of the term, protective systems intergrade with absorbing systems at the organism level (Sect. 3.1).

2.2 Dermal Tissues

The **epidermis (rhizodermis** in roots) is the outermost layer of cells in the primary structure of plant organs, except near the root tip, where the outermost layer is part of the root cap, and the rhizodermis begins to differentiate beneath the cap. The interface of epidermal and rhizodermal cells with the external environment makes them somewhat different from cells lying deeper within the plant body. An epidermal or a rhizodermal cell is necessarily different from an internal cell (Dormer 1980), in that it is asymmetric in shape and lives in an asymmetric environment. It has fewer neighboring cells than an internal cell has. These neighbors are mostly other epidermal or rhizodermal cells, but some are internal cells. We consider epidermis first, then appraise rhizodermis in comparison and contrast.

2.2.1 Shoot Epidermis

An essential function of dermal tissue on shoots of land plants is to protect against excessive water loss while still allowing controlled exchange of gases with the external atmosphere. The epidermis gives this protection while also protecting against potentially harmful invading organisms. A major physical component of the epidermal protective system is the layer of cuticle (Sect. 2.2.1.1) that generally covers all the external interface surfaces of the primary body of the shoot. Epidermis commonly consists of several types of cells: ordinary epidermal cells, stomatal guard cells, stomatal accessory cells (Sect. 16.3.4) and trichomes.

Effective epidermal control of gas exchange requires that there be no intercellular spaces between the ordinary epidermal cells. In leaf-blade epidermis, cells characteristically have considerable turgor, enabling this tissue to function mechanically as the hydraulically stiffened envelope of the blade. Because high turgor promotes cell rounding and consequent formation of intercellular spaces along the cell edges, it is not obvious how the ordinary epidermal cells can maintain a continuous envelope.

This becomes clearer if, keeping in mind the epidermis, we imagine a monolayer of bubble-like vesicles with their contents under positive gauge pressure and with their side walls glued together. All walls, including those that are glued together, are stretched. The pressure-induced tensional stresses in the walls generate mechanical forces that tend to separate the vesicles along edges where three vesicles meet. However, the stresses, as well as the mechanical forces, depend not only on the gauge pressure but also on vesicle radius. More specifically, wall stress is proportional to the product of the radius and pressure divided by wall thickness. Accordingly, the smaller the radius, the lower the stress. The situation in the turgid epidermis is similar. Wall stress is low if the radius of wall curvature is small. Thus, one would expect that epidermal envelopes consisting of small cells, or of cells having convoluted shapes, would be best able to sustain increased turgor pressure without pulling apart at the edges.

Surveys (e.g., Linsbauer 1930; Aneli 1975) have revealed that, in a wide range of genera, the anticlinal walls of leaf epidermal cells are sinuous, as observed from the outer leaf surface. Waviness of anticlinal walls is most evident just as the leaf blade is completing expansion. This suggests that sinuous cell walls allow epidermal cells to give adequate hydraulic support to leaves as they are expanding, before the veins are fully effective as mechanical stiffening tissue.

Obviously, morphogenesis of sinuous anticlinal walls requires a special distribution of symplastic growth rates at a multicellular or organ level. The seemingly simple structural pattern of these walls becomes complex if one tries to model the regulation of growth rates required to produce it.

Another structural adaptation for protecting the epidermis from desiccation is the abutting of ordinary epidermal cells on subjacent tissues. This arrangement provides little exposure of the inner walls of these cells to intercellular gas space. Because most epidermal cells are not active in photosynthesis, this is not a disadvantage.

The wall of an ordinary epidermal cell is usually thickest on the outer tangential side (Sect. 3.2.2). In some members of Poaceae, Cyperaceae, Palmaceae, Equisetaceae, and other families, the epidermal walls may accumulate grains or nodules of silica, which decrease their palatability to herbivores.

Most conifer needle-leaves, which differ greatly in morphology from laminar leaves, probably do not derive much mechanical support from a hydraulically stiffened epidermal envelope. In fact, in some conifers, notably *Pinus resinosa*, the epidermal cells of the needles become extremely thick walled and soon die (Gambles and Dengler 1982a). This dead layer presumably functions as protection, and the thickened walls substitute for hydraulic stiffening in mechanical support of the leaf.

Epidermal cells often produce trichomes, which, according to their structure and function, may be hairs (glandular or nonglandular), glands, scales, or papillae. These structures may intergrade with emergences arising from subepidermal cell layers (Uphof et al. 1962; Behnke 1984). Trichomes often are not mere evaginations of a dermal cell, as are root hairs, but may consist of one or more separate cells formed by a meristemoid that arises from an ordinary epidermal cell (Sect. 16.3.3). Epidermal trichomes, according to their structural adaptations, provide several kinds of protection to shoot surfaces (Juniper and Jeffree 1983).

Hairs and scales composed of dead, air-filled cells may decrease transpiration by increasing the depth of the boundary layer of still air near the epidermal surface. Even living trichomes, though they increase the surface susceptible to water loss, can still have boundary-layer effects that reduce losses. A dense stand of air-filled trichomes makes leaves appear pubescent, or "woolly". The trichomes' high albedo, by reducing the thermal load on the living tissue during periods of strong insolation, further reduces transpiration. In some taxa, long, air-filled trichomes are so dense that they form a felt-like layer that protects against desiccating winds.

2.2.1.1 Cuticle and Epicuticular Waxes. Cuticle is complex physically and chemically. Typically, it is multilayered, and overlies and melds with the outer walls of the epidermal cells. It is rich in waxes and cutin. Thickness of the cuticle typically is from 1 μm or less to about 15 μm.

The term "cuticle" in its broad sense includes all layers containing cutins and/or wax (Martin and Juniper 1970, Holloway 1982). The "cuticle proper" comprises the middle to superficial layers, which consist mainly of cutin (Fig. 2.1). The cuticle proper contains embedded wax crystals or bodies, and its surface is typically covered by an "epicuticular" layer of crystalline or amorphous wax. Beneath the cuticle proper is a "cuticular" layer that consists of the outer lamellae of epidermal cell walls incrusted with cutin. The cuticular layer typically is poor in cellulose, whereas other polysaccharides and pectin are well represented. The interface zone between the cuticular layer and the noncutinized part of the primary epidermal cell wall often stains intensely with ruthenium red and is interpreted as a pectin-rich layer. The process of cutinization may continue after epidermal cells cease to expand, giving rise to a secondary cuticle supplementing the primary one.

Cutins, which are high-molecular-weight, lipid polyesters of long-chain, substituted aliphatic acids, are the most characteristic components of cuticle. Monomeric composition varies with the taxon, and also with ontogeny and with position on the plant. Cutins are synthesized in epidermal cells, under the control of specific, localized enzymes. The soluble precursors of cutin pass through the plasmalemma into the wall, then through channels (possibly ectodesmata; see later) in the wall. Next, via some type of pore, they cross the cuticle already present, and condense, or polymerize, into solid form.

Cuticular waxes consist mostly of aliphatic constituents that are readily solubilized by lyophilic solvents. These constituents are based on even-numbered carbon chains from C_{12} to C_{32} and on odd-numbered chains from C_{17} to C_{33}. Some cyclic compounds, including triterpenoids, may also be major constituents of the waxes. Wax content in cuticle increases outwardly, with the outermost, epicuticular layer being almost entirely wax.

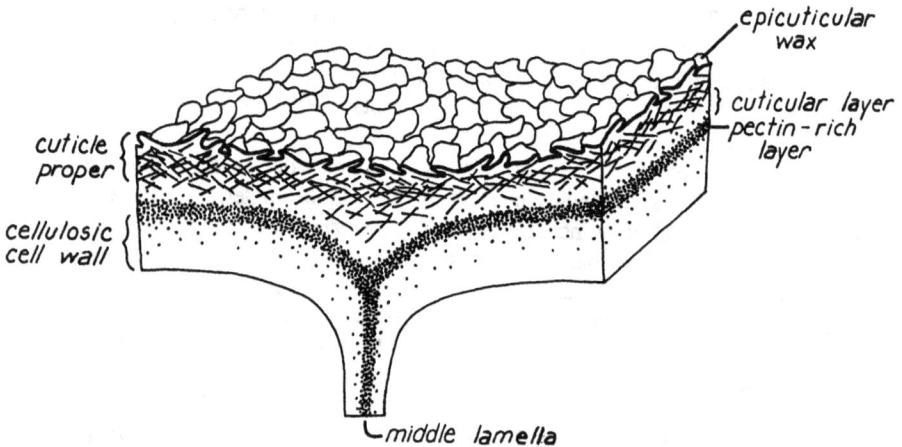

Fig. 2.1. Spatial relations of various layers of outer walls of typical leaf epidermal cells, emphasizing components associated with the cuticle. (Based on schematic interpretation of *Pyrus* leaf cuticle by Norris and Bukovac 1968)

The pectin content of normal cuticle of vegetative organs decreases outwardly. There is a pectin-rich layer, which is continuous with the pectin-containing middle lamella of the anticlinal walls. This explains why in some genera, including *Pyrus*, the cuticle can be detached from the epidermis by treatment with pectinase. The surfaces of the highly specialized cuticles of the stigmatic surfaces of angiosperm flowers tend to be higher in pectins and other carbohydrates than do the cuticles of vegetative organs. These substances probably function in pollen recognition and compatibility systems.

The precursors of epicuticular waxes, dissolved in a volatile solvent, migrate through the cuticle. As the solvent evaporates, these substances crystallize in ways that vary with the composition of the waxes and somewhat with environmental conditions that influence transport of the precursors and evaporation of the solvent.

The channels through which precursors of cutin and epicuticular waxes are transported from the plasmalemma to the surface of the cuticle need not be permanently open and perhaps need provide passage only for certain types of materials. Ectodesmata and cuticular pores may be components of these channels.

Wax surface patterns are so diverse micromorphologically that they may be taxonomically useful (Behnke and Barthlott 1983). The apparent texture of a waxy surface also varies with the degree of magnification at which it is viewed. Macroscopically, the surface is glossy or simply "waxy". Microscopically, the surface may be smooth, granular, grooved, or ridged. Ultramicroscopically, it may appear rough. Several properties of plant surfaces, including light scattering and reflectance, gloss or sheen, and water repellency (Sect. 2.2.1.2), are determined by epicuticular wax attributes (Hanover and Reicosky 1971), in conjunction with characteristics of epidermal hairs, if present.

Though epicuticular waxes are essentially insoluble in water, they can be weathered away. This weathering can be accelerated by wind and by air pollutants. Wax is produced mostly by young leaves, during early stages of leaf expansion. Soon after they mature, leaves may lose the capacity to replace epicuticular waxes that are degraded and weathered away. Thus, old leaves are not as well protected as young ones, and may have higher transpiration rates. At the same time, leaching losses to rainwater increase and resistance to pathogens decreases.

Even the surface cells of the shoot apical regions are protected by a rudimentary cuticle. If growth is rapid, however, the cuticle is likely to be too thin to give adequate protection by itself. This inadequacy is partly countered by developmental adaptations that allow the somewhat older leaf primordia to overarch the younger ones and the apical meristem. Often, too, the thinly cutinized meristems and embryonic shoot organs are enclosed in the sheaths of heavily cutinized scales that are the protective coverings of buds.

2.2.1.2 Cuticle and Water Relations. As noted in the previous section, the subsurface layers of cuticles on vegetative organs may include pectins and other hydrophilic carbohydrate components that are continually bathed in water and are continuous with the cell-wall apoplasm, whereas the cuticle proper and the epicuticular wax layers seal off the apoplasm from the external environment. These cutin and wax layers, being continuous and highly hydrophobic, exert control over water loss and also over absorption and loss of water-soluble materials through the epidermis (Sect. 3.2.2). The effectiveness of epicuticular waxes in restricting transcuticular movement of water and solutes depends on their surface properties (Martin and Juniper 1970), which in turn depend on their mode of deposition or accumulation.

Cuticular transpiration probably accounts for 5 to 10% of the total transpiration from leaves (Martin and Juniper 1970). Even if water stress is severe, some water continues to be lost through the cuticle and the plant eventually wilts. The water, and other small molecules, probably pass through cuticular pores (Sect. 3.2.2), which are polar in that they are lined with -COOH groups and thus have a pH-dependent function. The pores are dynamic structures, apparently not existing in dry cuticular membranes but appearing during hydration (Schönherr 1976). Extrapolation from the results of much research on isolated cuticular "membranes" to actual plant conditions is difficult because the naturally occurring waxes, which affect hydration of the cutin and its permeability to water, regardless of pH, usually are lost during preparation.

Xerophytes generally have thick cuticles and heavy waxy layers. However, a few have little surface wax and only thin cuticles, and apparently have evolved other means of controlling water loss. Both the thickness and the surface texture of waxy layers influence their effectiveness as barriers to water loss. However, permeability of the cuticle to water is not directly correlated with cuticle thickness (Riederer 1991).

Water repellency of leaves is attributable to characteristics of the epidermal-cell surfaces, including waxes and epidermal hairs, if present — or to a combination of these structures. For example, due to a combination of epicuticular waxes and short epidermal hairs, *Lotus* leaves are so unwettable that water drops falling on them behave as do mercury drops on a table top. The water repellency of the lower surface of *Acer pseudoplatanus* leaves has largely been ascribed to the papillose geometry of the epidermal cells rather than to wax or hairs (Holloway 1970). In some species, the rough texture of the surface waxes, revealed ultramicroscopically (Sect. 2.2.1.1), is responsible for a pronounced water repellency of the leaves.

If other surface characteristics are similar, unwettable surfaces function more efficiently than wettable ones as condensing plates that are below ambient temperature (Martin and Juniper 1970), as engineers can confirm. Thus, conifer needles, which are unwettable and have sharp edges along which droplets can "run", are effective condensers and collectors of fog droplets. The needles of *Pinus canariensis,* growing in its native habitat, function well in this respect.

Damaging a cuticular surface even very slightly, as by delicately stroking with a camel's-hair brush, can significantly increase the rate of water loss. Apparently, dust particles act as abrasives, cutting grooves in the wax and often becoming embedded. Once embedded, they act as microscopic wicks from which water evaporates.

Not only can water vapor be lost through the cuticle; but rain can leach organic metabolites and inorganic ions from foliage despite the cuticle (Tukey et al. 1958). Conversely, in polluted environments, some pollutants may accumulate in or pass through the cuticle and subsequently become a factor in animal and human nutrition (Riederer 1991).

2.2.2 Rhizodermis

The mode of origin of rhizodermis (Sect. 17.2.2) differs from that of epidermis, in that root apical-initial cells are deep-seated, whereas shoot apical initials are relatively superficial. That is, shoot epidermal cells differentiate from precursors that were themselves epidermal or protodermal, whereas precursors of rhizodermal cells were earlier overlain by other cells, usually those of a root cap. Nevertheless, in some taxa, deep-seated, rhizodermal precursor cells may constitute a discernable root **protoderm**.

Typically, rhizodermal cells have very thin walls, lack a cuticle, and are covered externally by a layer of mucilage (Chaboud and Rougier 1986). The lack of cuticle in these cells facilitates water movement from the soil to the interior of the plant (Sect. 3.2.3.1). Atypically, some rhizodermal cells produce a thin cuticle if exposed to humid air. This may be enough to reduce evaporative water losses and to make the roots somewhat water repellent. The ability of rhizodermis to produce a cuticle may be important in vegetative reproduction that occurs via adventitious roots that do not immediately contact the soil.

Because they are protected only by thin walls, root hairs (Sect. 3.2.3.1; 17.2.2) would seem to increase the vulnerability of a root to pathogenic invasion. This is not a serious problem, however, because root hairs, even if not invaded, usually function only a short time, then die and disintegrate as the root tip grows away from them. The fate of the whole rhizodermis is similar; it often serves only briefly as a protective layer, then defaults to a more persistent tissue.

2.3 Exodermis, Endodermis, and Metacutis

As the rhizodermis dies and disintegrates, the outer cortical layer may be converted into a second, less ephemeral protective tissue, the **exodermis**, via deposition of suberin and related substances within its cells (Sect. 17.2.4). Sometimes, though, all cortical tissue abaxial to the endodermis dies precociously and collapses. The protective function then devolves directly from rhizodermis to endodermis. This is common in gymnosperms.

Development of exodermis involves suberization and lignification (Sect. 17.2.4). However, even after lignified and suberized secondary walls have rendered the exodermis primarily protective in function, the protoplasts may remain alive. The suberin of exodermis and endodermis differs from the cutin of epidermis primarily in its predominance of longer-chain fatty acids and their derivative alcohols (Kolattukudy et al. 1975).

Exodermis varies in thickness from one to several cells. Its effectiveness in directly protecting against pathogens is not well known. We have observed that it protects only moderately against desiccation if roots are exposed to dry air. Thus, it apparently allows appreciable water loss despite its high suberin content.

Though less ephemeral than the rhizodermis, the exodermis and cortex in roots with secondary thickening eventually are variously ruptured and crushed. The innermost cortical layer, the **endodermis**, then becomes further suberized (Sect. 17.2.4), and assumes the protective function. The endodermal cells, of course, already had lignified and suberized Casparian bands, which made them a barrier to apoplasmic transport.

Thus, the surface of a young root is protected first by rhizodermis, then by exodermis, and finally by suberized endodermis. Close to the tip, just beneath the cap, the nascent external layer is rhizodermis. Farther back, it is exodermis, and still farther back, suberized endodermis. If extension growth stops, the rhizodermal and exodermal zones may become very short. The suberized endodermal zone may then even approach the root cap so closely that the white, succulent, rhizodermal zone, that is usually seen in growing root tips, is not visible.

The root tips of woody plants are more likely to grow episodically than continuously (Sect. 17.4.1). During periods of relative dormancy, the inner part of the root cap and the apical meristem underlying it are protected by **metacutization** of overlying cells (Sect. 17.4.2). In this process, fatty suberinic substances are deposited within the outer cells of the cap. This suberized layer may become continuous with the exodermis, or, through a bridging ring, with the suberized endodermis. Collectively, the bridging ring and the outer cells of the root cap form a protective tissue, the **metacutis** (Plaut 1918; Wilcox 1954).

Eventually, renewed elongation of the root bursts the metacutis. The new, white root segment that emerges is usually sharply distinguishable from the older, darker part. The new segment also is protected, successively, by rhizodermis, exodermis, and, again, metacutis. If this happens repeatedly, the root tip may assume a beaded appearance.

2.4 Periderm: Phellogen, Phellem, Phelloderm

All but the youngest shoots and roots of perennial woody plants are commonly protected by layers of **phellem** (cork). Phellem is part of, and a persistent product of, that complex of secondary tissues collectively called **periderm** (Chapter 21). Typically, periderm consists of a meristem — the cork cambium, or **phellogen** — and two classes of derivatives. By its meristematic activity, the phellogen produces on its adaxial side a parenchymatous tissue, the **phelloderm** (not present in all taxa), and, on its abaxial side, the phellem.

Phellem cells typically become suberized early in their development, and their protoplasts then die. Electron microscopy of secondary walls of phellem cells usually reveals a striking lamellar structure. This structure has been ascribed to alternating layers of suberin (electron dense) and waxes (electron lucent). It has also been ascribed to suberin of differing polarities arranged in alternating layers (Schmidt and Schönherr 1982). Commonly, a thin layer of cellulose, which may be lignified, is deposited over the suberin layers.

Sometimes, layers of phellem elements are separated by layers of nonsuberized, but otherwise phellem-like, cells, known as phelloid cells. In *Betula alba*, the phelloid cells typically contain a white substance, betulin, which gives the bark its familiar white appearance. The high albedo that this substance gives the bark may protect the stem from overheating during periods of strong insolation in winter.

Phellem is relatively impermeable to water and gases. A thick layer of phellem is an effective barrier against pathogens. It also protects against fires, because it is an excellent thermal insulator. Lenticels allow some gas exchange while only slightly reducing the protective qualities of phellem (Sect. 21.5).

In most woody plants, periderm develops during the first years of growth of axial organs, shortly after the vascular cambium has become active. The first periderm functionally replaces the primary protective tissues derived from the epidermis (or rhizodermis) and cortex. The phellogen cells in the sheath of first periderm may undergo anticlinal as well as periclinal divisions, thus allowing the circumference of the periderm to keep pace, at least initially, with diameter growth of the axis. The first periderm then produces multiple layers of phellem or phelloid cells externally, and usually some phelloderm cells internally. In many taxa, the first phellogen eventually becomes nonfunctional, and sequent periderms arise in successively deeper lying tissues of the cortex or secondary phloem. The developmental aspects of periderm are discussed in some detail in Chapter 21.

2.5 Rhytidome, or Outer Bark

In trees, the most evident protective tissue is the outer bark, which is thick, tough, impervious, insulating, and generally effective as a mechanical barrier. Technically, this tissue is the **rhytidome** (from the Greek *rhytidoma*, wrinkle). It includes all the tissues abaxial to the most recently formed (innermost) periderm. These tissues are the phellem, or cork; and phelloderm cells produced by the various phellogens (except the most recent one); and layers of dead phloem tissues that are included between successively initiated periderms. During their dying phase, these tissues commonly serve as disposal sites for excess salts, tannins, and other metabolic by-products, thereby removing these materials from the living parts of the plant. In the rhytidome, these materials still may serve a useful role, increasing the protective competence of this tissue. The formation of rhytidome and its relation to periderm are discussed in Chapter 21.

3 Absorbing Systems

3.1 General Perspectives

The term "absorption", as traditionally used in botany, refers to the processes by which substances from the external environment cross a boundary zone and enter a plant's spatial and physiological systems. In the light of present knowledge of plant structure, more rigorous definitions of absorption would be different for the different levels of the structural hierarchy. That is, at the organism level, absorption pertains to movement of materials, usually in water solution, from the ambient environment into the space grossly within the confines of the organism. At the tissue level, absorption pertains to movement across the plasmalemma from the apoplasm ("nonliving" organismal space) into the symplasm ("living" organismal space). At the level of the protoplast, absorption is not sharply focused because substances can enter from the ambient apoplasm by crossing the plasmalemma, or from neighboring protoplasts via plasmodesmata. Although the traditional use of "absorption" is not rigorous, rigor can be imparted by specifying the type and locale of absorption.

Because this chapter is about the tissue systems and other adaptations specific to absorption, we emphasize organismal, symplasm-apoplasm, and tissue perspectives, but we do not ignore subordinate, cellular aspects.

Several structural principles are common to tissues and to other adaptations that mediate absorption. The dominant principle is maximizing the effective interface surface. A second is location of the absorbing tissues or elements near the relevant interface. A third is the necessity for thin, permeable cell walls, though local mechanical thickenings are permitted. Some absorbing tissues produce and release enzymes or organic acids that dissolve materials not otherwise readily available for absorption, but this is not a general principle.

Of all substances absorbed by plants, water is paramount. The structure and physiology of living plant cells are water based, and water is the solvent vehicle for practically all other substances that are absorbed. The symplasm absorbs CO_2 for photosynthesis, not in gaseous form directly from the atmosphere, but after it has dissolved in the apoplasmic water in the walls of substomatal and other mesophyll cells. The absorption of other gases follows a similar pathway. (The movement of gases within the plant body is discussed in Sect. 9.5.)

Although the water incorporated during growth is at least equal in volume to that of the new tissue produced, hundreds to thousands of times more must be absorbed to replace water lost by transpiration. Typically, most of this water is withdrawn from the soil by peripheral root tissues. It is then transported inward to the super apoplasm of the central cylinder, within which it moves upward into the shoot.

3.2 Structural Aspects of Absorption from Outside the Plant

3.2.1 Embryonal and Cotyledonary Adaptations

Sporophyte ontogeny typically begins with a zygote, which develops into an embryo, enclosed in a seed coat along with available reserves. Because a young plant is not autotrophic until after it germinates, the developing embryo must either depend on reserves stored in its cotyledons or other embryonal storage tissues, or absorb solubilized reserves from nearby nonembryonic tissues in the seed. These tissues usually are the endosperm and/or perisperm in angiosperms, and the megagametophyte in gymnosperms. The embryo itself is an organism separate from the maternal sporophyte and constitutes a separate symplast — though in well-hydrated seeds the apoplasmic space of the two generations may be continuous.

If, during embryogeny, ample reserves have been stored in the cotyledons or in the embryonal axis, no special embryonic tissues are dedicated to absorption of reserves from gametophytic or maternal sporophytic tissues in the seed. In such seeds, water and solutes can be absorbed from the ambient apoplasm over much of the surface of the immature embryo by the protoderm or nascent rhizodermis.

However, if some or most of the reserves have been stored in the endosperm or perisperm or in other tissues that are not part of the embryonal symplasm, then the embryo during germination must have absorbing tissues in close contact with those tissues, in the manner of a parasite. During seed germination in many dicots, the protoderm or immature epidermis of the abaxial surfaces of the cotyledons probably absorbs water and solutes from outside the embryo.

In seeds of monocots, the embryo is laterally attached to a massive, single, cotyledon-like structure, the scutellum, which is especially well developed in grasses. The abaxial scutellar surface is adpressed to the endosperm and has many glandular cells, which, during germination, release enzymes that solubilize endospermal reserves. The same scutellar surface also includes groups of absorbing cells. In some taxa, these have flat or slightly papillose outer walls. Cells with such walls, though only moderately effective in absorption due to their limited surface area, function adequately during the slow germination that is characteristic of Palmaceae and Liliaceae, among other families.

In contrast, in the Poaceae, the scutellar absorbing cells constitute an absorbing epithelium. They are elongated perpendicular to the surface, in a palisade-like arrangement with many intercellular spaces. In *Zea* and *Triticum*, and especially in *Briza* (Haberlandt 1914), the scutellar absorbing surface resembles a rhizodermis with a dense complement of short root hairs, though the "hairs" are actually whole cells rather than evaginations of cells. The vast surface area of these thin-walled scutellar cells allows effective absorption of mobilized reserves during the rapid germination characteristic of grass seeds.

3.2.2 Shoot Tissues and Adaptations: Transcuticular Absorption

The major functions of typical epidermis are to protect mechanically and support hydraulically, and to minimize water loss to the external atmosphere (Sect. 2.2.1). In addition, however, there is plentiful evidence that foliar epidermis can function as an absorbing tissue. Water absorption by foliage has been known to botanists for over a century. Further, application of fertilizers by spraying onto foliage became common agronomic and horticultural practice decades ago. In these instances, empirically based practice preceded elucidation of absorptive mechanisms by basic research. The advent of radioactive isotopes for experimental use quickly allowed proof that various kinds of substances pass across the epidermal cuticle, though the exact route is still not clear.

In discussing the cuticle in this chapter, we emphasize the inward flow of substances absorbed from the external environment, whereas in the preceding chapter we focused on the protective function of cuticle in retarding outward flow of water and solutes. Discussion of the physical and chemical nature of the cuticle is divided between the two chapters. Here, we provide a general description of cuticle, including cuticular pores, through which water and solutes may be absorbed. A more detailed description of chemical structure and wettability properties of the cuticle were included in Section 2.2.1.1, because these properties are implicated in retarding water loss.

The outer walls of epidermal cells are usually thicker than the anticlinal and inner periclinal walls, even before secondary thickening begins. After cell expansion ceases, a secondary wall may be deposited, often not uniformly, and the whole wall may become lignified, especially in xerophytes. The layer of cuticle that overlies the epidermal cells is in turn overlain by epicuticular waxes.

Absorption may occur through a combination of pores in the cuticle and channels (ectodesmata) in the outer walls of epidermal cells. **Ectodesmata** are very slender, ribbon-like or columnar regions extending through the external walls of epidermal cells and trichomes, but not through the overlying cuticle (Franke 1967). In microscopic sections, they appear less dense than surrounding regions. Unlike plasmodesmata, they are not plasmatic, and thus are not extensions of the symplasm.

Great numbers of anticlinally oriented pores about 1 μm in diameter occur in the cuticles of many taxa (Miller 1985, 1986). If ectodesmata were to coincide with these pores (or permeable areas) in the cuticle, then ectodesmata and cuticular pores together could function — via diffusion and mass flow — as effective channels for movement of water and certain solutes (Franke 1969). In addition, because plant cuticles are polyelectrolytes with isoelectric points near pH 3.0 (Schönherr and Huber 1977), absorption via the cuticle and the epidermal cell-wall apoplasm may involve ion exchange. This would be particularly effective in the opportunistic capture of materials available as aerosols or dusts.

In addition to cuticular pores and ectodesmata, the needle leaves of conifers may have further adaptations that mediate absorption of water and solutes. In needle fascicles of *Pinus*, for example, the less mature tissues, at the needle bases, have thin epidermal walls that are cutinized only lightly, if at all (Leyton and Juniper 1963). The bases of the needle fascicles are surrounded by multiple layers of dead scales. Droplets of dew or rain that run down the water-repellant surfaces of the needles can be absorbed first by the basal scales and then by the living tissues of the needle bases. We have observed that in *P. sylvestris* the first epidermis of the leaf base and caulis surface soon dies. The hypodermal layer then produces a thin cuticle. This process may be repeated, and in older fascicles the bases of a needle pair tend to have wedged between them multiple layers of remains of dead cells. These serve as an absorbent pad that can store water until it is absorbed by the living tissues of the leaf bases. However, this adaptation, which allows *P. sylvestris* to grow on relatively dry soils if water is available from dew or mist, also opens a pathway for entry of toxins originating from air pollutants.

These observations provide a rationale for the known sensitivity of *P. sylvestris* to air pollution. In *P. nigra*, in contrast, resin fills the narrow space between the two adpressed adaxial surfaces of a needle pair and no absorbent pad of dead tissues accumulates. Thus, needle bases of *P. nigra* absorb less atmospheric moisture, and their sensitivity to air pollution is lower.

Ordinary, living trichomes of shoot surfaces, if their cuticles are wettable, can absorb some water, but the volumes absorbed are seldom significant to the plant's water economy. However, among the epiphytic Bromeliaceae, water absorption by the multicellular, scale-like trichomes densely scattered over the leaf surfaces has almost completely replaced water absorption by the root system (Uphof et al. 1962). In fact, in the "atmospheric" *Tillandsia usenoides* and closely related species, there are no functional roots, and complex trichomes, each consisting of dozens of cells, are the major sites of absorption of water and solutes. Some of these specialized trichome cells have many invaginations of the plasmalemma, which make absorption more rapid (Dolzmann 1964). Functionally, these cells are transfer cells (Sect. 3.3.1).

3.2.3 Root Tissues and Adaptations

3.2.3.1 Rhizodermis and Root Hairs. The tissue most specialized for absorbing water and solutes from outside the plant is the uniseriate outer layer of young root segments, the **rhizodermis**. This layer is comparable to the shoot epidermis in a topological sense and in lacking intercellular spaces. However, stomata are rare in the rhizodermis, and the external walls of rhizodermal cells, unlike those of epidermal cells, are not notably thicker than the anticlinal or inner tangential walls. Rhizodermal cells also generally lack a cuticle. In some taxa, however, the cellulose lamellae in the outer walls may be incrusted with fatty substances, in ways somewhat resembling cutinization, though waxes typically are lacking. A tendency toward functional cutinization is most evident in plants growing in water or in saturated soils, and is minimal in plants of mesic soils. The rhizodermal surfaces of some taxa, especially grasses, may be covered by a thin mucilage layer having an uncertain function in absorption (Chaboud and Rougier 1986).

The growth habit and growth rate of the root determine the degree of persistence, and the functionally effective area, of living rhizodermis. Generally, rhizodermis does not grow once it has formed root hairs. Accordingly, its ability to accommodate root-thickening growth is limited. Eventually, this growth bursts and fragments the rhizodermis. In dicots, an intact and functional rhizodermis commonly is present only near the tips of the youngest roots, where its longitudinal extent is determined by the relative rates of extension growth and acropetal progression of secondary thickening growth. Its area is thus variable, and this variation may have a seasonal component. In monocot roots, which typically lack secondary thickening, living rhizodermis may be quite persistent, even perennial, and its effective area may be extensive.

Though a rhizodermis efficient in absorption is likely to occur only on young or slender roots, the total rhizodermal area can be very large, partly because the number of such roots in most root systems is enormous. For example, a mature *Quercus rubra* tree may have more than 500 000 000 living root tips (Lyford 1975), and even a four-month-old *Secale cerale* plant may have more than 13 000 000 (Dittmer 1937). Because of these vast numbers of slender, living root tips, and the much smaller numbers of thicker roots, much of the total surface area of the root system is contributed by the rhizodermis. Root hairs further enhance the rhizodermal surface that these many root tips present to the environment.

A root hair (or root trichome) is a slender, thin-walled, tubular evagination from the outer wall of a rhizodermal cell (Sect. 17.2.2). Typically, each cell can form only one root hair, and only about half of the rhizodermal cells have hairs. Root hairs are mostly 5 to 15 μm in diameter and 100 to 1500 μm in length, with a wall thickness generally less than 1 μm. Quite typical root-hair diameters and lengths are: in *Fraxinus lanceolata*, 5 by 370 μm; in *Ulmus pumila*, 10 by 200 μm; and in *Avena sativa*, 13 by 1200 μm. Root hairs are simple protrusions and are never septate as epidermal trichomes may be. Though root hairs are often regarded as ephemeral, they tend to persist as long as the rhizodermis is alive and intact, and in some genera the rhizodermis can accommodate considerable secondary thickening before rupturing and sloughing off (Dittmer 1949).

Although commonly only about half of the rhizodermal cells produce root hairs, the total number present may be very large. A *Secale cerale* plant, for example, was determined to have more than 14 000 000 000 living root hairs, with as many as 100 000 000 new ones formed each day. These root hairs had a total area of about 400 m^2 — somewhat larger than that of all other components of the root system combined (Dittmer 1937). Total rhizodermal area, as enhanced by root hairs, may be several-fold greater than the shoot area from which water can be lost — even including the cell surfaces exposed to gas spaces within the leaves.

Root hairs do not develop in all higher-plant taxa, nor under all conditions in any one taxon. They seldom occur on submerged roots, and are probably less important in absorption in gymnosperms than in angiosperms. In the Pinaceae, root hairs develop as evaginations in the second, or even third, layer of cells beneath the root surface. These hairs are never abundant and have been found only on "long roots". Root hairs have not been observed at all in Araucariaceae, Taxodiaceae, and Cupressaceae (Sect. 17.2.2). In some taxa, the role of root hairs may be assumed by velamen or by the hyphae of root-associated fungi (Sects. 3.2.3.4; 17.8.2).

The rhizodermis meets all criteria of a highly efficient absorbing tissue. It has an extensive area and thin cell walls, and is located at an interface between plant and environment.

3.2.3.2 Exodermis and Cortex.

Rupture and sloughing off of the rhizodermis expose the developing exodermis beneath it. This layer is less well adapted to absorption than to protection. Exodermis, like rhizodermis, lacks intercellular spaces, but its walls are thicker and it typically forms no root hairs (Sect. 17.2.4). The exodermis, being a type of hypodermis, is the outermost layer of the root cortex. It is not to be confused with the endodermis, which has generally been considered the innermost layer of the cortex. Unfortunately, the two terms have not been used consistently in the literature (Guttenberg 1968).

The anticlinally-oriented walls of maturing exodermal cells in roots of most angiosperms develop Casparian bands similar to those in the endodermis (Peterson and Perumalla 1984; Peterson 1988). These bands then block the radial route of apoplasmic transport of absorbed substances. The Casparian bands in the exodermis are deposited in cells farther from the advancing root tip than are the bands in the endodermis. When the rhizodermis loses its continuity, the protoplasts of most maturing exodermal cells deposit their suberin lamellae over the entire walls. There is evidence that suberized root surfaces, which may account for a significant fraction of the total root surface, participate to some extent in absorption. In slowly growing root systems, they may mediate a substantial fraction of it during certain seasons (Kramer and Bullock 1966; Kramer 1969; Zimmermann and Brown 1971).

In young root tips, and possibly even after a suberized exodermis replaces the rhizodermis as the outermost layer, there are three possible pathways for water with solutes to move across the root cortex into the central cylinder:

1. Symplasmic, which involves crossing the plasmalemma in the rhizodermis and again in the central cylinder, at the boundary of the super apoplasm
2. Apoplasmic inwardly to the Casparian-strip barrier in the exodermis and/or endodermis, then symplasmic, at least through those layers, before returning to apoplasmic in the central cylinder, thus involving two plasmalemma crossings, as in (1)
3. Transcellular, which involves many plasmalemma crossings

Results of some studies suggest that the routes are predominantly apoplasmic and symplasmic, with few plasmalemma crossings. However, the low hydraulic conductivity of roots indicates a substantial transcellular component also (Steudle and Jeschke 1983).

3.2.3.3 Mycorrhizal and Cluster (Proteoid) Roots.

The rhizodermis, though ephemeral, is well adapted to absorb water and dissolved mineral nutrients from the rhizosphere, especially if its cells bear root hairs. However, the small volume of soil from which a root tip can absorb these essentials at a particular time can be depleted quickly, and replenishment by diffusion or capillary flow tends to be slow. Many plants counter local depletion by continually extending new root tips into soil volumes beyond those already depleted.

This stratagem, however, requires a continuing investment of biomass into root tissues, especially short or "fine" roots, which are readily shed. Indeed, the turnover of short roots of many trees is probably near 100% per year and constitutes a large sink of fixed carbon — so large that in some coniferous forest types more than half the net primary biomass production is invested in roots (Fogel 1983). The amount of organic matter returned to the soil by root turnover may exceed that contributed by leaf and branch litter (Fogel and Hunt 1979). Root contribution of organic matter is a major factor in producing a rich microflora in soils that otherwise would be deficient in available carbon. Furthermore, soil fungi, if they have access to metabolizable fixed carbon, can absorb and mobilize phosphorus — an essential nutrient, having low mobility — and other nutrients from large soil volumes via their extensive hyphal systems.

Under these conditions, it is not surprising that higher plants and soil fungi have coevolved some **symbiotic associations**, as well as many associations in which the fungi are pathogenic, parasitic, or saprophytic on roots. Rigorous classification of root-fungus associations into discrete types is hardly yet possible, because of incomplete information and many apparent intergradations. Different kinds of associations may supplant one another as ambient conditions change and as the ontogeny of each participating organism proceeds.

The various kinds of mutualistic symbiotic associations, commonly known as **mycorrhizae** (singular: mycorrhiza), between plant roots and soil fungi enable some plants to compete vigorously on sites where they could otherwise hardly survive. We will here discuss mycorrhizal root associations from the perspective of absorption, deferring discussion of more detailed structural and developmental aspects to Sect. 17.8.2.

In a typical mycorrhizal root, the fungal symbiont is so well integrated with host tissues that the whole structure constituting the functional association is regarded as a kind of modified root. The external hyphae associated with all types of mycorrhizae are, functionally, extensions of the host's root system. These hyphae enormously increase the volume of soil from which water and nutrients can be absorbed. The hyphae supplement or functionally replace root hairs and typically are longer, more persistent, and more numerous than root hairs. Unlike root hairs, they are branched. The arrangement and distribution of hyphae on and in the root are different in different types of mycorrhizae, as are the effects that the presence and activity of the fungus have on the root tip. On the basis of such criteria, two major types of mycorrhizae, **ectomycorrhizae** (Sect. 17.8.2.1) and **endomycorrhizae** (Sect. 17.8.2.2), can be distinguished. These types, however, are just convenient morphological categories and have minimal taxonomic relevance.

In ectomycorrhizal root tips, there is a dense, superficial mesh of fungal mycelium enveloping the rhizodermis as a mantle. From this mantle, numerous hyphae extend outward into the soil. Others grow intercellularly into the rhizodermis and cortex, where they usually form a **Hartig net** of coenocytic hyphae (or rarely hyphae septated into cellular strands). These hyphae, which typically are finely branched, envelope the cortical cells like fingers of grasping hands. Their diameters perpendicular to the cortical cell walls may be tenfold greater than in the tangential direction (Kottke and Oberwinkler 1986). The distal parts of such grasping branches contain dense parietal cytoplasm with abundant mitochondria. These hyphal branch tips attach themselves very intimately to the cortical cell walls. Groups of such hyphae may function as do the wall ingrowths in transfer cells of vascular plants (Massicotte et al. 1986; Kottke and Oberwinkler 1987). The cortical cells embraced by the Hartig net hyphae may be of the transfer cell type, especially in early stages of Hartig net formation. Presumably, carbohydrates from the host move into the fungal symbiont across such areas, and water and mineral nutrients derived from the soil are transferred into host cells at the same or other sites. Because of the fungal mantles and frequent radial enlargement of rhizodermal cells, or swelling of cortical cells, ectomycorrhizae are often easily recognizable under low magnification as short root tips that appear swollen. Ectomycorrhizae occur generally in Pinaceae, Salicaceae, Betulaceae, and Fagaceae, and in some genera of many other families. Their presence seems to enhance the ability of these plants to survive and compete in harsh climates and on infertile sites.

Endomycorrhizae, which occur on as many as 80% of all vascular plants, are more widespread than ectomycorrhizae. They occur in most plant families and in many taxa and individuals that also have ectomycorrhizae (Harley and Smith 1983). Endomycorrhizae typically have no distinct mantle and no Hartig net. The fungal hyphae, which are coenocytic, may form a loose weft on the surface and extend branches far out into the soil. Some hyphae directly penetrate living rhizodermal and cortical cells of the host, then indiscriminately form intra- and intercellular hyphal coils and clumps. They also form highly branched, dendritic, intracellular structures in scattered cells. These structures, **arbuscules**, may occupy a large fraction of a cell's volume. A host cell's protoplast surrounds such arbuscules with a highly convoluted plasmalemma which differs structurally and functionally from that at the cell periphery. The fungus, though it has perforated the host cell wall, remains outside the host's cytoplasm in what may be considered as an extension of the apoplast-interfacial matrix (Dexheimer and Pargney 1991). Exchange can readily occur across such a matrix (Bonfante-Fasolo 1984). Functionally, host cells containing arbuscules serve as transfer cells, but by a stratagem not involving wall ingrowths.

Many or most endomycorrhizae also form inter- or intracellular storage vesicles (often containing glycogen or lipids), as well as arbuscules. Accordingly, the term **vesicular-arbuscular (VA) mycorrhizae** has replaced "endomycorrhizae" in much recent literature. Individual arbuscules alter the cells they occupy relatively little. They live only a short time; then arbuscules are formed in other cells.

As a system, endomycorrhizae seem quite efficient. They may live longer than nonmycorrhizal root tips, but can also be shed and quickly replaced by new mycorrhizal root tips as conditions change. Although they sometimes have characteristic colors, VA mycorrhizae typically are not recognizable except by microscopic observation. This is why their very wide distribution was overlooked until recently.

Most members of the Proteaceae (and scattered representatives of other families) cope with relatively sterile soils and irregular water supplies by producing **proteoid**, or **cluster, roots** (Lamont 1982). These roots are ephemeral laterals bearing dense complements (dozens per cm) of short, higher-order laterals, which in turn bear great numbers of long root hairs. The total area available for absorption can exceed that of an equivalent mass of conventional roots by tenfold or more. Plants having proteoid roots are not known to have mycorrhizae.

3.2.3.4 Other Adaptations. In most epiphytic and terrestrial Orchidaceae, as well as in numerous genera of terrestrial and facultatively epiphytic Liliaceae and Amaryllidaceae, the rhizodermis has become modified into a mono- or multiseriate **velamen**. The outermost cells of this tissue may have root hairs (Pridgeon 1987). The cells are short lived and generally thin walled, except for spiral or net-like thickenings that prevent collapse after the protoplasts die. The walls also have many small slits or breaks. The boundary-layer effect of the dead cells, when they are dry and air-filled, reduces evaporative losses from aerial roots or from roots exposed to dry soils. These same cells are quickly saturated with water when it becomes available. The living cells of the exodermis, beneath the velamen, can then absorb water and solutes from the velamen over an extended time. In some orchids, the exodermis has groups of unsuberized, passage cells, which facilitate absorption from the velamen (Benzing et al. 1982). In effect, the velamen functions as a captive rhizosphere. It has a diverse microflora that sometimes includes algae (which is the basis of reports that the velamen can be photosynthetically active). Velamen as an absorbing tissue is most highly developed in epiphytic orchids having most of their roots exposed to the air or embedded in the bark of forest trees of the mesic or near-arid tropics (Sanford and Adanlawo 1973).

Another kind of modification for absorption occurs at the interface between a parasite and a host plant in the approximately 2000 species of hemi- or holoparasites, representing a range of angiosperm families. With a reduced capacity for autotrophic nutrition, or none at all, these plants must acquire most of their organic metabolites, and often water and mineral elements also, from the tissues of their hosts. Hemiparasites, of which the various mistletoes are examples, are often as green as their hosts, but holoparasites tend to have little or no chlorophyll and sometimes no leaves, or even no plant body in the usual sense.

Parasitic angiosperms have supplemented or replaced roots exposed to a rhizosphere with highly efficient absorbing tissues that can be actively positioned within a host in close proximity to living parenchyma and functional vascular tissue. These endophytic structures, collectively referred to as haustoria, are extremely varied and complex in structure (Kuijt 1977). Haustoria are commonly regarded as being highly modified roots, though there is no strong consensus on this point. Many, or most, haustoria undoubtedly are derived from root tissue (Baird and Riopel 1984), but some seemingly arise from hypocotyl (Weber 1987) or shoot tissues (Weber 1980).

Haustoria use combinations of enzymatic and mechanical processes to penetrate the tissues of the host (usually on stems or roots) and then to establish cell-to-cell contact between the xylem of the parasite and that of the host. Typically, a bridge between the super apoplasm of the host and that of the parasite mediates the major flow of water and solutes between the two organisms. Even if continuous xylem pathways from host to parasite are not established, extensive apoplasmic contact between cellulosic walls enables translocation from the host's to the parasite's apoplasm, from which absorption into the parasite's symplasm can occur. Although some phloem elements occur in well established haustoria (Kuijt and Toth 1976), the symplasm of the parasite seems to remain isolated from that of the host (Alosi and Calvin 1985).

The confusing complexity of haustoria arises in part from their polyphyletic origin and convergent evolution toward structures that, though they may be of diverse

histological origin, have similar functions. Another source of complexity and variability is the combined influence of two organisms, an aggressor and a defender, on
the morphogenesis of a haustorium. Added to this is the difficulty of interpreting
three-dimensional structures from two-dimensional sections, especially if tissues of the
host and parasite are not always clearly distinguishable.

In general, haustorial cell walls tend to be quite thick. Collenchyma may even
appear where mechanical function would not seem to justify it. Yet, other walls in
other tissues are extremely thin. A possible explanation is that the thick walls function
in apoplasmic absorption into the parasite, whereas the very thin ones occur in areas
where absorption by the symplasm is more significant (Dobbins and Kuijt 1973).

3.3 Structural Aspects of Internal Absorption and Transfer

3.3.1 Wall Ingrowths and Transfer Cells

In plants, substances are exchanged across barriers between: the outer apoplasm,
which is quite open to the environment; often, an inner apoplasm; and the symplasm.
All living cells probably participate in exchange between apoplasm and symplasm.
However, at sites where the exchange is especially active, special adaptations of the
plasmalemma and secondary cell wall may arise, in the form of **transfer cells** (Sect.
1.5.3).

The term "transfer cell" is well established in the literature, though in some
respects a transfer cell is not really a distinctive cell type. Rather, transfer cells
originate as specializations of other cells. Thus, one can speak rigorously of xylem-
parenchyma transfer cells, phloem-parenchyma transfer cells, and epidermal transfer
cells, but much less rigorously of transfer cells per se. Nonetheless, the general
features of the wall-membrane apparatus of a transfer cell transcend anatomical
differences derived from the origins of the cells themselves (Pate and Gunning 1972).
The salient feature in all is that the plasmalemma (through which the precursor
materials of the ingrowing walls must first pass) is molded by the ingrowths, whatever
their shape.

The stimuli that induce wall ingrowths and thus transfer-cell development probably
arise from a tissue or organ level in the control hierarchy. Most transfer cells show
a polarity in the distribution of wall ingrowths, which often develop on cell faces
confronting tissue, organ, or organism boundaries, but not on other faces. Because of
this asymmetry, it is sometimes more logical to refer to **transfer zones** (within cells)
rather than to "transfer cells". Within a transfer zone, the surface area of plasmalemma confronting the wall apoplasm may be 5- to 20-fold greater than that prevailing
before ingrowths developed on that face (Gunning et al. 1970).

The wall-membrane apparatus in a transfer zone is an exchange facility in a
transport route. Typically, the immediate exchange is between the apoplasm and the
cytoplasm of the transfer cell — which, however, is continuous with the greater
symplasm, via many plasmodesmata. Thus, either in absorption or in secretion, a
front line of boundary transfer zones can serve large regions of symplasm. A transfer-

cell protoplast can probably exercise absorptive and secretory functions for different substances, possibly at the same time (Pate and Gunning 1969; Gunning and Pate 1974).

The wall ingrowths of transfer cells have various configurations, from papillate to labyrinthine, to heavy bars similar to those in the walls of xylem elements. The ingrowths within a taxon may have characteristic structural features, but, in addition, transfer cells at different sites in the same plant may show tissue-specific characteristics superposed on those of the species.

The first transfer zones to appear during ontogeny are in phloem-parenchyma cells in the small veins of the cotyledons of epigeal germinators, or in xylem-parenchyma cells in the cotyledons of hypogeal germinators (Briarty et al. 1970). As the epicotyl grows, transfer cells typically develop in the minor veins of each leaf, especially in herbaceous dicots. These minor-vein transfer cells probably absorb, from the apoplasm, solutes produced by photosynthesis, and then load them into the phloem symplasm (Giaquinta 1983). The beginning of wall-ingrowth deposition in these cells coincides with the beginning of sugar export from the leaf (Pate and Gunning 1972). In variegated leaves, transfer zones generally remain poorly developed in the white areas, which cannot export sugar (Gunning and Pate 1969). The most complex labyrinthine wall ingrowths occur in the xylem- and phloem-parenchyma nodal transfer cells, which occur in higher plants generally. Transfer zones occur more often in shoots than in roots.

Epidermal and rhizodermal transfer cells mediate exchange of solutes between the symplasm and the ambient environment via the local apoplasm, which is hydraulically continuous with the soil solution in roots and with a liquid-gas interface in leaves. Anatomical features do not necessarily reveal whether these cells are specialized to mediate secretion or absorption.

Rhizodermal transfer cells occur in root tips of a range of species growing under iron-deficiency conditions. These may function in proton secretion as well as in absorption (Römheld and Kramer 1983). In addition to rhizodermal transfer cells, roots may also have transfer cells in nitrogen-fixing nodules (Pate et al. 1969) and in association with mycorrhizae (Allaway et al. 1985).

Because there is little or no symplasmic continuity between the parent and offspring generations during sexual reproduction in plants, one expects to find transfer cells along the surfaces of apoplasmic contact between the generations. This has been confirmed (Pate and Gunning 1972; Sect. 6.4.2), and in some plants solutes have been found generally to pass from the parent to the offspring via these cells. In some angiospermous seeds, sets of transfer cells of the parent and offspring confront each other across a common apoplasmic boundary. In effect, there is a secretory-absorptive system operating across an apoplasmic gap between one symplasm and another.

3.3.2 Plasmatubules

In addition to the wall-membrane apparatus of transfer zones, some plants have an alternative or supplemental method of increasing the plasmalemma surface available for solute exchange. Tubular evaginations (as distinct from invaginations) of the plasmalemma into the periplastic space and adjacent, loosened areas of wall occur in various tissues, at sites where solute flux between symplasm and apoplasm is believed to be especially great. These evaginations, which have been termed **plasmatubules** (Chaffey and Harris 1985), may be viewed as a short-term structural modification that is an alternative to more permanent, transfer-cell development (Harris et al. 1982). However, plasmatubules apparently can also form within transfer cells themselves. Thus, the plasma membrane can be extended both by invaginations and evaginations in the same cell (Harris and Chaffey 1985).

4 Supportive Systems

4.1 Introduction

The cells and tissues of a plant organ must withstand various static and dynamic mechanical stresses. These derive from the increasing weight that accompanies growth and from sudden forces imposed by gusts of wind, falling raindrops, and other external agents. The forces acting on a plant organ produce tensions, compressions, shear stresses, and torsions — and the organ must cope with all of them during development, as well as at maturity. Thus, interpretation of the structure of plant organs with reference to mechanical function is a logical part of studies of developmental anatomy.

Supportive systems to a great extent consist of multifunctional tissues. For example, tree stems are supported in large part by the water-conducting elements of the xylem. Parenchyma, which performs an important hydraulic support function in both herbaceous and young woody plants, also has roles in other vital processes.

It is difficult to appreciate fully the complex, organism-level hierarchical controls necessary to build a structural/supportive system composed largely of tissues that also are functional in other systems. These hierarchical controls must, among other things, insure that the tissues are arranged in a near-optimal configuration for performing their several functions. The complexity is probably greater than we can imagine, because the mechanical stresses on a plant seem to represent more than simple burdens that the plant must withstand or endure. Indeed, we suggest that the generation, transmission, and sensing of mechanical stresses are functions of the organism-level system by which the cells and organs communicate (Sect. 12.4.6). Mechanical stresses may, in a sense, transmit information. The frequency and other attributes of hydraulic pressure waves could be the basis of this communication (Sect. 11.3.3). This seems to be a promising area for future research.

4.2 Organs, Tissues, and Cells as Mechanical Components

It is often instructive to look at a plant organ as an engineer would view a work of construction. Long before engineers learned to manipulate structural materials, growing plants were using near-ideal materials for constructing complex beams. By studying plant structure, we can learn much about the relative efficiency and effectiveness of various supportive strategies. Those pioneers of plant anatomy, Sachs and Schwendener, were fully aware of this more than a century ago. However, during most of the twentieth century, these aspects of plant structure were rather neglected, with the notable exception of the work of Razdorskij (1955).

Stems, petioles, and leaf blades, with their complex internal structures, are not homogeneous beams. They generally consist of a system of "stiff" or "hard", frame-like arrangements of sclerenchyma or collenchyma embedded in a matrix of "soft" tissue, mostly parenchyma. These "hard" and "soft" tissues make up the **reinforcing** and **filling** components, respectively, that characterize complex beams. The relative structural importance of "hard" and "soft" tissues has been controversial. Some early botanists interpreted the term "supportive system" narrowly, restricting it to the structurally specialized, hard tissues — collenchyma, sclerenchyma, and secondary xylem (wood). Schwendener (1874), for example, believed that only the mechanical tissues per se actively functioned in support, embedded in a passive matrix or filling of soft tissues. If this were so, then a higher plant would be constructed on the same basic principle as a higher animal: of soft tissues supported by a skeleton of stiff, mechanical components. Contemporary modes and styles, of course, strongly influence prevailing thought: when Schwendener was working, construction engineers were supremely confident of the potential of latticework-beam construction, as exemplified by the style of the Eiffel Tower, built in 1889.

By the time Razdorskij was working, engineers were more enthusiastic about reinforced concrete than about latticework beams. Razdorskij (1955) viewed the stiff, sclerenchymatous or collenchymatous primary-tissue elements of plants as analogous to the steel rods used as reinforcing, and the parenchymatous tissue as analogous to the concrete matrix. Now, we might compare the hard structural elements of a plant organ to bundles of glass fibers embedded in synthetic plastic resins.

To consider plant structures as analogous to complex, reinforced beams, rather than to simple beams or latticework girders, requires a broad interpretation of the term "supportive system". In this broad usage, the term is applied at the organ rather than the tissue level and includes many, or most, of the parenchyma cells also. We now know that "soft" parenchymatous tissue, if turgid, can resist deformation and support a load. The principle is the same as that of a pneumatic tire. The walls of an inflated tire are under tension, and that tension is increased by putting the tire under load. The tire can support the load because its walls resist tension. An ordinary parenchyma cell, in effect, is a bag of fluid under pressure — and the walls of the bag are under tension. The walls can resist this tension because they are reinforced by cellulose microfibrils embedded in, and stiffened by, hemicelluloses and lignins.

Hydraulic-support systems may consist entirely of parenchyma-cell units. Hydraulic parenchyma is the common mode of support in young stem tips. Consequently, if water supply fails, the stem tips wilt and droop. Even later in their development, after most stems have developed mechanically competent vascular tissues, the hydraulic-support role of parenchyma is not negligible (Wainwright 1970). Because of its potential hydraulic-support function, parenchymatous filling tissue in a plant organ functioning as a complex beam is usually not passive.

4.3 Mechanics of Complex Beams: General Consideration

To understand the static and dynamic aspects of mechanical functioning of plant organs, one must first consider the attributes of ideal materials in ideal beams. We begin by invoking some basic mechanical concepts relevant to both statics and dynamics in ideal beams, with emphasis on statics. Later (Sect. 4.5), we discuss dynamic loading of plant organs. Although we appreciate that the functioning of structural elements of plant organs can be described mathematically, we have space here to give only a very abbreviated mathematical treatment.

4.3.1 Normal Stresses, Normal Strains, and the Modulus of Elasticity

Consider a block with two equal but oppositely directed forces, F_1 and F_2, pulling on it. The block is in tension, and the tension, T, is equal to the magnitude of either force — i.e., $T = F_1 = F_2$. If, instead of pulling, the two forces (still equal) are pushing, then the block is under compression. As compression is the opposite of tension, we will use the same symbol, T, for both. T is positive if the sample is in tension and negative if it is in compression.

The dimensions of a sample change if it is subjected to changing tension. Let L_0 be the length when tension is zero and L be the length when tension is other than zero. Then, the change in length $\Delta L = L - L_0$. The value of ΔL is positive when T is positive (tension) and negative when T is negative (compression). The magnitude of ΔL elicited by T depends on the dimensions and material properties of the sample. The ratio of the change in length to the original length is the **normal strain**, ϵ; thus, $\epsilon = \Delta L/L_0$. The ratio of the tension to the cross-sectional area (A) is **normal stress**, σ; thus, $\sigma = T/A$. Tensional stress can be called negative pressure. Conventionally, though, the term "pressure" is reserved for fluid systems, in which the force per unit area is equal in all directions, and the term "stress" for solids, in which the force per unit area is a function of the plane of orientation of the area being considered.

Within certain limits of T, the normal strain, ϵ, is reversible, implying perfect elasticity and direct proportionality between strain and stress (**Hooke's Law**). This behavior is expressed as $\epsilon = (\sigma/E)$, where E is **Young's modulus**. The value of E depends on the materials constituting the sample, not on size or shape. For a particular material, E represents that normal tensional stress, σ, that would in theory cause a doubling in length of the sample, if it could withstand that much stress. In practice, most samples fail at a maximal tensional stress, σ_{max}, much lower than E. The σ_{max} value is an *estimate* of the strength of a material in resisting tension, based on actual tests. The maximal strain, ϵ_{max}, is defined as L_{max} divided by L_0, where L_{max} is the length of the sample under maximal tensional stress, σ_{max}.

For most engineering materials, Hooke's Law is strictly applicable (i.e., strains are fully reversible) only within a restricted range of tensional stress, which *does not* include σ_{max}. For cell walls, however, Hooke's Law applies throughout almost the whole range of possible stress up to σ_{max} — provided that the stress is of very short duration. Some values of stress and strain parameters for walls of different cell types, tissues, and for other relevant materials, are listed in Table 4.1.

Table 4.1. Maximal tensional stress, σ_{max}, maximal tensional strain, L_{max}/L_0, and Young's modulus, E, for various fibers and tissues or tissue elements, as well as steel, aluminum, and cellulose for comparison. Values for fibers, tracheids, and collenchyma pertain to transectional area of walls only (i.e., excluding lumina). The values for parenchyma represent E_{tissue}

Material and source	σ_{max} N/mm^2	L_{max}/L_0 %	E N/mm^2	Reference
Fibers (phloem):				
Sabal, Pritchardia, or				
Livistona petiole	240	2.0	16 500	Razdorskij (1955)
Phormium tenax leaf	420	2.3	21 000	Razdorskij (1955)
Helianthus annuus stem	274	0.8	34 250	Razdorskij (1955)
Fibers (xylem) or tracheids:				
Robinia pseudoacaia	148		23 500	Kisser (1925)
Fraxinus excelsior	131		21 600	Kisser (1925)
Abies pectinata	112		29 700	Kisser (1925)
Collencyma (in non-growing organs):				
Heracleum spondilium	147	2.6	9680	Hejnowicz (1973)
Livisticum officinale	116	1.9	5500	Ambronn (1881)
Parenchyma tissue:				
Pisum stem			34	Burström et al. (1967)
Solanum tuberosum tuber			3	Nilsson et al. (1958)
Other materials:				
Steel	800	0.12	205 000	Razdorskij (1955)
Aluminum	93—120	3.0	67 000	Razdorskij (1955
Crystalline cellulose chains (experimental value)			137 000	Sakurada et al. (1962)
Crystalline cellulose chains (theoretical value)			300 000	Gillis (1969)

Pure, evenly distributed tensions or compressions are probably rare in plant organs. Rather, different parts of plant organs are simultaneously subjected to tension and to compression, as in simple bending, and the relative distributions of stresses across the organs are quite predictable.

4.3.2 Bending and Stiffness

Bending evoked by wind and/or gravity forces is the predominant mode of deformation of plant organs. Consider a simple, horizontal bar-type beam supported at both ends and under a load great enough to bend it appreciably. Intuitively, it is clear that during bending the upper surface of the beam is compressed and the lower surface is stretched. It is also clear that, somewhere between the compressed upper layers and the stretched lower layers, there must be a plane or axis that is unstressed. This is the **neutral plane** or **neutral axis**. Because it remains unstressed, the neutral plane or axis undergoes no length change as the beam is loaded. However, the planes above the neutral plane are shorter, and the planes below the neutral plane are longer, than they were before bending. These deformations increase in magnitude with distance from the neutral plane and thus are greatest at the surfaces.

The rules of geometry assert that the strain in a layer at distance x from the neutral plane follows the relation $\epsilon = x/R$, where R is the radius of curvature of the beam — which will be extremely long when loads are light. By convention x is positive on the tension side and negative on the compression side of the beam. By introducing ϵ from Hooke's Law and rearranging we obtain the relation, $\sigma = Ex/R$, where the values of σ and E are local. Necessarily, then, the local stress in a beam bent under load is inversely proportional to the radius of curvature of bending and directly proportional to the local Young's modulus and to the distance from the neutral layer.

It is now relevant to ask: how is the extent of curvature of the loaded beam related to the load, to the bending force? How does the extent of bending depend upon the shape and composition of the beam? And, what is the location of the neutral layer?

Assuming that the bending force, F, imposed by the wind, impinges upon the tip of an axial plant organ, and that the tissues are not compressed or stretched beyond their limits of stress, σ_{max}, the following relation holds:

$$F \leq \frac{\sigma_{max} \, I}{L \, x_{max}} , \qquad (4.1)$$

where I is the second moment of inertia of cross sectional area about the axis in the neutral layer and perpendicular to the longitudinal axis of the beam, x_{max} is the maximal distance of the element at the surface to which σ_{max} refers, and l is the distance along the longitudinal axis from the point of wind impingement at the level considered.

Further, $I = \int x^2 \, dA$, where dA is an element of cross-sectional area A at x. The I value of the beam depends directly on its cross-sectional geometry (Speck et al. 1990). In a plant organ with strands of mechanical tissue, the I of the strand system is the sum of the cross-sectional area of each bundle multiplied by the square of the distance of the strand from the presumed neutral layer. A useful geometric parameter in comparing bending properties of axial organs having mechanical tissues of different cross-sectional geometries and areas is the **section modulus**, Z, which equals I/x_{max} (Speck et al. 1990). By taking into account this modulus and also the product of force and distance of the bending moment, M ($M = F \cdot l$), we can write $M \leq \sigma_{max} Z$. To estimate the maximal torque that can be tolerated by axial organs with the same cross-sectional area of mechanical tissue, but with different cross-sectional geometry, one must determine the corresponding Z and introduce it into that equation.

Usually the bending force is distributed along the organ and evokes bending with a curvature of radius R. In such instances the relation $M = E\,I/R$ holds. Here M is the net torque of the external forces evoking the bending. The product of E and I is the flexural stiffness, S, and is an attribute of beams that depends both upon cross-sectional geometry and the constituent materials.

This discussion pertains to hypothetical "beams" having only one type of tissue. But, in fact, plant organs always consist of different tissues with different mechanical properties. Typically parenchyma, the ground tissue, is complemented by an armor of tissues having higher elastic moduli. The equation $M = E\,I/R = SR$ remains valid for a composite organ provided that an appropriate S value is used. The flexural stiffness (S) of an organ composed of n strands is expressed by the sum of the S values of the individual (i) strands. That is $S = E_i I_i$. The contribution of a single strand, i, to the flexural stiffness of the strand is given by the ratio of the flexural stiffness of the strand, $E_j I_j$, to the sum, $E_i I_i$.

We have considered here the cross-sectional geometry of plant organs as being symmetrical, but this does not mean that the neutral layer will necessarily be symmetrically positioned. This is so because the E value for a tissue is generally higher for tension than for compression. This relation predicts that in a bending organ the neutral plane or axis will be closer to the convex side than to the concave side. As a result, the strain will be relatively greater on the concave than on the convex side of a bent horizontal beam or organ.

Superficially, one might suppose the strength of a complex beam to be proportional to the amount of reinforcing material in it. According to beam theory, however, the distribution of reinforcing material is more important than the total amount (Wainwright et al. 1976). Reinforcing is most effective near the upper and lower surfaces and least effective in the middle, which is why "I" beams are so widely used in engineering work. The distribution of the reinforcing material determines the position of the neutral plane. If there are too few reinforcing rods on the tension side, the neutral plane is shifted toward the compression side, and the beam may fail due to rupture of these rods. If there are too many rods on the tension side, the neutral plane is shifted toward that side, and the beam may fail due to crushing of the filling material on the compression side.

Disregarding subsequent growth, we can say that construction of a complex beam is optimal if the reinforcing elements on the convex side just begin to break under tension as the filling material on the concave side just begins to collapse under compression. From simple experiments, we know that stems and petioles often seem to have attained this optimal construction at maturity — but not necessarily at every stage of development.

4.3.3 Shear Stresses and Strains

In real structures under loads, normal tensional or compressive stresses, evoked by forces perpendicular to a unit area or face, are always accompanied by **shear stresses**, derived from forces parallel to the unit area or face. Shear stresses in turn elicit shear strains. The magnitude of shear stress, τ, is given by the parallel force divided by the area on which it acts. The magnitude of the **shear strain**, Γ, is expressed as the tangent of the angle measuring the deformation of the unstressed square-unit face into a rhombus (Fig. 4.1). Thus, $\Gamma = \tan \theta = \Delta / L_0$.

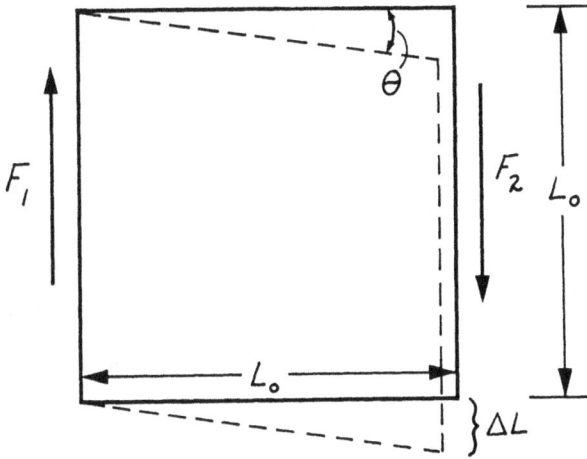

Fig. 4.1. Square representing the face of a cube being deformed by an upward force, F_1, along its left side and an equal downward force, F_2, along its right side. The magnitude of the deformation is a measure of the shear strain, Γ, which is defined as the tangent of angle θ, or $\Delta L/L_0$

Within the range of moderate stresses, shear stress, τ, is related to shear strain simply as $\tau = G\Gamma$, where G, a constant, is the **shear modulus**. This modulus, like Young's modulus (E), is a characteristic of the material constituting the sample.

In plant organs bending under moderate loads, shear stresses never become very high. But under extreme loading conditions, as when trees weighted with ice are further bent by wind, shear stresses may become critically high locally. Shear stresses are relatively greater along the neutral plane or axis, where normal stresses are minimal. In tree stems, some of the defects that foresters call "shakes" are probably failures due to very high shear stresses near the neutral plane during bending.

4.3.4 Buckling and the Slenderness Ratio

Buckling is severe deformation caused by excessive compressive forces in long, slender beams. In theory, "pure" buckling is caused wholly by compressive forces, but in practice buckling is generally preceded or accompanied by bending in which there are identifiable torques as well. Practically, there are two types of buckling. Buckling that leads to an elastic (reversible) bending that is distributed throughout the length of the beam, is **Euler buckling**. In contrast, a deformation that is local and consists of an irreversible and catastrophic "kink" is **local buckling**.

As a tree stem grows in height and diameter, the weight (static load) to be supported increases, requiring a change in relative dimensions of the stem. The classical **principle of similitude**, attributed to Galileo (Thompson 1942), proclaims that, "the larger the size of the structure, the greater the stresses per unit area". This is true because, as size increases, the weight to be supported increases as the cube of dimensions, whereas the cross-sectional areas of supporting members increase only as the square of dimensions.

From mechanics, we know that susceptibility to Euler buckling of a simple, homogeneous, cylindrical column increases with height and decreases with increasing diameter and with increasing Young's modulus (E) of the material. In construction, the requirement for resistance to Euler buckling determines the permissible ratio of height to diameter, the "slenderness ratio". The critical compressive stress, σ_e, for Euler buckling of a simple, homogeneous, cylindrical column is proportional to the E value of the material and inversely proportional to the square of its slenderness ratio (height, L, divided by diameter, D). This means that slenderness has a large effect on the critical stress for Euler buckling. It can be calculated that for a column to resist buckling, the critical slenderness ratio, L/D, must be inversely proportional to the square root of the height. (In engineering work, L/r, where r = radius, is often used rather than L/D to calculate the slenderness ratio; the coefficients, then, are twice as great.)

If one tries deliberately to cause Euler buckling of a column having a slenderness ratio (engineering, L/r) of less than 10, one must apply so large a compressive force that the column is crushed before σ_e, the critical stress for Euler buckling, can be attained. This always happens if σ_e is greater than the compressive strength of the material. Columns characterized by such a relation are "stocky". If the slenderness ratio is between 10 and 60, the column is "intermediate", and if the ratio is greater than 60, the column is "slender" (Wainwright et al. 1976). Stems of many tall grasses, and also some young trees, are "slender". Plants with slender stems are at some risk of Euler buckling, but there are stem structures that effectively cope with this risk (Spatz et al. 1990; Speck et al. 1990).

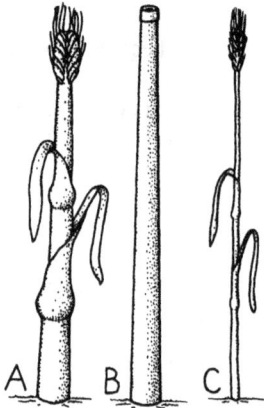

Fig. 4.2. Relative slenderness ratios. If a grass culm (A) were to be as tall as a 140-meter concrete chimney (B), it would also need to be relatively as thick at the base. The slenderness ratios (L/D) of such tall structures (here drawn as 13:1) seldom exceed 35:1. The ratio in a 1-meter-tall grass culm (C), in contrast, may have a ratio exceeding 300:1 (here drawn as 50:1). The range of practical L/D ratios is scale-dependent and is determined by the "principle of similitude" and other laws of geometry and mechanics as discussed in the text. (After Razdorskij 1955)

The marvelous slenderness and apparent efficiency of "design" of a tall grass culm make man's tallest artifacts, such as chimneys and skyscrapers, seem awkwardly stocky. Man's tallest structures are, indeed, much more stocky than the grass culm — not because of weaker materials, or bad design, but basically because of the demands of the principle of similitude. If a grass culm were to be as tall as a 150-m chimney, it would also need to be about as thick as the chimney at its base (Fig. 4.2). The slenderness ratio *(L/D)* of a meter-tall grass culm typically ranges from 100 to 330. If the culm were to be 100 m tall, and were to have the same resistance to buckling, its slenderness ratio would have to be reduced to 33. Its basal diameter then would be about 3 m.

Consideration of the observed slenderness ratios *(L/D)* of plants of different height classes confirms the validity of the preceding arguments. Plants ranging in height from 1.0 to 1.5 m can be the most slender, having *L/D* ratios of about 300. Plants ranging from 25 to 40 m are less slender: bamboos may attain ratios of 100 and palms about 60. If the height range is 50 to 150 m, as in the tallest forest trees, slenderness declines toward stockiness. Razdorskij (1955) gave examples: a 78-m *Tsuga mertensiana* that had an *L/D* of 42; a 128-m *Eucalyptus amygdalina* with an *L/D* of 28; and a 155-m *E. amygdalina* with an *L/D* of only 16.5.

As plant height increases from that of cereals to that of the giant bamboos, slenderness decreases approximately as the square root of the height increases. Among the palms, however, slenderness decreases faster than the square root of increasing height demands. This may be partly because palm crowns are heavy and concentrated compared with the relatively light foliage of a grass culm. In addition, young palms must be "overbuilt" to meet future support requirements (Sect. 4.6). Similarly, the tallest forest trees are more stocky than would be predicted from considerations of resistance to local buckling under most conditions. Such large, old trees have good margins of strength. Having survived the storms of centuries, which brought down trees of lesser strength, they are a highly selected sample.

4.4 Applications of Mechanics to Parenchyma Cells and Tissues

In the perspective of stress-strain relations and of the resistances to deformation of cell-wall materials and tissues, as they can be expressed by Young's modulus or the shear modulus, we can now consider the "strength" of the materials and structural elements from which plant stems are constructed.

4.4.1 Cell-Wall Tensions and Parenchyma-Tissue Strength

The walls of living, turgid, nonlignified cells are prestressed by the hydraulic pressure of the cytoplasm. Most of the mechanical strength of parenchymatous tissue is derived from the hydraulic property of its cells, because the walls, being unlignified, lack rigidity unless the cells are turgid. A flaccid cell cannot directly transmit physical moments. In contrast, outside forces acting on a turgid and prestressed cell evoke tensions in the microfibrils of its walls. For example, wall tension usually increases if a tissue is put under compression. The walls are quite resistant to rupture under tension because the tensile strength of the microfibrils is high. Turgid parenchyma tissue, unlike the walls of parenchyma cells, has appreciable elastic strength against compression and is effective in dissipating kinetic energy when transmitting a shock load (Sect. 4.5).

Hydraulic pressure in a cell is the same in all directions, regardless of cell shape, but the tensional stress in the walls of a cell that is not spherical is not the same in all directions. That is, the stress is anisotropic. It can be shown mathematically that, in the walls of a cylindrical container filled with fluid under pressure, the axial tensional stress is generally only half the tangential (i.e., circumferential) tensional stress. This relation provides a rationale for an observed bias toward an oblique rather than an axial orientation of cellulosic microfibrils in walls of parenchyma cells, which do not function as axial, tension-resisting, reinforcing elements (Sect. 4.4.2).

Both tensional and compressive forces, arising from gravity and wind, act along stems and other organs of free-standing plants. These forces are resisted by parenchymatous and other tissues. Though the cells constituting these tissues can be described as being on average 14-sided (Sect. 1.3), the cells nevertheless generally approach an irregular, stocky, cylindrical shape, with their long axes parallel to the axial compressive or tensional forces. Such forces, if compressive, generally decrease the meridional (axial) tensional stress in the wall and increase the tangential tensional stress. Because its contents are incompressible, a parenchyma cell under axial compression remains nearly unchanged in volume, though it becomes slightly shorter and wider. The cell may be deformed in this way to the limit allowed by the tangential strength of its wall. If the compressive forces are transient, the deformations are generally elastic.

We may now ask how the elastic compressive strength of a cylinder of parenchymatous tissues, as it might be expressed by Young's modulus (E), depends on turgor pressure. For potato-tuber parenchyma, the relation between E for elastic deformation and the turgor pressure has been both calculated from theory and determined experimentally (Falk et al. 1958; Nilsson et al. 1958). This work revealed that the E of parenchyma tissue increases with turgor pressure, as has been confirmed by Niklas (1988), and with the E of the cell walls. A careful analysis supports one's intuition that the elasticity of young organs depends heavily on turgor pressure. That is why the young, "hydraulically supported" parts of plants become limp and droop if deprived of water. Turgor is probably always a more important factor in the mechanical strength of parenchyma than are the physical attributes of the walls themselves. If turgor falls, the compressive strength of parenchyma declines drastically.

Although the strict theoretical relations between turgor, E value, and shear strength of parenchyma are not clear, it is clear that the shear strength of parenchyma, whatever the turgor, is low compared with that of so-called mechanical tissues. This explains why, for example, the neutral plane (where the shear stress is maximal), with respect to the bending of a petiole in response to wind from various directions, almost always lies within the mechanical tissues (if present) rather than within the parenchyma. In a *Populus* petiole, the direction of least resistance to bending is different in different segments along the length, and corresponding differences in distribution of mechanical tissues (vascular bundles) are seen in cross sections along the petiolar axis (Sect. 4.7.3). Thus, mechanical analysis can help us interpret some structural features.

4.4.2 Arrangement of Reinforcing in Cell Walls

In elongated cells that function as primary mechanical reinforcing elements, the thickened walls, whether collenchymatous or sclerenchymatous, typically contain some cellulosic microfibrils that are axially aligned, or nearly so. This is the most efficient microfibrillar orientation if the walls are to have maximal strength in longitudinal tension, which is the typical requirement for mechanical reinforcing elements in stems and petioles.

In other kinds of cells, especially in parenchyma functioning as compression-resistant filling tissue, the requirement is for cells to be more elastic than would seem to be allowed by axially aligned cellulosic microfibrils. In these cells, a helical arrangement of microfibrils is most effective. Helically arranged microfibrils, under tension, can efficiently cope with both axial and tangential components of wall stress.

In fact, we can treat parenchyma cells mathematically as cylinders of constant volume having walls reinforced by helical windings of microfibrils. The constant volume is that of the protoplast. Indeed, in fast deformations, such as occur during flexing of stems in the wind, cells do not cyclically change their volumes by uptake or release of water. Constancy of volume means that a cell that is stretched in length decreases in circumference, and a cell that is compressed increases in circumference. Parenchyma cells typically are extensively deformable and, as cells, may have relatively low E values. (Their walls, however, are complex structures reinforced by oriented microfibrils having a much higher E than the whole cells or the gross walls.)

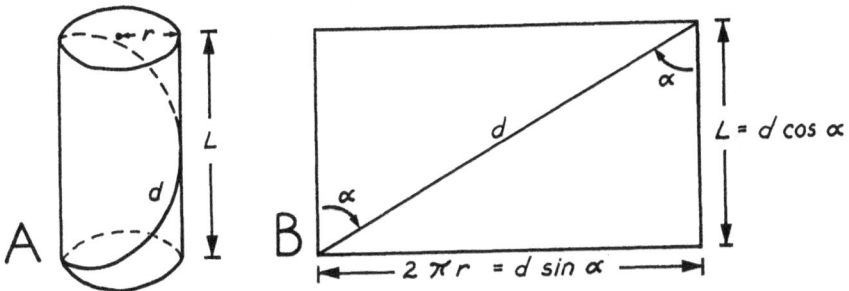

Fig. 4.3 A,B. Mathematical implications of fibrillar angle in the wall of a cell approximating a cylinder. A Segment of a cylinder of length L and radius r, delineated by one turn of the fibrillar helix measured along d. B Wall of the segment in A represented as unrolled onto a two-dimensional coordinate system. The length of one turn of the helix is the length of diagonal d; the angle of fibrillar inclination is α

What are the dynamics of deformation in fibril-wound cylinders, and what is the optimal angle of inclination of a microfibrillar helix relative to the major axis of the cell? If we accept the limitation that the cell be deformed at constant volume, we can derive the optimal angle mathematically. To begin, we can assume that cells having helically arranged microfibrils can be somewhat deformed with no appreciable change in the length of the microfibrils. Consider a segment of a cylinder, of length, L, and radius, r, subtended by one turn of a regular fibrillar helix having a length d (Fig. 4.3A). The volume, V, of the cylindrical segment, which is readily calculated from r

and L, must remain essentially constant, as must the length, d, of the fibrils along the helix. Yet, as stresses change, r and L change also. If one increases, the other decreases, while V remains nearly constant. Inevitably, as r and L change under stress, the angle, α, at which the helix is inclined relative to L must change also, however slightly (Fig. 4.3B). What is the range of α within which the volume of the cylindrical segment changes relatively little?

It can readily be shown that the relation between V and α is:

$$V = \frac{d^3}{4\pi} \times \sin^2 \alpha \, \cos \alpha \quad .$$

(4.2)

When this function is evaluated and plotted over a range of α (Fig. 4.4), the volume is seen to be relatively constant within a small range near its maximum, which occurs at an α of 54°44'. That is, the cylindrical segment becomes shorter and wider or longer and more slender as α ranges from its maximal value of 54°44' toward **A** or toward **B**, and very little change in volume results. How does this indicated value of α compare with actual fibril angles in mature parenchyma cells? We find that, though there is a range, angles near 54° commonly predominate. It seems that the limiting value of 54°44' will be automatically approached due to laws of physics, if a cylinder with deformable walls is helically wound with strong reinforcing strands.

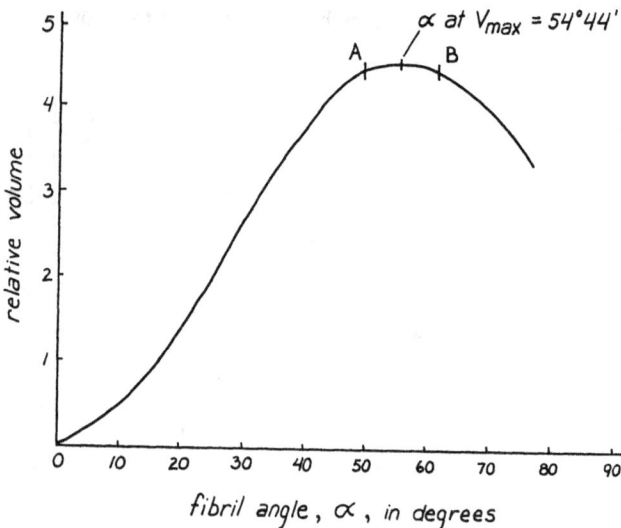

Fig. 4.4. The effect of changes of fibril angle on volume of a fiber-wound cylinder deformed by strain. See text for interpretation

As already mentioned (Sect. 4.4.1), in a cylindrical container with contents under pressure, the axial stress in the walls is only half the tangential stress. How closely does the optimal angle found above accommodate this 2:1 relation of tangential to axial stress? Note that the tangential stress is proportional to tangent α, while the axial stress is proportional to cotangent α. Thus, the ratio between the tangential and axial stresses equals the ratio of tangent α to cotangent α. When α is $54°44'$, that ratio is 2.00. The fibrillar angle of $54°44'$, then, optimally accommodates the 2:1 ratio of tangential to axial stress, as well as being the most efficient on the basis of volume enclosed per unit of reinforcing material.

4.5 Dynamic Loading and Organ Resilience

So far, we have almost ignored the rate of loading. For **static loads**, which in free-standing plants increase as the plant body grows, time is scaled in months and years. For **dynamic loads**, time is scaled in seconds and minutes. The strategies that provide the most efficient support of static loads may be inappropriate if the time scale is short, as when force is applied by a gust of wind or falling rain.

To survive and function, plant organs must resist these shock or dynamic loads without permanent deformation or rupture; they must be resilient. An essential attribute of resilient structures is their capacity repeatedly to absorb and release energy imparted by dynamic loading. Elastic deformations (strains) produced in the various structural elements of a system under dynamic loading represent stored energy that is immediately available to power restorative forces when dynamic loading declines.

What kinds of composite structures facilitate rebounding from short-term deformations? Through everyday experience, we have come to accept the remarkable resilience of materials made of glass or other fibers embedded in a pliant matrix. These composite materials have provided "high-tech" bows for archery, fishing rods, and skis, among others, with remarkable qualities based on reversibility of deformations imposed by short-term dynamic loads.

At both the tissue and the ultrastructural levels, plants long ago evolved structural members that perform mechanically in ways similar to modern, composite, fiber-and-matrix materials. At the tissue level, sclerenchyma strands embedded in an elastic parenchyma equip a plant to absorb the energy of short-term deformations and then reverse them, thus helping the plant tolerate dynamic loads. At the ultrastructural level, in the cell wall, there are cellulose microfibrils embedded in a matrix of hemicellulose and lignin.

4.5.1 Strain Energy

In systems under dynamic loading, the work done by external forces in generating deformations is temporarily stored within the deformed structures as **strain energy.** In ideally elastic materials, all the strain energy is recoverable during unloading. In real materials, there are frictional heat losses. Even without rigorous analysis, one can appreciate the relatively large amounts of strain energy that a plant, like a set of mechanical springs, can absorb from the environment as it is bent or twisted — and then give back again as it recovers seconds later. Resilience is the key to survival; the alternative may be permanent deformation or breaking.

The work that strain energy does during the input phase is calculable as a force operating through a distance. Considering, first, only the energy input producing deformation in a single direction, we can imagine the deformation (strain) of a small, square, cross-sectional element into a rhombus by a dynamic force imposing a shear stress (Fig. 4.1). The strain energy per unit volume, U_0, due to the shear, varies linearly from an initial zero to a final value, and we would expect the total to be quantifiable as one-half the product of stress and strain. Substituting, according to Hooke's Law (Sect. 4.3.1), we see that

$$U_0 = \frac{\sigma_x^2}{2\,E_x}\,. \qquad (4.3)$$

Here σ_x is the stress in the x direction and E_x is the Young's modulus in the same direction. Thus, the higher the E value, the less strain energy can be stored, as intuition would predict.

Dynamic loading is not limited to unidirectional forces, however. Torsional dynamic loading (Sect. 4.5.2) may evoke deformations in three mutually orthogonal planes, so that the strain energy may have a component along each of three mutually perpendicular reference axes (Ugural and Fenster 1975). An example is the energy input needed elastically to deform a small, initially cubical element so that each of its faces becomes a rhombus. Any three-dimensional, elastic tissue mass offers a large potential for temporary storage of strain energy because the components in the several directions are additive.

The maximal strain energy that can be absorbed by a system, U_{max}, is a measure of its resistance to dynamic loading, to mechanical shocks. The strain energy that a system can accept is higher if E is low, provided that the fracture strength of the material is high enough. Rubber, which has a very low E and is extensively deformable, readily accepts the kinetic energy delivered by shocks.

Parenchyma can perform similarly — if loading is not too rapid. Loading rate is important, because, for the preceding rationale to apply, there must be enough time for the energy-absorbing deformations generated locally to be propagated to and accepted by elements distant from the point of loading. In hyperdynamic situations, such as in shock loading, loading can be so rapid that U_{max} is exceeded locally and there is local breaking or shattering without propagation of deformations to more distant elements. The suddenness with which a shock is delivered, as well as its energy, determine the potential for damage. Even in shock loading, however, the tissue having the greatest U_{max} will fare best (Razdorskij 1955).

The strain energy that a specimen can rapidly absorb depends not only on its material composition, but also on arrangement of the mechanical elements relative to the applied stress, and on the extent of prestressing of the elements. Prestressing increases the ability of a large beam such as a tree stem to absorb and distribute the energy of rapid dynamic loading in various types of elastic deformations. Thus, U_{max} limits are not exceeded locally and the danger of breakage decreases.

Empirically, we know that prestressed beams, including tree stems, are more "live" than similar, unstressed beams and give a more clearly resonant sound if struck. This is the basis of a tradition of assaying a tree's growth vigor by the tone it emits when tapped with a hammer.

4.5.2 Torsions

A solid rod that is subjected, at its ends, to opposite torques about its axis is in torsion. There is then shear stress, perpendicular to the radius, in every cross section, such that if the rod suddenly were divided into wafer-thin, elemental cross sections free to rotate about the axis, they would do so, relieving the stress. The measure of the deformation induced by torsion is the angle of twisting, either α or ω (Fig. 4.5). In an ideally elastic material, the deformation, α, which represents the shear strain, Γ, is related to the shear stress, τ, by the shear modulus, G, such that

$$\alpha = \frac{\tau}{G} \tag{4.4}$$

and, if the angles are expressed in radians, such that

$$\alpha \approx \frac{AB}{L} = \omega \frac{r}{L} . \tag{4.5}$$

Hence

$$\tau = Gr \frac{\omega}{L} . \tag{4.6}$$

In any cross section of a rod in torsion, the torsional deformation ranges from nil at the center to a maximum at the surface. The distribution of stress is analogous to that in pure bending of a simple beam, in which the stress and deformation are zero along the neutral plane and maximal at the surfaces (Sect. 4.3.1).

If a torque twists a rod, energy is invested in elastic deformations (strains) within the wafer-like cross-sectional elements of the rod. The elastic deformations and the energy invested in them evoke a reactive torque, and, for an entire cross-sectional element, the reactive torque must equal the applied torque. This equality can be used to derive an expression for the attribute called **torsional stiffness**, which is the product of the shear modulus (G) and the moment of inertia (I) of the cross section. This attribute is fully analogous to flexural stiffness, which is the product of Young's modulus (E) and the moment of inertia (I) of the cross section (Sect. 4.3.2). Torsional stiffness theoretically is based on resistance to elastic shear deformations in which the face of each unit cube of affected material becomes a rhombus. In contrast, flexural stiffness theoretically is based on resistance of a unit cube to compression on one side and tension on the opposite side — that is, to elastic extension and compression — though, in a real beam, there is also a component of elastic shear deformation.

Even a superficial appreciation of the concept of torsional stiffness allows one to predict that, in a plant stem, the farther a tissue with a high shear modulus is from the axis, the greater will be the torsional-stiffness component of the stem's mechanical strength. Thus, torsional stiffness is relatively great in stems having a thick-walled epidermis, a continuous layer of peripheral sclerenchyma, or a thick phellem layer.

Another component of torsional stiffness in plant organs arises from separate strands of "reinforcing" elements embedded in a matrix of parenchyma tissue. These strands contribute to torsional stiffness only indirectly and less efficiently than do continuous layers of reinforcing elements. If a stem is twisted by external forces, these

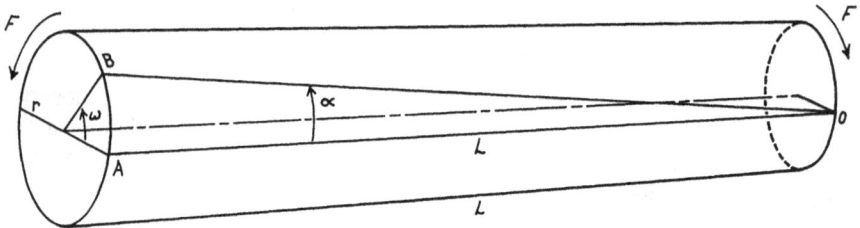

Fig. 4.5. A solid, rod-shaped beam subjected to opposite torques *(F)* about its axis at its two ends, producing torsion. Shear stress perpendicular to the radius exists in every section, and the material of the beam is deformed. The angle α is a measure of the deformation (and also a manifestation of shear strain, Γ). The relations between angle ω, radius, r, and length, L, also measure the deformation due to strain. See Eqs. (4.4) to (4.6) and related text for further interpretation

strands of mechanical tissues, being forced into longer helical courses, are brought under tensional stress. This stress in turn puts the ground tissue under compression, and the stem, upon being "wound-up", becomes slightly shorter. The internal stresses induced by the twist generate a reactive torque that opposes the externally applied torque. Most of the reactive torque (the "unwinding" force) arises from the tensile stress in the cell walls of the parenchyma — which is a consequence of the pressure being put on the incompressible contents of these cells by the tension in the strands of mechanical cells. Thus, tension in the walls of parenchyma cells is inevitably involved in resistance to torsion, even if strong strands of fibers are present.

4.6 Growth-Induced Tissue Stresses

Stresses in the form of tensions and compressions are evoked by differences in tissue growth rates. In fact, tissue stresses are probably generated by all normal growth processes. Such tissue stresses therefore can be termed **growth stresses**.

Many plant organs depend on growth stresses while performing their supportive functions, just as planned stresses are all-important in the dynamics of the air-supported "bubble" structures of cables, canvas, and plastic that engineers have developed. In plants, the supportive function of growth stresses is vitally important in fast-growing and short-lived "hydraulic" structures that include no fiber-like, mechanical elements. Growth stresses are also important in maintaining the resilience of long-lived organs such as tree stems (Archer 1986).

The inner part of a plant organ is typically in compression while the outer part is in tension. For example, in leaves and young stems, the epidermis usually is in tension, while the cortical parenchyma or mesophyll tissues are in compression. The epidermis, by functioning as do the sidewalls of a pneumatic tire, is an important component of the mechanical supportive system of a young shoot. This can be demonstrated by cutting a segment of the scape of a dandelion blossom into axial strips. The ends of the strips immediately curl outward. This phenomenon was well known to nineteenth century botanists, yet our understanding of it is still only superficial (Kutschera 1989).

Further, in stems of both angiospermous and gymnospermous trees having many annual increments, the outer wood is in tension, whereas the inner wood is in compression. Soon after a xylem increment is deposited, the thickened walls of its mature cells, through changing hydration states, tend to shorten slightly. This generates tension in the outer wood and compression in the inner wood (Jacobs 1945; Wilson and Archer 1977; Archer 1986). If there is to be no bending moment, the stresses must always be symmetrically distributed with respect to the axis (but see Sec. 10.3.2).

The magnitude and distribution of stresses vary with time during wood formation. Because the slow "creep" of strained elements relieves stresses, the maintenance of growth stresses (i.e., prestressing) in woody stems requires the continuing or episodic deposition of new wood. In steady-state conditions, the increase of stress due to the shrinkage of recently formed wood is in equilibrium with the decrease of stress due to the rheological "creep" of strained elements. In hardwood tree stems, compressive growth stresses may be as high as 20 MPa, whereas compressive stresses due to the weight of the crown are much lower — usually less than 0.1 MPa (Jacobs 1945; Kübler 1959b). Furthermore, the latter are distributed over the whole cross section of the stem, whereas the former tend to be localized subperipherally.

The prestressed state that is generated by growth has a large effect on a tree stem's mechanical properties. Prestressing promotes absorption, by elastic deformations, of the shock kinetic energy delivered by sudden gusts of wind to the crown (Sect. 4.5). A decrease in the rate of wood deposition as the tree matures, or as the environment changes, reduces the extent of prestressing, thus making the stem less resilient to shock dynamic loading. In fact, old trees in which the wood increment in the stem has been unusually small for several years are commonly overthrown by storms that are not particularly severe.

Growth-induced compressive stresses may be high enough to evoke localized compressive fracture zones in the older wood, a defect that the timber trade calls "brittle heart". More seriously, a seemingly inert log may make dangerously rapid movements if cuts or splits evoke a sudden release of growth stresses.

As a tree grows, the weight of tissues of all types increases. Thus, the static load becomes greater. The increase in foliage that accompanies growth also leads to greater dynamic loads evoked by the wind. Growth in branch thickness, however, increases flexural stiffness and thus gives the branch greater resistance to these increased dynamic loads. Mathematically, it can be shown that the flexural stiffness of a branch or stem is approximately proportional to the fourth power of its radius. Thus, branches and main stems can be considered reinforced beams that strengthen and stiffen themselves by adding new material as they grow radially.

Although the increased stiffness that accompanies radial growth reduces the bending that occurs under dynamic loads, it does not alleviate static-load stresses. If branches and main stems could not make active correctional movements, older and longer branches would be bent downward more and more by their own weight. This, indeed, occurs in *Picea abies*, for example, and is especially pronounced in "weeping" varieties of numerous species. Woody stems generally, though, have special physiological-anatomical systems that act primarily during thickening growth to increase stiffness and to generate restorative bending movements (Sect. 10.3.2).

Palms typically have no means of increasing stem diameter and stiffness as they grow taller. To compensate, a young palm produces a stem that is thick enough to meet future support requirements and that, consequently, is "overbuilt" (Rich 1987) until late in its ontogeny.

4.7 Design and Construction Strategies

4.7.1 Cantilever Beams and Stem Taper

The stems of free-standing plants, and also their petioles, typically are supported at only one end and function as **cantilever beams.** They are acted upon by bending moments resulting from the weight they support and from winds. The bending moments, being products of force and distance, are always greatest at the fixed end of the stem or petiole. We would therefore expect the ideal stem or petiole to be thickest at its base and then gradually to taper toward its free end. We would also expect that, for greatest efficiency of material use, the maximal bending stress would be the same all along the organ and that the organ would be circular in cross section and of uniform bending resistance. What would be the taper of such a beam, and what taper do we observe in actual tree stems?

Consider a coniferous tree in which the stem tapers uniformly from just above the basal buttresses to the vertex. Geometrically, this stem is a cone and has circular cross sections at all levels. The effective length of the beam is $H - h$, where H is the distance from the base to the center of mass of the crown (to which the wind may apply a bending force), and h is the distance from the base to the center of mass of the material in the beam (Fig. 4.6). It can be shown mathematically that, in a growing tree, if the material of the beam has a uniform Young's modulus, the diameter of the base (disallowing buttresses) must increase as the cube root of the distance $H - h$ increases. This relation can be considered to be an aspect of a D^3 **law of stem diameter**. Typically, angiospermous trees do not fit the model as well as do gymnospermous trees.

However, in coniferous trees growing in forest stands, the stems below the crown tend to be beams of uniform resistance to bending rather than beams of uniform taper. A beam that is circular in cross section and is of uniform resistance to bending has the form of a cubic paraboloid rather than a cone. Its diameter at any position is proportional to the cube root of the distance between that position and the point of application of the bending force. This manifestation of a D^3 law of stem diameter is most applicable to the form of the stem below the crown, in a mature but growing conifer. The stem within the crown tends to be conical (Büsgen and Münch 1931).

European forest scientists derived a D^3 law of stem diameter from empirical data more than a century ago (see review by Büsgen and Münch 1931). This relation has also been discussed more recently (Larson 1963; Wilson and Archer 1979). As we have outlined, a good case can be made for considering large coniferous stems as cantilever beams of uniform bending resistance.

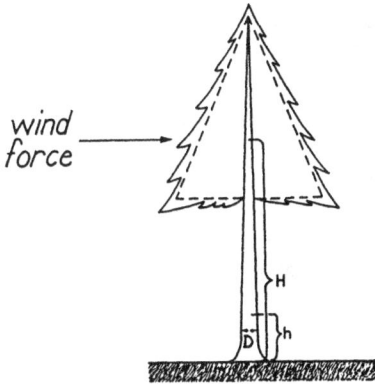

Fig. 4.6. A tree stem acts as a cantilever beam fixed at its base. As the wind applies force to the crown, bending moments are generated along the stem. The wind force, in effect, is concentrated at the upper limit of H, which approximately coincides with the center of mass of the crown. The effective length of the beam is $H - h$, where h is the distance from the base to the center of mass of the material in the stem. D is the stem diameter just above the basal buttress. (After Razdorskij 1955; Hejnowicz 1980)

A full analysis reveals that the compressed core wood that is a manifestation of natural prestressing in tree stems contributes to the bending strength by broadening the safe bending range. This entails no change in cross-sectional geometry or increased use of material. By considering prestressing when calculating the optimal taper of a tree stem for ensuring uniform strength, one can obtain a **D^2 law of stem diameter**. Indeed, in many cases, the D^2 law is more predictive than the D^3 law.

How different are these predictions for a tree that is, for example, 30 m high and has a basal diameter of 0.5 m? The interested reader is invited to calculate and plot profiles of stems of this tree, as predicted by both "laws" of stem diameter. From the plotted profiles, one can see that prestressing, such as occurs in vigorously growing stems, can lead to a stem that tapers somewhat more below the crown than does a less vigorously growing stem with less prestressing. At 10 m above the base of the stem in this example, the D^2 prediction of diameter is about 7% less than the D^3 prediction, and, at 20 m, 17% less.

4.7.2 Design for Resistance to Buckling

A large-diameter stem that is either hollow or has the most resistant tissue arranged around its periphery is quite resistant to bending, and hence to Euler buckling. At the same time, it may be precariously susceptible to local buckling (kinking). For example, whereas a slender, hollow grass stalk with a heavy seedhead can be bent to earth (Euler buckling) by wind and rain, and later recover, the more stocky, hollow scape of an onion seedhead may buckle locally (kink) if bent by the wind, and it does not recover. Local buckling is generally catastrophic and irreversible. The demonstrable difference in risk of local buckling between the slender and the stocky stems agrees with our observation that plants have been conservative in increasing the diameter of hollow stems or peripherally reinforced stems.

Note again that the higher the E value, the greater the critical stress for local buckling. Thus, it is not surprising that the tissue of hollow stems usually includes considerable sclerenchyma. To give adequate protection, sclerenchyma layers must be thicker in large- than in small-diameter stems. Compare a wheat or rice straw with a large bamboo stem.

Quite moderate amounts of peripheral sclerenchyma can protect thick stems against local buckling, if the internal space is filled with parenchyma. Even thin diaphragms of parenchyma, closely spaced, are effective in increasing resistance to kinking, though they contribute little to other types of strength.

To what relative extent is a hollow cylindrical stem protected against both Euler and local buckling, and what would be its geometry if resistance to the two types of buckling were equal? Without burdening the reader with mathematical details, we will briefly describe our approach to these questions. We set as equal the formulas for Euler and local buckling. Because E appears in the numerator on both sides of this equation, it cancels out and has no effect on the radius of gyration (for hollow stems) at which resistance to the two types of buckling is equal. Accordingly, the limiting dimensions are determined entirely by geometry and not at all by the material. The slenderness ratio under the limiting condition is likewise unaffected by E and is ultimately determined by a relation between the radius of gyration and wall thickness of the tubular stem or column.

These relations have been evaluated for several stems (Wainwright et al. 1976). From the values obtained, it is obvious that hollow stems are in fact much more slender than they could possibly be if they were as resistant to Euler buckling (bending) as they are to local buckling (kinking). The slenderness ratio of a dandelion scape, for example, is more than 20 times the hypothetical ratio for equal resistance. This means that hollow stems tend to be conservatively protected against local buckling, which tends to be catastrophically damaging, but can undergo considerable Euler buckling, which is benign and reversible. Knowing that natural selection ultimately imposes a cost-benefit analysis, we could anticipate this result.

4.7.3 Design for Simultaneous Resistance to Static and Dynamic Loads

Evolution has compromised between efficiency of material use and safety factors, with a resulting tendency toward mechanical optimization of "design". Any hypothetical, naturally driven system "designing" and controlling the construction of a tree would be under the same constraints as a human design engineer. In designing a complex beam, the engineer must reinforce it enough to avoid risk of failure, but must also avoid overly conservative design and attendant waste of materials. Similarly, in a tree, survival pressures imposed by competition for space, light, water, and nutrients penalize overly conservative design and inefficient use of materials. We must suppose that plants have evolved optimal compromises.

Can both high resistance to static loads and resilience with respect to dynamic loads be built into a structure without wasting materials? For strength in resisting static loads, the optimal cross section should have both high flexural (EI) and high torsional (GJ) stiffness. Both types of stiffness are promoted by increasing the cross-sectional area of *high-E* materials and distributing them around the periphery. Tubular construction thus seems appropriate. Twigs of *Paulownia tomentosa*, for example, approach this "design". However, if the diameter of the tube becomes too large, and the walls too thin, there is danger of catastrophic local buckling.

For effective absorption of the energy imparted during dynamic loading, the material of *low E*, which is able to accept large elastic deformations, should be located at the periphery, where such deformations are most likely. This seems to be contrary to the requirement for high-*E* materials near the periphery to resist static loads. Therefore, a compromise in use of peripheral space is required. Evolution has responded with stems having subperipheral axial strands of high-*E* material embedded in low-*E* material that extends to the periphery. The low-*E* material may be epidermis and parenchyma in herbaceous plants or in young organs of woody plants, and secondary phloem in older woody plants. The high-*E* material may be strands of phloem fibers, and sometimes of xylem as well, in young or herbaceous organs, and secondary xylem in older woody plants.

Because plant organs tend to be multifunctional, the demands of function on structure are diverse. For example, a stem not only must support, but also must provide translocation pathways. In addition, there are inevitable limitations during development. As the European cathedral builders understood quite well, even a perfectly designed vault cannot support itself during construction. Similarly, a stem must be able to respond to stresses elicited by static and dynamic loads during all of ontogeny, not just during one phase of it. Young organs may be weak until they develop the sclerenchyma that will reinforce them during maturity. Their structure seems to be more nearly optimal for static than for dynamic loads, as manifested by peripheral distribution of collenchyma. In a stem that has ceased elongating, fiber strands usually develop in the cortex and/or primary phloem. A woody cylinder develops, surrounded by a sheath of secondary phloem, which has a lower *E* value than has xylem. This structure can be interpreted as an adaptation to both static and dynamic loading, as well as to transport function.

The anatomy of a petiole can be viewed as a compromise between adaptation to the static load of the weight of the blade and to the dynamic loading imposed by the wind. In small leaves, adaptation to the shock dynamic loading caused by the impact of raindrops is also a factor. From considerations of the ability of a palm leaf, which is very large, to support a static load, it appears that the cross section of a petiole should be between semi-circular and kidney shaped, with blunt edges. Furthermore, a petiole should be oriented with its flat or concave side upward, to maximize both torsional stiffness and resistance to downward bending. The palm petiole approaches this shape. It can support the palm-leaf blade in a stable position relative to sunlight in a gentle to moderate wind and seldom kinks even in a strong wind (Wainwright et al. 1976).

5 Photosynthetic Systems

5.1 Structural Aspects of Photosynthetic Function

Green plants synthesize themselves mostly from water and carbon dioxide. Collectively, the synthetic reactions are heavily endergonic and must be driven by trapped energy derived from outside the plant. This energy is invested mostly in separating the hydrogen and oxygen of water and in producing high-energy organic intermediates that then drive the reduction of carbon and the synthesis of carbohydrates.

Radiant energy as sunlight, though diffuse, intermittent, and variable, is available over most of the earth's surface. Carbon as gaseous CO_2 is available everywhere in the atmosphere — at a reliable 340 to 360 ppm. Water is always present as vapor in the atmosphere, but on many sites liquid water in amounts needed to sustain land plants is often limiting.

During photosynthesis, plants "consume" small amounts of water, thereby liberating oxygen. However, photosynthetic land plants lose far more water through transpiration than they consume in photosynthesis. This loss occurs because the plants, with their living cells always bathed in water, must ultimately exchange gases with a relatively dry atmosphere. Structural and spatial arrangements that permit ingress of CO_2 to photosynthetic cells also permit egress of water from their surfaces.

Plants have no selectively permeable barrier that would allow CO_2 to diffuse from the atmosphere into the plant while preventing egress of water vapor. Because the cuticle (Sects. 2.2.1; 3.2.2) is nearly impervious, the predominant route for moving gaseous CO_2 into the plant is through at least partly open stomatal pores (Sects. 9.3; 9.4; 16.3.4). These pores also permit water loss, unless the relative humidity is 100%. If that water is not replaced, the affected cells become flaccid, photosynthesis is inhibited, other systems fail, and the plant succumbs. Water-management adaptations enhancing a plant's ability to cope with the often limiting availability of water confer selective advantages. Ultimately, the several photosynthetic pathways discussed here are in part water-management adaptations.

In this chapter, we focus on the many structural and spatial requirements of photosynthetic function at the organ, tissue, and cellular levels. For example, photosynthetic cells need good access to light, which implies near transparency of any overlying protective cells or tissue layers. They also must have ready access to CO_2, which is absorbed immediately from the apoplasmic water, but ultimately from the atmosphere. Photosynthetic cells must have water supplies adequate to replenish water transpired to the atmosphere. Photosynthetic tissues must be able to dissipate heat, to damp the wide fluctuations in carbohydrate production (due to changing light intensity) by temporary local storage, and to regulate export of products to other parts of the plant. At the organ level, there must be systems for supporting and positioning light-intercepting organs having chlorophyllous tissues, usually leaves, in order to trap optimal amounts of radiant energy. There must also be translocation systems, efficiently integrated with support systems, leading to and from each region of chlorenchyma. These systems must be able to supply water and solutes to the chlorenchyma cells and remove products from them.

Leaves of typical broad-leaved mesophytes are constructed in ways that maximize both light interception and the supply of CO_2 to the chloroplasts. Thus, leaf laminae tend to be thin and broad, which maximizes light interception, and the arrangement of their chlorophyllous cells promotes diffusion of gaseous CO_2 along intercellular channels to cell-wall surfaces close to the chloroplasts. (Recall that diffusion is about 10 000 times faster in the gaseous phase than in the aqueous phase.)

After moving from a stomatal pore through intercellular spaces, CO_2 moves along an aqueous pathway from the site of its solution in the water of the cell-wall apoplasm to its site of fixation. Solution and fixation together maintain a concentration gradient that promotes inward diffusion along the gaseous part of the pathway. However, photosynthesis tends to be accompanied by photorespiration, which produces CO_2 within illuminated cells. This decreases the inward diffusion gradient of CO_2 in the aqueous part of the pathway.

There seems to be continuing evolution toward photosynthetic systems that use CO_2 with great efficiency. This evolution involves both developmental changes in mesophyll and bundle-sheath anatomy, and biochemical-physiological changes in the photosynthetic process (Holaday and Chollet 1984).

Several alternate pathways of carbon fixation can coexist within the same taxon. The extent to which, during the development of an organ or individual, the ascendancy of a system or a "pathway" is genetically or epigenetically determined is not yet clear (see Nelson and Langdale 1989).

5.2 Alternate Pathways of Carbon Fixation

Via the sets of interrelated mechanisms collectively constituting photosynthesis, chlorophyll-bearing cells use trapped light energy to produce the high-energy intermediate, adenosine triphosphate (ATP), and strong reducing agents, such as reduced nicotinamide adenine dinucleotide phosphate (NADPH). Some of the energy is also used to split water, thus liberating oxygen. The ATP and NADPH are used almost immediately to drive reactions in which CO_2 is reduced and "fixed" into nonvolatile compounds.

There are two well-known pathways of CO_2 fixation. These are not mutually exclusive within a plant. Intermediates also occur. In the most common pathway, abbreviated as **C3** (for its cycle of 3-carbon compounds), CO_2 from the gas space within the leaf enters the aqueous segment of the diffusion pathway and migrates down concentration gradients into chloroplasts of the mesophyll cells. There, it is directly fixed by reactions of the reductive pentose phosphate (RPP) cycle (also known as the Benson-Calvin cycle), which synthesizes three-carbon organic acids and, ultimately, carbohydrates. This relatively simple C3 route, which is also known as the **direct pathway**, is probably always present as a "default" system, especially during early ontogeny, even though it may be complemented by an alternative route later in development (Nelson and Langdale 1989).

In the C3 pathway, CO_2 enters the RPP cycle, within chloroplasts, by carboxylating ribulose-1,5-biphosphate (RuBP). This carboxylation, catalyzed by ribulose biphosphate carboxylase (RuBPCase), produces an unstable six-carbon compound that decomposes into two molecules of a three-carbon compound, phosphoglyceric acid (PGA), per molecule of CO_2 fixed. The PGA and the photosynthetically produced, energy-rich ATP and NADPH are used partly to synthesize carbohydrates and partly to regenerate the RuBP needed to continue the cycle.

Because RuBPCase is able to catalyze the oxygenation as well as the carboxylation of RuBP, oxygen competes with CO_2 for fixation sites. Oxygenation of RuBP is the crux of the "photorespiration" phenomenon, which leads to decreased efficiency of CO_2 fixation unless the concentration of CO_2 is high relative to that of oxygen — but high CO_2 concentrations in chloroplasts inevitably slow the inward diffusion of additional supplies. This untoward situation is worsened by the further reaction of oxygenated RuBP, producing PGA and glycolic acid. In the cytoplasm outside the chloroplast, some of the glycolic acid is decarboxylated to serine. The CO_2 thereby released can reduce the inward diffusion gradient of CO_2 along the liquid segment of its pathway to fixation sites.

Plants in which photosynthesis operates via this flawed "direct" pathway are commonly referred to as **C3 plants** (Björkman and Berry 1973). To overcome the inefficiency of the inward diffusion of CO_2 along shallow gradients, C3 plants keep their stomata open for extended periods and consequently lose significant water vapor. This loss is irrelevant if water is plentiful, but can be a handicap if water becomes limiting.

There is a well-studied, alternate route of photosynthetic carbon fixation, usually abbreviated simply as the **C4 pathway**. Actually, though, this "pathway" is a two-stage system, in which an auxiliary module carries temporarily fixed carbon to the RPP cycle, where it is then fixed as in the C3 pathway. The innovative element in this system is the shortening of the aqueous segment of the diffusion route of CO_2 by the temporary fixation of CO_2 in the cytoplasm (not the chloroplasts) of the mesophyll cells. This occurs via carboxylation of phosphoenolpyruvic acid (PEP) to form a four-carbon compound, oxaloacetic acid. This reaction is catalyzed by PEP carboxylase. Oxaloacetic acid then is converted into malic and/or aspartic acids, which are not directly used to synthesize carbohydrates, but instead diffuse along gradients to specialized C3 cells located deeper within the organ, usually in the bundle-sheath region. These are the "Kranz" cells (Sect. 5.3.2). In these cells, the four-carbon acids are decarboxylated to yield CO_2 and pyruvic acid (or sometimes PEP). This CO_2, via a carboxylation catalyzed by RuBPCase, then enters the RPP cycle, as in the C3 pathway, and is reduced to carbohydrate. The pyruvic acid (along with any PEP so formed) diffuses outward to mesophyll cells and reacts with ATP to form more PEP, thus closing the cycle.

In this system, sites of the initial absorption of CO_2 from the gas space of the mesophyll are separated from sites of ultimate fixation into carbohydrate. In effect, temporarily fixed carbon is carried from mesophyll cells to nearby bundle-sheath cells by shuttles of C4 acids. Plants having this **two-stage photosynthetic system**, with the stages localized in different cells, are called **C4 plants** (Björkman and Berry 1973; Monson et al. 1984).

An important aspect of the C4 pathway is that PEP carboxylase, which catalyzes the initial fixation reaction in mesophyll cells, has a higher affinity for CO_2 than has RuBPCase, which functions in the C3 pathway. Therefore, PEP carboxylase fixes CO_2 more efficiently at low concentrations than does RuBPCase. Any CO_2 liberated by photorespiration is quickly recaptured by PEP in the cytoplasm and does not reduce the gradient for inward diffusion of CO_2. This minimizes stomatal opening and loss of water vapor to the air. Further, unlike RuBPCase, PEP carboxylase is not also an oxygenase, and its CO_2-fixing action is not inhibited by oxygen (Monson et al. 1984).

Thus, although the two-stage, C4-C3 photosynthetic system seems cumbersome and is not particularly energy-efficient, it often allows relatively uninhibited photosynthesis at a minimal internal CO_2 concentration of less than 1 ppm in C4 plants, compared with about 50 ppm in C3 plants (Krenzer et al. 1975). This implies that, during active photosynthesis, C4 plants can maintain much steeper inward gradients of CO_2 than can C3 plants. Thus, C4 plants, by regulating stomatal apertures, can control transpirational water loss more efficiently than can C3 plants.

The C4 innovation apparently arose independently many times during relatively recent stages of angiosperm evolution. It has been found in more than 100 genera of at least 16 families of monocots and highly evolved dicots (Edwards and Walker 1983). Some highly efficient crop plants have this photosynthetic system, including *Zea mays*, *Sorghum vulgare*, and *Saccharum officinarum*, as well as eight of the world's ten most troublesome weeds in crop fields. Most other field crops, though, are C3 plants. The C4 pathway seems to be absent from gymnosperms and from the more primitive orders of angiosperms, such as Magnoliales, Laurales, and Ranunculales. The relatively few known woody C4 plants include *Euphorbia forbesii*, a small subtropical forest tree, and *Atriplex hymenelytra*, an evergreen shrub that grows on extremely xeric slopes in Death Valley, California (Pearcy and Ehleringer 1984). Another woody plant that may use the C4 pathway is the mangrove, *Aegiceras majus*, which colonizes shallow salt water (Joshi et al. 1974).

5.3 Structural Implications of Photosynthetic Pathways

Though, in a sense, all cells that contain chloroplasts belong to the photosynthetic system, in higher plants most carbon fixation occurs in chloroplast-bearing cells that are arranged in masses of **chlorenchyma** tissue. In laminar leaves, the chlorenchyma commonly occurs between the upper and lower epidermis, where it constitutes the mesophyll and commonly also the parenchymatous bundle-sheath tissue.

Epidermal cells, except for stomatal guard cells, commonly have only poorly developed chloroplasts. Guard-cell chloroplasts function in relation to opening and closing of stomata, and thus are involved in regulating the efficiency of photosynthesis, though they do not contribute significantly to plant carbohydrate production.

In the mesophyll chlorenchyma, which typically has anastomosing intercellular gas spaces, more than half of the cell surfaces usually are in contact with the gas phase. The remaining surfaces are in cell-to-cell contact, and have numerous plasmodesmata. It is safe to assume symplasmic continuity among mesophyll cells, and between meso-phyll cells and bundle-sheath cells (Gunning 1976; Russin and Evert 1985).

Generally in laminar leaves, the mesophyll is traversed by many slender vascular bundles, minor veins, which functionally are extensions of the network of major veins. The total length of minor veins is large. There are usually from several to a dozen or more millimeters of minor veins per square millimeter of leaf area. In fact, the vascular system, which both supplies water and removes photosynthetic products, extends into the immediate vicinity of most chlorenchyma cells (Sect. 16.3.5 and next).

The vascular elements of minor veins are surrounded by one or two layers of parenchymatous bundle-sheath cells, which typically lack intercellular spaces. In most plants, these cells are somewhat specialized. In C3 plants, they are intermediate between chlorenchyma and vascular elements. In C4 plants, they constitute the intensely green, "Kranz" tissue (Sect. 5.3.2), which is the site of the C3 part of the two-stage, C4-C3 photosynthetic system.

Taxa vary widely in the frequency of plasmodesmata between bundle-sheath cells and phloem cells (Gunning 1976). That frequency is very high in the *Populus deltoides* leaf (Russin and Evert 1985), suggesting that there is symplasmic loading of phloem elements from the bundle sheath.

Usually, the cortex of young stems is chlorenchymatous, though stem stomata commonly are too sparse for efficient fixation of atmospheric CO_2. Stem chloren-chyma may serve not so much to synthesize carbohydrates from exogenous CO_2 as to fix local respiratory CO_2, with the liberation of oxygen, which can then be used in oxidative metabolism in developing tissues.

In young stems, the most appropriate location for chlorenchyma would seem to be the outer cortex. This location, however, is also appropriate for collenchyma or other mechanical tissues (Sect. 4.7.3). As if the plant had reached a compromise, alternating sectors or bands of collenchyma and chlorenchyma tissues are common in the outer cortex. In addition, photosynthetic and mechanical functions are not mutually exclus-ive: collenchyma cells may have abundant chloroplasts, thus exhibiting a dual function.

5.3.1 Leaf Structure in C3 Plants

In a typical, hypostomatous, C3, laminar leaf, and in some columnar leaves, the mesophyll tissues are well differentiated into palisade and spongy chlorenchyma. Palisade chlorenchyma consists of narrow, cylindrical cells oriented perpendicular to the leaf surface and usually arranged in one or two layers subjacent to the epidermis. Although compactly arranged, palisade cells have little mutual contact, due to long, narrow, intercellular voids along their anticlinal walls. These voids, which constitute about 10% of the palisade-tissue volume, are continuous with the much larger voids in the subjacent spongy chlorenchyma. This arrangement promotes gas circulation and exchange within the leaf. Because of their cylindrical shape and orderly arrangement, the palisade cells have extensive wall surfaces along which chloroplasts can closely approach gas space.

In contrast to palisade chlorenchyma cells, spongy chlorenchyma cells are irregularly shaped. They have lobes that commonly are arm-like and that allow the cells to maintain some mutual contact in spite of the large, anastomosing gas spaces.

Those spongy-mesophyll cells just beneath the palisade layer commonly extend laterally between the minor veins, thus constituting a three-dimensional, spongy reticulum that links the lowermost palisade cells with the bundle sheaths. These are the **paraveinal mesophyll cells** (Sect. 16.3.5). They seem to be an efficient cell-to-cell pick-up and delivery system radiating outward from the veins (Kevekordes et al. 1988; Fig. 5.1A). In *Glycine max*, all photosynthetic exports apparently must pass through the paraveinal mesophyll on their way to the phloem (Franceschi and Giaquinta 1983). The leaves of *Populus deltoides*, which have palisade chlorenchyma on both adaxial and abaxial sides, have a well-differentiated paraveinal mesophyll layer in the plane of the veins, between the two palisade layers (Russin and Evert 1984).

The sharpness of mesophyll differentiation into palisade and spongy morphological types varies with plant age and may, in part, be environmentally determined. The palisade layer is less well developed in juvenile than in adult leaves, and in shaded than in well-insolated ones (Sect. 16.3.5). In well-insolated leaves, the palisade chlorenchyma occupies a larger volume than does the spongy chlorenchyma and contains a much larger percentage of the chloroplasts (Schramm 1912; Turrell 1936). The spongy chlorenchyma contains a larger percentage of the chloroplasts in shaded than in well-insolated leaves. In general, the surface of palisade cells available for gas exchange is several times larger than that of spongy mesophyll cells. The total gas interface surface within a C3 leaf may be 7 to 30 times greater than the total external leaf surface (see data tables in Schramm 1912).

In palisade cells, many of the chloroplasts intercept incoming light edge-on rather than broadside. This is because these cells are very narrow, with most of their wall area oriented parallel to the direction of the incoming light, and because their typically lens-shaped chloroplasts are very numerous. This combination of cell shape and chloroplast orientation allows efficient interception of light of high, but not of low, intensity.

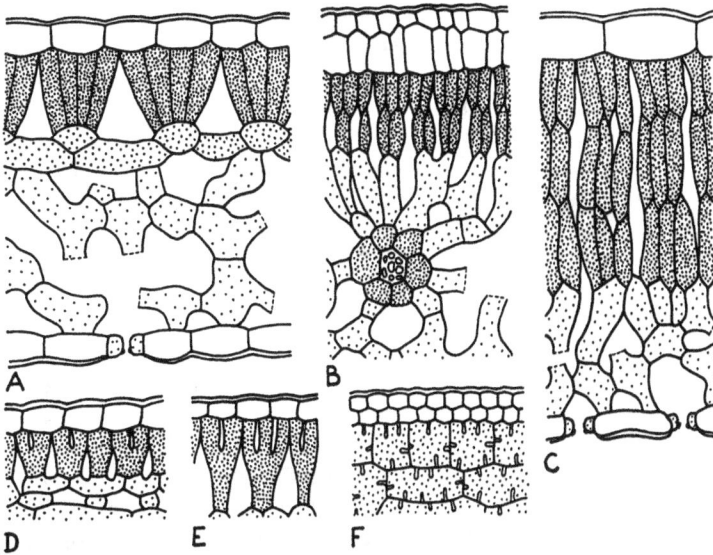

Fig. 5.1 A—F. Various arrangements of chlorenchyma in C3 leaves. *Density of stippling* indicates relative density of chloroplast complements. **A** Leaf with extensive development of paraveinal mesophyll cells, which form an anastomosing network between the palisade cells *(heavily stippled)* and the spongy mesophyll in a plane normal to the drawing. **B** Leaf having two tiers of palisade mesophyll and extensive spongy mesophyll. The upper epidermis is multilayered and lacks stomata. An extensive intercellular space system is ventilated by stomata in the lower epidermis (not shown). Bundle-sheath mesophyll *(moderately stippled)* is shown. It is not as clearly differentiated from spongy mesophyll in C3 leaves as in C4 leaves. **C** Hypostomatous leaf with three tiers of palisade cells and a spongy-mesophyll layer. **D**, **E**, and **F** Several shape adaptations that increase the surface area of mesophyll cells exposed to internal gas space. Cells that seem *Y-shaped* in **D** and **E** are of the "arm palisade" type. **F** Thin transverse sheets of mesophyll cells (in the plane of the drawing) separated by gas spaces occur in leaves of various members of Pinaceae and have numerous wall ingrowths which develop into slit-like gas spaces within their seeming "infolds"

In less specialized chlorenchyma, such as spongy mesophyll, there tend to be fewer chloroplasts per unit volume, and these can be redistributed by light-influenced chloroplast movements. As a result, the chloroplasts can assume positions near walls that are normal to the incoming light and can intercept it broadside. Thus, they efficiently use light of low, but not of high, intensity. Although the adaptive value of having more spongy and less palisade parenchyma in shaded leaves seems clear, it can be argued that, because light reaching the leaf surface is often diffuse, and there is appreciable scattering of light within the leaf, chloroplast migration in response to direction of illumination can make only a minor contribution to efficiency of light use (Britz 1979). However, in some plants, epidermal cells function as cellular "lenses", modulating the amount of light reaching the chloroplasts in the cells beneath, as discussed in Section 5.5.

In some C3 plants, the chlorenchyma is not differentiated into typical palisade and spongy types. For example, some *Populus* leaves have only palisade cells. In some leaves, especially the needle leaves of various gymnosperms (Sect. 16.3.5), mesophyll-cell-wall area available for display of chloroplasts is increased by development of apparent "infoldings" of the wall into the lumen. Transections of needles of *Pinus resinosa*, for example, show infolds (Fig. 5.1F), and suggest that the chlorenchyma has little intercellular space. However, longitudinal sections of such needles reveal that the chlorenchyma cells are arranged in transverse plates one cell thick, with each plate slightly separated from its neighbors by gas space (Gambles and Dengler 1982a).

5.3.2 Leaf Structure in C4 Plants

Leaves of C4 plants typically differ anatomically from those of C3 plants — even within a genus (Björkman and Berry 1973; Monson et al. 1984). These differences are not fundamental in a developmental sense. In a C4 plant, the separation of sites of preliminary (C4) and permanent (C3) fixation of CO_2 is evidenced by arrangement of the chlorenchyma into mesophyll and bundle-sheath types (Fig. 5.2). The mesophyll type typically is pale to medium green, whereas the bundle-sheath type is more intensely green, as is apparent in leaf cross sections at low magnification (Laetsch 1974). Haberlandt (1882) applied the German term "Kranz" to the wreath-like appearance, in cross section, of the compactly arranged, intensely green cells around the minor veins of leaves of certain species, which we now know are C4 plants. In three dimensions, this "Kranz" constitutes a sheath usually identical with a bundle sheath (Nelson and Langdale 1989).

The leaves of C4 plants are more thoroughly vascularized than are those of C3 plants. In C4 plants, neighboring veins may be separated by as few as four green cells, whereas in C3 plants there are apt to be ten or more (Crookston 1980). The bundle-sheath cells of C4 leaves, though similar in origin and development to those of C3 leaves, have thicker walls, with many pits and plasmodesmata. These cells also have larger and more numerous chloroplasts than do bundle-sheath cells in C3 leaves. In contrast, the mesophyll cells of C4 leaves are morphologically similar to those of C3 leaves, and seem to contain the same type of chloroplasts.

Presence of an intensely green bundle sheath, which is the basis of the **Kranz syndrome**, has been used as an indicator of the C4 system, but it is not infallible. *Suaeda monoica*, for example, though reportedly a C4 plant, has extensive water-storage tissue rather than bundle sheaths around the vascular bundles. Subepidermal layers of outer (C4) and inner (C3) chlorenchyma surround this water tissue (Shomer-Ilan et al. 1975), with no Kranz around the vascular bundles. In addition, the mesophyll cells of C4 leaves do not always conform to the common, spongy-mesophyll type. In some Kranz-displaying taxa of Amaranthaceae and Chenopodiaceae (which include arborescent forms), the mesophyll cells immediately surrounding the bundle-sheath tissues have a palisade arrangement (Laetsch 1968).

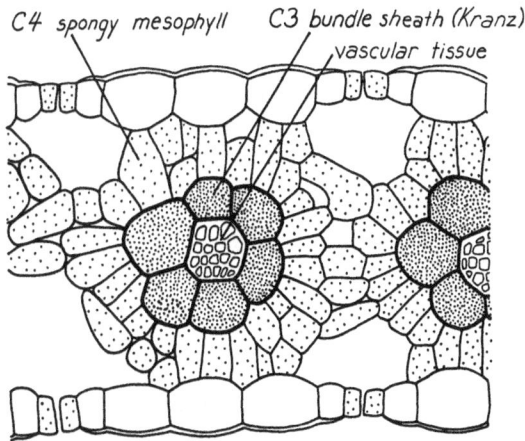

Fig. 5.2. Arrangement of chlorenchyma in a leaf in which "**Krantz anatomy**" indicates a C4 photosynthetic system. The bundle-sheath cells, having more and larger chloroplasts than do the spongy-mesophyll cells, are intensely green and also have thickened walls. This leaf lacks palisade mesophyll. The spongy mesophyll is aerated by stomata on both upper and lower laminar surface

The detailed structure of the bundle-sheath tissues also varies with the several biochemical variants of the C4-C3 scheme. Some of these differences are related to whether the bundle-sheath tissues are one or two layers deep (Nelson and Langdale 1989).

Although the Kranz bundle-sheath cells are a salient feature of C4 plants, carbohydrate is synthesized within them by the C3 pathway, as in C3 leaves — except that CO_2 is brought to them by the C4 "shuttle" innovation. Thus, the photosynthetic activity of these cells is independent of CO_2 diffusing from the gas spaces, and the compact arrangement of the cells, which slows diffusion to their chloroplasts, is not disadvantageous.

The Kranz cells seem to be connected with adjacent mesophyll cells by numerous plasmodesmata (Osmond and Smith 1976; Fisher and Evert 1982). It is likely that there is symplasmic transport of C4 acids to the Kranz cells, and of pyruvic acid (the precursor of PEP) from these cells to the mesophyll.

The chloroplasts of Kranz cells tend to be larger than those of ordinary mesophyll cells, and, during periods of rapid photosynthesis, they contain numerous starch grains. The walls of Kranz cells may be extensively thickened, which should increase their efficiency as apoplasmic transport channels.

Kranz tissues have been described as modified parenchyma sheaths of minor veins in the leaf. This may typically be so, but Kranz tissue is found in stems as well as in leaves, and may develop either from provascular tissue or from ground parenchyma. Further, within leaves, it may be associated with tissues other than mesophyll (Brown 1975). The term "Kranz syndrome" is sometimes used with reference to anatomical (as distinguished from biochemical) evidence of the presence of the C4 pathway. The term "Kranz mesophyll" has been applied somewhat loosely to the intensely green bundle sheaths in C4 leaves, and not always with reference to the C3 tissue only (Fisher and Evert 1982).

Examining the anatomy of chlorenchyma for the Kranz syndrome is a moderately reliable method of learning whether a plant is of the C4 or C3 type. However, there are many biochemically intermediate types that are also anatomically intermediate, or even typically C3 (Chaguturu 1981; Monson et al. 1984).

5.4 CAM and Other Variations

Besides C4-C3 intermediate photosynthetic types, there is another modification of the two-stage pathway of CO_2 fixation — i.e., **crassulacean acid metabolism (CAM)**. In this system, the initial fixation of CO_2 into acids and its final fixation via the RPP cycle are typically separated by some hours, and may be spatially separate also. The C4 part of CAM operates mainly at night, when temperatures are lower and the relative humidity is higher than by day. Stomata can be opened at night with minimal water egress relative to CO_2 ingress. The C3 part of CAM operates by day and requires light, but *no* stomatal gas exchange. Because stomata typically are closed by day and open at night, water use by these plants is highly efficient (Edwards and Walker 1983).

A CAM plant uses energy derived from oxidation of stored starch to fix CO_2 (in darkness) into C4 organic acids, which accumulate during the night. During the day, the plant decarboxylates the acids and, by a C3 system, photosynthetically fixes the CO_2 thus released.

CAM occurs in about 30 families of flowering plants, and, like the C4 pathway, does not seem to occur in gymnosperms or primitive angiosperms. It is rare in woody plants; the hemiepiphyte, *Clausia rosea*, is one of the few known examples (Ting et al. 1985). In some families, such as Crassulaceae and Cactaceae, almost all species have CAM, whereas in other families, including Liliaceae and Asclepiadaceae, only the succulent species do. Many orchids and bromeliads have CAM systems, though not necessarily in all developmental stages or under all environmental conditions. The pineapple, *Ananas sativus*, is facultative, using CAM when under water stress and otherwise using C3 (Ting 1985).

CAM plants, which typically are succulent, do not usually have Kranz anatomy. They have: (1) a marked diurnal fluctuation in organic-acid content — high in the morning and low in the evening; (2) a diurnal cycling of soluble and storage carbo-hydrates — in a pattern reciprocal to that in (1); (3) high activities of appropriate carboxylase and decarboxylase enzymes; (4) large, highly vacuolate chlorenchyma cells, with numerous large chloroplasts; and (5) a stomatal system that closes apertures by day and opens them at night.

Both CAM and C4 photosynthetic systems seem to be of polyphyletic origin. These systems may be an example of biochemical and structural convergence. Both systems have been regarded as adaptations to aridity or salinity. It seems possible that in both types of adaptation, rapid fixation of CO_2 into organic acids may ultimately have a salt-management as well as a water-management function (via photosynthesis), as anions other than those of organic acids and hydroxyl may not always be available in good supply.

Green cells commonly occur in deep-seated anatomical locations, where the weak light available would seem to preclude a significant contribution to photosynthesis. These cells, often considered an anomaly, usually are in or near sites of intense metabolism, where metabolically liberated CO_2, but little atmospheric CO_2, is available. For example, in young woody stems, there is often local greening in the cambium and nearby tissues. The important function of the chloroplasts in these cells may not be synthesis of carbohydrates per se, but the liberation of supplemental oxygen that can be used locally in respiration. Local CAM-like photosynthesis, using organic acids imported from elsewhere in the plant, could be occurring. We have no data on this, however.

5.5 Leaf Structure in Relation to Optical Properties

Leaf mesophyll and epidermis have optical properties that enhance light absorption for photosynthesis. The role of epidermis in controlling the internal distribution of light within a leaf was largely overlooked until relatively recently (Poulson and Vogelmann 1990). Haberlandt (1909), however, noted that the epidermal cells of some leaves are convex in shape, or otherwise modified such that light striking the leaf surface is focused onto certain regions within the leaf. In contrast to more recent investigators, he interpreted this focusing as providing the leaf with a mechanism for detecting light direction (Sect. 10.2.2.2), rather than for increasing photosynthetic efficiency. Nevertheless, the variety of epidermal-cell modifications that Haberlandt described is of some interest.

For example, in some leaves, the plano-convex, upper epidermal cells are the lenses. In other leaves, the outer epidermal wall may show a lens-like thickening. Haberlandt demonstrated, using pieces of peeled *Anthurium* leaf epidermis and photographic paper, that some walls are such effective lenses that they can be made to project recognizable images of objects "seen" by the leaf.

In addition, scattered among ordinary upper-epidermal cells of *Fittonia* leaves are many unusual, bicellular eye-like structures that Haberlandt called **ocelli**. An ocellus consists of a large, basal, epidermal cell having a dome-like external surface that protrudes above the general epidermal plane, and a small, bi-convex, lens-like cell at the vertex of this dome.

More recent workers have emphasized the role of epidermal focusing in intercepting light for photosynthesis. They have found that convex upper epidermal cells focus light onto the palisade layer, and that this focusing seems to result in local irradiances within the leaf that are significantly greater than those of the light incident on the leaf surface. In addition, they have found that chloroplasts at different levels within the leaf seem to be adapted to light of different intensities. In *Medicago* and *Oxalis*, focusing by epidermal cells reportedly channels light to a population of chloroplasts that are adapted to high light intensity (Martin et al. 1989; Poulson and Vogelmann 1990).

Focusing by leaf epidermal cells seems to be better developed in herbaceous than in arborescent plants. Whereas epidermal lens cells have been found in several herbaceous species, the leaf epidermis in the arborescent plants thus far investigated seems to be relatively flat (R. A. Donahue and G. Martin, personal communication to J. H.).

6 Storage Systems

6.1 Storage and Survival

Of the many metabolic reactions occurring in living plants, most yield only "intermediates" that are quickly used in the synthesis of other organic substances, or "metabolites" that are soon oxidized to yield energy. But, some products are not used immediately, including sugars, starches, and oils. These are "stored", or accumulated as reserves in a form that is protected from wastage but still available for quick remobilization and use. Reserves tend to be made and consumed at rates that vary greatly over time and in different plant organs. Conditions of *net* synthesis and *net* consumption of a particular plant product typically occur at different times and in different places. Spatial separation of these functions necessitates transport (Chap. 7). Temporal separation entails storage. Transport is essential even for hour-to-hour maintenance of the organism. Storage is essential for longer-term survival of the individual and for reproduction, and thus for survival of the species.

Though the water in a plant is not a "product" or "metabolite" in the sense that organic materials are, management of water reserves is included in the concept of storage. Absorption, storage, and use of water typically occur at different times and in different places within the plant. Thus, transport and storage of water, as well as of organic materials, is often basic to survival.

Superposed on the day/night rhythm of photosynthetic activity are irregularities due to variations in light intensity, temperature, and water availability. When conditions are favorable for photosynthesis, carbohydrate production in leaves often exceeds export. The excess is stored, usually as starch. At sunset, photosynthesis quickly drops to zero, but export of sugars derived from starch stored in the leaf can continue during darkness. Plants must also cope with more extended interruptions of efficient photosynthesis, as during long periods of heavy overcast, after defoliation by pests or hailstones, or, in deciduous trees, after the leaf senescence and abscission that mark the onset of the dormant season. When dormancy ends, many plants must be able to mount a burst of growth at the expense of stored reserves, before again being able to produce new assimilates efficiently. To cope with such fluctuations in photosynthate production and with unreliable water supplies, many perennial plants have evolved capacious storage systems.

In addition to reserves that are remobilized and used, a plant may accumulate some products that neither it nor its progeny use directly. Instead, these products have an ecological role, eliciting from insects or higher animals behavior that protects the plant or aids in pollination or seed dispersal. Thus, some reserves are not stored for the plant's own future consumption, but to promote, in other ways, its survival or the survival of the species.

A variety of plant tissues and organs serve as depots for storage. In typical mesophytes, including most temperate-zone and many tropical trees, the various structural and functional attributes are so highly integrated and interrelated that storage is apt to be a secondary rather than a primary function of tissues; no organ or tissue component seems specifically devoted to storage. Just as most living cells of a plant, being interconnected into a single symplasm, participate to some extent in symplasmic transport, so do most of them at times serve a storage function. Some exceptions may be sieve elements, companion cells, and Strasburger cells (Sect. 7.2.2), but even in these cells, slow transport may be de facto storage. The importance of the various components of the storage system may change with plant age and with seasonal or other external factors.

Typically in woody plants, reserve carbohydrate metabolites are stored primarily in parenchyma of the secondary xylem (Sects. 19.2.2; 19.3) and of the bark (Sect. 20.3.3). Storage proteins (which may occur at low levels in summer, but accumulate in autumn) have been found in *Fagus*, *Alnus*, *Betula*, and *Quercus*, but not in *Gleditsia*, or in *Robinia* (Wetzel and Greenwood 1991).

Tissues having storage as their major function are especially common in plants adapted to extreme habitats or to climates offering only strongly seasonal or irregularly episodic intervals of favorable conditions. In this chapter, we emphasize tissues and organs that have, as a major function, storage of reserve organic materials or of water.

6.2 General Characteristics of Storage Organs and Tissues

Within a plant, water is generally stored differently from organic materials. Organic reserves are typically stored within the symplasm, specifically in organelles such as plastids or spherosomes and in bodies such as aleurone grains. However, these reserves may also be stored apoplasmically, in the form of hemicellulose, typically in the walls of endosperm cells of certain seeds (Sect. 6.4.2). Water is often stored apoplasmically — in the cell walls, in the lumina of dead cells, or, rarely, in intercellular spaces. It is also stored in the vacuoles of living cells.

Many storage tissues consist of thin-walled parenchyma cells, but cells of other types (or the lumina of dead cells) can also become storage depots. Storage organs tend to be capable of rapid increase in volume. For storage of organic materials, this increase may arise from accessory vascular cambia or from proliferation of ray or other parenchymatous tissues. Parenchymatous water-storage tissues and organs typically are capable of notable swelling as they "fill" and then of shrinkage as water is withdrawn. If the water-storage cells are dead, they may become extensively embolized as water is withdrawn. Storage organs that sequester large reserves of organic compounds have relatively well developed-contact with sieve elements, but tracheary contact may be poor. This situation may be more balanced in water-storage tissues.

6.3 Systems that Store Water

We will discuss storage of water in the vacuoles of living cells first, then water storage in the apoplasm. Subsequently, we will consider a water-storage system based on mucilage that may be either apoplasmic or vacuolar.

6.3.1 Water Storage in Vacuoles

An important feature of living water-storage cells is that, upon giving up water, they shrink or even collapse, but in an orderly, noncatastrophic manner. They are able to lose water and regain it, repeatedly. The controlled absorption and release of water by such a cell depends on the presence of a living protoplast with an intact plasmalemma.

The walls of the large parenchymatous cells that function primarily in water storage typically lack secondary thickening and pits, and are extensible. Most of the cell volume is occupied by watery, vacuolar sap. There is a thin, peripheral layer of cytoplasm. These cells are nucleate and often have a few chloroplasts or other plastids.

When the cells are turgid, their contents are under pressure and their walls are in tension (Sect. 4.4.1). When water is withdrawn through the plasmalemma into the apoplasm, cell volume decreases and wall tension relaxes. The surface-to-volume ratio of the cells increases. The cells become flaccid or partially collapsed. Their walls assume undulating or folded contours, because, in living cells, water that is withdrawn is not replaced by gases. Consequently, when water is again available, turgidity of the water-storage cells can be restored. Shrinkage in the water-storage layers of leaves often causes curling. Entire water-storage organs may shrink, so that their outer layers become wrinkled or even pleated — as in some cacti.

In the stems and leaves of many taxa, the water-storage tissue is part of a **multiseriate living epidermis** derived from the protoderm by periclinal divisions. A nonchlorophyllous multiseriate epidermis from 2 to 16 layers thick is relatively common. The inner layers of this tissue often differentiate into putative water-storage tissue, which may include storage tracheids (Linsbauer 1930). Such water-storage tissue occurs in Moraceae, Pittosporaceae, Piperaceae, Malvaceae, and other families.

Succulent stems and leaves characteristically have internal (rather than epidermal) water-storage tissues. Succulence is typical of taxa that are adapted to xerophytic conditions and that may lack persistent, laminar leaves. These taxa include almost all Cactaceae and members of the genera *Agave, Aloe,* and *Mesembryanthemum.* In stems and other bulky organs, water-storage cells, when turgid, tend toward isometry. In contrast, in the laminae of broad-leaved taxa, they often are of the palisade shape.

In succulents, water-storage tissue usually has extensive contact surfaces with photosynthetic tissues. Succulents characteristically have CAM (Sect. 5.4), with its pronounced diurnal fluctuation in soluble organic acids. The acids are produced at night and stored in the vacuoles of the chlorenchymatous and nearby water-storage cells.

Water is stored in the stems and roots of some large trees. Cauline water-storage tissues are well developed in members of Bombacaceae, such as *Ceiba* (kapok tree), *Adansonia* (baobab tree), *Ochroma*, and *Bombax* (cotton tree), none of which are known to use CAM. *Adansonia digitata* is notable because it retains laminar foliage while having extensive water-storage tissues in its axes. Young *Adansonia* trees have large, turnip-like taproots that store water and organic substances. Once a tree is established, storage in the taproot becomes relatively insignificant as the tree develops a trunk consisting largely of water-storing parenchyma. With increasing age, the trunk may become grotesquely bottle-shaped and attain a girth of many meters. The stems and major branches of *Adansonia* trees, though they contain relatively little ordinary wood, are quite strong mechanically when well hydrated, because they are hydraulically supported. The water content of internal stem tissues may be as high as 400% (dry-weight basis), and cut blocks may shrink as much as 40% in volume upon drying (Wyk 1974). Stem girth can fluctuate with the seasons.

Aside from stems of trees or cacti, with their variously specialized tissues, the most capacious water-storage organs are fleshy tubers. Many of these are known primarily for their reserves of starch, inulin, or sugars (Sect. 6.4.1), and, indeed, many xerophytic taxa have subterranean tubers with large stores of both water and organic reserves. These enable the plants to live quiescently for long periods and then to mount a rapid burst of shoot expansion, flowering, and seed production after rains come.

6.3.2 Water Storage in Living or Dead Cells

Sapwood (Sect. 19.5.2) is a major water-reservoir in trees. For example, the total water storage in the trees of a *Pseudotsuga* forest has been estimated at 267 m^3/ha (or 26.7 ha mm units), of which 75% is in the sapwood. During summer droughts this reserve may be drawn upon to supply as much as 1.7 ha mm of water per day for short periods (Waring and Running 1978). Water storing tracheids and special tracheidal idioblasts, usually terminating veinlets (Tucker 1964), may occur in leaf mesophyll and in some young stem tissue and may also store significant water. Well-developed tracheidal idioblasts in the shoot of the halophyte, *Arthrocnemum fruticosum* may store water as well as translocating it to living, water-storing parenchyma from stomata adapted to absorbing dew (SaadEddin and Doddema 1986).

The **velamen** (Sect. 3.2.3.4) of terrestrial and epiphytic orchids and of certain other monocots consists of layers of dead cells that are air filled when water is not available. The velamen has been interpreted as a multiseriate rhizodermis. During rainy weather, its cells become saturated with, and then store, water. The velamen is typically underlain by a living exodermis (Sect. 2.3), which has passage cells that allow both apoplasmic and symplasmic transport of water to and from storage sites.

6.3.3 Water-Storage Systems Based on Mucilage

Another putative water-storing strategy involves **polysaccharide mucilages**. Cells and tissues rich in mucilages can function as water reservoirs because these substances, if dissolved or dispersed in the cell sap or in apoplasmic water, increase the water potential. Mucilages are common in cacti. The controlled absorption by, and release of water from, storage systems based on mucilage may or may not depend on the presence of living protoplasts.

In some taxa, the active polysaccharides of mucilaginous, water-storing cells may be derived from partial hydrolysis of the cell walls (Mauseth 1983). In other taxa, including *Echinocereus, Nopalea,* and *Opuntia,* mucilages are produced by the protoplasts of "secretory cells", which secrete these substances into the periplasmic space, between the plasmalemma and the cell wall. The mucilages accumulate there and gradually encroach on the cell lumen, often eccentrically. They eventually occupy the entire cell, as the protoplast gradually degenerates (Mauseth 1983, 1988).

Trachtenberg and Fahn (1981) found that the mucilage-producing idioblasts of *Opuntia ficus-indica* account for only about 3% of plant-body volume. Can such a small volume fraction be significant in water storage? Trachtenberg and Mayer (1982) suggested that the major function of mucilage cells may, instead, be dynamic storage of Ca^{++}, and hence involvement in CAM metabolism, because Ca^{++} is the major counterion to the organic-acid anions of that system.

Water held by mucilaginous substances in periplasmic or intercellular spaces is apoplasmic water, but because of barriers such as Casparian bands in the endodermis or exodermis, it is not necessarily continuous with the apoplasmic water of the root cortex or the rhizosphere.

6.4 Systems that Store Elaborated Reserves

6.4.1 Storage in Vegetative Organs

Storage of organic materials is the major function of masses of parenchyma in specialized, thickened regions of stems (fleshy rhizomes, corms, tubers), roots (fleshy roots, root tubers), or buds (bulbs, "heads"). The parenchymatous storage tissues in these organs are derived from primary and/or secondary meristems. Storage of organic reserves is typically symplasmic, although storage may also occur in the cell walls. Fleshy rhizomes, tubers, corms, or bulbs are often the only vegetative organs of a plant to survive during a dry or cold interval that separates successive growing seasons. Thus, they also function as organs of vegetative propagation. Their "capital reserves" give the propagules a competitive advantage.

Structural and functional details of tissues and organs that store elaborated reserve materials vary greatly with taxa and habitats. A few examples of the rich variety among familiar plants follow.

In *Beta* (beet: Chenopodiaceae), the storage organ is mostly root tissue. At an early stage of root development, an ordinary vascular cambium differentiates. Then, derivatives of a phellogen differentiate into a thin periderm that covers the organ's surface. Subsequently, the cells of the pericycle proliferate into parenchyma between the stele and the periderm. A concentric succession of accessory vascular cambia rapidly appears within this parenchyma. These cambia produce increments of nominal vascular tissue, having poorly differentiated xylem and phloem and copious parenchyma. The various cambia seem to grow simultaneously, which makes possible rapid growth in root thickness. A root transection reveals concentric rings of anomalous secondary thickening, consisting of parenchymatous xylem and phloem. All these rings have vascular connections to the leaves. The *Beta* root stores sucrose dissolved in the vacuolar sap.

Daucus (carrot: Apiaceae) and many other biennials develop fleshy taproots. In a young *Daucus* taproot, vascular-cambial activity quickly produces voluminous, highly parenchymatous phloem outwardly and somewhat "woody" xylem inwardly. Growth in the phloem region is augmented by proliferation of ray and pericycle cells — but there is no differentiation of accessory concentric cambia as in the *Beta* root. Reserves are stored in the carrot root as soluble sugars in the vacuolar sap and as grains of starch and crystalline carotene in plastids (Esau 1940).

Among the crucifers (Brassicaceae), the storage organ may be a tuberous root, as in mustard (*Brassica*) and horseradish (*Armoracia*); a tuberous hypocotyl, as in radish (*Raphanus*); or a tuberous stem, as in kohlrabi (*B. caulorapa*). In these structures, periderm is present; the secondary phloem is parenchymatous and thin; the vascular cambium is close to the surface; and the bulk of the organ is homologous with secondary xylem, though it consists mostly of parenchyma within which vascular strands are dispersed. The storage products are sugars and starch.

The storage organ of the common potato, *Solanum tuberosum* (Solanaceae), is the swollen tip (tuber) of a rhizome — a stem. This tuber, which has visible leaf and branch buds (the "eyes"), is obviously derived from ordinary rhizome structures. The rhizome and tuber have a central vascular cylinder and a thick cortex. In the tuber, the phloem is much more voluminous than the xylem. The rapid volume growth of the tuber is mostly ascribable to proliferation of parenchyma, particularly in the inner part of the "procambium" (Sect. 13.5.2), around the pith. The stored reserve is mostly starch, though some sugars and proteins are also present.

The root tubers (as distinguished from stem tubers) of many genera are formed as swellings of secondary roots. In *Ipomoea batatas* (sweet potato), the fleshiness of the root tuber is due to large volumes of parenchyma in the primary and secondary xylem. Some groups of vessels are surrounded by anomalous cambia that produce further parenchyma. In root tubers of *Dioscorea floribunda* and *D. spiculiflora* (yams), the storage parenchyma may be partly derived from activity of the ground meristem in a bulge between the root and shoot of the seedling (Martin and Oritz 1963).

In the *Manihot esculenta* tree (Sect. 6.6), root tubers are initiated along young, actively growing roots. This occurs locally after the cambium has already deposited a 10- to 20-cell-thick layer of ordinary secondary xylem, consisting mainly of vessel members and fibers. Tuber initiation is marked by a rapid increase in cambial activity, leading to formation of large-celled storage parenchyma instead of xylem fibers (Lowe, et al. 1982). Laticifers occur in the xylem parenchyma and the phloem. In some variants, the latex contains cyanophoric glycosides, and cyanide is liberated if the tissue is wounded.

In woody plants, the parenchyma of the secondary xylem, especially in the roots, and of the secondary phloem is significant as an episodically or seasonally depleted and replenished starch-storing tissue. Storage of proteins in special vacuoles within the ray parenchyma cells of stem sapwood has long been recognized. In *Populus*, these proteins are mobilized during bud break and moved to the xylem sap (Sauter, et al. 1988). The vascular rays of root wood tend to be larger and more abundant than those of stem wood (Sect. 17.9), and are important storage sites. Their storage capacity is enhanced by longevity. Because heartwood seldom develops in roots (Fayle 1968), rays of old roots have many living cells and can still have a storage function.

Although, in single microtome sections of wood of many species, xylem-parenchyma cells appear scattered, three-dimensional spatial studies confirm that in most species the axial xylem parenchyma is symplasmically continuous with the radially oriented parenchyma cells of the vascular rays. The symplasmic meshwork is continuous from the phloem across the cambium and into the xylem (Zimmermann and Brown 1971). Thus, it is likely that every living parenchyma cell of the wood and inner bark is in symplasmic contact with the conducting phloem (Ziegler 1964).

Although starch is the most common reserve substance stored in this symplasmic parenchyma meshwork in both roots and shoots, the proportion of starch relative to other reserve materials is highly variable, because starch is readily converted into sugars, and sugars can be converted into lipids. During the depths of winter dormancy, stored starch is at a minimum, whereas stored lipids are maximal. In areas where the soils of the root zone are rarely frozen, starch may be stored in the roots throughout the winter, but in very cold climates lipids may replace starch even in the roots (Ziegler 1964).

Storage tissues function as sinks when they are being loaded and as sources when they are unloaded. A widely accepted hypothesis is that solutes move from sources to sinks in response to gradients of simple concentration or of electrochemical potential. Each solute moves independently down its own gradient. This simple hypothesis may be valid for short-distance transport, but long-distance transport is more complex, involving the phloem and some active processes (Sect. 7.2.4).

6.4.2 Storage for a Later Generation

Endosperm (Sect. 13.1.3.2) and **perisperm** (nucellar tissue that has proliferated into a food-storage tissue) are specialized storage tissues that occur in seeds of some angiosperms. In gymnosperm seeds, most reserves are stored in a macroscopic, female-gametophyte generation, the **megagametophyte** (Sect. 13.1.2.1).

Materials stored in the endosperm, perisperm, and gymnosperm megagametophyte serve as initial "capital" for the new sporophyte and are not available to the maternal sporophyte, though the perisperm is part of the maternal sporophyte body. The predominance of seed plants on most land surfaces is partly ascribable to competitive advantages their offspring derive from inherited capital reserves in the seed.

Endosperm and perisperm, and the gymnospermous megagametophyte tissues that are functionally comparable to them, may accumulate carbohydrates and other reserves to the extent of 70% of their weight and then become relatively dehydrated and dormant in the mature seed. During germination, the cells of these tissues, in contrast to those of the embryo, are only partially reactivated, in that they generally do not again grow or divide.

Examples of endosperm are: the main, "starchy" part of the cereal seed; the coconut copra; and the stony part of the date seed. Examples of perisperm are: the hard tissue of the coffee seed and the principal part of the storage tissue of the pepper seed.

The reserve substances stored in endosperm, perisperm, and in gymnosperm megagametophytes typically are starch, oils, proteins, and hemicelluloses. Starch is stored in plastids (amyloplasts). Oils are held in spherosomes or simply as droplets in the cytoplasm. Oils and starch are rarely stored in the same cell, but oil-storing cells and starch-storing cells may occur in the same tissue. Proteins as well as amides and free amino acids may accompany starch in plastids and may also be stored in special protein bodies.

Appreciable reserves may be stored in the embryo itself, often in fleshy cotyledons specialized for storage, as in Fabaceae (Leguminosae), Fagaceae, Juglandaceae, and other families in which endosperm and perisperm are poorly developed or lacking. Fleshy cotyledons account for most of the volume of seeds of beans and peas, and of acorns, walnuts, and many other common nuts.

In addition to storage within the embryo and in endosperm, perisperm, or mega-gametophyte tissues, "arilloid" outgrowths from the ovules or associated tissues may also accumulate reserve substances. These substances are not again available to the parent plant or to the embryo, so in a sense they are not really stored reserves. The flesh of the berry-like fruit of *Taxus*, for example, is arilloid tissue (Singh 1978).

The parenchyma of fleshy fruits typically is rich in stored sucrose, other carbo-hydrates, and organic acids. Like the substances stored in arilloid tissues, these are not again available to the parent plant, but serve primarily as animal attractants. During the ripening of fleshy fruits, the middle lamellae of their large, highly vacuolated, thin-walled parenchyma cells become partially hydrolyzed and the cells then are easily separated. In addition, the cell contents often are somewhat autolyzed, and parts of the parenchyma become semiliquid or juicy. The edible sec-tions of a citrus fruit (a specialized berry) are filled with large spindle-shaped cells turgid with juice. These cells originate as multicellular emergences from the inner surface layers of the carpels and are distantly analogous to trichomes.

Perisperm is of the maternal sporophytic generation and may have symplasmic connections with other maternal tissues. Endosperm has no symplasmic connections with maternal tissues. Similarly, cotyledons, being of the new sporophytic generation (Sect. 13.1.3), have no such connections. There is also no symplasmic continuity between the embryo and the perisperm or endosperm. This means that storage material is transferred apoplasmically, either from the maternal sporophyte into the endosperm and then from the endosperm into the embryo, or directly from maternal tissue into the embryo. This transfer is probably facilitated by transfer cells (Pate and Gunning 1972).

6.5 Storage in Geophytes

Many plants subjected to episodes of unfavorable conditions can survive because they have stored reserves of energy-rich metabolites and water. Even mesomorphic, temperate-region plants must endure an annual period of cold weather, and some plants can survive and reproduce in habitats that offer only short, irregular episodes of favorable conditions separated by long periods that are harsh. An appreciable percentage of temperate-zone forest plants, and a much larger proportion of plants growing under especially austere conditions, survive because their perennating buds are borne on persistent, underground storage organs, where they are protected from the extremes of the atmospheric environment.

These plants, known as **geophytes**, are a major floristic component of dry steppe, chaparral, and other semidesert regions. They may also account for as many as 15% of the species in the temperate forests of western Europe (Géhu and Géhu 1977). However, geophytes commonly are difficult to recognize because careful excavation over much of a season is needed to establish a single-organism relation between a visible and familiar shoot and a relatively featureless underground storage and perennating structure.

During periods unfavorable for active growth, geophytes live on stored reserves. Some cytogenesis and organogenesis proceed, but volume growth and gross shoot morphogenesis are held in abeyance. When good growing conditions return, the geophyte can quickly send up a shoot, flower, and set seed. To survive, it must also synthesize and store metabolites against the next period of unfavorable conditions.

Pate and Dixon (1982) pointed out that there are great numbers of unrecognized geophytes in semidesert regions of western Australia. The subterranean storage organs of these plants range widely in size and morphology. Some form woody "lignotubers". The internal structure of these organs has been little studied.

6.6 Storage Organs of Some Lesser Known Crop Plants

World agriculture yields a vast array of plant products, or crops. Most significant are fruits, seeds, and vegetative storage organs that are rich in carbohydrates, oils, fats, or organic acids. Many plants that are presently cultivated in limited geographic areas have appreciable potential for wider cultivation. Geophytes have particular potential because of the considerable reserves of water and available organic metabolites in their large roots, rhizomes, or tubers. Their capacious storage organs allow these plants to survive in severely seasonal climates where ordinary forest trees cannot.

These large storage organs are characteristic of many plants relatively unfamiliar to inhabitants of the north-temperate zone. These plants belong to various families. The root tuber of *Manihot esculenta* (cassava tree), for example, known in America since ancient times, is of increasing economic importance as a source of tapioca flour and other starch products. *Manihot* has potential as a major crop plant in the tropics, in part because it does not have a narrowly limited harvest season. The *Manihot* tuber contains a ready reserve of starch available to the tree, and is also a self-storing food reserve available to humans.

Other plants with good potential include *Oxalis tuberosa* (Oxalidaceae), which was brought under cultivation in ancient America and is still second only to the common potato as a root crop in the Andes. Other potential (and former) crop plants include *Lepidium eyenii* (Brassicaceae), which produces good yields of radish-like roots even at very high altitudes; *Polymnia sonchifolia* (Asteraceae), with edible *Dahlia*-like tubers; *Ullucus tuberosus* (Bascellaceae), which has long been a tuber crop in the Andean highlands; *Canna edulis* (Cannaceae), which accumulates starch in its tuberous rhizomes and has been cultivated longer than *Manihot* or *Zea*; and *Tropaeolum tuberosum* (Tropaeolaceae), which has tubers that have been a staple in the Andean region for a long time because of high yields and easy storage (Vietmeyer 1984, 1986).

Even in areas not having a distinctly seasonal climate, many perennial plants store extensive reserves of starch or other metabolites. These reserves may be sequestered for long periods, as in monocarpic plants, which vegetate for many years, flower and fruit once, then die. For example, in *Caryota, Metroxylon,* and other monocarpic palms, the amount of starch stored in the ground parenchyma of the stem can become very large. The crushed and grated stem tissue of *Metroxylon* is the source of commercial sago (Tomlinson 1961). When a monocarpic palm begins floral development, sugar consumed in the rapidly growing floral tissue is replenished by sugar derived from starch stored in the stem. Thus, a continuous stream of "sap" rich in sugar flows toward the expanding inflorescence. This stream can be tapped by wounding the spadix. This allows much of the food value of the starch that had been stored in the stem for years to be harvested within a few weeks. Various species of *Arenga* have been used as "sugar palms" in this way (Tomlinson 1961).

7 Transporting Systems

7.1 Introduction

Within a plant body, water and solutes move through two spatial systems — the **apoplasm** and the **symplasm**. We have already introduced these concepts (Sect. 1.5) and used them in discussing several functional systems. Here, we discuss the special relevance of symplasmic and apoplasmic spatial systems to transport within the plant.

Symplasmic transport occurs through living protoplasts and through the plasmodesmata, which unite nearly all the protoplasts of a plant into a single spatial system (Sect. 1.5.1). Apoplasmic transport occurs through the lumina of certain kinds of dead cells and through the microspaces within the walls of most cells (living and dead). Both water and solutes can be transported either symplasmically or apoplasmically.

Certain cell types are specialized for rapid, long-distance symplasmic or apoplasmic transport. Most specialized for symplasmic transport are the sieve elements of the phloem. For apoplasmic transport, the specialized cells are the tracheary elements; transport occurs through their lumina, after their protoplasts have been autolyzed. These lumina collectively constitute the super apoplasm. The specialized symplasmic conduction system, the phloem, is concerned mainly with transport of assimilates, whereas the specialized apoplasmic conduction system, the xylem, is concerned mainly with transport of water and mineral nutrients.

In this chapter, we discuss the xylem and phloem as complex, organism-level, functional systems composed of both primary and secondary tissues. We also compare and contrast the various cell types that constitute these tissues in angiosperms and gymnosperms. We defer to Chapters 19 and 20 detailed discussion of the structure and development of the secondary vascular systems, and to Chapters 14, 16, and 17 most of our discussions of the development of the primary vasculature of stems, leaves, and roots, respectively. Here, we include some discussion of symplasmic transport outside the phloem and of apoplasmic transport that occurs through cell walls rather than through the lumina of dead cells.

7.2 The Phloem: Long-Distance Rapid Transport in the Symplasm

7.2.1 Phloem as a Tissue

Phloem is a highly specialized and complex tissue. Its complexity is ascribable to its dual origin — from secondary as well as primary meristems — and to its range of cell types, not all of which are directly involved in transport. Although phloem performs a transport function absolutely vital to large land plants, it constitutes a much less prominent part of a typical perennial plant axis than does xylem. The major phloem cell type that functions in long-distance transport is the sieve element. Enucleate when fully functional, a sieve element typically has close associations with one or more specialized, nucleate parenchyma cells, which often have the attributes of transfer cells.

Here, we focus on basic structural features and adaptations of the phloem as they relate to long-distance transport in the symplasm. Differentiation of phloem in the primary plant body is treated in Sections 14.3.2 (in shoots), 17.2.1 (in roots), and 16.3.6 (in leaves). A developmental account of the secondary phloem is given in Chapter 20.

7.2.2 Sieve Elements and Other Phloem Constituents

Sieve elements (Behnke and Sjolund 1990) have some structural features that increase their efficiency in transport in obvious ways, and other features that are of debatable function.

Sieve elements are elongated parallel to the transport route, which is usually along the axis of the vascular bundle or cylinder in which the elements occur. Sieve elements are of two types: **sieve cells**, which are not aligned into discrete tubes; and **sieve-tube members**, which are more specialized than sieve cells and are well-aligned cellular units composing **sieve tubes**. Sieve-tube members are characteristic of angiosperms, and sieve cells of gymnosperms.

There may be a 20-fold difference in the diameters of sieve elements within a single plant. In trees, and probably in herbaceous plants also, the largest occur in the secondary phloem of the main stem, and the smallest in small veins of leaf blades. The diameter of the sieve tubes of the secondary phloem of broadleaved trees is often no greater than that of sieve cells in secondary phloem of conifer stems, which are phylogenetically more primitive.

Sieve-element walls have **sieve areas** (Fig. 7.1). Cytoplasmic connections between the protoplasts of contiguous sieve elements pass through pores in these areas. Rigorously, the term "sieve area" refers to such a locality in the wall of one element only; thus, the cytoplasmic connections between neighboring sieve elements pass through a **sieve-area pair**. For simplicity, however, we and others often use the term "sieve area" to refer to the complementary pair functioning together, as well as to a single member of the pair.

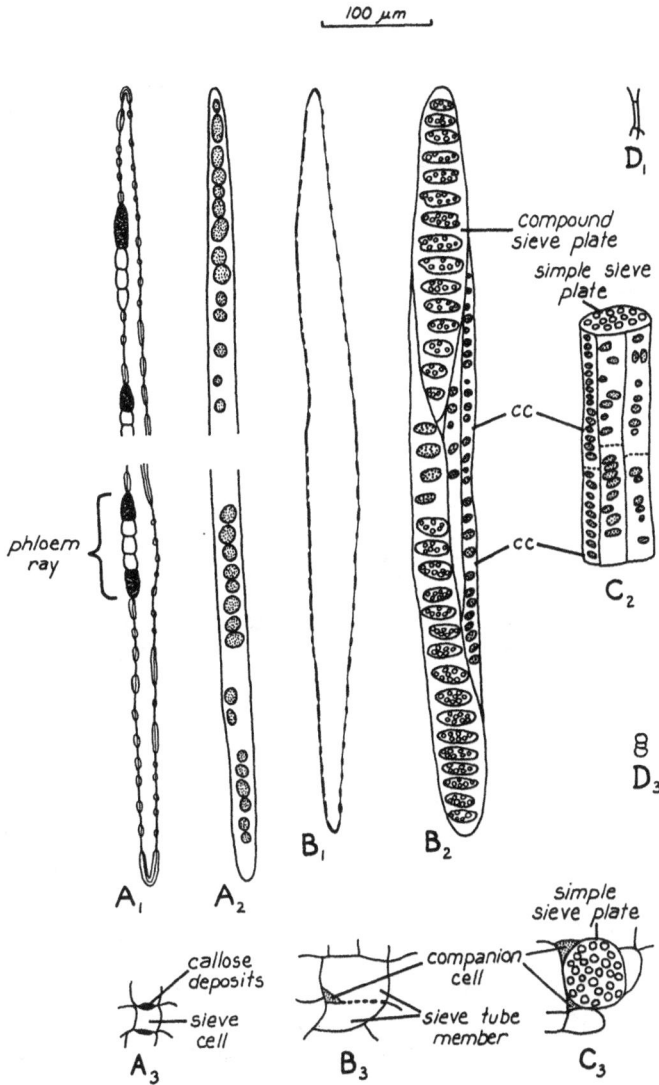

Fig. 7.1 A—D. Structural aspects of sieve elements. *Subscripts 1, 2,* and *3* indicate tangential section, radial surface, and transectional views, respectively. **A** *Pinus* sieve cells with mid-region omitted (typical length is 2—3 mm); A_1, *P. pinea* with Strasburger cells *heavily shaded.* Thin areas of wall are sieve areas. A_2, A_3, *P. sylvestris.* A_3 shows sieve areas thickened by callose deposits. **B** Sieve-tube member from *Betula verrucosa*, with highly oblique compound sieve plate and smaller lateral sieve areas. Cell faces *cc* were in contact with companion cells. **C** Sieve-tube member from *Fraxinus excelsior*, with nearly transverse, simple sieve plate and numerous lateral sieve areas; *cc* as in B_2. **D** Profiles of sieve-tube members from small vein of *Beta vulgaris* leaf in tangential and transverse view, to show small size when drawn to same scale as A, B, and C. The middle cell in D_3 is the sieve-tube member. (A_1 redrawn after Esau 1977; $D_{1,3}$ dimensions as in Esau 1972. Other drawings modified after Hejnowicz 1980)

Cytoplasmic connections traversing pores between sieve elements are much wider than plasmodesmata. Whereas the mean diameter of plasmodesmata is about 0.1 μm, the diameters of the cytoplasmic connections through sieve areas commonly are about 1.0 μm and may be as large as 1.5 to 2.0 μm. Usually, all the pores of a sieve area have similar diameters. Typically, they are simple cylinders, but may be enlarged in the middle-lamella region. **Callose** coats the surfaces of the sieve areas and also lines the pores.

In sieve cells, all sieve areas are similar, but in sieve-tube members, the sieve areas on end walls have larger, more specialized pores than do those on side walls. A wall portion bearing a specialized sieve area is a **sieve plate**. If there is only one specialized sieve area on that wall portion, the sieve plate is simple; if there is more than one, it is compound. In general, the more oblique the orientation of the end wall, the greater the number of sieve areas on the sieve plate.

If the phloem is wounded, massive amounts of additional callose may be deposited, often within a few minutes, on the sieve areas of the affected elements (Evert 1977). This **wound callose** seals off the pores of the sieve areas and isolates that part of the symplasm affected by the wounding (Lamoureux 1974; Sect. 20.3.1.2).

The protoplasts of sieve elements are structurally modified for function (Sect. 20.3.1.2). During differentiation, some organelles typically disappear or are modified, and some new structures appear. Functional sieve elements usually have no intact or recognizable nuclei, tonoplasts, dictyosomes, microtubules, or ribosomes, in spite of the seeming indispensability of these to living cells in general. Numerous exceptions, though, have been observed, both in angiosperms and gymnosperms (Srivastava 1974; Evert 1977).

Because a functioning element retains its plasmalemma, osmotic phenomena can continue, and mature sieve elements remain a part of the symplasm. The osmolarity of the sieve-tube contents — a mixture of cytoplasmic and vacuolar components — is quite high because of a high sugar content. This high osmolarity promotes influx of water and thus an increased hydrostatic pressure. The gradient of this pressure is considered an important motive agent in phloem transport (Sect. 7.2.5).

In many angiosperms, **P-protein** appears in sieve-tube members as they differentiate, and remains during the functional phase. This substance, which is usually of fibrillar or tubular form, has been found in sieve-tube members of practically all dicots investigated. It is less widely distributed in monocots, and occurs in few, if any, gymnosperms. The role of P-protein in the functioning of intact phloem is not well understood. It may serve to plug the sieve-area or sieve-plate pores after wounding. However, as it seems not to be universally present, it cannot be generally indispensable (Evert 1977).

In many species, sieve-element walls commonly become conspicuously thickened at an early stage and acquire a "pearly" luster. This distinctive appearance has led to the term **nacreous wall** (Esau and Cheadle 1958; Esau 1965a). This wall thickening does not occur over the sieve areas. Other wall regions may be thin, or so thick that most of the lumen volume is occluded. The nacreous wall contains cellulose microfibrils, polyuronides, and pectins (Cronshaw 1974b; see also Sect. 20.3.1.2).

In addition to these structural adaptations of sieve elements, adaptations of another kind of cell also contribute to transport efficiency. The loss of nuclei and some of the ordinary protoplasmic activities from sieve elements has apparently been compensated for by development of close relations with certain neighboring phloem-parenchyma cells. These neighboring cells have increased protoplasmic activities. Comparative studies have shown that parenchyma-cell associations with sieve elements are a major factor in the transport competence of the phloem (Behnke 1974; Eschrich 1980).

In contrast to sieve elements, which have reduced complements of organelles, the special parenchyma cells with which they are intimately associated have abundant organelles. These tend to be highly active physiologically (Sect. 20.3.2). Evidence of a close physiological association between the two cell types includes the alignment of primary pit fields in the parenchyma-cell wall with sieve areas of the sieve-element wall. Cytoplasmic strands in each sieve-area pore are connected with several plasmodesmata in the parenchyma-cell wall. The connections occur in an enlarged median cavity within the middle-lamella region. The close physiological association is also evidenced by the collapse of both the sieve element and the associated cells when the sieve element ceases to function.

In angiosperms, the sieve element and its associated parenchyma cell or cells, which are called **companion cells**, are derived from the same mother cell. Gymnosperms have a cell type that is functionally equivalent to companion cells, though there is no strict ontogenic relation as there is in angiosperms. Some writers call the associated parenchyma cells in gymnosperms **albuminous cells** (Sect. 20.3.2), though this term is inappropriate because the cells are not particularly rich in protein. We adopt the usage of some other authors, who instead refer to these cells as **Strasburger cells** (Sauter 1980). (For a discussion of phloem fibers and of phloem-parenchyma cells other than companion cells and Strasburger cells, see Sects. 20.3.3; 20.3.4).

In addition to the already-mentioned structural attributes of individual sieve elements and other phloem cells, the effectiveness of long-distance phloem transport depends on: (1) continuous sieve-element routes, with cell lumina oriented along the transport pathway between a source and a sink; and (2) transfer cells in contact with sieve elements or with their companion cells, at sites where transported substances cross the plasmalemma — i.e., at sources and at long-term sinks such as glands or storage parenchyma.

7.2.3 Transport Rates and Capacities

Velocity of transport in sieve elements is moderately high: about 20 cm/h in *Pinus* and 30 to 70 cm/h in dicot trees. The direction of transport is always from a source to a sink. During periods of photosynthesis, the sources usually are the leaves, and the major sinks are growing regions in roots and shoots. The major substances translocated are sucrose; other sugars; and the sugar alcohols, mannitol and sorbitol (Zimmermann and Brown 1971; Ziegler 1975).

The mass of solute material translocated through sieve elements has been estimated for a variety of plants. Various data can be made comparable by applying the concept of **specific mass transfer (SMT)**, introduced by Canny (1960). SMT quantifies mass transfer in grams of material per unit time per unit cross-sectional area of phloem (SMT_{ph}), or of sieve elements (SMT_{se}). The units are $g/cm^2/h$. Because sieve elements generally occupy only about 20% of the cross-sectional area of the phloem, the SMT_{se} values are about five-fold higher than SMT_{ph} values. For dicot stems, SMT_{se} is 10 to 25 $g/cm^2/h$ (Canny 1960), and for conifer stems, 2.2 to 2.5 $g/cm^2/h$ (original data are from Münch 1930).

7.2.4 Phloem as a Link Between Sources and Sinks

Most active sources and sinks are at the ends of slender branches of phloem strands, where every file of sieve elements is surrounded by parenchyma cells that have intense ATPase activity in their plasmalemmae. These plasmalemmae control loading and unloading of the symplasmic transport pathways (Eschrich 1980; Giaquinta 1983). In many of these parenchyma cells, the area of contact between symplasm and apoplasm is increased by wall ingrowths, as in transfer cells (Bentwood and Cronshaw 1978).

Along most strands of sieve elements, there are many small sources and sinks, consisting of the adjacent parenchyma cells, especially the ray cells in secondary phloem. Whether a particular contact is a source or a sink depends on external environmental (such as seasonal) factors and on internal physiological factors. Phloem seems to differentiate towards sites such as meristems or glands, where there is high demand for respiratory metabolites and materials for synthesis.

Transport changes with changing source-sink relations. For example, a developing *Populus* leaf first imports carbohydrates; then, as it matures, it begins exporting them. During intermediate periods, when the distal part of the leaf blade has stopped growing but other parts have not, the sieve tubes serving the distal part transport photosynthate basipetally while sieve tubes in the basal part are still importing from outside the leaf. Thus, local areas in a young leaf can be either importing or exporting, regardless of the net balance for the whole leaf (Larson and Dickson 1973; Turgeon and Webb 1973). The summation of many local supply-demand balances in organ-level sources and sinks determines the net gradients in the phloem and hence the direction of overall transport within it.

7.2.5 Probable Mechanisms of Transport Within Sieve Elements

The structural features of sieve elements become meaningful when we consider probable mechanisms of symplasmic transport through them. Transport seems to be based mainly on a flow of aqueous solutions rather than on movement of solutes independently of water. Of the various hypotheses proposed about the motive forces, the most prominent interpret the flow as: (1) a response to pressure gradients; and (2) electro-osmosis through the sieve areas separating sieve elements. Although these mechanisms are often considered mutually exclusive, they may possibly be superposed and function in concord.

The first of these hypotheses about the driving forces, often called the **Münch pressure-flow hypothesis**, is postulated to work as follows: a strand of sieve elements is loaded with sugars at a source, by an energy-requiring process. After transport to an adequately capacious sink, the sugars are unloaded, also by an active process. This active loading and unloading elicits a gradient of decreasing sugar concentration from source to sink, and hence, because of osmosis, also a gradient of hydrostatic pressure. Solution then flows down the pressure gradient. The flow itself — but not the loading or unloading — is basically passive, although further metabolic energy is needed to support the living state of the protoplasm of the sieve elements. Some experimental results do not support the Münch pressure-flow hypothesis, however, in that sugar concentration gradients may actually increase along the vein network from small or mid-sized veins within source leaves (Sect. 7.6.2; Roeckl 1949; Kursanov 1976).

Assuming that the flow is driven only by an osmosis-induced pressure gradient, and knowing the physical attributes of the system, one can calculate the required pressure gradient. The result is just on the limits of credibility — if one assumes that the pores of the sieve areas are completely open. But, if the pores contain filaments of P-protein or tubules of endoplasmic reticulum (ER), then the gradients required become unattainably high.

If flow in response to pressure gradients is, indeed, the prevalent mechanism of phloem transport, one can ask why sieve areas (or, at least, sieve plates) did not become completely open, or nearly so, during evolution, as did the perforation plates in vessel elements. An answer is that sieve areas, though they reduce the flow under a pressure gradient compared with what it would be in an open tube, also perform an emergency function. If a sieve-element strand is wounded, the cell contents rapidly shift in response to the wound-induced change in the pressure gradient, and certain structural elements of the cytoplasm collect on the sieve areas as "slime plugs", closing off the pores. The slime plugs are soon supplemented by deposits of wound callose (Eschrich 1975), which further protect the sieve-element strand against uncontrolled leakage.

Another question also arises: why, while retaining the emergency plugging function of the sieve areas, did evolution not generally increase the size of the pores? In fact, evolution has not only retained small pores in most lines, including some "successful" trees, but in some dicots has even increased the number of sieve plates along a sieve tube by allowing the mother cells to divide obliquely before the daughter cells differentiate into sieve-tube members. The oblique partitions then also bear sieve plates (Cheadle and Esau 1958; Evert 1960).

The small, but numerous, pores of sieve areas not only do not seem to decrease transport efficiency, but may even promote it. Further, the near-occlusion of a sieve-element lumen by thick nacreous walls does not seem to hinder transport. From biophysics, we know that, if transport in sieve elements were driven by electroosmosis (rather than mere pressure gradients), the presence of numerous small pores and many sieve areas along the transport pathway would be advantageous. The flow-efficiency ratio of large to small pores (taken in numbers such that the total pore area is the same) is just the reverse in electroosmosis from that in pressure-gradient flow. In pressure-gradient flow, a pore 10 μm in diameter will transmit 100 times more solution than will 100 pores, each 1 μm in diameter, though the total pore area is the same. Electroosmosis, on the other hand, is faster through the 100 small pores than through the single large one because the most active zone in each pore is the electrical double layer at the interface between the solution and the pore wall.

To operate, an electroosmotic system of flow requires a difference of electric potential across the pore. One way of maintaining this potential difference would be via an ionic pump in the plasmalemma around the sieve areas (Spanner 1974). The energy cost of operating such a system, however, is estimated to be higher than the cost of maintaining a pressure gradient. Conceivably, electroosmosis could be powered more efficiently by traveling waves of conformational changes in microfilamentous elements in the pore. These conformational changes could cause traveling waves of electrical change, by disturbing the surface double layer of charges, and thereby facilitate electroosmotic flow through the pores. This flow could operate in conjunction with pressure gradients (Hartt 1973).

If electrolytes are present, and there is an electric potential difference across the pore, the mobile part of the charged double layer — i.e., the solution in contact with the pore wall — will move by electroosmosis. In effect, electroosmosis through smaller pores is more efficient and, if there is a pressure gradient, an electric field in a small pore "lubricates" the pore and increases its effective diameter for pressure-driven flow by making the otherwise stationary boundary layer move. This may be why evolutionary change did not eliminate the small pores between sieve elements in so many taxa.

Calculations show that optimal pore diameter for electroosmosis is probably between 5 and 50 nm (Spanner 1974) — about five times the thickness of the double layer. Actual pores are 20 to 100 times that size — too large for efficient electroosmosis, yet too small for the most efficient pressure flow. If pressure flow and electroosmosis were functioning jointly in sieve tubes, there would be an optimal, intermediate pore diameter that would depend on the gradient of pressure and the electroosmotic factors involved. Evolution may have tended toward that intermediate, optimal diameter.

7.3 Non-Phloic Symplasmic Transport

Transport within the symplasm is not restricted to sieve elements, but also occurs through plasmodesmata, which interconnect most protoplasts of a plant (Sect. 1.5.1). Plasmodesmata connecting sieve elements to cells of adjacent tissues provide a symplasmic route for loading and unloading of the phloem (Sect. 7.6). They also provide a symplasmic route for radial exchange of solutes between phloem and xylem via the rays (Bel 1990).

Plasmodesmata, almost universally distributed in higher plants, are relatively uniform in ultrastructure (Fig. 7.2). If measured outside their plasmalemma linings, they range in diameter from 30 to 80 nm. A plasmodesmatal canal is as long as the thickness of the double wall it traverses. At each end, the canal may narrow and form neck constrictions. Nearly all plasmodesmata have an axial component, the desmotubule, 15 to 20 nm in diameter. The membranes of the desmotubule are continuous with the membranes of the ER on both sides of the double wall.

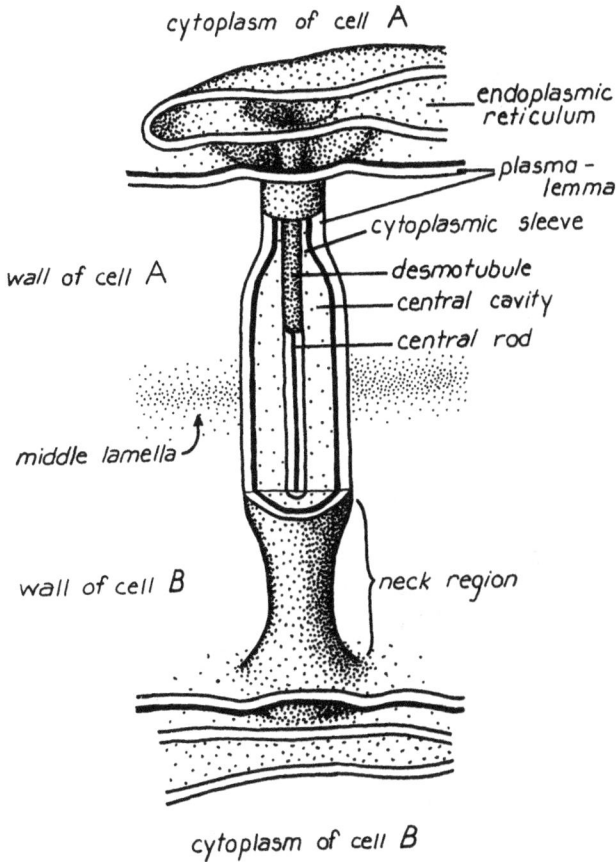

Fig. 7.2. Spatial relations and components of a simple primary plasmodesma traversing the compound wall between adjacent cells. Diagram is partly conjectural. (Based on information given by Robards 1976; Olesen and Robards 1990; Robards and Lucas 1990)

The sleeve-shaped lumen between the plasmalemma and the desmotubule within the plasmodesma has been called the "cytoplasmic annulus" because of its appearance in transection. If considered three-dimensionally, however, it is more appropriately termed "cytoplasmic sleeve" (Esau and Thorsch 1985). Cytoplasmic sleeves, as seen in electron micrographs, typically appear partially occluded, but have small open channels between the occluding bodies. It is possible that these occlusions are artifactual (Gunning and Overall 1983; Terry and Robards 1987; Robards and Lucas 1990).

Because the ER cisternae generally constitute a continuous system within a cell, and this system in turn is continuous with the desmotubules, the ER system seems to be continuous throughout the symplasm, though the desmotubules are quite slender. Nevertheless, the cytoplasmic sleeve, rather than the desmotubule, may be the major pathway for communication and symplasmic transport between cells (see Olesen and Robards 1990).

In relatively rapid symplasmic transport, a single plasmodesma conveys solutions at an estimated rate of about 0.02 μm^3/s (Robards and Clarkson 1976). This is equivalent to about 500 volume changes per second in the cytoplasmic-sleeve region — and to flow velocities of about 50 μm/s (18 cm/h). If these estimates approach reality, they are comparable to flow velocities in sieve cells of conifers (Sect. 7.2.3).

Transport through plasmodesmata occurs at least partly by diffusion. If driven by steep gradients over very short distances, diffusion can be quite rapid, but steep gradients *cannot* be maintained over long distances. The greater the distance, the less steep the gradient and the slower the diffusional transport. Bulk flow may also participate in plasmodesmatal transport. Diffusion through plasmodesmata can pro-duce fluxes of different molecular or ionic species in opposite directions, and, given adequate concentration gradients, the diffusional flux in one direction might even overcome a bulk flow in the opposite direction (Osmond and Smith 1976).

The diameters of the cytoplasmic sleeve and the desmotubule within it may restrict the size of molecules or particles that can traverse plasmodesmata. In *Abutilon*, for example, the outside diameter of the cytoplasmic sleeve is about 28 nm, and the outside diameter of the desmotubule is about 16 nm. Thus, the lumen of the cyto-plasmic sleeve is about 6 nm wide. It may be partially occluded. The "open" passages, in fact, may be as small as 3 nm (Terry and Robards 1987). This agrees well with the width of 3 to 5 nm inferred from studies of movement of fluorescent-labelled peptides of various sizes through plasmodesmata in *Elodea* (Goodwin 1983), and implies that peptides that diffuse rapidly through plasmodesmata must have a mass of less than 1000 daltons.

A generally unappreciated aspect of plasmodesmata is that, because of their great numbers, a large fraction of the plasmalemmal surface area of the plant is associated with them. In fact, in a cell having a wall 1.0 μm thick, with a plasmodesmatal frequency of 5/μm^2 and plasmodesmatal diameter of 60 nm, the area of plasmalemma associated with the plasmodesmata equals that lining the rest of the cell (Robards 1976).

Symplasmic transport through plasmodesmata can be studied directly by injecting hydrophylic fluorescent dyes (Terry and Robards 1987) and by experiments on electric coupling. Such studies indicate that the frequency of plasmodesmata ranges from less than 1 to about 100/μm^2. Typically, a higher-plant cell is connected with its neighbors by 1000 to 100 000 plasmodesmata (Clowes and Juniper 1968).

The frequency of plasmodesmata commonly differs on the various walls of a cell, as might be expected if plasmodesmata were functional in symplasmic transport or in signal transmission. For example, in the statenchyma of the root cap, plasmodesmata occur mainly on the transverse walls, suggesting a role in gravitropic signal trans-mission (Sect. 10.2.2.1). Further, in the companion cells of a *Populus deltoides* leaf, plasmodesmata are very frequent on the walls adjoining sieve-tube elements and are rare or absent on the walls adjoining other companion cells (Russin and Evert 1985). Transfer cells often have no plasmodesmata on cell faces having characteristic wall ingrowths.

7.4 The Xylem: Long-Distance Rapid Transport in the Super Apoplasm

The aqueous apoplasm encompasses all water-filled space within the plant body outside the protoplasts (Sect. 1.5.2). This space includes the ultramicrospaces between microfibrils in the cell walls. In this chapter, though, we focus on the tracheary elements, which become fully functional in transport after their protoplasts die and are autolyzed, leaving only their cell walls. Their lumina constitute most of the apoplasmic space. The microcapillary tubes formed by these lumina are vastly larger than the ultramicrospaces within the walls, and therefore are far less resistant to liquid flow. Because of the much greater transport capacity of these tubes than of the cell-wall component of the apoplasm, it is appropriate to refer to the tracheary-lumina system of liquid-filled space as the **super apoplasm.**

7.4.1 Tracheary Elements

7.4.1.1 General Aspects. Early in its differentiation, the cell wall of a tracheary element develops structural features that, after the protoplast dies, allow the wall to function effectively in water transport. These adaptations include secondary-wall deposits that strengthen the element even though not all of the primary wall is covered; lignification of the secondary wall and that part of the primary wall that underlies it; and at least partial hydrolysis of the uncovered primary wall. Another adaptation for transport is the arrangement of the tracheary elements in axial strands or files.

Tracheary elements vary notably in pattern of secondary-wall thickening. The thickening may be in the form of bars, rings, helices, or lattices. Deposits that over-arch adjacent, uncovered primary wall result in "bordered" primary-wall areas. In transection, a bar-shaped thickening thus may appear triangular, with the vertex of the triangle attached to the primary wall. Similarly, "bordered" pits result from the overarching of the primary wall of the pit area by adjacent secondary deposits (Sect. 1.4.2).

The lignified secondary wall, and the lignified primary wall underlying it, protect the tracheary element from collapse. This protection is necessary because turgid neighboring cells exert compressive forces on the element, and because, during rapid transpiration, there may be very great negative pressures in the water flowing through the element (Sect. 7.4.3.3). The advent of lignification of super-apoplasm walls was an important innovation in the evolution of large land plants (Barghoorn 1964).

7.4.1.2 Types of Tracheary Elements; Vessels. Two general types of tracheary elements — **tracheids** and **vessel members** — participate in water transport. Tracheids, the more primitive of the two, normally develop in strands or bundles. Within each strand, the ends of tracheids overlap. The elements of a strand function together, each tracheid receiving water from several of its basal and lateral neighbors and passing it on to several apical and lateral neighbors.

Secondary tracheids, especially those of gymnosperms, are well adapted for both transport and mechanical support. They are quite long and have moderately wide lumina. Sometimes it is convenient to classify tracheids according to their relative ability to function in mechanical support versus transport. Thus, one speaks of vessel-tracheids and fiber-tracheids (Sect. 19.2.1.3).

Tracheids are equipped with large, bordered pits (Sect. 1.4.2), which are specialized primary-wall areas that lack overlying deposits of secondary wall. Typically, each tracheid has multiple pit connections with a dozen or more neighboring tracheids, and, because water transport between tracheids occurs mainly through these pits, they are very important to organismal functioning. In secondary xylem of gymnosperms, pits generally occur on radial walls, but only in the latewood do they occur on tangential walls also. The pits on the tangential walls of the most recently formed xylem increment allow water to flow across the growth-increment boundary into the reactivating cambial zone in spring. This flow tends to be rather weak, because latewood tracheids, with their small pits and narrow lumina, are better adapted to support than to transport.

Pit size, shape, fine structure, and arrangement in the radial walls of tracheids vary widely in wood from different parts of a tree and from different species. The number of pits per tracheid, usually between 50 and 300, varies with tracheid size. Pits may be arranged in precise patterns of rows or lateral pairs. Little is known about the mechanisms that control these patterns.

The pits between tracheids and the parenchymatous cells of xylem rays are bordered on the tracheid side but usually not on the ray side (Jane 1970), and hence are half-bordered pit pairs. These pit pairs are variable among taxa and thus useful in wood identification, whereas the bordered-pit pairs between one tracheid and another are quite uniform. The area of contact between a ray parenchyma cell and an axial tracheid is termed a **cross-field**, and the pits that develop there are **cross-field pits**. They vary in number per field as well as in arrangement and morphology.

Radially oriented tracheids occur in the xylem rays of some woods, particularly in the Pinaceae, Cupressaceae, and Taxodiaceae. They also occur in some members of a few angiosperm families, notably the Proteaceae and Malvaceae (Chattaway 1948, 1951). Water flows radially through these **ray tracheids**, which commonly occur in strands and are slightly elongated along the ray. Ray tracheids are linked with neighboring axial tracheids through bordered pits in their side walls, and with other ray tracheids within a radial file through bordered pits in their end walls. These end walls commonly are oblique, providing a large area for pits. The tracheids usually are arranged in single or double rows along the ray margins; the mid-part of the ray consists of parenchyma cells. Data on the functioning of ray tracheids are scarce. We surmise that they permit fast radial flow between growth increments in the sapwood, thus integrating the increments into a single super-apoplasmic transport system.

Vessel members are characterized by **perforation plates** (Fig. 7.3) where one member adjoins another in an axial file (Sect. 19.2.1.3). An axial file of vessel members having perforation plates at the junctions, and thus arranged in the form of a pipeline-like conduit, is a xylem **vessel**. In contrast to tracheids, in which water transport between elements is mainly through pits in the walls, transport between vessel elements is mainly through the perforation plates, and is generally more rapid than that between tracheids. Flow rates in vessels with simple as well as scalariform perforation plates are nearly equal to those in ideal capillaries of appropriate diameters (Schulte et al. 1989). The morphology of vessel members also differs from that of tracheids in other ways (Fig. 7.3). Further, vessel members vary considerably among themselves in shape, pitting pattern, and dimensions. The end walls of vessel members that terminate a vessel are not perforated, though they usually have pits allowing the lateral exchange of water with other vessels or tracheids. A single vessel may include dozens or hundreds of vessel members.

In secondary xylem in the main stem of large trees vessel length ranges from a few centimeters to several meters in diffuse-porous wood (Sect. 19.5.1) and to about 10 m in ring-porous wood. Greenidge (1952), after measuring passage of compressed air through segments of stems, reported average lengths of the longest vessels of diffuse-porous hardwoods to vary from about 5 m in *Fagus grandifolia* to about 1.2 m in *Betula lutea* and about 0.9 m in *Acer rubrum*. He reported longer vessels in ring-porous woods: 8.5 to 15 m in *Quercus borealis*, 7.7 to 18 m in *Fraxinus americana*, and 5 to 8.5 m in *Ulmus americana*. In these three species, vessels typically were as long as 90% of total tree height. Differences in length of vessels may possibly affect the bulk flow of water more than do differences in diameter because of the large resistances to flow at the nonperforated end walls. Two or more vessels having their terminating members overlapping laterally and connected by bordered pit pairs form a **vessel file**.

Vessels in secondary xylem run, overlap and anastomose mainly in planes parallel to that of the cambium from which their cells derived. However, two or more vessels that are radially aligned, and therefore seemingly of different ages, may contact each other along their tangential walls for a short distance. In *Eucalyptus maculata*, vessels having a mean length of 50 cm were found to have an average of 10 lateral contacts with other vessels, and those contacts occupied 2 to 3% of the vessel surface. In the same wood, tracheids provided additional, indirect contact between vessels (Skene 1969). The vessel system of secondary xylem as a whole becomes a three-dimensional network of conduits (Burggraaf 1972; Zimmermann 1983).

Comparative studies of tracheary elements in extant and extinct plants have shown that: (1) the bar type of thickening of secondary walls is phylogenetically more primitive than the bordered-pit type; (2) tracheids evolved before vessel members; (3) tracheids with bordered pits assumed some support function in addition to transport function early in their evolution; and (4) vessels evolved independently in diverse taxonomic groups — among both monocots and dicots, in a few gymnosperms, and in *Pteridium* and *Selaginella*. Among extant dicots, vessels are missing only from Tetracentraceae, Winteraceae, and Trochodendraceae. Vessels are absent from most gymnosperms except *Welwitschia, Gnetum, Ephedra*, and close relatives.

7.4.2 Primary and Secondary Xylem

Xylem is a complex tissue having primary as well as secondary origins, and diverse cell types. Primary xylem arises from provascular strands during vascularization of the primary plant body. In arborescent plants, most primary xylem functions only briefly.

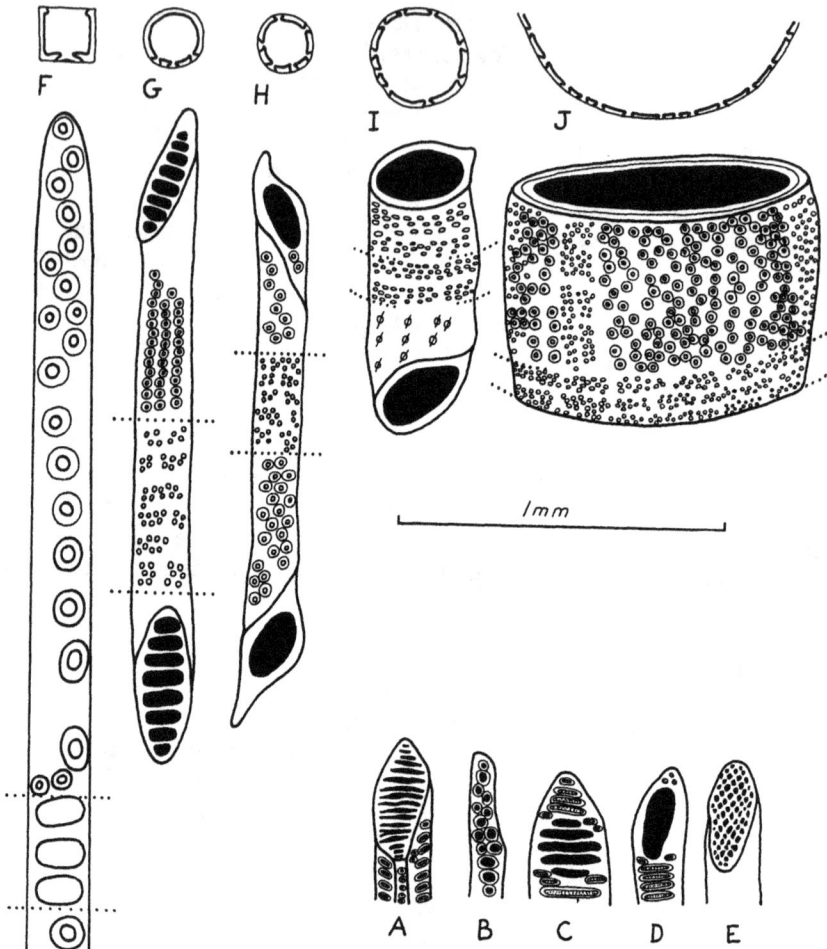

Fig. 7.3 A—J. Views of tracheary elements. **A-E** End walls of vessel members showing different types of perforation plates. **A** Finely scalariform (*Pteridium*). **B** Multiple circular "ephedroid" (*Ephedra*). **C** Coarsely scalariform, flanked by bordered pits (*Vitis*). **D** Simple, accompanied by pits (*Vitis*). **E** Reticulate (*Daemonorops*, Palmaceae). (**A,B,C,** and **D** modified after Esau 1965a; **E** drawn after descriptions given by Tomlinson 1961).

F-J. *Lower set of drawings* Radial faces of tracheary elements in whole or in part, showing range of shapes, pitting patterns, and approximate relative dimensions. Wall portions between *dotted lines* show cross-field pitting, indicating ray contact. **F** End of long tracheid of *Pinus*, with three large, unbordered, fenestral pits in a cross-field. **G** *Liriodendron tulipifera* vessel member, with scalariform perforation plates. **H** *Quercus alba* latewood vessel member, with simple perforation plates. Tips of original fusiform derivative extend beyond perforation plates. **I** *Acer negundo* vessel member. **J** Extremely wide earlywood vessel member of *Quercus alba*. Tips of original derivative have been lost. **F-J** *Upper set of drawings:* appearance of typical transections of cells represented in lower set. (**F** redrawn after Hejnowicz 1980; **G** to **J** redrawn after Eames and MacDaniels 1947)

It is much less extensive than secondary xylem. The characteristic cell type in both primary and secondary xylem is the tracheary element, but both parenchymatous and sclerenchymatous elements are also present. Parenchymatous elements generally retain their living protoplasts much longer than do tracheary elements (Sect. 19.5.2).

7.4.2.1 Protoxylem and Metaxylem.

Primary xylem is classified as either protoxylem or metaxylem, depending on the timing of maturation of the tracheary elements. **Protoxylem**, which is generally the first to mature, terminates the dendritic super apoplasm in the youngest root and shoot segments. **Metaxylem** matures after the axis of the organ has ceased to elongate.

A protoxylem tracheary element is adapted to exchanging water laterally as well as transporting it axially. Secondary-wall thickenings, in the form of annular or helical bars, allow the element to stretch after it matures, during the axial extension growth of adjacent, later-maturing tissues. The partial hydrolysis of unprotected areas of primary wall contributes both to extensibility and to lateral water exchange. As the organ extends axially, the distance between reinforcing bars of the protoxylem element increases. Finally, the element collapses under the positive pressure of contiguous turgid cells and negative internal water pressure.

A metaxylem tracheary element is better adapted to axial transport than is a protoxylem element. Its diameter is greater, especially in roots, and secondary wall characteristically is deposited as a network or as a pitted layer, making the element more resistant to collapse than an element with annular or helical wall thickenings. However, a metaxylem element is less efficient in lateral transport than is a protoxylem element and generally is not extensible.

7.4.2.2 Secondary Xylem Systems: Functional Types.

Secondary xylem of gymnosperms is less complex than that of angiosperms. In gymnosperm xylem, which generally has no vessels, the axial elements consist mainly of tracheids. These have a major mechanical as well as a transport function. Some gymnosperms have small amounts of axial parenchyma having a storage and a minor mechanical function. Radial elements include ray cells and sometimes ray tracheids (Sect. 19.3). Resin ducts may be present (Sect. 19.4).

In the secondary xylem of arborescent dicots, the transport and mechanical support functions are partially separated at the cellular level. Along with many intermediate types, there are two basic forms of thick-walled elements: vessel members (including vessel-tracheids) and fibers (including fiber-tracheids and libriform fibers). Numerous vessels, with wide lumina, serve as longitudinal transport pipelines, whereas most of the support is provided by fiber-tracheids and libriform fibers (Sects. 19.2.1.3; 19.2.2), which have little transport capacity. By partially separating the transport and support functions at the cellular level, dicot trees have developed generally stronger stems than have gymnospermous trees. Because of this greater strength, dicot woods are commonly known as "hardwoods".

The secondary xylem that functions in transport always contains some parenchyma cells (Sects. 19.2.2; 19.3), which form a continuous symplasmic system among the tracheary elements. This system is part of the greater symplasm. Vessels usually have relatively large contact areas with these parenchyma cells. There are two orientational types of parenchyma in secondary xylem: axial and radial (Sects. 19.2.2; 19.4).

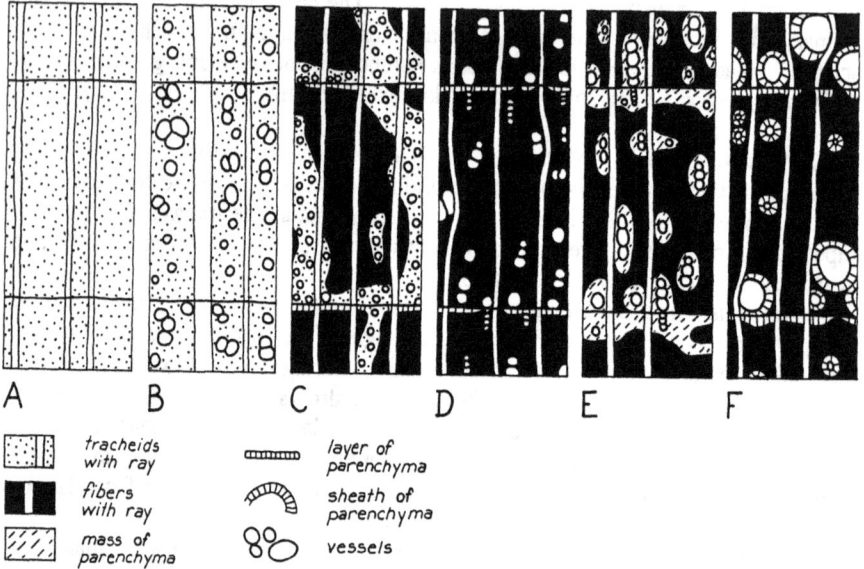

Fig. 7.4. A-F. Diagrammatic transections of functional types of secondary xylem systems. Horizontal lines across the panels indicate annual-increment boundaries. Vertical bands indicate vascular rays. **A.** Type 1. Typical gymnosperm xylem. Tracheids perform both transport and support functions. There are no vessels. **B.** Type 2. This xylem, having both vessels and tracheids, which may also include some fiber-tracheids, is characteristic of the relatively primitive angiosperm woods. Practically all elements have a dual function. **C.** Type 3. This xylem has distinguishable conductive hydrosystems and fibriform support systems. Three-dimensionally, each system is continuous and anastomosing. In transections, one system may seem dispersed in a matrix of the other. This type is exemplified by *Quercus*. **D.** Type 4. The hydrosystem consists entirely of vessels in contact with water-filled fibriform elements and is exemplified by *Aesculus*. The mechanical elements in this and the following types form a continuous matrix. **E-F.** Type 5. The vessels, as exemplified by *Fraxinus*, are surrounded by masses (**E**) or discrete sheaths (**F**) of parenchyma and have no direct contact with other tracheary elements or with fibriform elements. See text for further details. Modified after Braun (1963, 1970, 1982)

Five functional types of secondary-xylem systems can be distinguished in trees, based on three criteria (Braun 1970):

1. Does a single type of tissue perform both transport and mechanical support functions, or are the tissues functionally divided into a specialized hydrosystem and a specialized supportive system?
2. If a hydrosystem is distinguishable from a supportive system, how are the hydrosystem elements arranged with respect to the supportive-system elements?
3. If the hydrosystem consists primarily of vessels, do they directly contact the supportive-system elements or do sheaths of peritracheal parenchyma intervene (Fig. 7.4)?

Type 1. (Fig. 7.4A) This type is exemplified by typical gymnosperm xylem, in which tracheids both transport water and provide support. Water flows fastest in the most recently formed growth increment — or, early in the growing season, in its immediate predecessor (Sect. 7.4.4). Usually, several additional outer increments of the sapwood also participate. Pits in the tracheids in the mid-regions of a growth increment are mostly on the radial walls, whereas, near the abaxial border of the increment, pits are numerous on the tangential walls. This has functional implications for radial water flow, which may also occur through ray tracheids (Sect. 7.4.1.2).

Type 2. (Fig. 7.4B) Both tracheids and vessels function in transport. The vessels have contact with the tracheids and also with the fiber-tracheids, which have a largely supportive function. Generally, the vessel network of a growth increment makes no contact with that of neighboring increments, except indirectly, through tracheids. Relatively primitive angiosperm woods, such as *Alnus, Castanea,* and *Fagus,* are of this type.

Type 3. (Fig. 7.4C) In Types 3, 4, and 5, the hydrosystem is definitely separated from the fibriform, supportive system. The Type 3 hydrosystem consists of both vessels and tracheids, though most of the water moves in the vessels. As can be seen three-dimensionally, the hydrosystem and supportive system are each continuous and anastomosing. If viewed in transection, however, the hydrosystem elements in some taxa seem to be dispersed in a matrix of supportive tissue, whereas in other taxa, the "phases" seem to be reversed.

Type 4. (Fig. 7.4D) The hydrosystem consists entirely of vessels, in direct contact with the fibriform elements. Some of the fibers are alive; others are dead and water filled. The latter may make a small contribution to local water transport, but could hardly be significant in transporting transpiration water. Xylem of *Aesculus* is of this type.

Type 5. (Fig. 7.4E, F) Vessel elements, either singly or in groups, are completely surrounded by parenchyma in the form of irregular masses (Fig. 7.4E) or discrete sheaths (Fig. 7.4F). Vessel networks do not directly cross growth-increment boundaries, but there may be close radial contact between the narrow, terminal vessels produced during one growing season and the wider, initial vessels produced the next (Fig. 7.4E). The former may even serve as templates for the latter. This type, which occurs in *Fraxinus, Gleditsia,* and *Albizia,* seems to be the most advanced phylogenetically.

7.4.3 Hydrodynamics of Transport in the Xylem

7.4.3.1 Hydraulic Conductivity and Resistance in Tracheary Elements. To understand the dynamics of water movement in tracheary elements, we must consider flow in capillaries. The velocity of flow through a capillary can be expressed as a mean, but information about the profile of velocities across the lumen is more useful. From physics, we know that water immediately adjacent to the wall is a stationary boundary layer, whereas water in the center of the lumen outpaces the mean velocity. Both volume of flow and maximal velocity of flow depend on the capillary radius. In ideal capillaries, volume of flow is proportional to the fourth power of the radius and maximal velocity is proportional to the square of the radius. Interestingly, although the capillaries of the super apoplasm are not ideal, the flow rates in both wide and narrow vessels, with either simple or compound perforation plates, are essentially equal to flow rates in ideal capillaries (Hagen-Poiseuille flow) of the appropriate diameter (Schulte et al. 1989).

The radii of the virtual ultramicrocapillaries formed by the ultramicrospaces in the cell wall differ from the capillaries that are the cell lumina by perhaps three orders of magnitude. On this basis, the highest velocity of flow would be six orders faster in the lumina than in the wall microcapillaries, and the volume of flow would be greater by still larger factors. Obviously, resistance to flow in the super apoplasm is very much less than in the cell-wall apoplasm.

With suitable instrumentation, it can be shown that the specific resistance of the cell-wall apoplasm to water flow is about 10^6 times higher than the resistance of the super apoplasm, on the basis of equal cross-sectional areas of wall and lumen. Thus, resistance to flow through the super apoplasm is about 10^6 times lower than it is through the cell walls of parenchyma tissue having a similar cross-sectional area and path length (Raven 1977).

Water that is eventually lost from the plant via transpiration generally enters the plant through roots and flows first across the root parenchyma, then through the xylem in the root-shoot axis, and finally through the leaf parenchyma to the sites of evaporation. Thus, three successive segments of the pathway may be distinguished: root radial, root and shoot axial (longitudinal), and leaf radial. Nearly the same amount of water flows through each. The resistance of each part of the pathway is directly proportional to its length and inversely proportional to the effective cross-sectional area of the tissues involved. During evolution of the higher-plant body, dimensions have become adjusted so that the pressure drop along each segment of the pathway is appropriately moderate when water adequate to replace transpiration loss is transported from the roots to the leaves (Tyree and Ewers 1991).

In herbaceous plants, the segment of the water pathway having the highest and most variable resistance is that leading radially across the root (Raven 1977). Although comparable data from trees are not yet available, we know that the cross-sectional area of the water-conducting xylem in the stem, expressed in mm^2/g fresh weight of leaves, is similar in herbs and trees. Thus, one would suppose that the much longer xylem pathway in a tree would make the xylem contribution to total resistance higher than in herbs. Some such effect is unavoidable, but it is not large. Because of the relatively wide diameters of tracheary elements in tree stems, especially in taxa having vessels, resistance of the xylem per unit length is low compared with that of other segments of the pathway. It is likely that, in trees as in herbs, the major resistance to water flow is not in the xylem, but in the radial path across the root parenchyma.

What is the relative conductivity of the xylem segment of the transpiration path in various parts of a tree? The most readily obtainable conductivity measurement is based not on a unit of cross-sectional xylem area, but on a unit of fresh weight of leaves on the stem above the level being studied. This is the **leaf-specific conductivity (LSC)**, which is the amount of water flowing through an excised segment of stem, under conditions of gravity flow, per unit time per unit fresh weight of leaves supplied by the stem segment, usually expressed in $\mu l/h/g$ of leaf weight. Anatomically, variation in LSC values is ascribable primarily to differences in tracheary-element diameter. Thus, LSC is higher in the main stem than in the branches of a tree, and is especially low at the bases of branches and leaves in the taxa that have been investigated (Zimmermann 1983). The hydraulic constrictions at these junctions are important in regulating the distribution of water to leaves in the various parts of the crown, and are especially important in insuring enough water for the main leader (Sect. 7.4.4). Because the columns are continuous along the xylem pathway, the increased resistance in the constricted regions must be compensated for by a steeper gradient of water potential.

In a tomato plant, the water-potential gradient was found to be about 0.13 atm/m in the stem, whereas in the petioles it was considerably greater — 0.72 atm/m (Dimond 1966; Zimmermann 1983).

Junctional constrictions can be identified microscopically by the narrowness of the vessels. Constrictions have been found in, for example, branches a few centimeters from their point of attachment in several species (Zimmermann 1978); at the base of the *Populus* petiole (Isebrands and Larson 1977); and in the distal nodes of *Vitis vinifera*, all nodes of *Populus deltoides* (Salleo et al. 1982), and the proximal nodes of 1-year-old twigs of *Olea europea* (Salleo and Lo Gullo 1983). Thus, plants apparently are hydraulically segmented into: (1) internodal zones, having xylem conduits that are wide and therefore efficient in water conduction (but perhaps vulnerable to cavitation; Sect. 7.4.3.4); and (2) nodal zones, having xylem of opposite characteristics (Salleo and Lo Gullo 1986).

7.4.3.2 Motive Forces Driving Flow Through the Xylem.

Water flows through the xylem along gradients of water potential. In discussing this water potential, we use hydraulic-pressure terminology, and therefore express the potential in pascals, bars, and atmospheres.

What forces can move water up into the crowns of trees many meters tall and thus maintain the leaves in their turgid, hydraulically supported condition despite exposure to the sun and drying breezes? Simple capillary rise in the xylem is inadequate as a motive force in trees, because the height to which water will rise in a capillary with smooth, wettable walls and a radius typical of a tracheary element (about 2×10^{-5} m) is generally only about 0.75 m.

Two other kinds of forces may act jointly to raise water to the tops of trees. One force is **root pressure**, which is involved to a varying extent, depending on environmental conditions. Via the second force, the more important one, water is held, and is moved as a continuous column, by **cohesion** within the water and **adhesion** between the water and the cell walls of the xylem tubes and of the leaf mesophyll. The cell walls are hydrophilic, offering good adhesion for water. We will examine the interplay of these motive forces under various conditions.

First, consider the situation when there is no flow — as when there is no transpiration and no guttation. Assume that there is a height, h_0, to which water will rise in the stem because of root pressure (described more fully later). At a certain height, h_0, in the stem the water pressure equals the atmospheric pressure, the gauge pressure is zero, and there is no flow. The height, h_0, is usually less than h_{max}, the total height of the tree. Therefore, under most circumstances, root pressure alone cannot meet the crown's need for water.

If there is no flow through the system, h_0 is usually above the soil surface. (Capillary action is not a factor because gas-water menisci necessary for capillary movement are not present in functional elements.) The no-flow gradient of water pressure in the xylem equals about 0.1 atm/m. Because this pressure is entirely due to gravity, it increases downward. In the no-flow state, the pressures due to gravity are balanced by root pressure and cohesive/adhesive tensions, and the system is in equilibrium.

At any level, h, that is below h_0, the pressure in the xylem exceeds atmospheric pressure. The gauge pressure, P, at such a low h is:

$$P = DG(h_0 - h),\tag{7.1}$$

where D is the density of the xylem solution and G is the force of gravity. If the height h is above h_0, Eq. (7.1) is still valid, but the absolute pressure becomes less than 1 atm and the gauge reading is negative. For example, if h is approximately 10 m above h_0, the gauge pressure is -1.0 atm. Substituting h_{max} for h in Eq. (7.1) gives the gauge pressure in the xylem in the uppermost leaves. At heights up to h_0, the water molecules are in compression. Above h_0, they are in tension, with the water pulling inward on the cell walls.

If the relative humidity of the external atmosphere falls much below 100%, the relative humidity in the gas spaces of the leaf mesophyll also falls slightly. Water then evaporates from the water-saturated cell walls and diffuses along the gas space, eventually to be lost via stomatal pores. This transpirational water loss disturbs the no-flow equilibrium. The mesophyll cells, having lost some water, transmit their water deficit to the walls of neighboring cells through the cell-wall apoplasm. Because the water in all these walls forms a continuous, cohesive system with the water in the super apoplasm of the xylem, the water deficit in the mesophyll cells is transmitted apoplasmically into the stem as an increased tension — a more negative pressure. This steepens the gradient of pressures all along the xylem. Consequently, the level in the stem, h_0, at which water is at atmospheric pressure moves downward.

Because water flow through the plant is continuous from the soil to the air around the leaf, the total flux of water is the same through successive parts of the path. As already mentioned, the successive parts of the pathway are: root radial (r), from the soil to the root xylem; axial (x) within the xylem, from the roots to the leaves; and leaf radial (l), across the mesophyll into the leaf gas space and external atmosphere. A corresponding pressure differential, P, acts across each part of the pathway. Each part also has a characteristic resistance, R, to water flow.

Thus the amount of water (W) transported through the plant per unit time can be expressed as:

$$W = \frac{P_r}{R_r} = \frac{P_x}{R_x} = \frac{P_l}{R_l},\tag{7.2}$$

where the subscripts r, x, and l refer to the root, xylem, and leaf parts of the pathway, respectively. Note that P_l is the pressure differential resulting from the difference of water potential between mesophyll cell walls and the external atmosphere. This varies with humidity. Leaf resistance, R_l, is controlled mostly by stomatal pore apertures. The ratio of P_l to R_l is therefore quite variable and depends heavily on conditions prevailing at the moment. Xylem resistance, R_x, in contrast, is relatively constant. Hence, if W is to increase, P_x must increase. This happens if the pressure of water in the leaves becomes negative through transpiration. Quantitative adjustment of P_x to a greater W generally reduces water pressure at the bottom of the tree. Accordingly, the level, h_0, at which the water pressure equals atmospheric pressure moves downward, eventually into the roots. Pressure reduction in the roots in turn evokes a steeper r gradient across the roots, which promotes water absorption from the rhizosphere.

The velocity of water moving along the root-stem-leaf pathway depends on pressure gradients. Reported mid-day peak velocities vary greatly — from about 1 to 50 m/h. Values from conifers tend to be near the lower limit; those from ring-porous angiospermous trees are near the upper limit. In a survey by Huber and Schmidt (1936), the highest velocity, 43.6 m/h, was found in *Quercus pedunculata*, which has wide earlywood vessels (200-300 μm diam). Species with diffuse-porous wood tend to have peak velocities in the range of 1 to 10 m/h (Huber and Schmidt 1936; Daum 1967). Similarly, Zimmermann (1983) reported mid-day peak velocities for trees with narrow vessels (50-150 μm diam) as 1 to 6 m/h. Peak velocities for some conifers are in the same range as those for trees with diffuse-porous wood. This is not surprising, because only the widest vessels in diffuse-porous wood are wider than the normative tracheids of conifers. Thus, some coniferous xylem is highly efficient in water transport even though it lacks vessels.

Pressure gradients required to force water through xylem at peak velocities, disregarding gravity, are in the range of 0.05 to 0.1 atm/m, depending on the velocity (Zimmermann and Brown 1971). In standing trees, the force of gravity creates a further gradient of 0.1 atm/m, so that the total gradient needed to move water upward at peak velocity is 0.15 to 0.2 atm/m. This means that the pressures in leaves near the tops of trees under conditions of moderate to high transpiration will generally be negative, and that transpiration "pull" will be the major force moving the water upward.

Under conditions of little or no transpiration, in contrast, root pressure increases in importance in determining water status in the root-stem axis. In effect, root pressure is a kind of back-up system that can help maintain flow of water and solutes when the humidity of the external atmosphere is high.

What enables positive pressures to develop in roots? A key structural factor is the endodermis, with its Casparian bands. By dividing the root apoplasm into an internal and an external domain (Sect. 1.5.2), these bands bar outward apoplasmic water movement across the root. Water and solutes can enter (or exit) the internal cell-wall apoplasm and the super apoplasm with which the cell-wall apoplasm is contiguous, only by passing through the symplasm — that is, by crossing a plasmalemma, migrating within one or more protoplasts, and then crossing a second plasmalemma.

Root pressure may become positive due to this endodermal barrier and to ion-pumping activities and osmotic phenomena associated with the symplasm and its membranes. Ions and sugars are "pumped", and/or leak, out of the symplasm into the internal apoplasmic domain (adaxial to the endodermis), within the central cylinder. The presence of these solutes may lead to a positive osmotic pressure in the water of the root tracheary elements with respect to the soil water. This osmotic pressure can manifest itself as positive root pressure if the outflow of water is prevented by an endodermal barrier.

Root pressure increases the height in the plant (h_0) at which gauge pressure is zero, as discussed previously. If h_0 exceeds the height of the plant, the pressure in the central cylinder, under conditions of no transpiration, can manifest itself as guttation (Sect. 8.2.1), or, in decapitated root systems, as root exudation. Guttation occurs in many herbaceous plants and in some small trees.

If transpiration increases and water pressure in the xylem becomes negative, water flow from the symplasm to the super apoplasm in the root is facilitated, and the solute concentration in the xylem decreases. The efficiency of the salt-pumping activity of the symplasm then increases. However, the flux of water through the membrane separating the symplasm and apoplasm can be considered "normal" osmosis only if the osmolarity of the xylem solution in the root exceeds the osmolarity of the soil water.

If the osmolarity of the root xylem solution becomes lower than that of the soil solution, the flow may nevertheless remain inwardly directed, because the inward hydraulic-pressure gradient is steeper than the outward osmolarity gradient. This net inward flow of water along a hydraulic-pressure gradient, against the outward tendency of osmolar factors alone, is termed **reverse osmosis**. Reverse osmosis probably is always a part of the water dynamics of mangroves and other trees having root zones that are saturated with salty or brackish water. Reverse osmosis probably also occurs in most plants that are rapidly transpiring. As it is powered by the difference between water potentials inside and outside the leaf, reverse osmosis requires no energy expenditure by the plant. Energetically, reverse osmosis is the most efficient way of desalinating water (Cromer 1977).

7.4.3.3 Structural Adaptations of Tracheary Elements to Negative Pressures. As already discussed, negative water pressures are common in the xylem of trees during periods of rapid transpiration. Negative pressures cause compressive strains in the walls of the capillaries containing the water. From mechanics, we know that the strain depends not only on the pressure, but also on the radius of the container, and is inversely proportional to wall thickness. In the ultramicrospaces within a cell wall, the strain arising from the negative pressure of water is negligible because these spaces are so small. Their total cross-sectional area also is small relative to solid wall material.

The situation is very different in capillaries of the super apoplasm (lumina of dead cells). In these elements, the ratio of radius to wall thickness is high, and the compressive strain in the wall is great. An unlignified wall, though quite resistant to tension, is not resistant to compression (Sect. 4.4.1), and therefore is in danger of collapse. Lignin is the only common plant-produced substance that can substantially increase the rigidity of a cell wall and also offer high adhesion to water molecules. Water under negative pressure cannot easily break away from a lignified cell wall as it can from a suberized wall. In addition, the super-apoplasm capillaries are reinforced by deposits of secondary-wall layers on the inner surface of the primary wall.

In a cylindrical cell, the transverse compressive stress resulting from the negative pressure of internal water is twice as great as the longitudinal compressive stress. (This is analogous to the relation with respect to positive pressure; Sect. 4.4.1.) Protoxylem elements are effectively reinforced against transverse compression by annular or helically shaped secondary thickenings on the side walls. These thickenings cannot provide direct resistance to longitudinal compression. However, an entire strand of tracheary elements may shrink longitudinally under negative pressure; this shrinkage, even if only slight, would put shear and other stresses on neighboring cells, and the restorative forces generated thereby would help balance the compressive stress. Metaxylem and secondary-xylem tracheary elements are more resistant to negative water pressure, especially to longitudinal compressive loading, than are protoxylem elements, due to lignified secondary walls deposited in netlike patterns, or in layers that are continuous except for pits.

7.4.3.4 Embolism: Efficiency Versus Safety in Water Transport. Capillary water columns under negative pressure are unstable. Gas bubbles can appear and break the continuity of the water. The more negative the pressure, the more likely are bubbles to form. Once formed, a bubble expands, as water vaporizes into it. The bubble can continue to expand until the moving gas-water interface encounters a porous barrier having high adhesive forces in its pores. The interface will be stopped there if the negative pressures are not great enough to draw it through the pores. Viewed from the gas side, the meniscus in each pore curves away from the gas, and, in effect, is the meniscus of a small capillary. Forcing such a meniscus through a pore (capillary emptying) requires a negative pressure in excess of the oppositely directed pressure of the capillary-rise (capillary filling). It can easily be calculated that, for a pore 1 μm in radius, the capillary-rise force is equivalent to 1.5 atm. But if the radius is reduced to 0.1 μm, that force is increased to 15 atm!

It follows that, if the pores of the closing membranes of the pits had diameters of 0.1 μm or less, they could effectively contain gas embolisms against negative hydraulic pressures as great as 15 atm (equivalent to a water column about 150 m high). In gymnosperm tracheids, the bordered pits have a closure system consisting of a torus and porous margo. With colloidal gold suspensions, it has been found that the effective diameter of margo pores ranges between 0.1 and 0.2 μm (Frenzel 1929; Liese and Bauch 1964). Because the presence of even a few pores as large as 0.2 μm greatly decreases the effectiveness of the barrier, additional protection is needed. The very rapid local flow of water accompanying a spreading embolism provides this protection by causing the torus of a bordered pit to snap shut against the aperture.

The pores in membranes across the pits of vessel elements in angiospermous trees generally have diameters much smaller than 0.1 μm. These pores may thus be highly effective in confining a gas embolism to the vessel in which it originates. Though they protect against embolism, the small pores in vessel-to-vessel pits inevitably are a constraint on lateral, vessel-to-vessel flow. This constraint is minimized, however, by the extreme shortness of the capillaries formed by the pores.

In addition to the pit characteristics that help prevent embolisms from spreading from one tracheid or vessel to another, plants have other ways of preventing embolisms and of coping with them after they occur. These include other structural characteristics of tracheary elements, particularly of vessels.

For example, scalariform perforation plates, which are common in vessels of woody dicots in cold climates, may be important in confining gas bubbles that appear when the water in the vessels freezes (Scholander, et al. 1961; Sucoff 1969). These bubbles occur because gases are less soluble in ice than in water. When the ice melts in spring, the bubbles usually redissolve in the water before transpiration generates negative pressures, but, under some conditions, the small bubbles combine into large ones that pose an embolism threat. Zimmermann (1978, 1983) suggested that scalariform vessel-element perforations inhibit the movement of gas bubbles from one element to the next unless negative pressures become exceedingly high. Thus, bubbles are held in place, and most eventually redissolve rather than combining with other bubbles in the vessel to form an embolism so large that the vessel cannot be refilled. The refilling of tracheids in conifers is facilitated by lowering of water potential within tracheid lumens by capillary action. Indeed, in *Pinus sylvestris* the tracheid walls may become chemically active and reduce water potential below the values deriving from capillary forces alone (Borghetti et al. 1991). One can also appreciate that the cellular structure and organization of gymnospermous wood tends to restrain the spread of embolism, and that many gymnosperms are well adapted to grow in regions where water in the wood often freezes (see also Tyree and Dixon 1986).

Two other structural characteristics of vessels that affect susceptibility to embolism are diameter and length. These dimensions also affect the efficiency of water conduction. Zimmermann (1978), in fact, proposed that the two main factors directing evolution of the water-conducting systems of plants, such that large, arborescent angiosperms gradually became established in mesic environments, were (1) a requirement for efficiency of water conduction, and (2) an accompanying and seemingly incompatible requirement for "safety", in the sense of being able to cope with high negative pressures in the conducting elements.

The diameter of tracheary elements is an important factor determining the efficiency of water conduction. Under ideal conditions, wide vessels are much more efficient than narrow ones because the volume of flow increases as the fourth power of the diameter. Wide vessels, however, are much more susceptible to embolism than are narrow ones. Large-diameter vessels have evolved independently in various taxonomic groups, yet many arborescent species have small-diameter vessels.

Vulnerability to embolism seems not to be directly related to vessel diameter, but rather to be determined by the pit-membrane pore diameter. Within a species, the pore diameter is generally larger in wide xylem conduits than in narrow ones. Xylem elements of the same size, from different species, can have different vulnerabilities to embolism (Tyree and Sperry 1989).

Within the plant body as a whole, the spread of embolisms is limited by localized hydraulic constriction zones, attributable mainly to narrow vessel diameters (Sect. 7.4.3.1). Constriction occurs in, for example, the nodal regions and in the primary xylem at the bases of the lateral and adventitious roots. In addition, in some diffuse-porous trees, there is a high frequency of vessel ends in branch bases. Because of their high resistance to the passage of menisci, vessel ends may help limit the spread of embolisms into the branches (Zimmermann and Jeje 1981; Salleo and Lo Gullo 1986).

In vessels that are similar in diameter and structure, the danger of embolism increases as pressure becomes more negative. Thus, the problem of embolism would, in theory, be greater at the tops of trees than near their bases or in their roots (but see Sperry et al.). The decrease in mean width of vessels upward within each growth increment, from the roots into the crown, may lessen the danger of embolism in the treetops (Zimmermann 1978).

Vessel length affects both the efficiency of water conduction and the likelihood of embolism, but to a lesser extent than does vessel diameter. Long vessels (which tend to be wide) are very efficient in water transport but are highly vulnerable to gas embolism, whereas shorter vessels (which tend to be narrower) are less efficient and less susceptible to embolism. Therefore, to assess relative efficiency versus safety of water transport in a tree, it is important to know the relative frequency of long and short vessels. Commonly, individual trees have vessels of a wide range of lengths (Skene and Balodis 1968; Zimmermann and Jeje 1981), and thus have both the risks and the advantages of long and of short vessels.

Of greater interest than the length of the longest vessels is the relative frequency of vessels of various length classes. Probably every part of a tree has vessels of a range of lengths and probably has many more short than long ones (Zimmermann and Jeje 1981), but data are few. In *Acer rubrum*, average vessel lengths and diameters increase from twigs to branches to main stems to roots (Zimmermann and Potter 1982). In earlywood of *Prunus serotina* stems, about 75% of the vessels do not exceed 10 cm in length, whereas, in latewood, 95% may be in this category. In the stem of the palm, *Raphis excelsa*, almost 80% of the vessels are less than 5 cm long, but a small percentage may attain lengths of 30 to 50 cm (Zimmermann et al. 1982).

Embolism of tracheary elements need not be permanent. Embolized elements may be refilled with water during periods when the xylem water is under positive rather than negative pressure. They may then again function in transport (Milburn and McLaughlin 1974; Zimmermann 1983; Sperry et al. 1988). For example, in some species, freezing-induced embolism during winter is followed by spring recovery as root pressures become positive (Tyree and Sperry 1989).

If an embolized tracheary element cannot be refilled with water, it must be replaced functionally. In stems of woody dicots and gymnosperms, the vascular cambium provides replacement tracheary elements during most growth episodes of the plant. However, palms and some other groups have attained an arborescent growth habit without having evolved a vascular cambium or other lateral meristem system capable of adding new axial vascular elements that can functionally replace the old. The life spans of cambium-less trees, though shorter than those of large dicots or gymnosperms, nonetheless extend into scores of years. How do these trees surmount the embolism problem? In palms, one factor may be high root pressure (Davis 1961).

The xylem parenchyma may provide another means for some trees to cope with the problem of embolism in tracheary elements. Commonly, the embolizing gas is mainly CO_2 (Chase 1934; Carrodus and Triffett 1975), which possibly can be absorbed in the xylem parenchyma by CAM-like dark fixation (Sect. 5.4) and thus be removed from the tracheary elements. In this way, xylem parenchyma may be indirectly involved in water transport (Höll and Meyer 1977). However, xylem parenchyma seems not to be involved in refilling embolized tracheary elements in *Pinus sylvestris* (Borghetti et al. 1991).

Permanently embolized tracheary elements may serve a transport function in some trees. For example, parts of the xylem in the pneumatophores of roots of *Taxodium* and some other genera seem specifically adapted (Sect. 9.5.2) to undergo embolism and to transport gas rather than liquid. Theoretically, the permanently embolized heartwood (Sect. 19.5.2.3), which constitutes a continuous system in the inner parts of secondary wood in axes of older trees, could function in gas transport to roots. However, at least in angiosperms, an important feature of heartwood is the occurrence of tyloses in embolized vessels. Tyloses are outgrowths of neighboring parenchyma cells, through the pits into embolized vessels where they block the lumina and inhibit gas transport.

7.4.4 Water Movement Through the Stem: Distribution in the Crown

In woody plants, tracheary elements form strands or files extending from the proto-xylem and metaxylem in the root, through the secondary xylem of the root and stem, to the protoxylem and metaxylem in the vein endings in the leaves.

Flow in the xylem takes the path of least resistance, which is along the axial direction of the vessels and tracheids. In the secondary xylem, this is approximately the major direction of all the fusiform elements — the direction of the wood "grain". There may, however, be noticeable local differences between the direction of the vessels and of the grain, especially where the grain angle deviates considerably from axial (Burggraaf 1972; Zimmermann 1983).

Data about routes of flow through the xylem of trees come mainly from studies using dye solutions injected into small, bored holes or through hollow, steel chisels driven into the sapwood. Treated trees are felled and sectioned to reveal dye distribution patterns. Additional data come from severing selected roots, then immediately immersing the cut surfaces of the remaining parts in dye solutions. These studies show that water ascends through a much smaller part of the thickness of the sapwood in ring-porous than in diffuse-porous dicots or in gymnosperms. In fact, in ring-porous trees, rapid upward transport is largely confined to the outermost growth increment. Dye solution injected directly into the heartwood has revealed no upward transport in this region in any species (Kozlowski and Winget 1963; Vité 1967; Zimmermann 1983).

In trees in which water moves in more than just the outermost increment, the ascent of dye solutions may deviate from the axial in three ways, as seen in transections. These are: (1) a sectorial pattern, in which the dyed sector becomes tangentially wider above the point of injection; (2) a spiral ascent (indicating spiral grain); and (3) a reversing spiral ascent (indicating interlocked grain; Sect. 18.5.3), in which the dyed sapwood tissue forms progressively larger zig-zags with increasing height (Zimmermann and Brown 1971). These types may also intergrade. Because the various ascent patterns are correlated with wood grain, we must seek an ultimate explanation for them in cambial dynamics (Sects. 18.5.3; 18.5.4).

Water transported upward through the xylem is distributed to all living parts of the crown. The leaves at the top of the crown are the most important in competing with neighboring trees for sunlight. These leaves, however, are the most "expensive" to maintain in terms of water requirements. Being more exposed to air movements, these leaves transpire more water than do lower leaves. How can upper leaves compete successfully with lower ones for water (Zimmermann 1983)?

Variation in vessel length and diameter, and thus in LSC values of xylem, in different parts of a tree (Sect. 7.4.3.1) suggests an answer. In lower branches, vessels generally are narrower than in higher branches, and thus have lower LSC values. This variation within the crown can automatically regulate the distribution of water to branches at different heights. When transpiration is rapid, the high flow resistance of the narrow vessels in lower branches produces a large pressure drop along these branches. The resistances in the pathways leading to leaves in the upper crown are lower, thus allowing the upper leaves to compete for water with leaves in the lower crown (Zimmermann 1978, 1983).

However, even if the pathways leading to the upper crown have lower resistances than do pathways to the lower crown, the pressures in the upper crown are still quite negative, especially if transpiration is rapid. In fact, because the super apoplasm has no semipermeable membranes, any negative pressures within it extend to all parts of the functional xylem, including the protoxylem. Cells in growing tissues must obtain water from the protoxylem despite the negative pressures. How is the turgidity of the hydraulically supported meristematic tissue maintained? Only one force, the osmotic pressure of the living cells, can explain the flow from the protoxylem into growing cells under conditions of negative pressures generated by transpiration. Even that force, however, is often not great enough to allow an unimpeded rate of growth. Rates of shoot-extension growth are, in fact, reduced during periods of rapid transpiration, and accordingly are often greater by night than by day (Kozlowski 1971, p. 278).

To insure that water reaches the crown, even if some embolisms occur, xylem always has a "back-up" system of narrow conduits, which are less susceptible to embolism than are wide ones (Sect. 7.4.3.4).

7.5 Primary Vascular Bundles and Rapid Long-distance Transport Routes

Xylem and phloem elements differentiate within discrete vascular bundles that have an orderly three-dimensional distribution pattern within the primary plant body. This distribution pattern determines the routes of import into and export from leaves, and the interrelations of different leaves and axillary buds.

Primary vascular bundles usually include both xylem and phloem, arranged with the xylem adaxial to the phloem. These bundles are called **collateral.** A bundle in which there is phloem both adaxial and abaxial to the xylem is **bicollateral.** Strands of xylem unaccompanied by phloem are unknown in stems, but may occur in leaves (Sect. 7.6.2).

In primary tissue, phloem develops in or toward cell masses in which growth and cell division are intensive (Wardlaw 1970) or that perform photosynthetic or storage functions. Physiologically, one can think of primary phloem as developing toward metabolic sinks or sources.

An axial bundle, if followed acropetally and basipetally through the entire primary stem, is found to be associated with several branch-like leaf traces. If we assume an apical viewpoint, a leaf trace always leads from a leaf into a stem. Materials exported from an older leaf, **A**, to a younger leaf, **B**, in the same sympodium (Sect. 14.3.1) first move down along the trace of leaf **A**, then upward along the trace of leaf **B**. The axial part of a trace may be in lateral contact with neighboring sympodia, or may be connected to them by branches. Thus, if **A** and **B** belong to different sympodia, several steps of lateral exchange may be involved. In many taxa, there is considerable transport autonomy within sympodia (Kursanov 1976).

Transport from a leaf to the bud in its axil follows the same basipetal, lateral, apical sequence as that from older leaf to younger leaf. The provascular stage in axillary bud development occurs very early. Differentiation and maturation of protoxylem and protophloem from provascular strands, however, are not necessarily continuous with those of the parent stem and may be much delayed in dormant buds (Tucker 1963; Sachs 1970).

Primary vascular bundles tend to be somewhat inclined relative to the axis of the stem. Thus, the primary vascular system as a whole typically has a twisted configuration, as is manifested by the helical movement of injected dye solutions. The observed twisting of traces may be a consequence of phyllotaxy, or of torsions accompanying extension growth.

7.6 Structure in Relation to Transport Within Leaves

7.6.1 General Perspectives

Transport systems within leaf laminae have generally evolved in such a way that transport routes to and from mesophyll cells are as short as possible (Sect. 5.3). Haberlandt (1924) recognized this as the fundamental "Bauprinzip" of leaves. Comprehensively viewed, transport includes both vascular and other-than-vascular routes — that is, not only the specialized vascular tissues of the veins, but also the nonphloic symplasm, and the cell-wall apoplasm, which consists of the walls of both vascular and nonvascular cells. In this view, essentially all the cells of the leaf are part of the transport system. Although we will discuss these different modes of transport separately, they must, of course, function as an integrated system.

7.6.2 Vascular Pathways in Leaves of Dicots

In dicots having pinnately veined leaves, the leaf traces and their subsidiary bundles extend from the stem through the petiole into the leaf blade, where they become the midvein. The midvein "branches" pinnately into first-order lateral veins. In palmately veined leaves, there is no midvein. Instead, the bundles that compose the trace diverge near the junction of the lamina and petiole. The veins thus formed then branch and rebranch (but see Sect. 16.3.6.2, with reference to "branching" of veins). In both palmately and pinnately veined leaves, the small veins form a usually closed network that divides the entire leaf blade into small (near-microscopic) fields called **areolae.** Generally, each areola is surrounded by veins, partly sheathed (discussed later). In addition, an areola may have one, several, or many blind vein endings within it. Thus, a mesophyll cell is unlikely to be more than a few cell diameters from vascular tissue.

Small veins usually are entirely primary. Large veins, because of the meristematic activity of ribbons of vascular cambium within them, may include some secondary tissues also.

Typically, large and medium-sized veins include vessels and tracheids on the xylem (adaxial) side, and sieve tubes with associated parenchyma cells on the phloem (abaxial) side. In the smallest veins, the xylem elements are tracheids with spiral or annular thickenings, and the phloem elements are sieve-tube members with associated companion cells (Turgeon et al. 1975). Some of these companion cells may be connected with their associated sieve elements and with bundle-sheath cells via numerous plasmodesmata. This suggests that these cells function as "intermediary cells" in loading the sieve elements with photosynthate via the symplasm (Fisher 1986). Minor veins may or may not have recognizable transfer cells (Pate and Gunning 1969).

The most distal phloem is often near, but not at, vein endings. It may consist of single files of sieve elements or only slightly differentiated parenchyma. The vein endings themselves often consist of several slender files of tracheids, or of pairs of tracheids, or even single tracheids, unaccompanied by phloem.

Most small and some larger veins may be surrounded by compactly arranged, parenchymatous **bundle sheaths**, which prevent direct exposure of the vascular elements to intercellular gas space. These sheaths are often only one cell thick, but may be thicker, and, in the larger veins, may also contain collenchyma and sclerenchyma (though, rigorously, these constituents are not part of the bundle sheath). Bundle-sheath cells are typically elongated along the vein in C3 plants and are thin-walled, but in C4 plants often have obviously thickened walls and (when viewed in transections) are radially elongated in a "Krantz" arrangement (Sect. 5.3.2).

In some taxa, bundle sheaths may have localized apoplasmic barriers, such as Casparian bands or suberin lamellae, somewhat as in endodermis. The occurrence and distribution of Casparian bands in bundle sheaths have been documented for some members of the Primulaceae and Plantaginaceae (Gunning 1976). Putative Casparian bands also occur in analogous cells in monocots (Van Fleet 1950, 1961) and in some gymnosperms (Sect. 7.6.3). This means that in some leaves, as in roots, there is an inner and an outer apoplasm, at least locally.

The larger veins, isolated from nearby mesophyll by the nonvascular, bundle-sheath tissues, may be analogous to express highways, which, having been designed for long-distance travel, provide no direct access to the communities they traverse. In contrast, in the sheath surrounding the fine veins, much of the transpiration water supplied via the super apoplasm enters the symplasm, soon again to be transferred to the outer apoplasm and lost (Canny 1990).

In leaves of many dicots, panels of parenchyma, known as **bundle-sheath extensions**, may connect the bundle sheaths with the upper or lower epidermis, or with both. Rather than being elongated, as the term "extension" suggests, the cells of this tissue are similar to those of bundle sheaths, and are compactly arranged. The term "extension" actually refers to the appearance of the tissue in transection. Bundle-sheath extensions can serve as conduction routes to the epidermis (Wylie 1947, 1951; Sect. 16.3.6.2).

The major function of the phloem of small veins is to collect and export sugar produced during photosynthesis. The concentration of sugar in the small veins may even be higher than that in mesophyll. In *Beta vulgaris*, for example, it is 4 to 16 times higher (Kursanov 1976). In addition, because the phloem accounts for no more than half of the vein volume, sugar concentration in the sieve elements must be considerably higher than this. Sugar concentration in *Beta* leaves increases from small to medium-sized veins (Kursanov 1976). Gradients in *Robinia* leaves are in the same direction, but less steep (Roeckl 1949). Thus, in these taxa, loading of phloem with sugars and longitudinal transport in the small veins appear to proceed against a concentration gradient. This raises a question about the general applicability of the Münch pressure-flow hypothesis (see Sect. 7.2.5).

Two types of **intermediate cells** occur between sieve elements of minor veins and the mesophyll. One type has plasmodesmata on the interface with mesophyll cells; the other lacks them. Accordingly, all small veins may be classified into two general types: the open type having numerous plasmodesmata at the interface, and the closed type lacking them. Most woody plants have intermediate cells of the open type and utilize symplasmic transport for phloem loading. In contrast, most herbaceous plants have intermediate cells of the closed type and are dependent upon apoplasmic transport for phloem loading. The closed type may be subdivided into subtypes differing in structural adaptations promoting phloem loading from the apoplasm (Gamalei 1989).

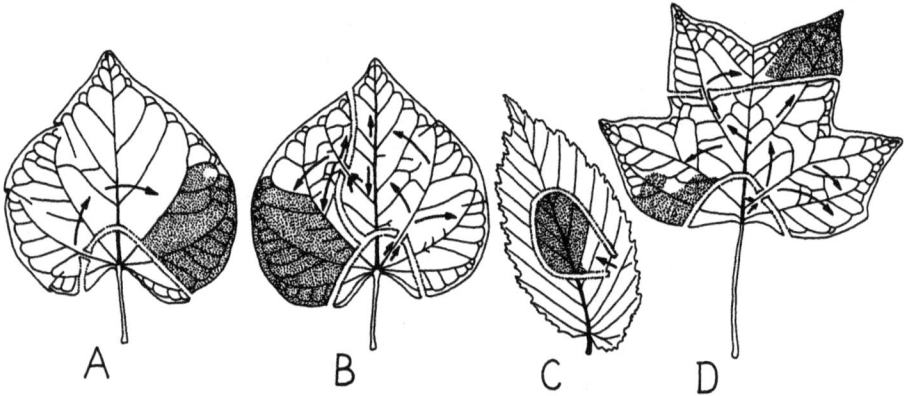

Fig. 7.5 A-D. Experimental cuts demonstrate the capability of the closed network of veins in dicotyledonous leaf blades to supply the same area by various alternate routes, though not without limits. The *stippled areas* became desiccated and died. **A** and **B** *Cercis canadensis*; **C** *Ulmus americana*; **D** *Liriodendron tulipifera*. (Redrawn after Plymale and Wylie 1944)

Collectively, small veins usually account for more than 90% of total vein length in a leaf and for about half the total vein volume. In dicotyledonous trees, the total length of veins per unit leaf area, ranging from 40 to 170 cm/cm^2, is similar to that in herbs (Schramm 1912; Plymale and Wylie 1944; Kursanov 1976). Total vein length per unit leaf area is greater in leaves that developed in full sunlight than in leaves that developed in shade. For example, in *Ulmus campestris*, vein length per unit area was found to be about 170 cm/cm^2 in sun leaves, but only 105 cm/cm^2 in shade leaves. In *Alnus glutinosa*, the lengths in sun and shade leaves were 81 and 36 cm/cm^2, respectively. Typically, the volume of vein tissue is about 25% of the total mesophyll volume, though the proportion was found to be greater than 30% in *Tilia* and *Ulmus* (Plymale and Wylie 1944).

Because veins form a generally closed network, an area of leaf blade can be supplied with water by several alternate routes. The technique of deliberately cutting veins and then observing the movement of dye solutions during transpiration, and also noting the areas that die for lack of water (Fig. 7.5), allows demonstration of the extreme flexibility and overcapacity of the xylem water-supply system (Plymale and Wylie 1944). This flexibility and overcapacity have survival value because of the continual danger of damage to leaf blades by, for example, leaf-eating insects.

Evidence suggests that the phloem export system is less flexible than the xylem system, or that the adjustment time is greater. For example, in apple leaves damaged by a leaf miner, large areas were found to have impaired ability to export carbohydrate (Schneider-Orelli 1909). Preventing exports may lead to degenerative changes in the photosynthetic system (Kursanov 1976). Hence, whether an area beyond a severed vein can continue to contribute photosynthetic products to the plant may depend on the time it takes for the photosynthetic apparatus to degenerate, relative to the time it takes for alternative phloem export routes to be established.

7.6.3 Vascular Pathways in Leaves of Gymnosperm

Compared with the elaborate networks typical of dicot leaves, vascular systems of gymnosperm leaves seem, on first consideration, to be poorly developed. Though large veins may "branch" into smaller ones (but see Sect. 16.3.6.2), the smaller ones typically are not directly interconnected. Thus, the leaves generally lack a closed network of veins, except in Gnetales. For example, in *Ginkgo*, the two vascular bundles of the petiole diverge upon entering the blade, to become the right and left veins. Each of these veins then forks dichotomously many times, providing each half of the blade with independent venation. There are no lateral vascular connections. After a large vein is cut, the part of the blade that it had supplied will die.

The needle leaf of a gymnosperm such as *Pinus* or *Cedrus* usually has a single central vascular bundle or two closely aligned bundles (Gambles and Dengler 1982b) that are formed by early dichotomous branching of a leaf trace in the stem, before it enters the petiole. Venation of scale leaves of gymnosperms differs developmentally from that of needle leaves, but otherwise is similar. In some broad-leaved members of *Podocarpus*, there may be further dichotomous dividing of the veins within the blade, somewhat in the manner of *Ginkgo*. Vascularization of gymnosperm leaves is discussed from a developmental perspective in Section 16.3.6.1.

The veins of most gymnosperm leaves contain a vascular cambium. In needle leaves of many conifers, this cambium produces secondary phloem, but no secondary xylem (Ewers 1982b). The primary xylem apparently functions as long as these needles live, which is typically 30 or more years in *Pinus longaeva*.

Water and nutrients are transported laterally between a vein and the mesophyll via **transfusion tissue**, which may entirely ensheathe the vein and/or extend laterally into the mesophyll, as in *Cycas* leaflets. Transfusion tissue includes transfusion tracheids and transfusion parenchyma. Some of the transfusion parenchyma cells function as Strasburger cells. The tracheids have slightly thickened, lignified walls with bordered pits. They are often located lateral to the tracheids of the vein xylem, and may be oriented transverse to the vein. Transfusion tracheids vary in structure and dimensions depending on their position in the vascular bundle. This suggests that their function may also vary with position (Gambles and Dengler 1982b). Transfusion Strasburger cells have dense cytoplasm and are located adjacent to the phloem of the vein. Other transfusion parenchyma cells are mostly adaxial to the vein xylem.

In some conifers, the transfusion tissue is surrounded by a distinct layer of compact cells that has been called endodermis. Though the endodermal cells of *Pinus* needles reportedly have Casparian bands, these are not comparable to the narrow, discrete bands in root endodermis. Rather, the entire radial and end walls are heavily impregnated with a substance that is probably lignin rather than suberin (Gambles and Dengler 1982b). These walls, nonetheless, can function somewhat in the manner of true Casparian bands (Carde 1978), forming barriers between the internal and external fractions of the apoplasm (Scholz and Bauch 1973). Water moving from the xylem across the leaf endodermis must pass through the symplasm and is therefore susceptible to osmolarity effects and possibly other physiological controls.

Huber (1947) reported that in *Pinus sylvestris* the transfusion tracheids abut on the leaf endodermis opposite its anticlinal walls, whereas the transfusion parenchyma cells are opposite the middle parts of the endodermal cells. In view of present knowledge of apoplasmic and symplasmic spatial systems, we interpret the transfusion parenchyma cells as being well positioned for symplasmic transport of assimilates from the mesophyll to the phloem of the vein. Similarly, the transfusion tracheids are well positioned for apoplasmic water transport from the xylem of the vein to the anticlinal walls of the endodermal cells and then to the mesophyll walls. The quantity of water moved outward could be significant only across an immature (i.e., unlignified) endodermis, however. Further information is needed on the interrelation between development of photosynthetic function, with production of assimilates, and development of a lignified barrier that prevents the leaching outward of these assimilates.

7.6.4 Nonvascular Routes of Translocation in Leaves

We are here concerned with the structural aspects of intrafoliar, nonvascular, short-distance translocation of: (1) photosynthetic products from chlorenchyma cells (a source) to the phloem (a sink), and (2) water from xylem elements to mesophyll and epidermal cells that are losing water by transpiration.

The translocation of photosynthetic products from chlorenchyma to phloem may include both symplasmic and apoplasmic routes. In leaves of water plants such as *Vallisneria*, in which apoplasmic transport would be susceptible to large leakage losses, empirical observations support the expectation that transport is largely symplasmic (Gunning 1976).

In plants that use an apoplasmic route from mesophyll to phloem, transport leads from mesophyll protoplasts to the cell-wall apoplasm, and then back to the symplasm as the products enter the phloem. The plasmalemmae of green mesophyll cells are, in fact, more permeable to sugars and amino acids moving outward (into the apoplasm) than inward, and, according to Kursanov (1976), an appreciable fraction of the total sugar in the lamina may be in the apoplasm. Rapid translocation of assimilates from the mesophyll symplasm to the cell-wall apoplasm may be a means of isolating them from biochemical systems that could degrade them. The sugars and other metabolites that concentrate in the cell-wall apoplasm can be leached from leaves by rain and mist, as has been demonstrated by many studies (Tukey 1970).

Mesophyll cells have structural adaptations that increase membrane transport capability. For example, surface area is increased by the elongate shape of palisade cells and by the "arms" of spongy-mesophyll cells (Sect. 16.3.5). In fact, the surface area of a green mesophyll cell considerably exceeds the total surface area of the chloroplasts it contains and thus would seem to be large enough that the protoplast could readily pump out the assimilates to the cell-wall apoplasm. The large plasmalemma surface area also promotes CO_2 uptake.

It seems likely that in some aquatic plants such as *Vallisneria*, and in some terrestrial plants such as *Cucurbita pepo* (Madore and Webb 1981), the symplasmic route predominates. In contrast, in *Beta*, apoplasmic transport is of prime importance (Kursanov 1976). In many plants, both routes contribute significantly. The symplasmic route may be more ancient in an evolutionary sense, but the apoplasmic route may become dominant under unfavorable environmental conditions (Gamalei 1991). The contribution of the apoplasmic route is probably related to the occurrence and activity of transfer cells next to the small veins. The contribution of the symplasmic route probably depends on the number, size, and distribution of plasmodesmata (Gamalei 1989). Accordingly, in *Fraxinus* and *Populus* leaves, the reportedly high frequency of plasmodesmata linking certain cells in sequence — i.e., mesophyll, bundle sheath (including the "intermediate cells" already mentioned), phloem parenchyma, and sieve elements in small veins — suggests that symplasmic transport of assimilates is important (Russin and Evert 1985; Gamalei 1989).

Unloading of photosynthate from the phloem has been less studied than has loading. In roots, photosynthate is unloaded via the symplasm (Giaquinta 1980) and may be unloaded by a similar mechanism in some leaves (Eschrich 1986). In bean stems, sucrose is reportedly unloaded into the apoplasm (Minchin and Thorpe 1984).

In addition to nonvascular transport of assimilates, there is also nonvascular translocation of water, from the xylem to transpiration sites. These sites include all cells exposed to the internal or external atmosphere. The water may flow to these cells symplasmically, but, if apoplasmically, then the route of flow may differ from the route of the diffusive flow of assimilates. In broad-leaved taxa, most of the assimilate flows through the adaxial part of the blade, from the palisade mesophyll to the phloem, which is in the abaxial part of the vein. In contrast, most of the water flows from the xylem, which is in the adaxial part of the vein, through the apoplasmic space in the tissues in the lower (abaxial) part of the blade, to the lower epidermis, where stomata are commonly numerous and the transpiration rate high.

Superficially, it may seem that the leaf epidermal layers, which are the most remote tissues from the xylem water supply and are exposed to the external atmosphere, would be the most susceptible to desiccation. However, cuticle layers protect against rapid water loss from the epidermis (Sect. 2.2.1.2), and efficient routes of water transport help insure an adequate supply. The walls of epidermal cells, which are a part of the cell-wall apoplasm, are in continuous, though indirect, water contact with the xylem, except where there is a functional endodermal barrier. In many angiosperms, parenchymatous bundle-sheath extensions provide a relatively direct route of water transport from the xylem to the epidermis (Wylie 1952).

8 Secretory and Excretory Systems

8.1 Distinguishing Secretion from Excretion in Plants

Distinguishing secretion from excretion is more difficult in plants than in animals. Excretory systems of animals are well known. They rid the body of useless or harmful substances. Animal secretory systems, which are well known anatomically and physiologically, produce substances that either perform a useful function within the individual or, after release to the exterior, benefit the individual or the species through their effects on other organisms.

In plants, though, neither the structures involved in the release of substances, nor those substances, can be easily categorized as excretory or secretory, because it is commonly unclear whether a substance is useful, harmful, or superfluous to the plant releasing it. Some substances that serve no apparent purpose for the plant may, in fact, have long- or short-range ecological functions affecting survival of the individual or the species. Arguably, many "secondary" products released by a plant protect against pathogens, foraging animals, and competitors (Swain 1977). Some of these substances may also be metabolic waste products. Closely related taxa often release different substances, or similar substances in greatly differing amounts. This suggests that these products are not essential to plant life, but rather result from minor differences in metabolic byways that give members of one taxon an advantage over those of others in a particular ecological situation.

Because of the difficulty of ascertaining that a product released by a plant serves no useful function, we generally refrain from using the term "excretion". Instead, to describe the disposal of substances in cells, tissues, or organs that are "programmed" to die, such as heartwood, we use the more neutral term "deposition", which requires us to make no judgement about the possible purpose of these substances. In addition, we apply the term "secretion" — nonrigorously — to the various processes by which individual cells or multicellular structures do any of the following: (1) remain alive and discharge their products outside the plant body or into special, nonliving ducts or cavities within that body; (2) store the products within the living cell that elaborated the products, or between the protoplast and the wall of that cell; or (3) undergo lysis and rupture, releasing their contents, often into a duct.

8.2 Deposition in Dead Cells, Tissues, and Organs

There are various processes by which plants deposit metabolic products in cells, tissues, or organs that soon die. For example, great quantities of metabolic products are deposited in the heartwood, which constitutes a large fraction of the mass of a mature perennial plant such as a tree. These products, which include tannins, oils, gums, resins, and salts of organic acids, sometimes accumulate to the extent of 25% of the dry weight of the wood. These deposited materials, collectively known as "extractives" (Hillis 1968), are mostly synthesized in the xylem-parenchyma cells of the older sapwood, from precursor substances derived from elsewhere in the plant. The site of deposition is neighboring tracheary elements and senescent parenchyma cells in the sapwood-heartwood transition zone. This deposition, which generally does not occur in young plants, may begin in stems that are 5 to 10 years old in *Eucalyptus*, 15 to 20 years in *Pinus*, and 80 to 100 years in *Fagus*. Some taxa seem to form no heartwood (Dadswell and Hillis 1962).

Dead or dying tissues or organs other than heartwood also are sites of deposition of metabolites. Many of these plant parts, unlike heartwood, are shed by the plant. During its lifetime, a typical tree produces and sheds many sets of leaves, and most trees shed many layers of dead phloem and sequent periderms, as bark flakes. Furthermore, metabolically active, meristematic shoot organs commonly produce and shed trichomes, scales, or stipules. Like heartwood, these tissues and organs are often rich in products such as tannins.

Many heartwood extractives, such as tannins, though they are in a sense metabolic "waste" products, are functional in imparting decay resistance to the wood. Substances deposited in leaves or bark may have allelopathic effects, increasing the competitiveness of the species after the organs or tissues are shed.

8.3 General Aspects of Secretion

Some structures secrete, in unmodified or only slightly modified form, substances supplied by the vascular system. These structures include hydathodes, salt glands, and nectaries. Other structures secrete substances synthesized in their constituent cells (Fahn 1979).

If substances are discharged onto the external plant surface, we speak of **exosecretion.** If discharge is to an internal site, or if the substance is stored in the protoplasts, as in laticifers, the process is **endosecretion.**

Further, in **merocrine secretion,** a protoplast releases substances through its intact plasmalemma (or tonoplast), whereas in **holocrine secretion** substances are released as a consequence of cell disintegration (Schnepf 1969). In merocrine secretion, if the substances cross the plasmalemma (or tonoplast) in soluble form, we speak of **ecrine secretion**, and if they accumulate in vesicles of a different physical phase, which are then eliminated by exocytosis, we speak of **granulocrine secretion**. In granulocrine secretion, the membranes of the vesicle fuse with those of the plasmalemma (or tonoplast), or the vesicle becomes surrounded by an evagination and is ejected from the protoplast. In either case, the vesicle contents are discharged outside the cytoplasm.

In some glands, an early phase of merocrine secretion is followed by a terminal phase of holocrine secretion (Schnepf 1974). Most of the glandular trichomes that have been studied have well-expressed phases of granulocrine secretion, as exemplified by essential-oil production in leaves and resin accumulation in buds. A large fraction of the resin produced in these buds may be released in a final, lysigenous, holocrine phase.

Hydathodes are essentially passive pores that secrete water, with solutes (Sect. 8.4.1). In contrast to substances secreted by the merocrine and holocrine modes, the solution secreted by hydathodes is continuous with that of its source, the xylem apoplasm (the super apoplasm).

Synthesis of hydrophilic substances is associated with dictyosomes, whereas synthesis of lipophilic substances is associated with the endoplasmic reticulum (ER). Glandular cells that release lipophilic substances have extensive smooth ER. Cells in which there is a final holocrine release of lipophilic substances often accumulate the substances in vesicles within ER cisternae (Schnepf 1974).

Lipophilic substances, in contrast to hydrophilic ones, are not generally compatible with the aqueous solution in the cell-wall apoplasm and do not readily pass through the wall. Hydrophilic substances easily pass through an ordinary cellulose-pectin wall and appear in the intercellular spaces, or in the subcuticular spaces of epidermal cells. Movement of substances through the epidermal cuticle is limited unless there are cuticular pores (Sect. 2.2.1.1). Substances therefore may accumulate as blisters between the wall and its cuticle, forcing a separation. The accumulated substances are liberated when the cuticle ruptures (Fahn 1979; Boughton 1981).

The cells near the bases of some secretory structures have suberized or cutinized walls, some of which are equipped with Casparian bands. These bands, by separating the cell-wall apoplasm of the secretory tissues from the remainder of the cell-wall apoplasm, could, in theory, limit and direct the flow of secreted substances. However, because cells with Casparian bands occur only in secretory structures exposed to the external atmosphere, they may be as important in limiting water loss through a ruptured cuticle as in preventing backflow of secreted substances into the larger apoplasm (Schnepf 1974).

8.4 Structures of Exosecretion

8.4.1 Hydathodes and Glands Secreting Aqueous Solutions

Hydathodes, which release water through special pores in the epidermis, are usually associated with vein endings near leaf margins (Fig. 8.1). When transpiration is minimal, and root pressure causes xylem sap-pressure to be high (Sect. 7.4.3.2), water commonly exudes through these pores. Hydathodes have been reported in more than 30 families of dicots (Metcalfe and Chalk 1979). Mainly a feature of herbs, they are present in some shrubs, but not in tall trees. They tend to occur at the tips of leaf laminae or at the tips of laminar lobes or marginal teeth. In a few species, they are scattered over the leaf surface, as in *Ficus diversifolia* (Lersten and Peterson 1974).

A hydathode may have several or many water pores, which seem to be modified stomata having simple guard cells. Some or all of these pores may be occluded by cuticle and incapable of guttation (Lersten and Curtis 1982; Curtis and Lersten 1986). The number of pores per hydathode is inversely related to pore size. If the water pores are larger than ordinary stomata, there are only a few pores per hydathode. If the pores are smaller than stomata, there are numerous pores per hydathode, as in *Sanguisorba*. The guard cells of water pores are generally less responsive to environmental stimuli than are the guard cells of ordinary stomata. They may, nevertheless, open and close with changing water status.

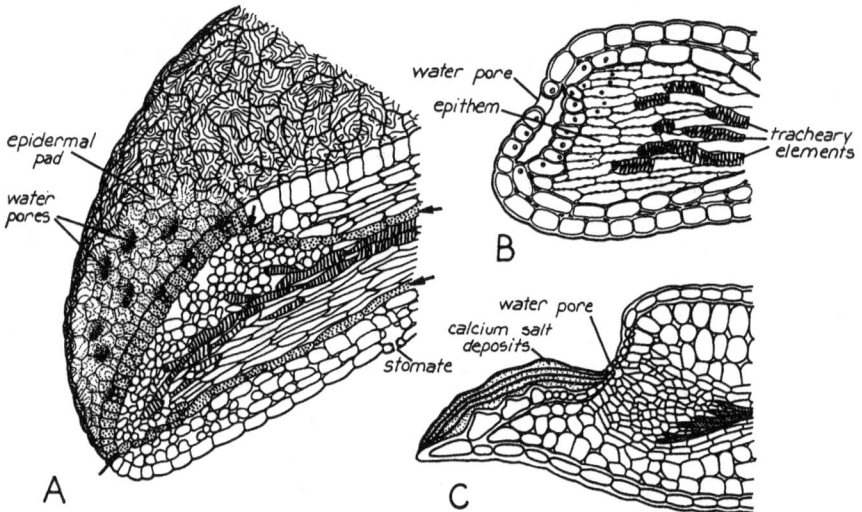

Fig. 8.1 A—C. Hydathodes. **A** Cutaway rendering of hydathode at margin of *Physocarpus opulifolius* leaf. *Arrows* indicate bundle sheath (*stippled*) that is continuous with cells of an epidermal pad (stippled) with numerous stomata modified as water pores. These are underlain by epithem. Tracheary elements are drawn with scalariform thickenings (redrawn after Lersten and Curtis 1982). **B** Epithem hydathode with single pore at tip of leaf-margin tooth in *Primula* spp. The walls of cells located between tracheary elements and water pore are wettable. (Modified after Haberlandt 1904). **C** Structure functioning as both hydathode and "chalk gland" at tip of leaf- margin tooth in *Saxifraga* spp. Walls of cells between vascular elements (*shaded*) and pore are wettable, as in **B**. Note "chalk" deposits. (Drawing based on information and photographs published by Kurt 1929)

Many hydathodes have loosely packed parenchyma cells in the region between the pore area and the ends of the hydathode-associated veins. These cells have wettable surfaces, in contrast to the unwettable surfaces of cells lining ordinary substomatal chambers and other intercellular gas spaces (Sect. 9.2). Collectively, the parenchymatous hydathode cells constitute an **epithem**, and the hydathodes containing them are therefore termed **epithem hydathodes** (Fig. 8.1A,B). The wettability of the epithem-cell surfaces facilitates water flow between the ends of the xylem elements and the pores. The tracheary elements of the vein ends commonly interdigitate with the epithem cells, but they do not contact the epidermis unless the epithem cells are very loosely arranged. Circumferentially, the epithem may be bounded by cells that can be interpreted as a terminal extension of a vascular bundle sheath (Fig. 8.1A). The cells of this sheath may become suberized or may deposit Casparian bands. Usually, the epithem cells are relatively passive physiologically. If they secrete sugars into the passing solution, the hydathode functions as an extrafloral nectary (Sect. 8.4.3).

Some structures that function as hydathodes lack epithem cells and differ little from ordinary stomata. These simple, stomate-like hydathodes commonly occur at the tips of leaf-margin teeth. The cell surfaces of the substomatal chambers of these hydathodes, like the cell surfaces beneath ordinary stomata, are unwettable, in contrast to epithem-cell surfaces, which are wettable. Thus, moving solutions do not fill the adjacent intercellular spaces, but flow directly out through the pore.

The details of structure, development, and function of hydathodes are so diverse that several classification schemes have been proposed, as summarized and documented by Metcalf and Chalk (1979).

8.4.2 Salt Glands

Salt glands, which occur in or on the epidermis of stems or leaves, are similar to hydathodes (Fig. 8.1C), in that the main source of their secretions is the transpiration stream (Hill and Hill 1976). These glands may occur in large numbers, with $1000/cm^2$ not being unusual. Salt glands collect and release excess ions unavoidably acquired from the saline soil water that the plant absorbs in response to transpiration losses.

The morphology and anatomy of many salt glands suggest that these structures originated from multicellular trichomes adapted to glandular function. In this respect, salt glands differ from hydathodes, which originated from stomata. Nevertheless, in the literature, salt glands have been referred to as hydathodes, active epidermal hydathodes, and trichome hydathodes.

Salt glands are not always raised above the epidermal surface. For example, in *Tamarix aphylla*, they are embedded in the epidermis (Thomson et al. 1969). The major functional cells of these glands, as seen in microscopic sections, have dense cytoplasm containing many small vacuoles, or "vesicles". In *Tamarix*, the walls of salt-gland cells confronting either the external atmosphere or nonglandular neighboring cells are heavily cutinized, except for the so-called transfusion areas on internal or lateral walls. Through these areas, the secretory cells maintain symplasmic continuity with neighboring cells via numerous plasmodesmata. The cuticle on walls confronting the external atmosphere is perforated by pores that are much smaller than the apertures of stomata. Through these pores, the salty solutions in the periplastic spaces can flow to the exterior. During periods of severe water stress, turgor-based movements may occlude the pore-bearing part of the external surface of a salt gland (Thomson et al. 1969). In this respect, some salt glands are similar to the water-absorbing trichomes of *Tillandsia* (Sect. 3.2.2).

The obviously trichome-derived salt gland of *Atriplex spongiosa* differs anatomically and functionally from the glands just described (Osmond et al. 1969; Lüttge and Osmond 1970). It consists of a stalk of one or several small cells, surmounted by a single, enormous, bladder-like cell with a capacious vacuole. Numerous plasmodesmata connect the stalk cells with the bladder cell and with the underlying parenchyma. By an apparently energy-consuming process, salt is transported to the vacuole of the bladder cell, where it accumulates in soluble form (Lüttge 1971). Salt glands of this type, which also occur in other taxa, have no evident means of actively secreting salt from their bladder cells. Holocrine secretion occurs when the salt glands or the leaves bearing them senesce.

This type of salt gland, like that of *Tamarix*, commonly has cutin-like deposits that isolate a local region of apoplasm around the salt-depot cells from the greater cell-wall apoplasm of the underlying parenchyma. In some glands, there are also features similar to Casparian bands (Fahn 1979, 1986). Apoplasmic isolation is essential if salt is to accumulate against gradients before being released.

8.4.3 Nectaries

Nectaries are specialized secretory structures that release a relatively concentrated solution of sugars, often with an admixture of other substances. **Floral nectaries** occur within flowers and are directly associated with the pollination process. **Extrafloral nectaries** occur on vegetative structures or on peripheral floral parts and have no direct relation to pollination or reproduction (Elias and Gelband 1976; Elias 1983). Some extrafloral nectaries seem to function in attracting insects, primarily ants, which, by their presence and actions, protect the host plant against grazing mammals and foliage-eating insects (Keeler 1977; Majer 1979). While defending the nectaries that secrete their food, these ants may seem to be preying on other insects (Bentley 1977a,b).

We suggest that some extrafloral nectaries have a function only incidentally related to insect attraction: they serve as safety valves for releasing excess sugars when photosynthetic activity is especially rapid. We have observed nectar dripping from tips of *Ailanthus* leaflets during periods of rapid photosynthesis. On sunny, humid days, *Tilia*, *Liriodendron*, and other trees also lose droplets rich in sugars, which may be deposited as "sap" on automobiles parked nearby. Aphids commonly are present on the leaves from which the droplets are lost, but we have also observed release in the absence of aphids. "Safety-valve" release of excess sugars may involve local ruptures of the cuticle.

The modes of sugar translocation in nectaries are diverse. In some nectaries, the sugar-rich solution exuded by the phloem may be transported along the cell-wall apoplasm without again entering a protoplast. Yet, as the solution moves along the wall near a plasmalemma, the sugars may be acted on by enzymes, and some of the nonsugar components may be resorbed by the symplasm. The floral nectary of *Acer platanoides* functions in this way (Vasiliyev 1969).

Operating very differently are nectaries having specialized tissue in which sucrose unloaded from nearby phloem undergoes enzyme-catalyzed changes in the symplasm and is then secreted into the cell-wall apoplasm. If this secretion is of the ecrine type, the cells assume the structure and function of transfer cells. Granulocrine secretion, however, is the more common mode (Fahn 1988). In short-lived flowers, nectar may be released in a holocrine manner, by cellular breakdown of the secretory tissue, as in *Turnera* (Elias et al. 1975).

Depending on the taxon, a nectary may consist simply of an area of secretory epidermis or a population of glandular trichomes, or it may be more complex, consisting of both nectariferous and non-nectariferous tissue. The nectar may flow to the surface through stomata, as in the floral nectaries of *Campsis*, or through cuticular pores, as in *Abutilon*. In the foliar nectaries of *Turnera*, there are no release pores. Rather, nectar is secreted beneath the cuticle, which is pushed out from the cell walls, thereby forming blisters. These finally rupture, releasing the nectar (Elias et al. 1975). (In contrast, release from floral nectaries in this taxon is holocrine; see earlier discussion.)

8.4.4 Stinging Emergences

Stinging emergences, less accurately called stinging trichomes, are multicellular, glandular structures occurring on leaves, and sometimes on stems, of at least four families of dicots: Urticaceae, Loasaceae, Euphorbiaceae, and Hydrophyllaceae. These emergences usually are outgrowths of both epidermal and subepidermal cell layers, and thus are not trichomes, which are outgrowths of the epidermis only. Generally, a stinging emergence consists of a glandular stinging cell of specialized structure that is partially sheathed and basally supported by a pedestal of less specialized cells of epidermal or hypodermal origin. A stinging cell may have a thin, brittle, siliceous wall, as in Urtica, or may contain a needle-like crystal within a thin cellulosic wall, as in *Tragia*. In *Urtica*, the slightest contact with the stinging cell breaks its brittle, glasslike tip, releasing irritants (Thurston 1974).

The stinging emergence in *Tragia* is an unusual example of specialized evolutionary development and finesse in control of crystallization of inorganic components. The central stinging cell contains a compound crystal of calcium oxalate. Usually, a single, tetrahedral, needle-like crystal projects from a cluster of shorter tetrahedral crystals at its base. These crystals are embedded in the cell walls near the basal end of the stinging cell and are braced by cellulose beams higher up. A remarkable feature is a deep groove along one side of the needle crystal, extending toward its base (Thurston 1976). Slight pressure on the tip pushes back the wall, exposing the needle crystal. Then the secreted proteinaceous toxins can presumably flow along the groove, into the victim's skin. Thus, a plant cell can direct the growth of an inorganic crystal into the form of an injection needle, while the remainder of the glandular structure serves as a syringe.

8.4.5 Glandular Epidermis

In addition to glandular emergences specialized as stinging cells or salt glands, which release aqueous solutions of toxins or salts, there are other kinds of glandular trichomes or glandular epidermal cells. Some of these release aqueous solutions of hydrophilic substances, such as the precursors of polysaccharide gums, but most of them release lipophilic secondary plant products, including essential oils, fatty oils, resins, flavonoid aglycones, and quinones. Glandular trichomes with resin-filled secretory cavities occur in *Rhododendron* and *Ledum*, and in various genera of legumes, including *Bauhinia*. Embedded epidermal glands (gland dots) with schizo-genous cavities occur in the leaves of some other legumes (Turner 1986).

Among the relatively few arborescent plants in which glandular epidermis has been carefully studied, there is much variation in anatomical and functional detail. We summarize information about the glandular epidermis in *Populus deltoides*, as illustrative of the diversity that occurs even within a species.

Populus deltoides has extrafloral secretory structures on the adaxial sides of the modified stipules that serve as bud scales; on leaf-margin teeth; and at the bases of leaf blades. Each of these three kinds of glands is active at a particular stage of leaf development (Curtis and Lersten 1974). In this species, when a dormant vegetative bud forms in late summer, the stipular scales develop a secretory epidermis on their adaxial surfaces. Early in the dormant period, the cuticle over this secretory layer ruptures, initiating holocrine release of a clear yellow resin that fills the spaces between the various appendages. In the spring, the pair of stipules around each emerging leaf also develop adaxial, palisade-like, secretory epidermis. Before a leaf emerges, the epidermis of its stipules releases resins that cover younger stipules and leaves. This release is also holocrine. Secretion from each pair of stipules begins about three days before the internode above the pair begins to elongate. It continues for about ten days, or until the petiole of the associated leaf stops elongating. The stipules then senesce and abscise. There is no whole-bud cycle of resin production and release. Instead, each pair of stipules goes through its own cycle in sequence, as part of a developmental process that continues throughout the growing season.

Although the first leaves of *Populus deltoides* to appear in spring have no glands on their marginal teeth, later leaves have glands on most teeth except near their tips. These glands consist of epidermis covering vein endings within the mesophyll.

The basal glands in the *Populus deltoides* blade differ among early, mid-season, and late leaves. The first leaves to unroll usually have two, stalked, basal glands, covered with secretory epidermis at the tips. One or two nectar-secreting stomata, capable of guttation, occur on the petiolar sides of the glands. Leaves emerging about a week later often have no basal glands at all, but during the main part of the growing season, emerging leaves generally have two, three, or four basal glands. A few have none. Late leaves have an average of four, and these are smaller than the glands on earlier leaves. They are only slightly elevated and their areas of secretory epidermis are small. Generally, no nectar-secreting stomata are associated with them.

Bud and leaf resins in *Populus* probably protect against leaf-eating insects (Levin 1973; Curtis and Lersten 1974). The function of the nectar that is often secreted by the same glands as the resins, though at different sites, is unknown.

8.4.6 The Root Cap as a Secreting Organ

The root cap, though primarily protective in function, is also secretory. The cap consists of parenchyma cells that commonly contain starch. The cell layers at the periphery of the cap typically secrete polysaccharide mucilage, which presumably lubricates the root tip as it advances through the soil. The mucilage also promotes the growth of soil fungi (Sect. 17.8.2). While the cells are alive, secretion of mucilage is a merocrine process (Mollenhauer et al. 1961). In *Lepidium*, for example, secretion is granulocrine, via exocytosis of mucilage-transporting vesicles, chiefly in regions of the anticlinal walls near the surface (Volkmann 1981). Peripheral cells of the cap are shed while alive and metabolizing, but in *Zea*, at least, there is no evidence that they secrete mucilage after detachment (Guinel and McCully 1987). Generally during root-cap development, cells are slowly pushed by growth processes from the inner to the outer regions, where they become separated, die, and disintegrate. Their final contribution of lubricating secretions is thus holocrine.

8.5 Structures of Endosecretion

8.5.1 Diversity of Structures

Endosecretory structures and modes of function are diverse. Substances may be released by either the merocrine or holocrine mode, or they may be retained within the protoplast until they are released by wounding of the cell. Substances secreted by the protoplast may be deposited in the periplastic space, or outside the wall in special intercellular spaces. If release is by the holocrine mode, the secretory cells undergo lysis, forming lysigenous cavities or ducts. As in exosecretion, a phase of merocrine secretion often precedes the terminal, lysigenous phase. The structures of endosecretion may be initiated directly by derivatives of primary or secondary meristems, or may develop by further differentiation of parenchymatous tissue. The major types of endosecretory structures of endosecretion found in woody plants are briefly discussed next.

8.5.2 Ducts and Cavities with Merocrine Secretions

Many woody plants have complex merocrine endosecretory structures. These usually include specialized, secretory epithelial cells that form schizogenous ducts or cavities. The epithelium, in turn, may be surrounded by a sheath of auxiliary cells. The substances that are released accumulate in the ducts or cavities. Volatile ("essential") oils are produced in this way in Hypericaceae, Myrtaceae, and many other families. Similarly, resins or balsams accumulate in Araliaceae, Anacardiaceae, and various families of Coniferales. Mucilaginous polysaccharides collect in Bombacaceae and Sterculiaceae, among other families.

The ducts or cavities that receive merocrine secretions (but not holocrine secretions) do not seem to be isolated from the greater apoplasm of the plant by layers of cutinized or suberized cells (Schnepf 1974). Nevertheless, there is no evidence that the secretions spread into the greater apoplasm. Secretions that are even slightly lipophilic could hardly diffuse through the water-saturated apoplasm, and diffusion of hydrophilic, mucilaginous polysaccharide secretions would be strongly inhibited by their large molecular size.

The **resin ducts** that occur in many conifers are the best known examples of ducts or cavities containing merocrine secretions. In many genera, they are normal features of the primary body of both root and shoot. In *Pinus halepensis*, for example, resin ducts have one terminus near the protoxylem poles of the root, and the other terminus at or near the bases of the cotyledons; in the primary stem, ducts in the cortex extend branches through the petioles into the mesophyll of the leaves (Werker and Fahn 1969).

In addition, some genera, including *Picea*, *Pinus*, *Pseudotsuga*, and *Larix*, always form resin ducts in their secondary xylem. Other genera, such as *Cupressus*, never form them, and still others, such as *Abies*, *Sequoia*, and *Cedrus*, form them only in response to injury (Sect. 19.4).

In those genera in which resin ducts are a normal feature of the secondary xylem, the ducts are present in each growth increment. The ducts may be axially or radially oriented. Axial ducts are approximately aligned with the axial direction of the fusiform elements.

Although axial resin ducts do not occur in the phloem, radial ducts, located within enlarged, fusiform rays, are continuous radially across the xylem, through the cambium, and into the phloem, where their outer ends are enlarged into vesicles. The lumen of a radial duct may not be open for flow across an active cambium. The lumen of each radial duct is connected, at its inner end, with the lumen of an axial duct. Radial ducts typically, but not exclusively, occur in those genera in which resin ducts are a normal feature of the secondary xylem.

In *Pinus halepensis*, the innermost axial resin ducts of the first xylem growth increment branch and unite tangentially (Fahn 1974), whereas the ducts in the older increments apparently do not. This suggests that the duct system consists of many separate sets of radially aligned axial ducts, with the ducts within each set interconnected by a radial duct but not connected with the sets of ducts along other radii. Such a system would be two- rather than three-dimensional (Fahn 1974). However, topological study of the xylem structure of *Picea abies*, *Larix decidua*, and *Pinus sylvestris* revealed that axial and radial resin ducts can constitute a three-dimensional network (Sect. 19.4), because axial ducts contact radial ducts on each side, and, similarly, radial ducts contact axial ducts on each side (Bosshard and Hug 1980).

The dimensions of axial resin ducts vary widely. In *Pinus halepensis*, these ducts are generally less than 10 cm long (Werker and Fahn 1969), whereas in *P. contorta*, lengths ranged from about 4 to 43 cm in a 75-year-old tree and from about 1 to 12 cm in a 30-year-old tree. Diameters in *P. contorta* varied less — from 60 to 105 μm — and were smallest near the pith (Reid and Watson 1966).

Radial ducts may be continuous across several growth increments. Although there are few data on their dimensions, they are generally shorter and narrower than are axial ducts in the same species. In *P. strobus*, the diameters of axial and radial resin ducts are about 150 and < 60 μm, respectively (Panshin and Zeeuw 1980). Diameters are larger in older than in younger trees.

Both radial and axial ducts are ordinarily filled with **resin**, which is under positive pressure due to the turgor pressure of the epithelial cells. If a duct is cut open, the resin slowly begins to flow out. It flows slowly because it is highly viscous and because the severing of the duct causes the local pressure to drop. Expansion of nearby epithelial cells by osmotic water uptake then tends to constrict the opening.

8.5.3 Ducts and Cavities with Holocrine Endosecretions

Although ducts and cavities containing holocrine endosecretions may appear superficially similar to structures associated with merocrine endosecretions, the origins of these structures are commonly different. Whereas merocrine secretory ducts and cavities are formed schizogenously, holocrine structures are formed by lysis of the protoplast and rupture or dissolution of the cell wall, resulting in a cavity containing a mixture of secretory and autolytic products. In the lysigenous oil cavities of the flavedo (exocarp) of *Citrus* fruits, essential oils are synthesized and accumulate as droplets in the plastids of the secretory cells. These cells, when mature, are lysed, and the droplets then fuse into larger droplets within the cavities.

A newly formed lysigenous cavity may enlarge and accumulate additional endosecretions as cells around it undergo lysis. In addition, a schizogenously formed intercellular space that is at first filled with merocrine secretions may later enlarge and accumulate holocrine secretions as adjacent cells undergo lysis.

Though ducts containing terpenoid resins are mostly confined to gymnosperms (Sect. 8.5.2), some angiospermous trees have cavities with holocrine endosecretions that are distantly comparable (Sect. 19.4). These **gum ducts** may be oriented axially or radially (within rays), but both types are seldom present in the same wood. They occur in the secondary xylem and secondary phloem. In the wood of some taxa, gum ducts are a normal feature, whereas in other woods they form in response to wounding. In *Ailanthus*, there is extensive development of gum ducts and cavities.

One of the functions of gum seems to be plugging of wounds. In addition, deposits of gum in embolized vessels may help prevent the vessels from being channels for the spread of fungi. In this way, the gum may perform a function similar to that of tyloses.

Though lysigenous gum-filled cavities typically occur in the wood, they also occur in the bark of some trees. For example, in various *Acacia* species, gum-arabic or gum-acacia are formed and accumulate in the bark. In addition, in Prunoideae (Rosaceae), the gum-filled cavities of the xylem may extend completely through the cambium, phloem, and bark, and, as a result, the gum may be externally visible.

Eucalyptus has **kino veins**, which are lysigenous canals of a different kind, initiated in layers or strands of parenchyma cells formed by the vascular cambium in response to injury (Sect. 19.4). Some of the cells in these layers or strands accumulate large amounts of polyphenols, then undergo lysis. Subsequently, neighboring cells also undergo lysis, and a canal filled with polyphenol deposits is formed. This canal may be a part of the xylem or the phloem.

A special feature, absent in gummosis, is the peripheral "cambium" that differentiates around a developing kino vein. The inner derivatives of this cambium accumulate polyphenols, then undergo autolysis and contribute to the kino accumulation. The outer derivatives eventually become suberized and form "periderm", which limits further development of the kino vein (Skene 1965).

8.5.4 Lithocysts

Inside of an idioblast called a **lithocyst**, calcium carbonate accumulates, as a **cystolith**. The cell wall of a lithocyst has club-shaped ingrowths, around which the secretory cytoplasm is arranged. This cytoplasm, which appears as a mass of radiating strands, mediates the growth of the mass of calcium-carbonate crystals that becomes the cystolith. The calcium carbonate, which the protoplast deposits outside the plasmalemma, becomes attached to the end of a wall ingrowth.

Lithocysts commonly occur in the epidermis of *Ficus* and other genera of Moraceae, and in trichomes and various types of parenchyma, though seemingly not in xylem parenchyma. Lithocysts may provide deposition sites for excess calcium, but the significance of this is not well understood.

8.5.5 Oil Cells

Oil cells are idioblasts that synthesize lipophilic substances. These substances accumulate outside the plasmalemma, usually in a vesicle formed by a single invagination of the plasmalemma (Kisser 1958). Oil vesicles are readily misinterpreted as vacuoles, though they typically have a higher index of refraction. Oil cells occur in leaf mesophyll and fruit mesocarp and, in many families, in the stem and root cortex (Fahn 1979). In a few families, they occur in xylem rays, giving woods such as *Tectona* and *Guiacum* "oily" properties. Lehmann (1925) described the structure and mode of development of oil cells in *Liriodendron*, *Magnolia*, and various other genera.

8.5.6 Laticiferous Systems

The specialized secretory cells known as **laticifers** are derived from parenchymatous cells and generally have some or all of the following characteristics:

1. The secretory activities and products are ordinarily confined to the cells of the laticifer system itself, with no programmed release of secretions to other tissues or to the outside, though there may be release after wounding.
2. A central vacuole contains an emulsion or suspension of latex, unless the tonoplast has degenerated after cell maturation.
3. They form an anastomosing system associated with vascular bundles and especially with the secondary phloem.
4. The cytoplasm, including organelles, may either become specialized or degenerate.

5. There is commonly formation of starch that has the unusual property of being nonmetabolizable under ordinary conditions (Spilatro and Mahlberg 1986).

In some *Euphorbia* species, grains of nonmetabolizable starch may function in plugging wounds in laticifers (Biesboer and Mahlberg 1978). The latex of *Hevea brasiliensis* contains high levels of chitinase and lysozyme, which may enable laticifers to perform a defense function (Martin 1991).

The latex that laticiferous cells synthesize and store is a somewhat viscous emulsion that is usually milky-white, but may also be colorless or a clear yellow, amber, or brown. The droplets of the hydrophobic, dispersed phase of the latex characteristically are rich in polyisoprene hydrocarbons, triterpenols, and esterified fatty and aromatic acids. The aqueous phase may include enzymes, sugars, mucilages, and the usual vacuolar-sap complements of sugars and organic acids.

Latex occurs in more than 12 000 species in about 20 families of dicots; in a few monocots; and in at least one genus of pteridophytes (Metcalfe 1967). Latex or latex-like secretions occur in the cortical tissues of some gymnosperm embryos and seedlings (Werker 1970), but are otherwise generally absent from gymnosperms. In a few plants, notably *Parthenium argentatum* (guayule), latex is produced and stored in ordinary parenchymatous cells that *are not* differentiated as laticifers (Bonner and Galston 1947). Characteristically "laticiferous" families that include numerous arborescent species are Euphorbiaceae, Apocynaceae, Sapotaceae, Moraceae, and Caricaceae.

There are two types of laticifers: nonarticulated and articulated (Bary 1877). This classification has little taxonomic significance, as the two types commonly occur among species of the same family. Nonarticulated laticifers are multinucleate bodies developed from single cells that grow symplastically and intrusively. These laticifers may be simple, rather straight, unbranched tubes, or they may be repeatedly branched. Articulated laticifers consist of chain-like files of elongated cells and may be unbranched or more or less elaborately branched. The end walls of these laticifer elements may either have primary pit fields and remain intact at maturity, or disappear (as do the end walls in vessels), forming a virtual syncytium similar to that of nonarticulated laticifers. At points of anastomosing, side-wall segments as well as intervening end walls may disappear. Thus, although laticifer systems are not true transport systems, they may function as such when there is flow toward a wound-related opening.

Nonarticulated laticifers, which are common in Moraceae, Apocynaceae, Euphorbiaceae, and Asclepiadaceae, seem to occur only in the primary plant body. They may originate in the embryo, as in *Nerium oleander* (Mahlberg 1961), and then grow symplastically and intrusively, with repeated branching. Alternatively, they may arise in the subapical region of the shoot tip, as in *Cannabis* (Schaffstein 1932). During the very extensive elongation growth of nonarticulated laticifers of *Euphorbia marginata*, the repeated nuclear divisions (without accompanying cytokinesis) are not randomly distributed, but have a wave-like distribution in time and space (Mahlberg and Sabharwal 1967).

Articulated laticifers may occur in both primary and secondary tissues. For example, in *Hevea brasiliensis*, they occur mostly in the primary parenchyma of the stem, leaf, and root, and as axially oriented tubes in the secondary phloem of the trunk (Vischer 1923). In this species, the end walls of the original laticifer cells disappear completely during differentiation, and numerous anastomoses form. The laticifers in the successive incremental layers of secondary phloem form concentric networks that are not necessarily interconnected.

Opening of a laticifer network by wounding allows the latex to begin to flow out. Normally, the flow soon slows and eventually stops because of several types of constriction and plugging. In leaf abscission layers, callose plugs often close off the laticifers. Loss of latex from the system, as in the tapping of rubber trees, stimulates synthesis, until the approximate pre-wounding supply is established. Control of latex synthesis is poorly understood (Fahn 1979). Mature laticifers of *Hevea* have no functional plasmodesmata in the walls confronting parenchyma sheath cells. However, many plasmodesmata occur in the common walls of the sheath cells and the parenchyma cells of phloem rays. Loading of *Hevea* laticifers probably requires first a symplasmic pathway, but ultimately an apoplasmic one for transport of metabolites from sieve cells (Fay et al. 1989).

9 Aerating Systems

9.1 Introduction

Intercellular gas spaces enable a plant to absorb CO_2 from, and release oxygen to, the external atmosphere, while simultaneously minimizing water loss. In addition, these spaces supply oxygen to subterranean plant parts. Without some provision for transporting at least minimal oxygen to organs embedded in the soil, these parts could not survive the episodic water saturation that is characteristic of soils of many otherwise mesic areas.

9.2 Basic Features

The aerating systems of large plants consist mostly of intercellular spaces, and may also include the lumina of dead cells that are not filled with water. The various channels and gas spaces are usually interconnected throughout an organ or even a whole organism, and allow flow, as well as storage, of gases. The natural input and output terminals of the aerating system are the stomata (Sects. 9.3; 16.3.4) and lenticels (Sect. 21.5).

The cell surfaces confronting the intercellular gas spaces, though they seem to consist mostly of water-saturated cell walls, are completely lined with a thin, water-repellent, cuticle-like layer (Martin and Juniper 1970; Ende and Linskens 1974), and thus are unwettable (except in some hydathodes; Sect. 8.4.1). Their water repellency is so effective that pressure differentials of about 1 atm are necessary to infiltrate the intercellular gas spaces with liquid water (Sifton 1957; Pitman et al. 1974). Of course, if these small capillary spaces were not water repellent, they would spontaneously fill with water during periods of high humidity and could not subsequently function in aeration.

The intercellular gas spaces originate mostly by **schizogenous** separation of cells along their edges. This seems to happen as follows: where the middle lamella of a newly formed partition wall comes into contact with the wall of the maternal cell, it thickens, so that in transection a triangular mass can be seen. This mass, by processes that are not yet understood (but see Sect. 12.2.1), becomes continuous with the middle lamella of the mother cell. Small cavities may appear in this newly extended middle lamella, especially at or near the newly created cell corners. Because of turgor pressure, all the walls are in tension, and for mechanical reasons cell edges tend to round off by pulling apart around the newly formed cavities. As cell-rounding continues, the cavities enlarge and gradually become continuous with the rest of the intercellular-space system (Martens 1938; Priestley and Scott 1939; Kollöffel and Linssen 1984).

Commonly, the surface and volume growth of cells is so distributed that the intercellular spaces become larger as the adjacent walls are "pulled apart" by growth stresses. Leaf spongy mesophyll and other tissue specialized as **aerenchyma** may develop in this way. Some special systems may operate at a tissue or organ level, controlling weakening of the middle lamella and thus influencing the shapes of intercellular spaces that develop. Just how the surfaces of the spaces, originally formed in highly hydrated middle-lamella materials, become lined with an unwettable, cuticle-like layer is not understood (see also Sect. 16.3.5 and literature cited there).

An additional component of the intercellular gas-space system is the space relinquished by dying parenchyma cells. Typically, when a parenchyma cell dies, its protoplasm is autolyzed, its cell wall is partly digested, its liquid residue is resorbed by its neighbors, and its space is filled with gas from the intercellular gas-space system. If whole groups of cells die and disintegrate as a normal part of development of intercellular spaces, we speak of **lysigenous** intercellular-space formation. These spaces commonly develop in plants growing in marshy areas and also in ordinary banana leaves (Skutch 1927). The central cavities of internodes of grass stems begin in this way, and are then further increased by stretching and tearing of tissues, yielding **rexigenous** spaces.

In plant organs that do not have aerenchyma, the aerating system usually comprises only a small fraction of the total volume, generally less than 10%. Nevertheless, in such organs there may be a very large internal surface lining the gas spaces. For example, the internal surface of the aerating system in the mesophyll of mesomorphic leaves is 7 to 30 times greater than the external leaf surface. This ratio may be even higher in xeromorphic leaves (Turrell 1936).

9.3 Stomata and Stomatal Function

The epidermis of leaves and other aerial plant parts generally is perforated by numerous stomatal pores. These also occur in some underground organs, particularly rhizomes (Fahn 1974), and have even been found on some primary roots, notably in *Ceratonia siliqua* (Christodoulakis and Psaras 1987).

A stomatal pore is a minute opening between a pair of guard cells. Commonly, subsidiary epidermal cells, which are morphologically distinct from other epidermal cells, are functionally associated with, and adjacent to, the guard cells. The stomatal pore, together with the pair of guard cells and any subsidiary cells, constitute a stoma or stomate. This structural unit is also called a stomatal apparatus.

If viewed from above the leaf surface, a guard cell usually appears kidney-shaped, though other shapes also occur. The shape, as seen in a section transverse to its longest dimension, varies widely with the taxon. In addition, serial sections through a single guard cell reveal great variation, because the walls are not uniformly thickened. Thickening patterns vary widely with the taxon, but the wall is usually thinnest at the poles. The wall that forms the margin of the stomatal pore usually has a thickened ridge or lip. The ridges of two guard cells can overlap or abut tightly enough to seal the pore.

It has often been assumed that when turgor changes guard cells change their shape *because of* their nonuniform wall thickness. It is now clear that these fine details of sculpturing of the guard-cell walls are generally secondary to the main operating movements of the cells (Sect. 10.2.1.2). This may explain why the sculpturing can be so variable (Aylor et al. 1973).

Stomata are essential to the survival of land plants because they minimize transpirational water loss from green tissue while still allowing the quantities of CO_2 needed for photosynthesis to enter. They do this by regulating the effective pore diameter as an integrated response to: the CO_2 concentration in the intercellular gas space; the saturation deficit of the air; the water stress in the leaf; and the general water status of the plant. The stomata thus constitute a highly variable resistance to movement of gases between the interior atmosphere of the organ and the ambient air.

Open stomata are not very effective in barring diffusion of water vapor. In fact, the rate of diffusion between the interior atmosphere of a leaf and the ambient air can be almost as great as if the epidermis were not even present (Ting 1982). In general, diffusion through small pores, unless they are very closely spaced, is more nearly proportional to the sum of pore perimeters than it is to the sum of pore areas. In most higher plants, the stomatal pores are arranged such that the "perimeter law" is most nearly valid if the stomatal apertures are only slightly open. If, instead, they are fully open, diffusion is more closely related to pore area. As the pores close, diffusion is reduced only slowly at first, then declines rapidly as full closure is approached (Bange 1953; Lee and Gates 1964). If the stomatal guard cells can close the pore quickly, they can protect the leaf from the severe water deficits that otherwise would accompany drying winds (see next).

Some surfaces of the stomatal guard cells are exposed to the atmosphere of the substomatal chamber (Mansfield et al. 1990). The guard cells respond directly to changes in CO_2 concentration in this chamber. They open if the CO_2 concentration in the chamber becomes very low, as it does when CO_2 is removed and fixed by photosynthesis. If, because of failing light or other factors, the CO_2 concentration increases, the pores close. However, the wind, which moves fresh supplies of CO_2 to the leaf surface, has a large effect, steepening the inward CO_2 gradient and shortening the diffusion distance. As a result, the CO_2 concentration in the intercellular space within the leaf increases markedly with wind speed. This tends to cause the mean stomatal aperture to decrease.

The extent to which the stomata close in response to wind-induced increases in CO_2 concentration is influenced by the water balance in the guard cells. Because wind carries water vapor away from the leaf surface, thus steepening the water-vapor gradient near the surface, wind usually increases water loss. The stomata of many plants close if there is a steep gradient of partial pressure of water vapor between the interior gas space and the external atmosphere (Bunce 1985). In many species, stomatal conductance falls in response to increasing water-vapor-pressure differences between the leaf gas space and the external atmosphere (Jarvis and Morison 1981). The difference is determined mostly by leaf temperature and by the relative humidity of the ambient air. The stomatal response to these factors is evoked by the change in bulk leaf water. In at least some plants, however, stomatal closure can be ascribed to guard-cell water deficits that develop directly from loss of water through areas of thin cuticle overlying these cells (Appleby and Davies 1983).

Stems of Carboniferous-Era trees, such as *Lepidodendron*, had neither stomata nor lenticels. Instead, they had cavities — parichnos — which extended from the leaves into the stem and are assumed to have been filled with gas enriched with oxygen derived from photosynthesis in the leaves. In *Lepidodendron*, these cavities extended into the roots, where they are known as stigmaria. There are also no stem stomata in most extant gymnosperms, though some have parichnos, or vestiges of them accompanying the leaf traces through the cortex (Jeffrey and Wetmore 1926). In *Ephedra*, stomata do occur on the stem, which is the major photosynthetic organ.

In some woody dicots, stem stomata may be much larger than leaf-blade stomata. Lenticels often originate beneath these large stomata (Sect. 21.5). In *Fraxinus excelsior*, the large, pre-lenticel stomata are the only type that occurs on the stem, whereas stems of some other woody plants (e.g., *Platanus*) have both the large, pre-lenticel type and the smaller, leaf type. In still other woody plants (e.g., *Gleditsia*), all stem stomata are of the same size and type as those on the leaves.

The stomata of leaves are usually not distributed uniformly. In many plants, they occur primarily in the abaxial epidermis. Frequency of stomata is variable among leaves — usually about 500 to 1000/mm^2 in mesomorphic ones. The frequency is higher in sun-adapted than in shade-adapted leaves. Stomatal densities are about 40% lower on recent leaves than on leaves from 200-year-old herbarium specimens of the same species from similar sites. This may be related to increases in atmospheric CO_2 since the advent of worldwide industrialization (Woodward 1987). If so, the relation predicts increasing efficiency of water use by crops.

9.4 Stomata, Water Economy, and Photosynthetic Efficiency

Because water molecules are small and light compared with CO_2 molecules, and because the water-vapor gradient between the gas space within the leaf and the external atmosphere is often steeper than the CO_2 gradient in the opposite direction, many water molecules are lost for each CO_2 molecule absorbed. For example, for *Xanthium strumarium* in good light, at an external relative humidity of 55%, water loss per CO_2 molecule fixed was calculated to be 156 to 336 molecules (Raschke 1979), depending on the hydration state of the plant and microenvironmental conditions. Though much of this loss is inevitable, C4 plants seem to have been able to minimize it via CO_2-fixing reactions that are highly efficient even at low CO_2 concentrations (Sect. 5.2). In addition, some plants have epidermal structural adaptations, such as hairs, that reduce water loss (see later discussion).

The ratio of the amounts of two substances diffusing through a porous barrier is given by:

$$\frac{M_1}{M_2} = \frac{\delta C_1}{\delta C_2} \frac{D_1}{D_2} , \qquad (9.1)$$

where M is the amount of substance diffusing, δC is the difference of concentration across the barrier, and D is the diffusion coefficient of the moving substance.

At the porous barrier between the leaf and the external atmosphere, the δC for water depends on relative humidity, temperature, and illumination. Stomatal aperture has only a slight influence as the gas space in the leaf is practically saturated with water vapor under almost all conditions. In contrast, δC for CO_2 is strongly influenced by stomatal aperture. If the aperture is small and photosynthesis is rapid, CO_2 concentration within the leaf may fall rapidly, resulting in a large δC. If the pores open, the difference, δC, between the nearly constant external CO_2 concentration and the concentration within the leaf decreases. Thus, the ratio of δC for H_2O to δC for CO_2 is large when the stomata are widely open and decreases with a narrowing of the pore. If the pores are narrowed gradually, they will reach some aperture at which the ratio of molecules of CO_2 entering to the number of water molecules leaving will be maximal. Plants having the C4 photosynthetic pathway (Sect. 5.3.2) have minimized the pore aperture size at which this ratio is maximal.

Many mesophyll cells are not directly adjacent to a substomatal chamber. Therefore, CO_2 that passes through a stomatal pore typically must diffuse a distance equivalent to several cell diameters along intercellular gas-space passageways. For efficiency, this path must be short, even though diffusion in the gas phase is rapid compared with diffusion through liquid. This requirement provides a rationale for the usual peripheral location of the chlorenchyma that takes up CO_2 from the gas space in C4 plants (Sect. 5.3.2). C3 leaf structure in typical broad-leaved taxa promotes CO_2 diffusion to deeply-lying fixation sites. This route is usually from the substomatal chambers across the large intercellular spaces of the spongy mesophyll, then through the narrow spaces between the palisade cells, within which CO_2 is rapidly fixed. The passage of this gas through wide, then narrow, spaces allows development of steep gradients that increase the flux of CO_2 to the palisade chlorenchyma.

Changes in the turgor of mesophyll cells imply changes in their volume. These changes, expressed as a percentage of the total cell volume, are small, but the percentage change of the intercellular-space volume is larger, because gas spaces compose a smaller volume fraction of the leaf than do the cells (Sect. 9.2). These episodic or periodic changes in gas-space volume are probably physiologically significant.

9.5 Intercellular Space and Respiratory Gas Exchange

9.5.1 General Perspectives

An organ that supports photosynthesis is unlikely ever to be deficient in oxygen, because a gas-diffusion system that can supply adequate CO_2 for photosynthesis can also supply enough oxygen for respiration. This is so because the diffusion coefficients and the gradients are more favorable for oxygen. Even a bulky organ, unless in an oxygen-poor environment, should not suffer internal oxygen deficiency if it has a system of continuous intercellular spaces throughout its volume. This is because the diffusion of small molecules through gas is about 10 000 times faster than through water — and, in this respect, living cells are similar to water. Diffusion of oxygen over a relatively long path is sufficient for respiration, even though diffusion of CO_2 along a similar path would be inadequate to support photosynthesis, as already discussed.

By techniques developed by Burton (1950), the rate of oxygen diffusion was calculated to be about 430 times higher through potato-tuber tissue than through water. Even though intercellular spaces make up only 0.62 to 1.34% of tissue volume of the tuber, they obviously contribute enormously to oxygen diffusion. At these diffusion rates, the oxygen level of the interior gas space of an average-sized potato tuber would be about 3%, despite continuous removal of oxygen for respiration; and, with 1% intercellular space, even a potato tuber of larger diameter — 14 cm — could maintain an oxygen concentration of 2% at its center. If there were no intercellular gas-diffusion space, the oxygen supply would become critically deficient in much smaller organs.

Intercellular gas spaces occur in almost all tissues. A gas canal system consisting of narrow spaces along cell edges pervades even the vascular cambium (mostly via rays) and the secondary xylem. The volume fraction of this space system in the latter tissue is usually less than 0.5 %, but this is adequate for respiratory metabolism of the xylem parenchyma. In addition, some taxa have evolved specialized aerating systems. Several of these are discussed next.

9.5.2 The Special Significance of Gas Space in Roots and Rhizomes

In soils that are water-saturated or otherwise deficient in rhizospheric oxygen, a plant's intercellular gas-space system can carry significant amounts of oxygen from the above-ground atmosphere to the root-tip region. In waterlogged soils, there is, in fact, measurable movement of oxygen from the roots to the rhizosphere (Armstrong 1982). Even in roots of typical mesophytes, there are indications that not all oxygen consumed in root metabolism typically comes from the soil atmosphere; some comes from the shoot.

Many herbaceous plants growing in poorly-aerated soils have cortical aerenchyma extending from the aerial organs into the rhizome or root system. This tissue has very large, anastomosing intercellular spaces. In some of these plants, such as Oryza sativa, aerenchyma forms in the roots as part of a general developmental program, even if the roots are growing in a well-aerated medium. In contrast, in other taxa, genetic capability alone does not insure development of root aerenchyma. Inductive conditions are needed. For example, in roots of Zea mays, the stimulus for systematically distributed groups of cortical cells to collapse and form aerenchyma is provided by local increases in ethylene, in response to oxygen deficiency in the rooting medium (see Webb and Jackson 1986).

Some aquatic plants circulate gas internally between leaves and rhizomes, via mass flow through intercellular lacunae. In Nuphar lutea and several other species, a purely physical, flow-through system reportedly is driven by temperature gradients between young leaves and the external atmosphere. In these plants, the stomata are located mainly in the adaxial surface of the leaf. In a young leaf, air flows through these stomata when the leaf is warmer than the external atmosphere. Gas moves from the warmer side to the cooler side of the leaf by thermodiffusion, which generates a positive pressure in the leaf aerenchyma. Because this system does not operate as effectively in older leaves, in which spatial relations are different, the positive pressure in the young blades can drive an internal circulation system, in which air taken in by the young blades flows down the petioles to the rhizome, then returns and escapes via older petioles and leaf blades. This results in greatly increased oxygen supply to the rhizomes and roots and also in increased CO_2 supply to the interior of the leaves (Dacey and Klug 1982; Grosse and Mevi-Schütz 1987; Mevi-Schütz and Grosse 1988).

Some woody plants growing on wet sites also transport oxygen from organs above the water table to those below it (Armstrong 1968). **Pneumatophores**, highly specialized structures containing aerating tissue (Sect. 7.4.3.4), occur on the roots of some tree species growing in swamps. There are several kinds of pneumatophores. One kind is exemplified by the erect, conical, negatively geotropic roots that project above the mud as branches of long, horizontal roots in species of *Sonneratia*, a mangrove. The spongy cortex of the erect, conical roots functions as aerenchyma. These roots, including their apices, are covered with cork perforated by numerous lenticels (Sect. 21.5). Fibrous absorbing roots occur at increasingly high levels if the muddy substratum is progressively raised by sedimentation (Goebel 1886).

Another possible aerating structure, more common than the pneumatophore of *Sonneratia*, is anatomically and morphologically quite different from it. This kind of structure, which includes the "knee" of *Taxodium*, is produced by local proliferation of derivatives of the vascular cambium on the upper sides of horizontal roots. It consists mostly of xylem. Preliminary studies of the *Taxodium* knee by one of us (Z. H., with coworkers and students at the University of Bonn) revealed that most of the tracheary elements of the older wood are embolized — that is, gas-filled. The gas space of these elements is continuous with the external atmosphere through narrow intercellular spaces in the rays extending through recent xylem that is not yet embolized, across the cambium, and into the secondary phloem and periderm, which are highly fissured.

The extensively embolized, older xylem of *Taxodium* knees seems more suited for conduction or storage of gas than of liquids, though there is no universal agreement that its function is aeration (Kramer and Kozlowski 1979). The possibility that the xylem in analogous structures of *Amoora*, *Carapa*, and *Heritera* belongs to the aerating system was hinted at in older research (Groom and Wilson 1925). We regard the xylem of *Taxodium* similarly.

Strands of gas-filled tracheary elements in roots of arborescent taxa that do not form pneumatophores may also transport gas. This function would be furthered by the radially oriented intercellular spaces within xylem rays, which connect the relatively oxygen-rich xylem with the oxygen-consuming cambium. In addition, the vascular cambium in the trunks of some species is not a barrier to gas exchange between xylem and phloem, because the intercellular spaces along the rays are continuous across this meristematic layer (Hook et al. 1972). Further research is needed to determine the extent to which gas-filled xylem elements supply oxygen to roots of such trees.

Gas movement in the intercellular gas spaces in tree stems and roots probably does not depend entirely on simple diffusion or thermodiffusion. Several types of diurnal cycles could exert a pumping action on the aerating system. For example, there usually are notable diurnal changes in water pressure in the xylem, due to fluctuations in rate of transpirational water loss. These pressure changes, which can even be measured as changes in diameter of tree trunks, are accompanied by changes in gas-space volume. Added to this variation are pressure changes due to diurnal temperature changes in the stem.

9.5.3 Lenticels

As already noted, the anastomosing gas-space systems of stems and other massive organs are not confined to a particular tissue. For example, those in the xylem or even the pith may be continuous across the vascular cambium, phloem, and any cortex, with gas spaces in substomatal chambers or lenticels, which are adapted to gas exchange with the external atmosphere. A lenticel is a local, specialized area of periderm (Sect. 21.5). Some lenticels on young stems develop from stomata, and there is not always a sharp distinction between these two structures. However, lenticels can arise independently of stomata, especially on roots.

Lenticels typically are much larger than stomata, and usually have no single, large, pore-like openings or internal chambers. They have no physiologically responsive closing systems, and thus do not actively regulate gas exchange as do stomata. Rather, they passively facilitate it via the looseness of their cellular structure.

Because it is not feasible to discuss fully the structure and function of lenticels before the developmental aspects of the phellogen and peridermal derivative tissues have been discussed, we defer further consideration of lenticels and their functions to Section 21.5.

10 Movement Systems and Positional Perception

10.1 The Phenomenology of Movements in Higher Plants

There is no single, integrated movement system in higher plants — nothing remotely comparable to the nerve-muscle-bone system of higher animals. Yet, plant organs react to some stimuli by moving. Plant movements are mediated by a variety of systems. Some of these systems involve participation of living protoplasts, whereas others are based on physical principles and operate with only passive or indirect involvement of living cells. Some systems operate via special structural features of the walls of dead cells. These features result from highly controlled activities of the protoplasts while they were alive (Haupt 1977).

Because some plant movements do not fit neatly into convenient categories, we classify movements somewhat arbitrarily, according to a two-stage, dichotomous system. By our first dichotomy, we divide plant movements into those that depend on the active participation of protoplasts and those that do not. We then subdivide those plant movements that depend on protoplast participation into **turgor movements** (Sect. 10.2.1) and **growth movements** (Sect. 10.2.2). Those movements not depending on protoplast participation are divided into **cohesion movements** (Sect. 10.3.1) and movements based on **reaction wood** (Sect. 10.3.2).

However, it is often expedient to use established terminology that does not always conform to a dichotomous system of classification. For example, if an organ moves in response to change of stimulus over time, and the direction of movement is determined by the structure of the organ regardless of the direction of the stimulus, it is of the **nastic** type. But, as nastic movements may be powered either by growth or by turgor changes, they do not fit well into any of the categories we have adopted. For example, seismonastic movements, encompassing short-term responses to contact, are driven by turgor changes (Sect. 10.2.1.1), while longer-term nastic movements are driven by growth (Sect. 10.2.2.4). Further, specific types of nastic movements, such as nyctinasty, are powered by turgor changes in some plants but by growth in others.

In **tropic** movements, unlike in nastic movements, the direction of movement is determined by the spatial orientation of a stimulus. Tropisms are growth responses, not ascribable to reversible changes in turgor, and thus may generally conform to our classification scheme.

An established classification of tropic and nastic movements is based on the source and sometimes also the duration of the primary external stimuli that evoke them. If the external stimulus is gravity, we speak of gravitropism; if it is light, we have phototropism and photonasty (or nictinasty); if the stimulus is mechanical and of short duration, the movement is seismonastic; and if it is mechanical and of long duration, the movement is thigmotropic.

Still other movements are autonomous. These are not a direct response to external stimuli, but rather are modified or entrained by the stimuli. The mechanism of these movements is not well understood. There is an autonomic component to many nastic movements, such as epinasty, hyponasty, and nyctinasty. Another autonomic movement is **circumnutation**, or seeking movements, of tendrils, as discussed in Section 10.2.2.3 (see Haupt 1977).

Plant movements can also be classified on the basis of the rate at which they occur. This rate is related to the immediate energy source that powers the movements. In this scheme, there are three types of movements: (1) catastrophic, or explosive; (2) transient; and (3) slow to long-term.

Catastrophic and explosive movements are exemplified by various seed and pollen dispersal mechanisms in which stresses accumulate during maturation or drying, and by rapid movements of traps of some carnivorous plants. The stresses eventually become so great that a small shock can trigger catastrophic failure of the resisting tissues. When these tissues suddenly rupture, an explosive release of strain energy hurls out the seeds or pollen grains. A related, large-scale phenomenon is the unpredictable, dangerous splitting and lashing out of some tree trunks as they are being felled or split (Sect. 4.6). Catastrophic movements usually are a terminal response of the structures powering them, and are not repeatable or reversible. Explosive movements result from a disturbance of the balance in turgor tensions, as, for example in anthers of *Berberis*, or from a disturbance of the balance of accumulated tissue stresses in traps of some carnivorous plants. These movements are repeatable.

Transient movements generally involve participation of living protoplasts and are powered by osmotic energy made available by differential turgor pressures of cells or tissues. These movements usually occur within minutes, hours, or days after the stimulus, and, in theory at least, are reversible and repeatable. Leaf-orientation movements such as seismonasty are typically of this type. We discuss transient movements from the perspective of the involvement of protoplasts, via turgor changes (Sects. 10.2.1.1; 10.2.1.2).

Slow and long-term movements, as in phototropism, gravitropism, or the reaction wood of trees, depend primarily on differential growth. These movements may not become evident until days or even months after the stimuli are received and may continue for long periods, especially in tree stems. Osmotic phenomena power photo- and gravi-tropic bending of stems (and ordinary growth as well), whereas movements due to reaction wood are mostly powered by forces associated with swelling and shrinking of cell-wall materials during certain stages of xylem maturation. Movements due to reaction wood have a definite structural basis in the specialized properties of particular cell-wall layers. The protoplasts that deposit these specialized wall layers soon die and are not directly involved in the movements.

10.2 Movements Involving Direct Participation of Protoplasts

10.2.1 Turgor Movements

Rapid changes of turgor are involved in seismonastic movements and in movements of stomatal guard cells.

10.2.1.1 Seismonastic and Related Movements. Some plants, such as *Mimosa pudica*, are "excitable", in that they can make rapid movements in response to touch. Although most plants do not respond in this way, it is unlikely that the mechanisms underlying rapid response to touch arose de novo in each species that displays this trait. More likely, these mechanisms developed as dramatic expressions of some widespread, if often unrecognized, phenomena in plants.

Turgor-powered, **seismonastic** movement requires a plant to have: (1) receptive structures that, if touched, become locally distorted, thus initiating a signal; (2) "motor" cells capable of differential volume changes on two sides of an organ; and (3) a signal-transmission system that couples the receptive structures with the motor cells and can elicit volume changes in them.

Motor cells, if defined as cells capable of changing their volume rapidly upon receiving a specific signal, also include cells involved in nonseismonastic turgor movements of plant organs. There is a broad range of operational modes and positional arrangements of motor cells. These cells may occur on only one side of an organ — or on both sides, with different response patterns on each side. Motor cells on opposite sides may be induced to change in volume in divergent directions, or in the same direction but to a different extent. Water released from cells that decrease in volume may accumulate in the cell-wall apoplasm or in intercellular spaces, or it may be taken up by other protoplasts.

Anther filaments of *Berberis* exhibit seismonastic explosive movements. If the adaxial surface near the base of a filament is touched, the filament is stimulated to bend. The contact stimulus is received by specialized, epidermal "feeling cells", which protrude above the general surface. The walls of these cells have localized thin areas near the cell bases — a kind of hinge. When the shape of such a cell is changed, the walls of the "hinge" region bend and the protoplast is locally distorted, evoking an **action potential** (Sect. 11.3.1), which is propagated along and across the filament.

The elongated cortical cells that are the "motor" cells of the filament have thickened primary walls rich in hydrophilic carbohydrates (Ziegenspeck 1928; Guttenberg 1971). These walls are elastic and can be greatly extended as turgor increases. Before excitement, the walls of the cells on the adaxial side of the filament are more extended than are the walls of those on the abaxial side, but, on the two sides of the filament, osmotic and imbibitional forces are in equilibrium, as are the tensions in the cell walls. When these equilibria are disturbed by propagation of the action potential, cells on the adaxial side lose turgor and shorten, while cells on the abaxial side elongate. As this happens, the filament bends toward the style, thus bringing the anther into contact with the pistil or with an insect visiting the flower. Despite its seeming complexity, this system operates rapidly. The filaments can bend to touch the style within one second after excitation (Fleurat-Lessard and Millet 1984).

Many leaves provide examples of complex movement systems driven by turgor changes occurring in **pulvini** — specially structured organs at the bases of leaves, pinnae, or pinnules. The literature on the physiology of these movements is extensive (Satter and Galston 1981). We focus on the structural adaptations underlying them, and on *Mimosa pudica* because of the depth of information available about it.

The sensitive structures of the *Mimosa* leaf are the numerous bristles. These are multicellular and widely distributed. Each is up to 2.5 mm long. They have circumbasal or axillary bolsters of thin-walled cells. If a bristle is touched and slightly displaced, the cells of its bolster are deformed, thereby generating an action potential that is propagated to the pulvinus. However an action potential can also be generated by wounding, quite aside from bristle displacement. Structurally, the pulvini of *Mimosa* are similar to those of other genera of Fabaceae. They are cylindrical to spindle-shaped and surround the base of a petiole or petiolule. The central vascular bundle of the petiole or petiolule in the pulvinar area is surrounded by sclerenchyma, which in turn is surrounded by a thick, parenchymatous cortex — which except for the inner layer represented by the starch sheath (Sect. 10.2.2.1) is the motor tissue (Fleurat-Lessard 1988). The outer cells of this cortex are smaller than the inner cells and can change markedly in size (Haupt 1977). There are appreciable tissue stresses in the pulvinus. The sclerenchyma is under tension and the motor tissue is under compression. The turgor powered tendency of the motor cells to expand is balanced by tensions in the underlying sclerenchyma. The cells of the cortex are interconnected by numerous plasmodesmata — providing the electrical coupling necessary to propagate action potentials (Sect. 11.3.1).

The rapid, active phase of motor-tissue operation during seismonastic leaf movement typically involves loss of turgor and of cell volume on one side of the pulvinus, in tissue that initially was under compression. Simultaneously, the turgor of the cells on the opposite side is maintained or increased. The leaves recover their original orientation as the motor cells slowly regain their previous turgor status. Seismonastic movement is especially complex in *Mimosa*: in first- and second-order pulvini, the bending is downward, whereas in third-order pulvini, it is upward.

The walls of motor cells are thin and elastic, especially on the side of the pulvinus that shrinks during rapid leaf folding. These cells have a large central vacuole that contracts quickly, causing rapid decreases in cell volume. Water expressed from this large vacuole apparently first collects in numerous small vacuoles in the parietal cytoplasm, and is then secreted into the apoplasm (Satter 1979). From there, it may be exuded into intercellular spaces. In the seismonastic downward bending of a first-order pulvinus of *Mimosa*, the water released from the shrinking abaxial motor cells moves through the intercellular spaces to hydathodes (Sect. 8.4.1) in the adaxial epidermis of the pulvinus or nearby stem tissue.

The ultimate control of this set of finely regulated movements is not understood. However, it seems increasingly likely that, whereas transmission of action potentials by ordinary cells involves transient losses of K^+ (Sect. 11.3.1), arrival of the action-potential signal in the especially excitable motor cells triggers a catastrophic loss of K^+. This is largely balanced electrically by concomitant loss of Cl^-. Water then flows out in response to the changed osmotic gradient and the cells shrink (Kumon and Suda 1984).

10.2.1.2 Movements of Stomatal Guard Cells.

Rapid movements powered by turgor changes include the opening and closing of stomatal guard cells. These movements help to minimize water loss by transpiration while interfering minimally with the uptake of CO_2 (Sect. 9.4).

With regard to the cellulose microfibrils in their walls, guard cells, like parenchyma cells, are constructed in a manner analogous to that of a radial-ply automobile tire (Sect. 4.2). Also as in an automobile tire, the cross-sectional shape varies with the pressure, though the circumference does not. The chief mechanism of opening of guard cells is localized swelling and bending as the turgor increases. These increases are due to the activities of ion pumps and channels in the plasmalemmae and tonoplasts of the guard cells (Raschke et al. 1988). It is now clear that, in guard cells that are kidney-shaped (as seen in surface view), the swelling and shrinking occur near the ends of the cells, where the walls are thin (Sect. 9.3).

Raschke (1979) used optical-sectioning microscopic observations of living, operating guard cells to compare changes in guard-cell volume with stomatal aperture in *Vicia faba*. He found that a 17.5% decrease in volume was accompanied by a 67% decrease in aperture. The most effective guard cells are those that transduce small changes in volume into large changes in aperture.

In open guard cells, the main osmotica are potassium salts. The major anions are those of malic and other acids of the tricarboxylic-acid cycle, and chloride. Apparently, vacuoles of the guard cells accumulate anions during the opening phase; potassium ions then migrate inward along the electrical gradient (Outlaw and Lowry 1977).

Significantly, mature guard cells of at least some species seem to have few or no plasmodesmatal connections with the leaf symplasm (Willmer and Sexton 1979; Willmer 1983) and depend largely on their own resources for immediate energy. Accordingly, the primary function of chloroplasts of these guard cells may be to supply high-energy compounds to their own protoplasts.

In many taxa, the subsidiary epidermal cells, which immediately adjoin the guard cells, exchange water and ions with the guard cells via the cell-wall apoplasm, and thus play a role in guard-cell movement. Not surprisingly, areas of infolding characteristic of transfer cells occur on the walls between guard and subsidiary cells in some species. The morphology of guard and subsidiary cells is, of course, highly variable, as are the extent and manner of co-functioning of these two kinds of cells.

There are also other kinds of relationships between guard and subsidiary cells. For example, increasing pressures in guard cells and increasing pressures in subsidiary cells affect stomatal aperture in opposite ways (Wu et al. 1985). However, stomatal aperture is not a simple function of the pressure difference between guard and subsidiary cells. It probably is determined by the difference between a complex function expressing pressure in the guard cells and another complex function expressing pressure in the subsidiary cells. A significant feature of this relation is that a specific aperture size can be produced by various combinations of guard- and subsidiary-cell pressures (Cooke et al. 1976). Details of mechanical interactions between guard and other epidermal cells have been incorporated into a model for stomatal behavior by Delwiche and Cooke (1977).

10.2.2 Growth Movements

As do turgor movements, growth movements involve differential volume changes on two sides of an elongating organ. In growth movements, however, increases in cell dimensions are accompanied by *plastic* rather than *elastic* deformations of the walls. That is, control of growth movements is exerted at the level of wall plasticity, and elastic deformations are only incidental. Growth movements include tropisms, such as photo- and gravitropism, and some nastic movements.

Growth movements may superficially resemble turgor movements. For example, the leaves of *Carica papaya* have no pulvini, yet they move upward at sunrise and downward at sunset (Yin 1941). Although in some other plants movements such as these are caused by changes in turgor, in *Carica* these movements are caused by small, rhythmically fluctuating differences in growth rates on the adaxial and abaxial sides of the petiole (Sect. 10.2.2.4).

By extrapolating from these nyctinastic movements in *Carica*, one can imagine some of the mechanisms that may be involved in the growth movements that occur during shoot development, such as folding of primordial leaves inside the bud and leaf unfolding as the bud opens and the shoot expands.

10.2.2.1 Gravitropism. One can mistakenly assume that all roots are fully gravitropic and grow downward, without noticing that this is true mainly of the primary roots of seedlings. Some secondary roots do, indeed, grow downward, but most have strong lateral-growth tendencies. Indeed, graded responses are more typical than are fully gravitropic ones. We emphasize here the structural and physical bases of gravitropism in those organs in which it is strongly evident.

The most thoroughly investigated site of gravity perception in higher plants is in the root cap. There, gravistimuli are registered by specialized cells containing **statoliths**, which are amyloplasts having especially large starch grains that sediment rapidly. Cells containing statoliths are **statocytes**, and tissue consisting of statocytes is **statenchyma**. In the positively gravitropic root of *Lepidium sativum*, in which the nature of graviperception has been intensively studied, the statocytes are confined to the central part of the root cap, where they form a precise array of four tiers.

Statocytes of the root cap may show strongly polar organization, the nucleus being proximal and the ER complex distal. In vertically growing roots, the statoliths are sedimented onto the ER complex, where they form a radially symmetrical pattern with respect to the cell axis (Volkmann and Sievers 1979). If the root is deflected from the vertical, the statoliths sediment toward the new "bottom" of the ER complex. Their distribution is then no longer symmetrical. Disturbance of the tension in cytoskeletal elements during statolith movement is involved in gravistimulation (Sievers et al. 1991).

Studies of changes in the pattern of Ca^{++} transport and flow of electric currents around the root tip following gravistimulation have revealed that rapid changes in ionic currents through membranes are early events in graviperception. The regulation of an ER-localized Ca^{++} storing compartment in statocytes and polar movement of Ca^{++} across root tips may be especially important (Sievers et al. 1984; Moore 1985).

If a root cap is removed, amyloplasts immediately begin to develop from proplastids in the cells on the margins of the quiescent center (Sect. 17.1.2) and neighboring regions of the root apex. As these new amyloplasts become enlarged, gravitropic responsiveness returns, and then a new root cap forms. After the new root cap is well developed, the cells behind it no longer show enlarged amyloplasts (Barlow and Grundwag 1974).

Gravitropically directed growth, as distinguished from graviperception, occurs not in the root cap, but in the root-elongation zone, which is usually a few millimeters basal to the cap. The kind of signal that is transmitted from the statenchyma to the growing cells is unknown, as are its route and mode of transmission (Behrens et al. 1985), but there are some clues. The cell walls transverse to the shortest route from the statocytes to the cortex of the elongating zone have more plasmodesmata than do walls having other orientations (Volkmann and Sievers 1979).

Gravistimulation changes the distribution of growth in the root elongation zone. Geometrically, it is obvious that if the root is to curve downward, growth must at first be greater on the upper than on the lower side. However, during the course of the gravitropic response, the relative growth rates on the two sides may vary, with the concave side even growing faster than the convex side at certain times (Selker and Sievers 1987).

The range of graviresponsiveness shown by lateral and secondary roots appears to be due to the graded uncoupling of gravicurvature from gravistimulation in the columella (Sect. 17.1.5; Moore 1985). Some root systems, especially "fibrous" ones, form a beautifully arranged, three-dimensional, "fan-like" array consisting of roots growing in every direction from horizontal to vertically downward. Each root grows in a direction determined by its position in the system.

Statocytes occur not only in root caps, but also, it seems, in some primary stems. In these stems, the cells in an easily distinguishable inner cortical layer, called the **starch sheath**, have conspicuous amyloplasts, similar to those in the root cap. It seems probable that these amyloplasts are statoliths, but our understanding of the functioning of the starch (or statocyte) sheath has increased only moderately since the time of Haberlandt (1924). In some stems, the putative statenchyma does not form a continuous sheath, but occurs in separate strands accompanying primary vascular bundles.

The putative statocytes in the stem are relatively short and contain mobile amyloplasts. The mobility of putative stem statoliths is similar to that of root statoliths. Deflecting a stem even 10° from the vertical causes an observable change in their distribution. The amyloplasts appear very early in stem development, when the frusta (Sect. 14.1.1) of the bud-borne embryonic shoot begin to expand. After the internode has elongated and secondary thickening is underway, these bodies may be digested (Wright and Osborne 1977).

10.2.2.2 Phototropism. Whereas gravity is continuous in time and relatively uniform in intensity and direction, light is highly variable in all of these attributes and also in quality. In response to this variability, higher plants have evolved movement systems that adjust leaf and branch positions in ways that optimize light interception by the leaves. These directed-growth responses to light constitute phototropism.

By simple phototropic adjustment movements, many shoots can respond to sustained directional differences in intensity of illumination. For example, if a large object obscures the light received on one side of a shoot but not on the other, stem growth is inhibited on the strongly illuminated side, thus causing the stem to bend toward the light. This differential growth may involve unequal transport of, inactivation of, or responses to, auxin.

More difficult to understand are the relatively rapid growth movements that keep leaves or flowers oriented perpendicular to the direction of the brightest illumination. Such movements enable some leaves and flowers to "follow the sun" (sun-tracking, or heliotropism), thus providing maximum illumination of the upper leaf surface throughout the day (Wainwright 1977; Ehleringer and Forseth 1980). Although growth controls sun-tracking movements in some species, in other species changes in turgor in pulvini control the movements.

10.2.2.3 Thigmotropism. Tendrils, which are modified leaves, stem tips, or branches, provide good examples of thigmotropism. They move in response to contact with a solid object. Because this movement is due to irreversible expansion of the cell walls, it is a growth movement. Tendrils are organs of support common on plants lacking rigid supportive stems including some woody vines.

Before a young tendril contacts a solid object, it makes autonomic, **nutating** movements. These movements are due to cyclically modulated differential growth on opposite sides of the tendril. If the tendril contacts a suitable object, growth is inhibited on the side of contact and the tendril begins making tight coils. A tendril can respond in this way rapidly, wrapping around an object one or more times within an hour (Satter 1979). Tendrils of some species are quite selective about the objects they encircle. Tendrils of some will not wrap around unsteady objects.

In this discussion, we have considered tendril movement to be thigmotropic. That is, we have considered the direction of curling to depend on the direction of the stimulus. However, some tendrils are **thigmonastic**, their direction of curling being predetermined by asymmetry in the tendril structure. If a tendril — whether nastic or tropic — makes only momentary contact, it usually does not begin coiling and retains its irritability. If permanent contact is made, secondary walls may be deposited. The tendril then becomes somewhat woody and loses its irritability.

10.2.2.4 Nastic Movements. Some common types of movements powered by growth are responses to stimuli, though the direction of the responses is not related to the direction of the stimuli. These growth movements are called "nastic", as are some movements based not on growth but on changes in turgor. Some nastic growth movements result from differential growth on the upper and lower sides of an organ that is held at a large angle to the gravity vector. A prerequisite for these movements is that the two sides of the organ differ anatomically, though not necessarily morphologically. For example, the upper and lower sides of the hypocotyl hook of seedlings, or of bent pedicels of flowers, may be similar in gross structure but not in cell dimensions. Movements are generated as those differences are diminished or accentuated by differential growth. If growth is more rapid on the adaxial side, as in many developing leaves, curvature is outward and downward, and the movement is **epinastic**. Conversely, if growth is more rapid on the abaxial side, the movement is **hyponastic**. Whereas auxin probably is the major regulatory hormone in tropisms, ethylene may be the principal one controlling nastic movements (Harvey 1915; Kang 1979).

Leaves of some plants undergo diurnal nastic ("sleep") movements, or **nyctinasty**. These leaves move downward at sunset and upward at sunrise, due to greater growth on the upper or lower sides of the petioles, respectively. Similar movements occur in some other plants, but are powered by turgor changes in pulvini rather than by growth. Rhythmic growth movements of a petiole can continue only until the leaf matures.

In some plants, nyctinastic growth movements are correlated with localized diurnal changes in auxin production in the leaves. For example, Yin (1941) found that if a *Carica papaya* leaf blade is excised, the petiole stops growing and the nyctinastic movements cease. If auxin is supplied to the stump of the petiole, slight growth occurs, but at the same rate on both sides and there is no bending. However, applying exogenous auxin to various parts of the blade of an intact leaf causes differential growth in the petiole, and thus petiolar movement. For example, supplying auxin to lobes *L* and *B* in Fig. 10.1 elicits downward movement, whereas supplying it to lobes S and T elicits upward movement. These variations arise because different parts of the blade are connected with different vascular bundles in the petiole, as indicated by matching the letters in the two parts of Fig. 10.1.

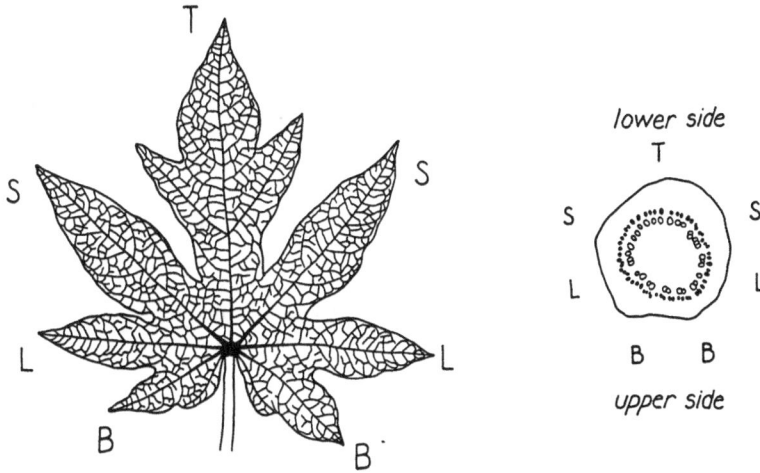

Fig. 10.1. Diagram of typical leaf of *Carica papaya* and corresponding petiolar transection. The phloem strands of lobe T are continuous with those of the lower side of the petiole, those of lobes B with those on the upper side, and those of lobes S and L with those of the lateral sides of the petiole. Thus, rhythmic cycling of rates of endogenous auxin production in different parts of the leaf can cause rhythmic upward and downward curvature of the petiole and, consequently, leaf-blade movement. (After Yin 1941)

Yin (1941) also demonstrated circadian cycles of auxin production in the different parts of the blade, with entrainment of the production in the various parts to different phases of the cycle. These endogenous rhythmic phase shifts in auxin production in different parts of the leaf cause cyclical changes in the slow elongation growth along the top and bottom of the petiole, and movement results. Many plants reorient their leaves photonastically and/or phototropically by bending and/or torsion movements involving pulvini. Nondirectional light signals are perceived in pulvini, but perception of directional signals may involve other parts of the leaf also (Koller 1990).

10.3 Movements with No Direct Participation of Living Protoplasts

Some movements are driven by hydration changes with no direct participation of living protoplasts. These changes may involve the uptake and loss of water by the walls of dead cells, or the dehydration of the walls and protoplasts of senescing and dying cells.

Here, "dehydration" of a protoplast implies drastic and terminal change, far beyond a mere loss of turgor. Turgor movements involve relatively small, reversible changes in the hydration states and volumes of living protoplasts, whereas dehydration of a protoplast typically results in loss of a large fraction of protoplast volume and usually an irreversible disorganization of its living systems.

Hydration changes involving only cell walls may occur in cells in which the protoplasts, before they died, deposited specialized wall layers. These hydration changes can cause a variety of movements. For example, just as unequal growth on opposite sides of a living primary stem can cause it to bend, so differential hydration changes of cell walls on opposite sides of a dead or dying organ can lead to bending or other movement. Likewise, differential swelling or shrinking of two layers of a laminar organ can cause curling and twisting movements not dependent on living protoplasts. The type and extent of these **hydration-related movements** depend on the relative amount and direction of swelling or shrinking in the two regions of the organ. Familiar, small-scale movements of this kind include the opening of fruits, opening of plumes of air-borne seeds, and tight closing of bud scales (Sect. 10.3.1). Many large-scale correctional and orientational bending movements of woody stems are also of this type, though they may be less familiar because they are very slow (Sect. 10.3.2).

10.3.1 Cohesion Movements

Movements resulting from loss of water depend on the cohesion between water molecules and on the adhesion of water to the walls of dead cells. If, as water is lost, cohesion between water molecules causes severe elastic stresses in the cell walls, at some moment either the stresses will become too great for the walls to sustain, or the cohesive forces will become too great for the water to sustain. Then, either the walls will suddenly yield and break, or gas embolism will form in the water, leading to rapid release of stresses and, as a consequence, rapid movement. Movements of this type may occur in drying pods and other fruits, and in drying wood.

The **contractile roots** of some perennial herbs, particularly those having corms or bulbs, provide another example of cohesive movement (Davey 1946). Seeds of these species germinate at the soil surface, yet as a result of root contraction, the junction between stem and root ultimately moves beneath the surface. Further, the effects of frost heaving seem to be counteracted by root contractions that pull the root collar back into the soil in the spring.

Contractile roots, which tend to be straight and vertical, typically have abundant cortical or secondary-phloem parenchyma arranged in plates transverse to the root axis. As contraction begins, cells of alternate plates die and collapse. Cohesive forces then bring together those plates remaining turgid. This prevents radial collapse. The tissues of the central cylinder (and often the rhizodermis also) become wrinkled or wavy in appearance as the root contracts (Thoday and Davey 1932; Reyneke and Schijff 1974).

Detailed information is not available on the occurrence of contractile root movements in tree species. During early years of development, however, some contraction in taproot length seems to be common. Some tree-root contraction may be powered by tension-wood systems (Sects. 10.3.2; 19.5.2) rather than by the parenchyma-based cohesive contraction systems of herbaceous roots. For example, the contraction of aerial roots of the banyan tree, *Ficus benjamina*, is evidently based on tension wood (Zimmermann et al 1968).

10.3.2 Movements Based on Reaction Wood

Various slow movements of woody stems are powered by changes in the cell walls of **reaction wood**. This wood is formed by most woody species in response to certain gravity-derived stimuli. Movements powered by reaction wood enable a tree to adjust branch or stem positions, and hence crown form, in response to changing competitive conditions or catastrophic damage. Reaction wood is also important economically because it can cause severe bending and warping as sawed lumber dries.

The general term "reaction wood" includes **compression wood** in gymnosperms and **tension wood** in angiosperms (Sect. 19.5.2). These specialized woody tissues produce the same gross kinds of orientational movements, but use different physical means and surely must have evolved separately. Compression wood typically develops on the lower sides of leaning stems or lateral branches, whereas tension wood develops on the upper sides. The walls of reaction-wood cells differ structurally and chemically from the walls of ordinary wood cells, and tend to undergo greater volume changes during maturation. Because of the typically asymmetric distribution of reaction wood, these dimensional changes generate high internal bending moments and consequently produce movement.

Compression wood and tension wood differ from each other in chemistry and microanatomy (Scurfield 1973). Tension wood, the reaction wood of angiosperms, has more fibers and fewer vessels than does normal wood, and the fibers have a specialized, "gelatinous" wall layer, the "G" layer. As they mature, tension-wood fibers tend to shorten, due partly to changes in the G layer. This shortening generates bending moments that produce slow, corrective movements of stems. The mechanism of shortening is only partly understood.

In a leaning *Populus* stem, tension wood does not begin to shorten immediately after it is deposited. Instead, the tension wood formed during the vegetative season tends to shorten as a whole after leaf fall in autumn. If a stem containing tension wood is cut before autumn shrinkage has begun, the tension wood will shrink rapidly and irreversibly during drying, perhaps causing the wood to warp severely.

Tracheids of compression wood, the reaction wood of gymnosperms (Westing 1965), have secondary walls that are thicker than those of ordinary tracheids. The wall thickenings are composed of materials that have a great potential for swelling, in contrast to the shrinking that is characteristic of the G layers of tension wood. Swelling generates the internal bending moments that produce the slow, corrective movements of gymnospermous stems.

The asymmetric deposition of compression wood does not guarantee that bending moments will be generated, because the cells of compression wood do not necessarily swell. We have observed apparent compression wood in the lower main stem of a 10-year-old *Pinus sylvestris* tree, but effective bending moments were produced only in the upper part of the stem.

Boyd (1972, 1978) considered lignin deposition to be basic to the swelling of compression-wood cells. However, because the swelling to a large extent reverses upon drying, it seems likely that a hydrophilic substance, such as laricinan, is more directly involved than is lignin. Laricinan fills the spaces that in dehydrated compression tracheids appear as interfibrillar fissures (Włoch 1975). The swelling of this substance, after the tracheids mature, probably generates the bending moments (but see Timell 1986).

The reversible shrinking and swelling of compression wood with changing atmospheric humidity is illustrated by the up-and-down movements of dead side branches of conifers as the weather changes. Wood from such branches can be used to make a hygrometer.

In an experiment on the formation and function of compression wood, young main stems of Pinus sylvestris were bent just as they stopped elongating (Hejnowicz 1967a). They were fastened to supports and thus fixed in the bent position. After a few weeks, during which some compression wood was formed, the bark was removed from the bent regions of the stems and the trees were freed from the supports. The stems straightened immediately, indicating that the compression wood that had formed while the stems were in the bent position had been under compressive stress, which was released by the freeing from the supports. However, as the exposed compression wood subsequently dried, it shrank, and the stems gradually resumed the bent shape they had had while fastened to the supports. This suggests that the length of the zone of dry compression wood corresponded to the length of the arced cambial zone that had produced the wood. The hydrophilic polysaccharides in the thickened secondary walls of the newly formed compression wood had imbibed water and swelled, thus generating the compressive stresses that tended to straighten the stem.

Similar experiments with naturally bent leading shoots of Chamaecyparis pisifera gave similar results. There is evidence that the straightening of leaders that are suspended (i.e., having their tips recurved) toward the end of a growing season in various Cupressaceae and other conifers, such as Tsuga, probably operates by the formation and action of compression wood.

Münch (1938) pointed out that the static as well as the dynamic properties of reaction wood are important. Compression wood is adapted actively to develop compressive stresses, but it is also adapted passively to resist these stresses if they are imposed by heavy static loads, as near the base of the lower side of a leaning stem. Compression wood has higher compressive strength than ordinary wood. Extensive lignification of compression wood is probably more closely related to its static, load-bearing function than to its dynamic, bending-moment function.

Not all tree species produce reaction wood. In lateral branches of young, leaning, main stems of Tilia and Liriodendron, the stresses that power upward correctional bending develop indirectly, by networks of fiber bundles in the secondary phloem, rather than by reaction wood. An underlying mechanism is greater production of secondary xylem on the upper than on the lower side of the stem. This results in greater lateral distention of the overlying network of phloem-fiber bundles on the upper than on the lower side. This greater lateral distension in turn elicits longitudinal shortening forces that are greater along the upper than the lower side, and the stem bends upward (Böhlmann 1984).

11 Intra-Organismal Communication Systems

11.1 Some Fundamentals

A fundamental principle of higher-plant structure is the interwoven coexistence of two spatial phases: the symplasm and the apoplasm (Sect. 1.5), both of which are continuous and constitute communication routes. The symplasm can serve as a communication route in two major modes: by translocation of a signal substance and by transmission of nonsubstantive signals. The apoplasm also can serve as a communication route in two major modes: by diffusive or facilitated transport of signal substances, and by transmission of nonsubstantive signals, such as mechanical stresses. There also seem to be routes of communication by which a signal repeatedly crosses the membrane separating the symplasm and apoplasm. For example, symplasm and apoplasm jointly participate in transmitting action potentials (Sect. 11.3.1). Both also participate in polar "secretion-absorption" of signal substances, in which there is polar secretion of a substance from a protoplast into the apoplasm and absorption of the substance from the apoplasm by the next protoplast along the route, and so on.

An intra-organismal communication provides developmental and functional integration of the various parts of the plant. It also encompasses an "alarm" system that can integrate defensive responses against various stress agents (Chessin and Zipf 1990).

11.2 Translocation of Substantive Signals

In Chapter 7, we discussed rapid long-distance transport in the symplasm (i.e., in the phloem) and in the apoplasm (i.e., in xylem-element lumina, or the super apoplasm). Obviously, these transport routes are involved in communication if they transport **signal substances**, such as phytohormones or morphogens, which produce effects in cells remote from their sites of origin (Sect. 12.4.2). The parts of the symplasm most actively involved in communication by phytohormone and morphogen distribution probably are the protoplasts of meristematic and associated parenchyma cells, and possibly the protoplasts of sieve elements. The parts of the apoplasm most probably involved in this communication are the cell walls surrounding the just-mentioned parts of the symplasm, and lumina of the mature tracheary elements (the super apoplasm).

If phytohormones or morphogens are synthesized in definite centers, then there may be concentration gradients of those substances along the routes they traverse while moving from those sources, as is more fully discussed in Section 12.4.2. Concentration gradients could: (1) divide the cell society into physiological regions, with regional borders delineated by threshold concentrations; or (2) integrate the activities of cells by presenting graded intensities of a consistent signal throughout the effective domains of the signals.

At least three modes of signal propagation by translocation of substances are plausible. One is diffusion of a substance between a source and a sink. Diffusion through living plant tissues is slow, typically only about 0.1 mm/h, but it can still be effective over short distances. A second mode is propagation of concentration waves of signal-substances. Waves provide several possibilities for communication: phase, phase velocity, amplitude, and synchronization (Sect. 12.4.3). A third mode is facilitated transport, in which the signal substance is carried, as a solute, along with mass flow of a solution. Facilitated transport, which occurs in the xylem, or super apoplasm, and in the phloic part of the symplasm, is generally much faster than diffusion.

Propagated waves of signal substances have been observed in various biological systems and described using general theoretical models (Prigogine et al. 1975). Of special interest to botanists is the propagation of auxin waves along the vascular cambium (Wodzicki et al. 1979; Zajączkowski et al. 1984; see also Sect. 12.4.3). The length of an auxin wave is about 20 mm. If, as has been suggested, the period is about 20 min, then the velocity of wave propagation is about 60 mm/h.

In *Pinus sylvestris* stem segments, the amplitude of auxin waves is affected by exogenously supplied regulating substances. Specifically, applying exogenous auxin (IAA) to the apical end increases wave amplitude several fold, and that enhanced amplitude is propagated much faster than the applied auxin molecules. The enhancement of amplitude can be completely suppressed by applying abscisic acid (ABA) to the apical end, but the suppression is prevented by zeatin and by gibberellic acid (Wodzicki and Wodzicki 1981).

The structural basis of the polar transport of auxin is problematic (but see Jacobs and Gilbert 1983). Although there may also be some nonpolar transport of auxin, polar transport is more prevalent and is almost certainly the single most significant example in the plant world of intra-organismal communication and integration via transport of a signal substance. Polar transport of auxin is based on movement through parenchymatic, undifferentiated, or partly differentiated tissues. It is especially well developed in the vascular cambium and in the adjacent, immature vascular tissues.

The route of polar auxin transport between cells is probably not entirely symplasmic (Goldsmith 1977). Some evidence suggests that the two ends of a cell differ in functional permeability of membranes to certain anions, and that polar transport of auxin is related to this difference (Fuente and Leopold 1966). Using indirect immunofluorescence in pea stems, Jacobs and Gilbert (1983) found that an auxin-anion carrier is localized in or on the plasma membranes at the basal ends of the parenchyma cells ensheathing the vascular bundles. They did not see specific fluorescent labeling of this kind in pith or cortical parenchyma cells. It is important to recognize that many cytoskeletal elements (Marchant 1982) are polar, having, in effect, a memory of orientation as to apical and basal directions. This memory of polarity goes back through the cell lineages to the embryo and egg (Cohen 1979a,b). Could these ultra-structural polarities elicit end-to-end asymmetries in the permeability attributes of the membranes of cells (see Sect. 12.4.2.1)?

11.3 Rapid Transmission of Nonsubstantive Signals

At least three modes of transmission of nonsubstantive signals are known or postulated in higher plants. The first mode is based on **action potentials** (APs) which are bio-physical and electrochemical phenomena (see later). This mode is a distant counterpart of nerve transmission in animals. The second mode is quite obscure, but possibly involves rapid changes in ionic equilibria between ER tubules and ambient cytoplasm (Sect. 11.3.2). The third mode depends on mechanical stresses, which, at the cellular level, may modulate permeability properties of cell membranes (Sect. 11.3.3).

11.3.1 Action Potentials and Their Structural Basis

There are two electrochemical prerequisites for an AP. First, there must be, in undisturbed cells, a trans-membrane electric potential difference (E_m), which arises partly from passive diffusion of ions to which the membrane is somewhat permeable (I_p). This E_m is sustained by an ATP-driven electrogenic ion pump in the membrane (Bentrup 1980, 1982). Second, there must be a trans-membrane concentration gradient of ions to which the membrane is relatively less permeable (I_u), and this gradient must be such that, because of active processes in the membrane, the ions would not be in electrochemical equilibrium even if the membrane were more permeable to them. Under these conditions, if the membrane is perturbed by a stimulus such that perme-ability to I_u is increased, the flow of I_u across the membrane increases. If the stimulus is above a threshold level, the electrochemical equilibrium suddenly collapses. This collapse is locally catastrophic and is propagated along the membrane in undiminished intensity as an all-or-nothing phenomenon. After a lag period, the electrochemical equilibrium can be reestablished — then another collapse can be triggered (Davies 1987).

As a consequence of oxidative metabolism, the inside of a plant cell is generally negative relative to the outside. This is manifested by a difference in electrical potential across the plasma membrane. Although the trans-membrane E_m, typically about 100 mV, is small in absolute magnitude, the electric field of about 10^7 V/m is actually quite strong. This trans-membrane field can affect the orientation of some molecules in the membrane, presumably along ion channels that regulate or maintain the low permeability of the membrane to I_u. In some plants, I_u is represented mainly by Cl⁻, which is at a higher concentration inside than outside the cell (see Kikuyama et al. 1984). However, the generation and propagation of an AP is not necessarily coupled with a large Cl⁻ efflux. It may depend more closely on an induced flux of K^+ and some other ions across the membrane, which lowers the E_m. This decrease of the electric field further increases the permeability, which quickly leads to the catastrophic collapse of the E_m. This collapse, along with the much slower process of restoration, constitutes one instance of an AP.

A plant cell develops an AP as a unit. Contiguous cells, however, can be excited by the current flow that accompanies an AP. If they are excited, a propagated AP appears. In a propagated AP, the current flows in ionic form through the plasma membrane of the excited cell into the apoplast (i.e., the membrane is then permeable), and also through the plasmodesmata to the protoplast of the neighboring "target" cell. This transient current is possible because the circuit is closed externally by ionic flow through the apoplast. The transient current across the membrane of the cell that is to be excited initiates the catastrophic collapse of the E_m in that cell. This second excited cell then excites another cell, and so on.

The continuity of both the symplasm and apoplasm is necessary for long-distance propagation of APs (Sibaoka 1966). The electrical continuity of the symplasm via plasmodesmata (Overall and Gunning 1982) allows flow of a current accompanying an AP inside the membrane; the continuity of the apoplasm allows flow on the outside.

Various kinds of living cells can initiate and propagate APs. In a *Lupinus* stem, for instance, an AP can be evoked in the outer stem tissues, largely cortical parenchyma, so that the potential difference measured on the surface of the stem is several tens of millivolts (Zawadzki and Trebacz 1982). Long-distance propagation of APs has also been reported in the parenchyma of vascular bundles and in phloem, and probably also occurs in protoxylem (Sibaoka 1966; Sinyukhin and Gorchakov 1968; Kursanov 1976).

The attributes of an AP moving along a plant vascular bundle — velocity of propagation, amplitude, form, and frequency of repeatability — are all variable, as are the characteristics of an AP in an animal nerve-and-muscle system. Propagation velocities, for example, have been reported to be 20-60 cm/min along pumpkin stems (Sinyukhin and Gorchakov 1968); 120-180 cm/min along excitable parenchyma cells associated with phloem and protoxylem in *Mimosa pudica*; and 1200 cm/min across an insect-trapping leaf of *Dionaea* (Sibaoka 1966). Thus, communication can be much faster by APs than by translocation of signal substances in the phloem, where the velocity of flow is usually less than 2 cm/min.

In carnivorous and seismonastic plants (Sect. 10.2.1.1), APs transmit signals between the sites of stimulus reception and of overt responses. Responses in both these plant groups include rapid movements. In flowers of *Incarvillea* spp, an AP signal moves through the style to the ovules if a pollen grain germinates on the stigmatic surface. This evokes elevated metabolic activity (Sinyukhin and Britikov 1967).

APs may mediate the integration of plant functions to a greater extent than is yet appreciated (Davies 1987). If so, primary and secondary vascular tissues may be routes not only of transport of substances, but also of fast communication by APs. Some of the same principles apply to the propagation of APs along vascular and adjacent tissues as to the transmission of similar signals along nerves of animals (Paszewski and Zawadzki 1974).

If vascular bundles and associated parenchyma are routes of AP signal transmission, then the Casparian-band barrier in the endodermis may have functions beyond those usually ascribed to it (Sect. 2.3). Excitation of a root tip may produce APs detectable on the stem surface, but not along the root surface. This is probably because surface detection of an AP, which depends on the flow of ionic currents between the sites of electrode attachment and the walls of cells propagating the AP, requires an electrically conducting continuum in the cell-wall apoplasm. The cell-wall apoplasm of the root does not meet this requirement. Casparian bands of the endodermis divide root apoplasm into an internal part, including the vascular tissues, and an external part, composed of the cortex and rhizodermis (Sect. 3.2.3.2). The bands also electrically insulate the vascular tissue, in which the AP is propagated, from the cortical tissues.

Electrical insulation by the root endodermis may have important implications if transmission of APs along the root-stem vascular routes is vital to the plant. It has commonly been observed that setting plants so deeply that the stem base is buried in the soil may lead to severe physiological disturbances. Kursanov (1976, p. 479) cited evidence suggesting that, in these plants, the currents accompanying APs that are transmitted from root to stem or vice versa are "short-circuited" in the buried stem segment, which has no endodermis.

11.3.2 Unidentified Rapidly Transmitted Signals

Rapidly transmitted, unidentified signals in a plant can be produced by, for example, irradiating some of the leaves on a spinach plant for short periods with red or far-red light. Seemingly, a "fast signal" moves from irradiated to nearby, shielded leaves (Karege et al. 1982), resulting in the quick modulation of peroxidase activity in both irradiated and shielded leaves, in a near-parallel manner. Rapid transmission of the signal is inhibited by LiCl and other agents affecting ionic equilibria in cells (Penel et al. 1985). Thus, the signals could involve perturbations of ionic equilibria that are rapidly transmitted from irradiated to shielded leaves. Available data suggest that living protoplasts are involved and that signal transmission depends neither on transport of a signal substance nor on APs of the type discussed previously (Sect. 11.3.1).

Though they involve the apoplasm and the symplasm, and the plasmalemma that is the boundary between them, APs are nonetheless based on cellular units. Because the symplasm contains compartments much smaller than cells and separated by membranes, one might expect that a wholly intrasymplasmic phenomenon, similar to APs, could occur. An obvious intrasymplasmic compartment, the endoplasmic reticulum, is continuous within a cell, and, via the desmotubules of the plasmodesmata, also forms a continuous system throughout the plant (Sect. 7.3; Fig. 7.3). It is possible that the desmotubules, as a consequence of this continuity, can transmit or relay rapid signals from cell to cell.

11.3.3 Mechanical Stresses as Rapidly Transmitted Signals

Partly as a consequence of growth, plant tissues are under mechanical stresses (Sects. 4.6; 12.4.6); some are in tension, others in compression. Superposed on growth stresses are other stresses resulting from static mechanical loading. For example, bending of an organ by static loading produces compression on one side and tension on the other.

In response to such stresses, a plant organ behaves as a unit rather than as a population of individual cells. For example, a stimulus that affects the turgor in a tissue will disturb the equilibrium of tissue stresses, and, via this change of stresses, information about local turgor changes may be transmitted rapidly to other parts of the organ. Because of rapid transmission of stress-related stimuli, mechanical stresses may be integrating factors in growth and development.

Stresses in a cell wall may be transmitted, via strains, to the protoplast because the plasmalemma is intimately associated with the wall; or the stresses may influence the protoplast only more indirectly through a stress-mediated phenomenon such as a piezoelectric effect in the cellulose micelles (Fukada 1968).

Cells may also possibly detect tissue stresses, and disturbances of these stresses, via stretch-activated ionic channels in the plasmalemma. The frequency of opening of these channels, and the duration of the open state, depend on the tension applied to the plasmalemma via a submembrane network of cytoskeletal elements connecting the channel proteins (Edwards and Pickard 1987; see also Schroeder and Hedrich 1989). It is possible that stretch-activated channels register and transduce signals resulting from changing mechanical stresses.

Although little is definitely known about the relevance of mechanical stresses to the integration of plant development, some information is beginning to emerge (Sect. 12.4.6). In Section 4.7.1, we mentioned a principle of mechanical design that many forest conifers seem to follow while maintaining uniform resistance to bending in the stem below the crown. That principle is the D^3 law of stem diameter (or the D^2 law, if the trees are vigorous and natural prestressing is effective). Many trees also seem to follow a design principle requiring the slenderness ratio to change inversely with the square root of tree height (Sect. 4.3.3). How can a local set of cambial cells regulate stem diameter growth according to stem height and distance of the cambial region below the crown or above the base?

In response to this question, McMahon (1975) suggested that the xylem-ray parenchyma may register a gradient of stress along the radial course of the rays if there is local bending, as might be caused by the wind. These cells may then transduce the stresses to signals that can affect the activities of local cambial initials. This idea, which assigns living xylem-ray cells a role in stem morphogenesis, seems to merit further investigation.

Mechanical stresses probably are propagated not only through cell walls and cell membranes, but also via the hydraulic continuity of the liquid phase of the symplasm and the liquid phase of the apoplasm, especially of the super apoplasm. Rapid physical signals (pressure changes) induced by osmotic changes in the apoplasm at one end of a phloem pathway can apparently be propagated to the other end of the pathway as hydraulic pressure waves. The frequency and other attributes of the waves could carry information, making the phloem a channel of rapid communication between source and sink (Grusak and Lucas 1986).

11.4   The Significance of Plasmodesmata in Communication and Integration

Empirical studies have shown that plasmodesmata, which unite most of the protoplasts of a plant into a single symplasm (Sect. 7.3), help integrate growth and development (Carr 1976; Robards and Lucas 1990). The fact that plasmodesmata are not present between all contiguous living cells at maturity, and are not distributed equally over the walls of cells in which they occur, suggests that symplasmic transport is greater along some pathways than along others (Fisher 1990).

For example, the large numbers of plasmodesmata in the cross walls of the cells located between the statenchyma in the root cap and the root-elongation zone suggest that there may be intense symplasmic transport along this route. This transport may be involved in the transmission of gravitropic signals (Sect. 10.2.2.1). Similarly, the high frequency of cytoplasmic connections between sieve elements and their associated companion cells in some roots suggests that a symplasmic pathway of phloem unloading is feasible (Warmbrodt 1985, and references cited therein). In many species investigated, there is a high frequency of plasmodesmata between mesophyll and bundle-sheath cells of the leaf (Fisher 1990, and references cited therein).

Some cells have no plasmodesmatal connections. Indeed, symplasmic isolation appears necessary before a cell or a group of cells can express latent totipotency by undertaking a radically different developmental pathway. For example, plasmodesmatal connections between generations (parent sporophyte, gametophyte, and embryonic sporophyte) are either absent or ephemeral (Rodkiewicz 1973). Each generation thus constitutes a separate symplasm and is a separate organism, even if not independent. For example, in *Zea mays*, plasmodesmata are absent between maternal sporophyte reproductive structures and the gametophyte tissue within (Diboll and Larson 1966). In *Capsella*, although plasmodesmata were reported between the egg or zygote and cells of the ambient gametophyte, these disappeared by an early stage of embryogeny (Schulz and Jensen 1968; Sect. 13.6).

In addition, an early step in the differentiation of an archesporial cell in many higher plants is deposition of callose between the protoplast and the cell wall (Heslop-Harrison 1966; Rodkiewicz 1973), with consequent plugging of plasmodesmatal connections. This isolation seems necessary for the reproductive cells to adopt a separate developmental program.

It also seems that a cell in culture must break plasmodesmatal connections with other cells before it can begin to develop into an embryo. It is even possible, as Steward (personal communication from F. C. Steward to J. R. 1975) speculated, that extreme physical shock can so greatly disrupt a plant's plasmodesmatal connections that it loses some organism-level control and develops many competing centers of meristematic activity. Steward had noted that the shock waves produced by a wartime parachute bomb that exploded just above the ground had stripped all the twigs and small branches off nearby *Acer pseudoplatanus* trees, and that, after a few months the standing trunks seemed to be covered with a green moss, which, on close examination, proved to be, not moss, but myriads of small adventitious shoots. Steward suggested that the plasmodesmata among many cambial and parenchymatous cells had been broken or occluded by the shock waves, thus preventing the cells from receiving positional and other developmental information from the rest of the organism. Individual cells, or small communities of cells, were then free to organize themselves into adventitious shoot meristems. Information carried by cell-wall arrangements and properties, and in the cytoskeletal ultrastructure, would, of course, persist, and the original polarity would therefore be unaffected. This interpretation, though speculative, raises some questions that are open to experimental investigation.

Part II Developmental Anatomy

12 Basic Concepts from an Organismal Perspective

12.1 Introduction

Organismal development requires structural and functional integration and coordination among cells, in order for the small and simple to become large and complex and to assume a form and structure characteristic of the taxon. The developmental process must be integrated at the tissue, organ, and organism levels. Individual cells, though in some senses totipotent, are only pliant constituents of development.

An organism can be likened to a cybernetically directed machine (Calow 1976) running through a sequence of programs: embryonic development, juvenile stage, mature reproductive stage, senescence, and death. In most higher animals, development of structural detail and complexity occurs mainly during a period of embryonic and fetal growth within the maternal body, or in an external egg under some environmental control; the development of structural detail is mostly completed during infancy. Juvenile stages involve an increase in size, and during later life emphasis is on maintenance and, episodically, on reproduction, not on continuing structural development of the individual.

Higher plants follow sets of developmental programs that differ from those of higher animals in two ways. Firstly, they produce new tissues and additional organs throughout their lives. In particular, woody plants — the most complex and long-lived of plants — are always developing. Their bodies are never completed. When development stops, the organism dies. Development and anatomy are inseparably integrated. Secondly, developmental programs in plants are flexible; many differentiated cells can revert to a meristematic state and reinitiate the developmental cycle.

The morphogenesis of a woody plant, in contrast to that of most animals, is not limited in time, and, therefore, is susceptible to continuing modification in response to environmental changes. A woody plant cannot isolate itself from external influences to the extent that an animal can, nor can it relocate itself. If threatened, a woody plant must adapt. Commonly, it responds by further development, by growth.

This continuing development of woody plants, a kind of persistent embryogeny, is localized in **meristems** and their progeny cells and tissues. Collectively, meristems may constitute only a small fraction of total plant volume. Because meristems have the potential to produce new cells and organs even after most older tissues are senescent or dead, the concept of age in plants is blurred. Different organs of the same plant, or different parts of the same organ, commonly are of different ages.

The complex ontogeny of woody plants is due in part to the multiple centers of cytogenesis and morphogenesis. The major routes of development are: (1) from the zygote and embryo, (2) from the shoot and root apical meristems, (3) from the vascular cambium, and (4) from the phellogen(s). Activities along these and other routes are coordinated, and the activity along each is subordinated to the whole.

If we disregard the various vegetative and other nonsexual modes of reproduction of higher plants, we can consider that sporophytes originate as zygotes. **Embryogeny,** the common route of early development, normally leads from a zygote to a mature embryo in a seed. However, many differentiated plant cells can become embryogenic under defined conditions.

An embryo has an established root-shoot polarity (Sect. 13.1.4). One end of the embryo is focused around a superficially located, **shoot apical meristem,** which produces the progenitors of the cells that will construct the shoot. Within the other end is a deep-seated, **root apical meristem**, which produces progenitors of all the cells of the root system. Thus, the apical meristems indirectly construct the remainder of the plant body. In effect, the root and shoot apical meristems maintain a virtual continuing embryogeny throughout primary and secondary growth.

Development of the secondary body is based mostly in the **vascular cambium,** which arises in the provascular tissue of the primary plant body and persists as an extensive, almost sheath-like, secondary meristem (Sects. 18.1; 18.6). Located between the secondary xylem and secondary phloem, this cambium extends throughout the plant, except in the very youngest root and shoot regions. By producing secondary vascular tissues, the vascular cambium causes the stems and roots to grow in thickness. The cells produced on the adaxial side of the vascular cambium develop into secondary xylem, or wood. Those on the abaxial side develop into secondary phloem, or bast. The bast eventually becomes part of the bark, which also includes derivatives of the phellogen(s).

Sexual reproduction of a seed plant yields a zygote, which then becomes an almost completely meristematic, seed-borne embryo. The embryo develops into a germinating seedling, with a single shoot meristem and a single root meristem. As the plant grows, there is a progressive decentralization of the meristematic function, until, if the plant becomes a large tree, it has thousands of root meristems, thousands of shoot meristems, an extensive sheath of vascular-cambial meristem, and one or more phellogens. Each individual meristem has some local autonomy, though each also depends on the rest of the organism for most of the essentials of survival. In some ways, this dependence of an individual on the whole has a counterpart in social structures of humans and some other animals. In a sense, a tree is a society of meristems as much as it is an individual organism. Further, the distinction between an individual organism and a "colony" is not always clear; for example, a tree may produce shoot sprouts from roots (*Ailanthus, Robinia*), or branches may produce roots (*Ficus, Rhizophora*). In either case, an individual may become a colony. Obviously, a tree is not an individual in the same sense that a cat or a human is.

In addition to development along the major meristematic routes, higher plants having extensive secondary growth tend to have various facultative developmental routes. These commonly are activated through injury or exposure. For example, within wound callus, parenchyma cells may undertake a series of divisions that lead to an adventitious apical meristem or a vascular strand. Similarly, in response to increased exposure, a phellogen (Sect. 21.2.1) may differentiate from parenchyma cells in nonfunctional secondary phloem.

12.2 Cellular Aspects of Development

12.2.1 Cell Division and the Cell Cycle

Multiplication of cells during development of the higher-plant body is based on mitotic nuclear division, which usually is followed quickly by cytokinesis. This sequence is so common that the term "mitosis" is often used synonymously with "cell division", even though mitosis is not invariably followed by cytokinesis. Mitotic division followed by cytokinesis produces cells that, in theory, have identical complements of nuclear genes. However, sister cells are always somewhat unequal (Sect. 12.2.2.2).

The series of cellular events that occur, and the time interval, between one cell division and the next in the same lineage is a **cell cycle**. This cycle includes changes in various cellular components, most significantly in the nuclear DNA and in the orientation of microtubules.

Within the nuclear-DNA cycle (the chromosome cycle), several phases can be distinguished. The terms applied to these specifically nuclear phases are often considered, loosely, as applying to phases of the whole cell cycle, not just to phases of the nuclear cycle. The phases of the nuclear cycle are G_1 (gap one), S (synthetic, with reference to DNA), G_2 (gap two), and M (mitotic). Noncycling cells are considered to be in the G_0 phase. The S phase, in which the chromosomal DNA is replicated, commonly lasts 2 to 10 h, which is usually less than half of the total cycle time. The S phase is separated from the preceding mitosis by G_1, which is highly variable in duration. During the S phase, the amount of DNA increases from the 2C amount present in G_1 to the 4C amount. During the G_2 phase, which follows S, the nucleus is potentiated to divide. Nuclear division occurs during the M phase, which consists of four subphases: prophase, metaphase, anaphase, and telophase. During anaphase, the DNA level per chromosome set is reduced from 4C to 2C.

A microtubule-reorientation cycle consists of a sequence of microtubule arrays (Fig. 12.1): a cortical array (during interphase), a preprophase band, a mitotic spindle, and a phragmoplast (Schnepf 1984; Lambert et al. 1991). The formation of microtubule arrays is thought to depend on microtubule-organizing centers (Lloyd and Barlow 1982), which often, but not always, are concentrated along cell edges.

The preprophase band is a loose, peripheral annulus of microtubules. The position of this band predicts the location of the edge of the partition wall (Wick 1991). Preprophase bands have been recognized in a wide range of taxa, but their presence may not be general. Highly predictable divisions, without prior display of preprophase bands, may occur during sporogenesis (Buchen and Sievers 1981) and in the vascular cambium.

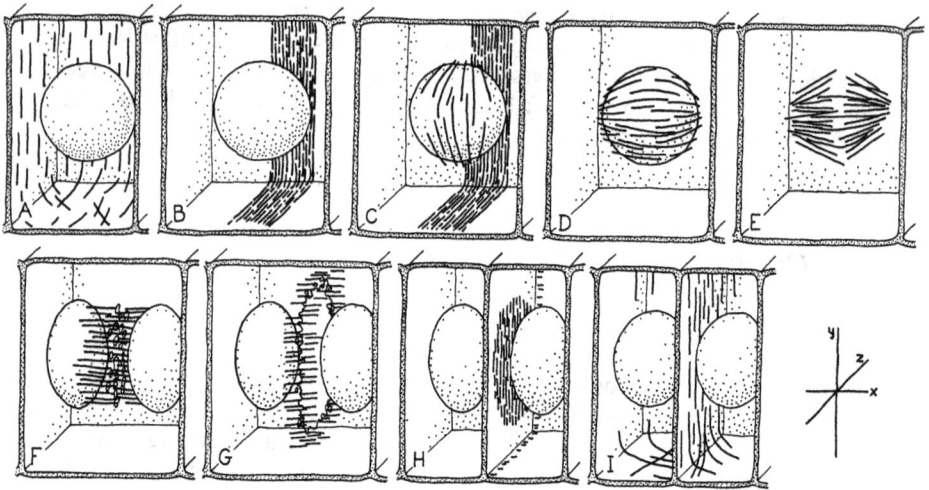

Fig. 12.1 A—I. Schematics of successive stages of the microtubule (MT) reorientation cycle during a mitotic cycle. Based on cytological studies of *Sphagnum* leaflets (Schnepf 1984), but probably applicable in principle to many higher taxa also. **A** Interphase MTs are in peripheral (cortical) positions. **B** In preprophase, MTs form a band. **C** During the transition between preprophase and prophase, MTs begin moving to the nuclear envelope. **D** Early in prophase MTs become aligned near the nuclear envelope, mostly in the directions of the future spindle "fibers". The nuclear envelope disintegrates later in prophase (not illustrated). **E** During anaphase and metaphase, aggregations of MTs constitute the spindle. **F** Late in telophase, MTs form a phragmoplast near the center plane of the maternal cell. **G** During cytokinesis, MTs are most obvious in the phragmoplast, near the cell equator. **H** Early in postcytokinesis MTs become aligned in layers on both sides of the new partition wall in a *yz* plane (*see coordinate axes at right*). **I** During transition from postcytokinesis to interphase, MTs become distributed in the peripheral cytoplasm near all walls of the daughter cells. (After Schnepf 1984)

The spindle is most prominent during the M phase and cytokinesis. It is a barrel-shaped network of meridionally oriented microtubules and microfilaments (Cohen 1979a). Usually, but with some exceptions, the spindle is structurally continuous during mitosis and subsequent cytokinesis. The equatorial part of the mitotic anaphase spindle becomes an aggregation of vesicles — the phragmoplast — within which the cell plate is subsequently organized (Venverloo et al. 1980).

Cytokinesis begins with this aggregation of vesicles — the phragmoplast complex — in the plane of the future partition. These vesicles gradually coalesce into two membranes, with a disc-shaped region between them. The membranes become the respective plasmalemmae along the partition faces of the two daughter cells as the disc-shaped space between the two membranes becomes the **cell plate**. This space is filled with liquid derived from the contents of the vesicles that coalesced to form it. The cell plate has pores lined with membranes of the same type and origin as the new plasma-lemmae along the partition wall. The protoplasts of the daughter cells remain in contact through these pores, which develop into plasmodesmata. From its inception each plasmodesma contains a tubule of ER.

The cell plate usually appears in the central part of the phragmoplast during telophase of karyokinesis. It extends itself centrifugally by accreting vesicles at its edges, and soon reaches the walls of the mother cell. Meanwhile, on both surfaces of the cell plate, the first layers of the primary partition wall are deposited by the respective daughter protoplasts. The first layers contain cellulosic microfibrils. The originally liquid middle layer of the cell plate gradually is dehydrated and becomes a gel-like middle lamella rich in pectic substances.

Daughter protoplasts typically deposit primary wall layers not just on the new partition, but also on the walls they inherited from the mother cell (Mahmood 1968, 1990). This nondiscrimination between old and new walls contributes to the asymmetry of each daughter cell, in that one wall consists only of new material, whereas the opposite wall consists of old and new material. Meanwhile, the mother-cell wall "loses" cellulose locally along the annulus where the new partition wall intersects it (Roland 1978a,b; Jeffree et al. 1986). This "loss" may be due partly to stretching, accompanying intensive local area growth of the wall, and partly to hydrolytic processes that culminate in a juncture between the middle lamella of the new partition and the middle lamella of the intersected mother-cell wall (Fig. 12.2).

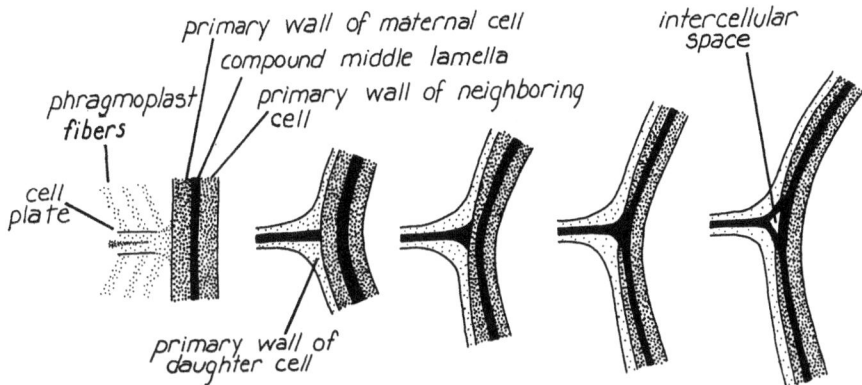

Fig. 12.2. Successive stages (*left* to *right*) of merger of middle lamella of a newly formed partition between daughter cells with compound middle lamella between maternal cell and an adjacent cell. Middle lamellae are *solid black*; older primary walls are *densely stippled*; primary walls of daughter cells are *lightly stippled*. An intercellular gas space soon opens along the junction between new and old middle lamellae

A cell cycle that consists of karyokinesis without subsequent cytokinesis is **coenocytic.** Coenocytic cycles are a normal part of development of nonarticulated laticifers and of the early phases of embryogeny in many gymnosperms and some angiosperms (Sects. 13.2; 13.3). In some septate fibers and xylem-parenchyma strands, there seem to be several successive coenocytic cycles, after which phragmoplasts form and cytokinesis occurs between the linearly arranged nuclei.

If a cell cycle includes the S but not the M phase, the chromosome number is doubled, but the nucleus fails to divide. This is an **endomitotic** cycle, resulting in **endoduplication** or **endoreplication.** Endomitotic cycles are common in normal development, and can be induced artificially by chemicals that interfere with the normal dynamics of microtubules and microfilaments. It is becoming clear that endoreplication normally occurs in cells differentiating along certain developmental pathways, but not necessarily in all cells of a particular tissue. Thus, part of the symplasm may have cells with endoduplicated DNA and part may not.

12.2.2 Cytodifferentiation

In animal ontogeny, a significant early event is the **"determination"** of cells to take a specific route of development. Cells that have been determined proceed along the same developmental route whether grown in situ, in isolation, or in a new site after transplantation. In some cases, however, they receive a specific signal before they become **competent** to proceed with that development.

Although the concepts of determination and competence are used in discussing plant development (Mohr and Sitte 1971; McDaniel 1984), these concepts have not been demonstrated to operate in plants to the extent that they have in animals. In plant cells, the determined state is less stable than it is in animal cells, and determination is more likely to be gradual.

Some meristems can be interpreted as being in a determined state. For example, after excision and culturing in vitro, the cells of apical meristems may remain determined as either root- or shoot-meristem cells, for a limited number of cycles (Wareing and Al-Chalabi 1985). In some tissues, the first stage of development is the differentiation of so-called "mother cells", which become determined by their position in the meristem. In situ, but not necessarily after isolation or relocation by grafting, the progeny of mother cells differentiate into cells constituting one specific type of tissue rather than another.

A shoot-apical meristem may be determined in a further sense, in that, after it has given rise to a number of cell generations, it may produce derivatives that are no longer in a "juvenile", or purely vegetative, phase, but rather are in a mature, or "adult", reproductive phase, during which they produce flowers. Evidence that shoot-apical-meristem cells can be determined either as "juvenile" or "adult" was provided by the grafting of "adult" *Citrus* shoot tips (with three leaf primordia), onto the epicotyls of young seedlings. The grafted apices retained their "adult" status (Navarro et al. 1975).

The term **cell differentiation** has been loosely applied to two somewhat different concepts. Firstly, there are the processes whereby a particular, unspecialized cell becomes specialized in structure and function. Secondly, there are the processes by which cells in an initially uniform population become different from each other. The first concept is temporal only and focuses on the differentiation of a single cell and its progeny. The second is spatiotemporal and focuses on diverging paths of differentiation among the constituent cells in a population. We use the term **cytodifferentiation** in referring to changes in single cells.

Cytodifferentiation probably depends more on an increase of certain cellular capabilities and reduction of others than on development of new capabilities. Commonly, some minor aspect of a cell's activity not critically important to its own life-sustaining systems is expanded into a major activity. Molecules related to that activity are a "luxury" at the cellular level (Holtzer 1970), but may have an essential function at the organismal level.

Cytodifferentiation is somewhat analogous to the "differentiation" of colors during passage of white light through a prism. Nothing is added to the light by its differential diffraction. What is new is the spectrum, the order, the differentiation. Similarly, differentiation at the organismal level can result in something new relative to active meristematic cells. This includes new cell types, new tissues, and new organs.

The process of cytodifferentiation culminates in maturation, during which the cell becomes functional. The cells of some tissues attain functional maturity only after their protoplasts have died. This **programmed death** comes after a protoplast has finished its synthetic contributions to, and pattern-making impositions on, the cell wall. Cell death is essential to the functioning of xylem tracheary elements.

12.2.2.1 Cytodifferentiation and the Cell Cycle. Encoded in the genetic material of a meristematic cell is all the information needed to construct, operate, and reproduce the plant. Most of that information is repressed at all phases of the cell cycle (Sect. 12.2.1). Effectuating a genetic program for cytodifferentiation requires derepression of relevant groups of genes. This is often called "switching on", but perhaps a better term is "activation".

Cytodifferentiation is the fate of all cells that are pushed away from, or left behind by, a plant's meristems. Two types of cell cycles (or divisions) can be distinguished in these cells: (1) those that do not influence determination of the cells, and (2) those during which determination changes or is fixed. We follow the zoologists and call the first type **proliferative cycles** and the second **quantal cycles** (Holtzer et al. 1975). "Quantal" in this sense implies two alternatives, as "on" or "off", or "yes" or "no".

In plants, proliferative cycles commonly occur mostly in so-called "mother cells" — that is, in cells that, because of their position in the organism and their lineage, are determined to follow a particular program of cytodifferentiation. Xylem- and phloem-mother cells produced by the vascular cambium are examples. The number of cells that arise from a mother cell depends on the number of proliferative divisions that occur. The number of proliferative cycles thus is a major determinant of the volume of the tissue or organ.

Cytodifferentiation of idioblasts or of small multicellular structures, such as stomata and trichomes, is often preceded by divisions yielding distinctly unequal daughter cells (Sect. 12.2.2.2). Perhaps the cell cycle ending with the unequal division is quantal, and the unequal division is already an expression of a determination that a particular developmental program is to be followed. However, in some cases of idioblastic cytodifferentiation, especially those in which the larger and smaller daughter cells undertake different developmental programs, there is evidence that switching on of the separate programs for the daughter cells depends on factors operating after the unequal division. In these cases, the cycle after the unequal cycle may be quantal.

Can a genetic program for cytodifferentiation be activated at any time in the cell cycle, or only during a limited time of susceptibility? It seems likely that the phase of most complete repression coincides with the M phase of the nuclear cycle, during which the chromosomes are tightly condensed. After mitosis, there is probably a trend toward derepression — or at least a susceptibility to derepression.

Cell division commonly occurs early in regeneration processes. Typically, parenchyma cells near a wound resume division before cytodifferentiation occurs. Cell division may erase previous developmental programs and make possible the activation of new ones. Furthermore, in normal (not wound-induced) development, specific programs of cytodifferentiation are determined as the cells undergo multiplicative growth, during which each phase of the cell cycle becomes open to epigenetic stimuli that could induce cytodifferentiation. From this superficial analysis, it would seem that initiation of cytodifferentiation depends on recent occurrence of cell division, and that there is a limited time of susceptibility.

Cytokinesis per se, however, may not be essential for activating developmental programs. For example, cells at a wound surface of a potato tuber become suberized even if the divisions that are a normal consequence of wounding are inhibited (Kahl et al. 1969). Wheat seedlings developing from seeds irradiated so heavily that cell division is completely inhibited still show seemingly normal cytodifferentiation of many cells (Foard and Haber 1961; Foard 1971). In wheat roots treated with colchicine at concentrations that inhibit cell divisions, groups of pericycle cells following the developmental program for initiation of lateral root primordia still grow essentially normally, though they do not divide (Foard et al. 1965). The primordia consist of a few, large, polyploid cells rather than many small, diploid cells, because colchicine inhibits cytokinesis but not DNA replication or chromosomal division.

Thus, some phase(s) of the cell cycle other than cytokinesis may be essential to determination of cytodifferentiation. For example, those phases of the cycle that occurred before seed irradiation, or those not inhibited by colchicine, may be essential. This is especially true for the replication of DNA during the S phase. The occurrence of endomitotic DNA replication in differentiating cells (Sect. 12.2.1) is evidence of some connection between DNA synthesis and cytodifferentiation.

12.2.2.2 Unequal Cell Divisions in Differentiation. Divisions in which the two daughter cells differ greatly in size and shape occur early in the developmental programs of various cell types (Bünning 1952; Wareing 1976). These patently unequal divisions occur in zygotes; during embryogeny; during development of trichoblasts, stomata, and sclereids; and in the genesis of phloem sieve elements and their companion cells. Determination that a cell will divide unequally occurs before mitosis begins and is often detectable at that early stage, from the position of the preprophase band of microtubules or from the position of the preprophase nucleus.

Unequal divisions are especially prevalent where the surrounding cells are already highly differentiated, as in developing stomata. Subsidiary cells as well as guard-cell mother cells may arise from unequal divisions of the guard-cell precursors (Sect. 16.3.4). Unequal divisions are also common where organ polarity is strong, as in the rhizodermis, in which trichoblasts commonly arise from such divisions.

The smaller daughter cell of an unequal division usually follows the more specialized program of cytodifferentiation. It may become an idioblast or may divide further and produce some small, specialized, multicellular structure. Its developmental pathway depends primarily on its position. For example, in rhizodermis, the smaller cell may become a root hair, whereas in leaf epidermis it may give rise to a stomate or a trichome.

Though our discussion here has focused on patently unequal divisions, all divisions are somewhat unequal, in that the apportioning of the cytoplasm during cytokinesis is unequal in volume and in number and types of included organelles. Mitotic daughter cells are also unequal in their shape (geometry) and in the structure of their walls (Mahmood 1990).

The asymmetry of cell division is due not only to the asymmetry of cells themselves, but also to the at least slight asymmetry of all cellular environments. For example, geometric considerations ordain that periclinal divisions of cells at an organ surface will always be asymmetric, and most divisions transverse to the axis of polarity are necessarily asymmetric because the two ends of the mother cell differed in polarity.

12.2.3 Cell Growth

12.2.3.1 General Perspectives. Newly formed daughter cells, except for the smaller of the cells derived from obviously unequal divisions, tend to grow until they reach the approximate volume of the mother cell. Meristematic cell division, combined with intervening cellular growth, constitutes **multiplicative growth**. In contrast, **extension growth** occurs at the margins of the meristematic regions, where cell divisions are infrequent. Growth that is not limited to one dimension at these margins can be considered as expansion growth.

Cell growth and division have opposite effects on cell dimensions. Growth increases the average cell dimensions; division decreases them. The relation between growth rate and division frequency determines the mean cell dimensions (Green 1976).

In meristems, mean cell dimensions may depend, indirectly, on various internal or external factors. These factors have their primary effects on cell growth rates or division frequencies. However, even if a variable factor has an immediate effect on, say, growth rate, and cell size begins to change, a new steady-state size cannot be attained quickly. Rather, the new steady-state size will be approached asymptotically over time. To illustrate the conservatism of this response, consider a simple example, in which elongated cells have a mean steady-state length of 1 mm under prevailing conditions. If growth conditions change such that the cells extend in length 3 mm per time unit and divide transversely once per time unit, the new *steady-state length* will be 3 mm, but that length will not be attained immediately. After one cell generation (one time unit), the average cell will be 2 mm long (because to the original 1 mm of length we add 3 mm, then divide by 2 to account for the division). After another time unit, the length will be 2.5 mm, then 2.75 mm, etc. Eventually, the length will be indistinguishable from 3 mm.

Cell growth entails wall extension. A key factor in this extension is wall plasticity, which may be anisotropic. This anisotropy can determine the direction of greatest growth and thus the final shape of the cell (Green 1980). Growth of the wall of an expanding cell has various aspects, including (1) deposition of new wall layers on the inner surface of the older layers, (2) insertion of new structural elements among the older, and (3) areal extension of walls as they yield under turgor-induced stresses (Cosgrove 1985). The genesis and ultrastructure of cell walls are discussed in Section 1.4.5.

To what extent does areal growth of a cell wall resemble its strain under mechanical tension? It is well known that putting an isolated strip of primary wall under moderate tension generates an elastic strain that is roughly proportional to the tensional stress. The amount of elastic strain is a function of the stress and of the elastic moduli of the cell walls (Sect. 4.3.1). Elastic strain is reversible, however; it is not growth. In contrast, if the stress surpasses a threshold value, the cell will slowly expand as the wall creeps under the stress induced by turgor pressure. This is plastic strain, which increases with time. Plastic strain is a component of growth. The amount of growth depends on the stress and on the rheological properties of the wall, which in turn depend in part on the physiological activities of the protoplast.

There is evidence that, with some possible exceptions (as in certain leaf mesophyll cells; see Sect. 16.3.5), positive turgor pressure within a cell is a precondition for wall growth. However, although the growth rate has been described as proportional to the margin by which actual turgor pressure exceeds the threshold pressure for growth to begin, the relation between growth and stress is more complex than a Hooke's Law analogy suggests. Although turgor pressure inevitably produces tensional stresses in cell walls, the stress in a specific wall area depends not only on the pressure within the cell, but also on cell geometry (Sect. 1.3; 4.4.1; Dormer 1980).

Somewhat confusingly, there is evidence that, at a cellular level, maximal growth does not occur where turgor-induced stress is maximal. For example, cylindrical cells in a stem generally elongate faster than they increase in diameter, though the axial stress is only half the circumferential stress (Sect. 4.4.1). Indeed, observations of growth, especially of the intrusive type (which is localized at cell edges and tips; Sec. 12.2.3.2), suggest that, in general, *within a cell*, growth is maximal where the wall curvature is maximal (and the stress is minimal), at a particular turgor pressure.

If areal growth of cell walls is viscoelastic creeping, that creeping is probably controlled via modulation of the viscoelastic properties of walls. It seems probable that auxin may affect these properties and growth by modulating both metabolic degradation and synthesis of cell-wall polysaccharides (Masuda and Yamamoto 1985).

Constraints imposed by the cell-wall network prevent cells of higher plants from slithering past one other, as animal cells do. For relative movement between cells to occur, the middle lamella must first be dissolved. Even in intrusive growth, however, there is no actual slippage, though there is dissolution of the middle lamella (Sect. 12.2.3.2). In many meristematic tissues, the growth of contiguous cells is closely coordinated at the tissue and organ levels and is commonly symplastic. However, at the cellular level, especially during later developmental stages, as in the formation of the extensive intercellular gas-space systems of leaf mesophyll (Sect. 16.3.5), the constraints of symplastic growth are lifted.

During the cell-wall expansion that is a part of symplastic growth, what is the fate of the plasmodesmata that were laid down during cell-plate formation (Sect. 12.2.1)? Are they merely "diluted" — that is, dispersed, or spread out — but otherwise essentially unchanged in distribution? The answer seems to be "no". Overall plasmodesmatal frequency tends to be maintained or increased, indicating that new plasmodesmata are added.

Such changes in frequency and distribution occur in, for example, seedling root tips of *Trifolium, Raphanus, Zea*, and *Sorghum* (Seagull 1983). These changes may be due to formation of additional plasmodesmata near already-formed ones (Jones 1976). Somewhat similarly, in the cambial fusiform cells of *Sorbus aucuparia*, the plasmodesmata are widely separated in the pit fields of the radial cell walls, whereas, after cell enlargement and secondary-wall deposition in the differentiating fiber-tracheids, they occur exclusively in densely packed, eccentrically located, clusters on the pit membrane in local thickenings of primary-wall material. Most radial cambial walls have undergone repeated periclinal divisions and radial expansion growth, and thus have been rebuilt many times; therefore, the plasmodesmata in these walls are unlikely to have arisen primarily during cell-plate formation during the relatively infrequent, anticlinal divisions. Rather, these plasmodesmata arise secondarily (Barnett 1987).

In the cambial zone (Sect. 18.1) and its periphery, where fibers differentiate (Sect. 19.2.2), plasmodesmata eventually seem to develop between nonintrusively growing cells and adjacent areas of wall newly laid down by intrusively growing cell tips (Carr 1976). It is unclear how these, and other secondary plasmodesmata, are formed. How some new plasmodesmata may be formed was revealed by studies of graft unions between partners chosen such that the graft interface could be located by means of species-specific markers (Kollmann et al. 1985). Such studies with *Vicia* and *Helianthus* indicated that many "half-plasmodesmata" traversing the wall of only one "partner" cell occur at the first step of the formation of new plasmodesmata. The second step is union of opposite, appropriately aligned, halves. Misaligned halves may remain as permanent half-plasmodesmata.

12.2.3.2 Symplastic Growth and Intrusive Growth.

The coordinated growth of cells within organs has been called **symplastic growth** (Priestley 1930). Unfortunately, the terms **symplast** and **symplastic** are also widely used — sometimes interchangeably with **symplasm** and **symplasmic** — to refer to the continuum of protoplasts that are interconnected through plasmodesmata (Sect. 1.5). To promote clarity, we have adopted the rigorous usage advocated by Erickson (1986), whereby "symplastic growth" is a kind of coordinated growth, and "symplasm" and "symplasmic" refer to the interconnected protoplasts. Symplastic growth is discussed further in Section 12.3.

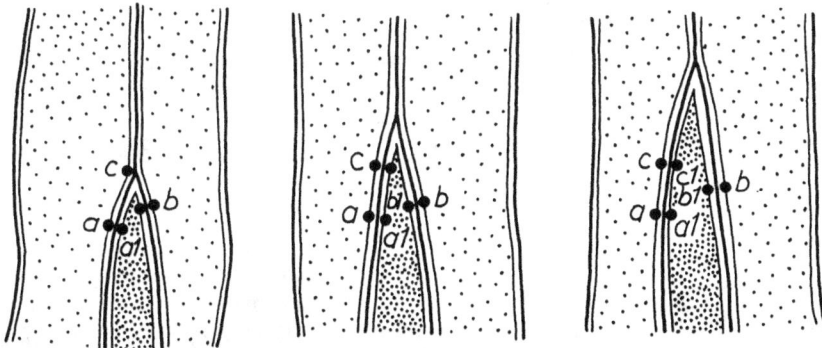

Fig. 12.3. Successive stages of intrusive growth of a cell tip between two neighboring cells. Temporal progression is from *left* to *right*. Note that pairs of points identifying material features, such as *a* and *a1*, which are initially opposite each other, maintain the same relative positions as the intruding cell tip advances. Similar intrusion can occur along cell edges

Neither in symplastic growth nor in **intrusive growth** do contiguous walls move (slide or slip) with respect to each other, and though new contacts can be made during intrusive growth, they do not result from slippage of existing opposite walls one past the other. Intrusive growth occurs only at cell tips or edges (Fig. 12.3). An intrusively growing cell tip or edge is directed toward the middle lamella between two neighboring cells, and is commonly assumed to act as a wedge (Wenham and Cusick 1975). In fact, however, there is probably no wedge-like action in a mechanical sense. Rather, a precondition for intrusive growth seems to be that the tissue be under tensional stress in a direction perpendicular to the direction of the incipient growth, at least locally. The cells then tend to separate if the middle lamella between them is dissolved. For example, in laticifers, the intruding edge may grow between neighboring cells after the edge has dissolved the middle lamella enzymatically by secreting pectinase (Wilson et al. 1976).

The narrow band of new wall continually being formed by the advancing edge of an intrusively growing cell almost immediately becomes glued to the walls of the neighboring cells by pectic materials. These substances form new middle lamellae of ordinary appearance. Secondary plasmodesmatal connections can be established across them (Carr 1976).

Cells in many tissues grow intrusively. A major mode of increase in diameter of vessel elements is intrusive growth along their edges (Sect. 19.2.1.3). Intrusive growth also occurs in the cells surrounding the vessel elements. Intrusive growth of cell tips is important in development of wood fibers and tracheids, as well as of other fibers, sclereids, and laticifers, and in various aspects of cambial dynamics (Sect. 18.4.3).

12.3 Patterns of Symplastic Growth at Supracellular Levels

12.3.1 Symplastic Growth and Cell Patterns

Because meristematic cells generally are cemented together along their walls, the structure of the meristematic tissue may be visualized as a three-dimensional meshwork in which the filaments of the mesh are the contact edges of the cells. Less easy to visualize in this way are differentiating tissue systems such as mesophyll or aerenchyma, in which there is extensive intercellular space. Instead, such tissue forms a loose meshwork, within which the "rules" of symplastic growth seem to have been suspended or modified.

During symplastic growth of meristematic tissues, the cell-wall meshwork is extended and, commonly, deformed. Obviously, the meshwork does not submit passively to this growth as sponge rubber does to bending and stretching. Rather, active cellular processes are involved in growth: new molecules are introduced between the old in the walls, and new cells form within the spaces previously occupied by mother cells in the mesh.

Even after several generations of cells have been formed and have grown symplastically within the mesh, lineage groups of cells can often be recognized by analyzing local geometric relations in the meshwork of cell-contact edges. Even if cell-lineage relations cannot be recognized with certainty, the wall structure can be used to confirm probable lineages ("families" or "packets") of cells. Because each daughter protoplast deposits a new lamella of primary wall on all of its faces, the generational status of a wall in a lineage is revealed by its thickness and lamellar structure (Mahmood 1990).

In an organ that is growing primarily in one direction, the meshwork of cell edges is being extended primarily in that direction, as in the elongation zones of primary roots and shoots. If the shape of a root or shoot tip, and the shapes of its constituent cells located in equivalent zones, do not change with time, then most new partitions must be oriented transversely to the direction of growth. As a result, the cells tend to be arranged in rows or files (sometimes called ribs), as can be seen in longitudinal sections of primary roots. In stems, if internodes continue to elongate after cells in files have stopped dividing transversely, as is typical of phloem fibers, then cell length will show a positive correlation with internode length.

If, on the other hand, meristematic cells grow and divide in two dimensions rather than in one, the progeny cells form layers rather than files. Such development commonly occurs, for example, in the formation of bud scales from cataphyll primordia (Sect. 16.6). Because of such relations, the arrangements of cells in tissues and organs contain information about the cell divisions and cell growth that produced the arrangements. The cell-wall network is a record of where the organism is in its ontogeny.

Attempts have been made to explain tissue geometries in terms of the physical laws governing bubble arrangements in foams, and also by formulating developmental algorithms. However, these approaches may be valid only for some tissues under certain conditions.

A system of bubbles is known to assume a state of minimal surface energy. If the bubbles are of uniform size, the surface energy is minimal if: (1) the bubbles are tetrakaidecahedrons (14-faced solid figures); (2) the various partitions form angles of 120° with each other (i.e., three cells meet along a common contact edge); and (3) the partitions between two bubbles have minimal area, and hence are flat (Lord Kelvin, as Thomson 1887; Dormer 1980). In contrast, if two bubbles of unequal size come into contact, the partition wall between them is concave on the side of the smaller bubble, though the walls still meet at an angle of 120°.

Although the resemblance of cell arrangements to bubbles in a foam is notable in certain tissues, such as stem cortex and pith, the systems for regulating cell division are generally much more complex than they would be if they depended only on surface-energy considerations. Thus, although the partition wall between cells of different sizes is usually concave on the side of the smaller cell, as in bubble systems, many dividing cells behave quite differently from bubbles in other respects. For instance, the area of a new partition wall, though minimal in many tissues, is maximal between cells in the early stages of provascular differentiation, and the shape of these cells rapidly changes from isodiametric to elongate. In addition, cambial fusiform cells divide periclinally in a plane of maximal area, and divide anticlinally (pseudotransversely) in a plane of intermediate area. Division in a plane of minimal area is rare in fusiform initial cells (Sect. 18.4.2).

In addition, the principle of only three cells meeting along a common edge, and hence forming an angle of 120° between walls, is violated in some tissues. For example, in developing aerenchyma, the cells commonly divide such that the partitions in two layers are superposed — and four cells briefly seem to meet along a common edge. These cells, however, soon round off their corners and pull away from the initially common edge.

Some developmentalists have formulated algorithms in an attempt to explain tissue geometries. This approach, as used by Lindenmayer (1975, 1984) and Korn (1984), emphasizes rules acting at the cellular level. A rigorous algorithm of symplastic growth should also invoke rules of growth and cell division at the tissue and organ levels. These seem more difficult to formulate, however. Tensor analysis offers possibilities (see later discussion).

Because the regulation of symplastic growth is so complex, we need further ways to study it. One approach is to study the relation between stress trajectories in tissues (Sect. 12.4.6) and planes of cell division. The plane of minimal shear stress may be the most general determinant of the plane of cell division in symplastically growing tissues, such as shoot tips (Lintilhac 1974b). Other stress-related factors probably are involved in regulating volume growth of neighboring cells in a coordinated or "symplastic" way.

12.3.2 A Description of Symplastic Growth in Terms of a Growth Tensor; Periclines, and Anticlines

Symplastic growth occurs most notably in root apices, in dome-shaped shoot apices, and in some fruits. It may involve growth of different parts of the cell-edge meshwork in the same direction at different rates, as well as parts growing in different directions at different rates. The different rates are coordinated with, and subordinated to, the realization of the overall pattern and shape of the organ that is being generated.

Tensor calculus is an orderly, rigorous method of describing these different growth rates. This calculus allows the growth of an organ to be represented as a second-order tensor quantity, which may be called a **growth tensor** (Hejnowicz and Romberger 1984). A brief introduction to this approach follows:

A Note on Tensors — Many physical quantities and relations can be expressed as tensors of the zero, first, or second order. A **scalar quantity**, which is a zero-order tensor, is fully determined by a number alone, without reference to a **coordinate system**. A **vector quantity**, which is a first-order tensor, is specified jointly by a number giving a magnitude and by a direction with reference to a defined coordinate system. In a specific three-dimensional coordinate system, both the length (magnitude) and direction of a vector are defined by a set of three numbers. The values of these numbers depend on the coordinate system, and they "transform" according to strict mathematical rules if the coordinate system is changed. A generalized vector, in which the field includes all magnitudes and all directions from any point, is a **second-order tensor,** or simply a **tensor**. Such a tensor is defined in a specific three-dimensional coordinate system by a 3 x 3 array of 9 numbers. These 9 numbers constitute a matrix, A_{pq}, in which both p and q successively assume values of 1, 2, and 3. Readers not familiar with the meaning and use of tensors and tensor terminology may wish to consult Feynman et al. (1964, Vol. 2, Chap. 31), and Appendix A in Hejnowicz and Romberger (1984).

Specification of a growth tensor requires the choice of a coordinate system and a knowledge of the **growth displacement velocity (V)** of points, such as cell-wall intersections, in growing organs, as a function of their positions.

In an orthogonal coordinate system, each of the three terms along the **major diagonal** (which includes the indices 11, 22, and 33 of the 3 x 3 matrix representing the growth tensor) is the relative elemental growth rate (linear) of a line element ($RERG_l$) in the direction tangent to the corresponding coordinate line at the considered point. That is, the three components of the major diagonal of the matrix represent $RERG_l$ in three orthogonal directions. It is noteworthy that the sum of these components is the same for all orthogonal coordinate systems. That sum, at a specified point, represents the relative elemental rate of growth in volume, $RERG_{vol}$, at that point. It is a scalar quantity.

Specification of the field of growth displacement velocities, V, of selected points in an organ is simplified if the coordinate system is chosen appropriately — i.e., with reference to the form of the growing organ, as follows: if $RERG_l$ at a point is not isotropic, then there must be three mutually orthogonal directions along which $RERG_l$ attains extreme values (i.e., the rate in a particular direction is higher or lower than the rates in all neighboring directions). These are the **principal directions of growth**. Between successive points along a coordinate line (which in natural coordinate systems is usually a curved line), these directions change in a continuous way. Thus, one can draw lines tangent to the principal directions. These tangents are the **trajectories of the principal directions**. Three such mutually orthogonal principal trajectories pass through every point in the organ. The surfaces tangent to pairs of principal trajectories can be denoted as **principal surfaces**. There are three orientational types of principal surfaces — and one of each type passes through every point.

We now advance an empirical statement, to be substantiated later: at every point on an organ surface, two principal trajectories are tangent to the surface and a third trajectory is normal to it. It follows that the organ surface coincides with the principal surface of one orientational type. This principal surface is **periclinal**. The principal surfaces of the two remaining orientational types are normal to the organ surface. According to their orientation relative to the organ axis, one of these latter principal surfaces is **anticlinal longitudinal**, the other **anticlinal transverse**. By analogy, the two types of principal trajectories that are tangent to the organ surface are **periclinal longitudinal** and **periclinal transverse**. The third type of trajectory, which is normal to the principal surface, is **anticlinal**. These meanings of periclinal and anticlinal encompass the traditional meanings. Note that there are two types of periclinal trajectories, but only one anticlinal type.

The growth tensor is least complex if the coordinate lines of the chosen system coincide with the principal directions of growth. In such a system, the major diagonal components of the corresponding matrix express $RERG_l$ in the principal directions.

In the cell-wall meshwork visible on the surface of, for example, a shoot apical dome, there usually are indications of two types of principal directions of growth — periclinal longitudinal and periclinal transverse. Usually, also, enough additional evidence can be deduced from the cell-wall pattern visible in an axial median section to select a three-dimensional orthogonal coordinate system that will yield an applicable

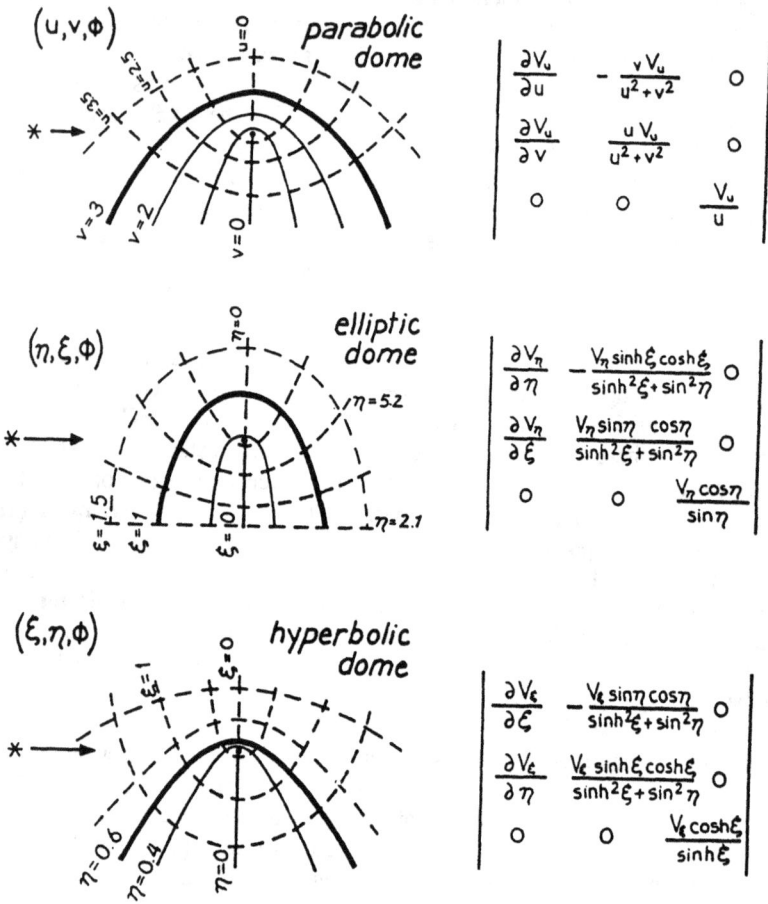

Fig. 12.4. Dome-shaped shoot-apical meristems, orthogonal curvilinear coordinate systems applicable to them, and corresponding growth tensors. Several curvilinear coordinate systems, including paraboloidal, and elliptic and hyperbolic variants of prolate spheroidal types, are superficially compatible with dome-shaped organs. Further, hemispheroidal domes may be treated as special cases of elliptic domes. The assigned order of coordinate variables (for example, u, v, ϕ in the paraboloidal system) is such that the first variable for each coordinate system (u, η, or ξ) applies to position along the periclinal coordinate curves, the second (v, ξ, or η) to position along anticlinal coordinate curves, and the third (ϕ in all cases) to position along the latitudinal coordinate circles (azimuthal around an axis and not represented in these two-dimensional drawings). The *boldly drawn curves* represent dome surfaces. The geometric focus of each dome (the focal point of the parabolas, ellipses, or hyperbolas of a system) is on the axis at the level marked by an *asterisk* and *arrow*. Applicable 3 x 3 matrix forms of the growth tensor are written to the *right* of each dome type diagram. As mentioned in the text, the sum of the three terms of the major diagonal (*top left* to *bottom right*) of each tensorial form represents the relative elemental growth rate in volume (RERG$_{\text{vol}}$) at any point specified by the coordinate variables. If each tensor is multiplied by a scaling factor derivable from its form, the indicated RERG$_{\text{vol}}$ values are scalar and comparable for all forms. (Redrawn after Hejnowicz et al. 1984)

form of the growth tensor. We note, however, that the shape of a specific *surface* profile, as seen two-dimensionally in an axial median section, does not in itself specify an orthogonal system that will generally coincide with principal directions inside the organ. That is, the organ surface does not unequivocally define the system having coordinate lines coinciding with the trajectories of the principal directions of growth at points within the organ — though the systems may coincide at the surface. One can imagine *several* orthogonal curvilinear coordinate systems that are compatible with, for example, a dome-shaped organ surface (Fig. 12.4). To determine the most appropriate coordinate system, one must know the course of periclines and anticlines inside the organ.

From tensor analysis of organ growth, we already know that if the pattern of principal trajectories is steady over time, then segments joining recognizable points of the wall meshwork, and chosen so as to be tangent to the principal trajectories (hence mutually orthogonal), will maintain their orthogonality during growth. This means that, if a set of points that delimit mutually orthogonal segments change their position during growth, the segments delimited after growth will still be orthogonal, whereas other segments, initially orthogonal but not tangent to the principal trajectories, will lose their orthogonality during growth. It is also known that patterns can be recognized in the cell-wall meshwork of every plant organ in which growth was, or is, distinctly anisotropic, and that the orthogonality of the elements of this pattern is maintained during growth. It was, in fact, for the express purpose of describing such patterns that Sachs (1874) and coeval botanists introduced the terms "pericline" and "anticline".

It follows that surfaces and pattern elements that show a persistent orthogonality represent principal patterns or trajectories. It also follows that the planes of cell division typically are normal (or nearly so) to the principal directions of growth; consequently, periclines and anticlines can be recognized in the arrangement of cell walls. Of course, we are here speaking of the general, the normative, features of cell-wall arrangements — not of the orientation of a particular face of a cell.

Orthogonal curvilinear coordinate systems having coordinate lines that, at any point, coincide with trajectories of principal directions (that is, that are periclines and anticlines) are called **natural coordinate systems** (Hejnowicz 1984). In these systems, the growth tensor assumes a relatively simple form (Fig. 12.4). Further, vector fields necessary for the calculation of growth rates throughout the organ can be experimentally determined in a natural coordinate system (Sect. 14.1.6). This approach to growth analysis seems promising.

12.4 Positional Information and Regulatory Factors in Differentiation and Pattern Formation

A large, complex organism is a structural hierarchy, in which some of the levels are: organism, organ, tissue, and cell. The development of such an organism requires that the differentiation of its constituent cells, tissues, and organs be determined mostly by epigenetic information coming from other cells, tissues, and organs — that is, by

positional information. If there were no such information, there could be no orderly anatomy; and higher organisms with chaotic or random anatomy are inconceivable. Thus, it seems to be a principle of plant development that a cell's fate is a function of its position. This principle is summed up in the aphorism: "The plant forms cells, not cells the plant", which is often attributed to de Bary (in lectures given in about 1880), though its exact origin and original wording are unclear (Barlow 1982).

By influencing cell determination, position within the organism exerts control over pattern. There are patterns at every level of the structural hierarchy. For example, the distribution of xylem rays is a tissue-level pattern; the shape of guard cells a cellular pattern; and the distribution of secondary-wall thickening in xylem elements a subcellular pattern. Is there in living organisms a set of general principles that controls genesis of patterns, or are different patterns formed in different ways?

Note that some kinds of patterns in nature, if viewed developmentally, are not as different from one another as they at first appear. For example, although each local landscape pattern differs in detail from all others, and obviously is a product of its unique geographic position, we know that a set of basic factors governs landscape development everywhere. These factors include erosion and other geological factors, temperature, precipitation, vegetation, and human activities.

Similarly, the basic unity of all life forms suggests that there may be a common set of principles underlying **pattern formation** in all organisms, though the particular subset of principles followed in forming subcellular patterns may differ from those followed at higher levels of the structural hierarchy. The same principles of pattern formation could apply to, for example, the vascularization of leaves and the innervation of mammalian limbs or bird wings (see Wolpert 1981).

Results of experiments on epithelial mesenchyme in insect limbs have been interpreted as indicating that developing cells receive positional information relative to a virtual two-dimensional coordinate system of the polar type (French et al. 1976; Sect. 12.4.2.1). In such a system, one coordinate gives position on a circumference; the other gives distance from a center. Successive sectors in a clockwise or anticlockwise direction around the circumference are somehow distinguished one from another — "numbered", in a sense — and the direction of numbering imparts a chirality. Pattern formation in developing organs of other taxa may also be understandable by reference to polar-coordinate systems (but it is not obvious how these systems could obtain positional information from a concentration gradient of a morphogenic substance; Sect. 12.4.2.2).

In many plant species with spiral phyllotaxy (Sect. 15.2.1), the "handedness", or chirality, of the spiral in a lateral branch is commonly opposite that in the main shoot. That is, lateral shoots tend to be antidromous (Raunkiaer 1919; Wardlaw 1965; Dormer 1965) relative to parent shoots. This is easily explained on the basis of polar coordinates, and is what we would expect of lateral shoots that were initiated within the morphogenic field of a terminal shoot apex. However, no such relation applies to lateral shoots that are adventitious or initiated from axillary residual meristems. These laterals, initiated outside the morphogenic field of the shoot apex, may be either antidromous or homodromous.

The response to the positional information received at one stage affects the information available during the next stage. That is, the hierarchical level of detail of positional information to which differentiating cells respond generally decreases as developmental decisions are made and the remaining options become increasingly limited. A corollary is that specification of the pattern is at a higher hierarchical level than are the responses to the information. The specificity of different patterns results mostly from genomic specificity of response to the positional information. Certainly, there is not a basically different map for each stage of development, though the maps at different stages may differ in included detail.

Evidence supporting these views includes the harmonious development of periclinal chimeral shoots formed from species of different genera (Sect. 14.1.5). Examples are *Crataego-Mespilus*, *Laburno-Cytisus*, and *Pyro-Sorbus*. The developing cells in the chimeral apex respond to positional information in their normal, species-specific ways. For example, in *Crataego-Mespilus*, in which the outer one or two layers of the tunica consist of *Mespilus* cells, the epidermis is indistinguishable from that of typical *Mespilus*. Nevertheless, the chimeral shoot apex develops in a coordinated, symplastic manner, indicating that the map for this apex is very similar to the maps for the separate genera. We can speculate, therefore, that the high hierarchical level of the map provides a basis for the obvious superficial patterning similarities of the organisms within supraspecific systematic groups. What kinds of systems specify positional information? Several hypotheses are discussed in what follows.

12.4.1 Axiality, Polarity, and Differentiation

Morphological, developmental, and physiological data indicate that at least one axis is assumed by, or imposed upon, almost every cell, on the basis of its position in the plant. Figuratively, establishing polarity adds a directional arrow to such an axis.

Usually, there are three mutually orthogonal axes, and almost every cell eventually behaves as though it were responding to direction along at least one of these axes. However, many functional and developmental events, including cell growth and division, are oriented in relation to an established axis without being limited entirely to one direction.

In a hierarchy of spatial ordering, from randomness up toward increasing order, axiality may be at a lower level than polarity. Can there be polarity before there are poles of an axis? In most developing systems, however, an axis probably assumes a polarity almost at the moment it is formed. Which comes first, axis or polarity, may be only a semantic question.

A plane normal to an axis can be constructed through any point on that axis. Additional straight or curvilinear axes orthogonal to the first axis can also be drawn either within or tangent to this plane at that point. In plants, the orientations of the axes along which a variety of morphogenic events occur can often be recognized in organ symmetry and in the orientations of the major dimensions of the cells. One of the cell axes usually coincides with the axis of organ polarity, which may indicate the direction of the principal stress acting on or in the cell (Kotenko 1986).

Most cell divisions are oriented with some reference to cell polarity (Bünning 1952; Kotenko 1986). In fact, the polarity of the mother cell inevitably is transferred to the daughter cells, because most of the faces of the daughter cells are inherited from the mother cell, and because a part of the maternal cytoskeleton (Lloyd and Barlow 1982), with its arrays and strands of microtubules, microfilaments, etc., is also passed on.

Polarity may also be manifested by unequal distribution of cytoplasmic components during cytokinesis, particularly if the partition wall is transverse in relation to polarity. Unequal distribution of cytoplasmic material may be important in allowing the cells within a lineage to take divergent developmental paths. A mechanism that might lead to unequal, but controlled, distribution of cytoplasmic material during cytokinesis is an electric field strong enough to cause electrophoretic migration of charged molecules or small organelles. Electric currents and oriented electric fields have been reported in egg cells of *Fucus* and *Pelvetia*, in some animal eggs (Jaffe 1981; Robinson 1985), and in embryos of *Daucus carota* (Brawley, et al. 1984). The current may be a consequence of unequal membrane permeability at the opposite poles of the cells.

Though various types of polarity may coexist in a cell, tissue, or organ, an axial, root-shoot polarity is primal in higher plants. This polarity is manifested by a basipetal flow of auxin (Kaldewey 1984; Sects. 11.2; 12.4.1) and by physiological and developmental responses to that flow. However, root-shoot polarity cannot be explained simply as a manifestation of a concentration gradient. As the poles of an axis of a tissue or organ become increasingly distant, separated by many thousands of cell lengths, the concentration or electric gradients along individual cells become vanishingly small. Only a local vectorial property of the cells, in the form of a cytoskeleton of oriented and polar elements or other differences between opposite ends of a cell, can account for undiminished manifestations of polarity. The concept of feedback based on the divergence between the vector of signal flow (auxin flux) and long-established polarity (as borne, for example, in the cytoskeletons of the cells) invokes in each cell a dynamic polarity vector, rather than just a static, inherited polarity (Sect. 12.4.3).

12.4.2 Morphogens, Phytohormones: Their Possible Modes of Action

In higher animals, a nervous system provides a means of rapidly transmitting bioelectric signals to practically all living tissues, and a system of circulating blood and lymph permits hormones produced by specialized cells or organs (glands) to be distributed throughout the organism. In contrast, plants have no endocrine glands and no true circulatory or nervous systems. The origins of the plant regulator molecules and the means by which they are translocated to their target tissues are obscure.

Despite the inadequacy of knowledge about phytohormones, a great mass of barely interpretable lore concerning, for example, the multiple effects of auxin, has accumulated over the decades — and from this, a broad, auxin-based dogma has evolved. But, are the "plant growth regulators" and phytohormones really "regulators" in themselves, or only the "servants" of more basic systems that control the synthesis and translocation of these morphogens, and the sensitivity of receptor tissues to them? Phytohormonal regulation requires not only an emitter (a synthesizing site) and a receptor, but also a set of systems for controlling synthesis, transport, and inactivation. All these systems must be coordinated if overall organismal development is to be regulated. The hierarchical level at which phytohormones act on cells and tissues may be lower than the level at which plant morphogenesis is coordinated and controlled.

It is at this possibly lower level — the level of action — that phytohormones conventionally are discussed. Thus, a phytohormone is generally considered to be an endogenously produced substance that: (1) after arriving at a particular locality, seems to regulate cellular processes, (2) produces its effects on cells remote from those in which it originated, and (3) characteristically acts at very low concentrations. Some substances meet all three of these criteria and thus qualify as phytohormones. Other substances meet only two. For example, a substance may have effects in the cell that produced it, and these effects may not be distinguishable from those evoked by the same substance coming from elsewhere in the plant. In addition, a soluble, diffusible substance such as sucrose occurs at high concentrations, yet can influence vascular-tissue differentiation at sites remote from its origin. To accommodate these substances, the broader term "morphogen" is often used. Phytohormones such as auxin or cyto-kinin can be considered as morphogens of a particular type.

There are, moreover, many synthetic substances — commonly, artificial analogs of natural phytohormones — that partially fit the definition of a phytohormone. Without attempting to be fully rigorous, we apply the term "phytohormone" to naturally occurring substances and the term "plant growth regulator" to synthetic analogs that have phytohormone-like activity.

There are at least five classes of phytohormones: auxins, gibberellins, cytokinins, abscisic acid, and ethylene. Most living cells have the genetic information necessary to synthesize all of these, and, in theory, could produce the substances for their own use. However, local production and use of phytohormones by individual cells would contribute nothing to the whole organism's morphogenesis unless that production and use were coordinated by systems at the organ and organismal levels. Regulation of phytohormones at the organismal level is based on differences among cells in the use of genetic information. Most cells use only a small, but closely regulated, fraction of the genetic information relating to synthesis of, or responses to, phytohormones. Some cells, on the basis of spatiotemporal factors, synthesize and export a particular phyto-hormone, while other cells do not synthesize the substance but may become "potentiat-ed" to respond developmentally to the arrival of the substance from elsewhere.

12.4.2.1 Auxin Flow and Differentiation. Auxin is much involved in differentiation, though it is unclear where in the hierarchies of control it exerts its effects. One effect of auxin is to stimulate elongation growth of plant organs, via a metabolically con-trolled increase in extensibility of primary cell walls. This increased extensibility allows a plastic deformation of the walls (Taiz 1984), which may be why in both monocots and dicots the epidermal cell layer has a crucial role in growth responses to auxin. Some emerging concepts of this aspect of auxin action emphasize auxin's control of orientation of microtubules, and thus, indirectly, of microfibrils deposited in the cell wall (Bergfeld et al. 1988). In addition to its role in cell-wall extension, auxin is also involved in inducing differentiation of vascular tissues.

Morphogens such as auxin are usually considered as mediating or directing morpho-genesis, either by being present at a suitable concentration, or via a gradient of their concentration. However, studies have shown that auxin-induced differentiation of vessels (Sachs 1981a, 1984) is related not to the mere presence of a gradient of auxin, but rather to a polar flux of auxin. In addition, auxin flux is required during most or all of the process of vessel differentiation, and this process neither directly uses, nor produces, auxin.

Several types of evidence indicate that the rate and direction of signal flow, rather than the presence of auxin or a concentration gradient, induces vessel differentiation. For example: (1) as seen in serial tangential sections, auxin-induced vessels or vessel elements in some secondary xylem form closed rings (circular polarity: see later); (2) auxin can induce differentiation of vessels if it is polarly transported in the absence of a gradient; and (3) vessels many meters long are formed in tree stems in short time periods, even though gradients along individual cells can hardly be significant (Sachs and Cohen 1982).

How flux could control differentiation is not evident. Sachs (1984) proposed that if a polar flow induces in cells an enhanced capability for transport along the pathway of that flow, a positive feedback loop between flow and polarity will result. In this loop, flux causes an increased polarity, which in turn facilitates flux through the cells. Those strands of cells that develop this feedback loop more rapidly than do neighboring ones may drain flowing auxin away from these neighbors. Thus, flux and differentiation could become channelized, and strands of vascular conducting elements could develop (Sect. 16.3.6).

Polarity, which is a manifestation of directional components in the organization of an organ (Sect. 12.4.1), provides the rudiments of a coordinate system, with a primary axis. The major vectors of "channelized" auxin flux, which may be regarded as a signal flow, may form the basis of a coordinate system relative to which other types of differentiation are established. The hypothesis that signal flow is a basis of orderly morphogenesis seems to have its greatest explanatory potential in large plants, such as trees, in which the distance from root apices to shoot apices is long.

In a shoot, the differentiation of new primary and secondary vascular tissue is controlled by stimuli emanating from the leaves. The main component of these stimuli seems to be a basipetal flux of auxin. For this signal-flow system to operate, a continuous route from leaves to the considered site is necessary. In experimental systems, leaves can often be replaced by an exogenous source of auxin. In incision experiments involving application of exogenous auxin, the "drainage" character of the system of primary vascular strands induced by auxin becomes quite apparent. Polarity is stable if the applied auxin flows with the established polarity; but, if auxin flow assumes a new route, polarity may follow it and assume a new direction. In extreme cases, the polarity of some differentiating fusiform xylem cells is reversed (Sachs 1981b, 1984).

One can imagine bringing together the ends, or poles, of a polarity axis of a tissue to form a circular "axis". Such an "axis" could be "polar" in either a clockwise or an anticlockwise direction. Polar-coordinate systems can be based on such axes (Sect. 12.4). Sachs and Cohen (1982) attributed the development of circular vessels to a circular flow of auxin, somewhat akin to eddy currents. If ordinary, one-way polar flow of auxin is impeded, as in a cambial zone at the basal end of a cut stem segment, circular vessels may be formed. These are often small and may consist of only two vessel members joined at their perforation areas. Circular vessels also occur in some intact stems above axillary buds, and in branch junctions of trees (Hejnowicz and Kurczyńska 1987; Lev-Yadun and Aloni 1990). Circular signal flow provides the simplest explanation for the formation of these vessels.

Circular polarity has the unusual property of being centered around a point of singularity, where there is no flow and "direction loses its meaning". (At the North Pole, all directions are south). Such points may possibly serve as centers of new morphogenic fields — especially those that involve new "polar" coordinate systems.

12.4.2.2 Gradients of Morphogens. Most differentiating cells are strongly influenced by substances produced both locally and elsewhere in the plant. Collectively called morphogens, these substances may move short distances by diffusion along gradients from sites of origin, release, or activation (see also Sect. 11.2) to sites of action. Gradients of morphogens, if they were sustained and were to interact with each other, could constitute a plausible basis for formation of morphogenic patterns in time and space; that is, they could determine positional value.

How could gradients of morphogens be sustained? One can postulate systems regulating synthesis or release of morphogens at source sites, along with diffusion processes that would result in concentration gradients of the substances as they moved from source to site of action. A set of gradients of several morphogens could specify a pattern of some complexity. Meinhardt (1984), for example, speculated on the pattern-forming potential of a hypothetical system of synthesis and depletion of two morphogens, of which one, an activator formed autocatalytically, promotes synthesis of a second, an inhibitor. This inhibitor suppresses synthesis both of itself and of the activator. It also diffuses away more rapidly than the activator. Thus, the field of activator concentration lies within a larger field of inhibitor concentration, which acts to stabilize the activator and to maintain a centrifugal gradient.

In theory (Turing 1952), such systems of autocatalysis and inhibition could generate regular patterns of activator concentrations (or ratios of activator/inhibitor concentrations) capable of self-regulation, self-regeneration, and migration. Activator maxima could serve as signals evoking determination of receptive cells. And, if the cells were already determined, the signal would be shifted to contiguous cells, causing them to become determined. Thus, a strand of cells in a primordial leaf blade along the path of a moving activator maximum could be determined to become provascular, then vascular (Meinhardt 1984).

It is possible that the ratio of effective concentrations of two morphogens at a site, rather than the concentration or gradient of one, defines positional value in relation to a pattern. If so, environmental fluctuations that evoked changes in absolute concentration of the morphogens might have little effect on the *ratio* of concentrations, and thus little effect on positional values. Such a system involving auxin and sucrose may operate in the vascular cambium, with a high auxin/sucrose ratio prevailing in the differentiating xylem and a low ratio prevailing in the differentiating phloem (Warren Wilson and Warren Wilson 1984). Such proposed systems, though speculative, illustrate how interactions of gradients of several morphogens might specify patterns.

12.4.3 Wave Phenomena in Differentiation

As already mentioned, biochemical feedback systems, along with diffusion, can, in theory, generate gradients of two morphogens that interact in ways that may produce stable spatial patterns. This concept addresses only part of the problem of control of morphogenesis, however, because events in differentiation happen sequentially at the same location in the organism. It is possible that the sequence of events is coupled to a cyclic process, in which time is specified by the phase of the process and also, on a broader scale, by the number of cycles that have elapsed. Such a cyclic process is inherent in spatiotemporal patterns based on waves. Wave-based processes could specify not only a regular pattern in space at a given instant, but also a cyclic change in pattern with time at a particular site. The concept of wave-based patterns is well founded in theoretical studies (Prigogine et al. 1975), and there is some evidence that these patterns occur in both plant and animal morphogenesis, as a consequence of nonlinear biochemical feedback and transport processes.

Results of a series of studies in our several laboratories have led us to conclude that in many, perhaps even all, trees that have well-developed vascular cambia, there are "behavioral" morphogenic waves, though not necessarily of the type just postulated. These migrate along the sheath of cambial fusiform initial cells, and are based not on changing concentrations of substances, but on changes in behavior of the cells in time and space. These waves are manifested by the nonrandom spatial distribution of the rightward (Z) and leftward (S) orientations of several types of morphogenic events in the cambium. The events include oblique anticlinal divisions and overlapping of oppositely directed, intrusively growing tips of fusiform initial cells. Local areas in which directional morphogenic events are predominantly S or Z are called S or Z **domains**, respectively (Sect. 18.5.1). Typically, these domains migrate acropetally along the cambium.

Research indicates that several cambial waves of differing length and velocity, but of the same period, can be superposed in the same locality. Superposition results in complex patterns of nonrandom distribution of S and Z morphogenic events. Common, but indirect, manifestations of these complex wave phenomena are the wavy grain patterns in the wood traditionally used for violin backs and for some fine cabinetwork. Formulation of a theoretical basis for understanding and interpreting these waves is of interest from a basic-research viewpoint (Hejnowicz and Romberger 1973, 1979). It seems likely that the slowly migrating waves manifested by cambial domains are merely indicators of more rapid, underlying wave phenomena that we can only begin to appreciate.

The superposition of waves of constant period but variable velocity results in the formation of a stationary pattern of amplitudes (an envelope) (Fig. 12.5). Now, let us consider an assemblage or bundle of longitudinal elements that can actively pump certain ions from outside to inside. In addition, some passive leakage of these ions can occur in the opposite direction.

If we assume that a cyclical change in properties of the elements is propagated along the elements such that the net movement of the ions into the elements is altered, it follows that cyclical changes in rates of ion pumping inward or ion leakage outward will be propagated along the elements (as happens in excited nerves). The concentrations of certain ions in the medium around the assemblage will then change cyclically — over space at an instant and in time at a point. There is a propagated wave of concentration change. This wave can be understood by analogy to the surface waves of water: at any instant, we can see cyclic changes of water level in space. But, if we stand in the water, we also become aware of its changing height over time.

If waves of change in net ion flux are propagated along the elements, the external ion concentration at a specific position will reflect the summation of changes in the several individual elements nearby. Let us assume that the propagated waves of change have the same period but different velocities. Their lengths will then be different.

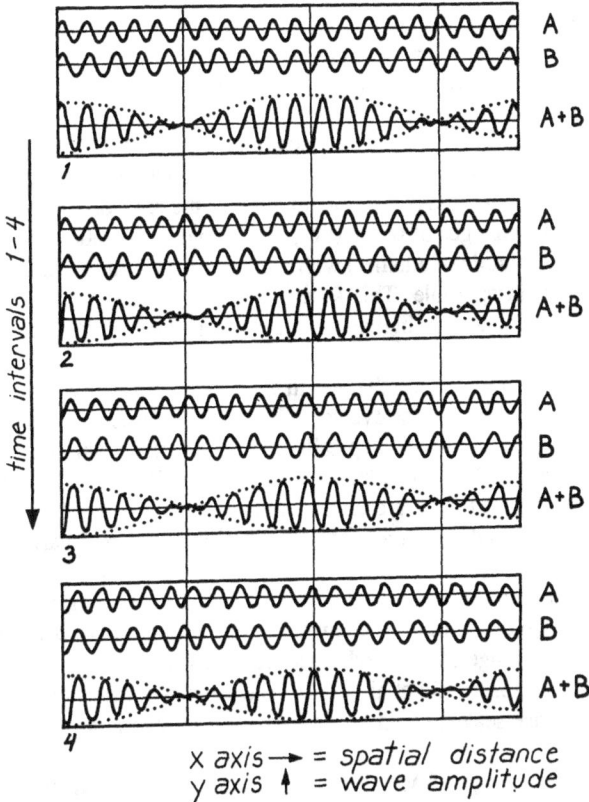

Fig. 12.5. Superposition of waves of constant period but variable velocity results in formation of *stationary* patterns of wave envelopes. Here, waves *A* and *B* have the same period, but move from left to right at different velocities. The composite wave *A*+*B*, formed by superposition, has a stationary pattern of amplitudes, within a wave envelope, as is illustrated by the time sequence (indicated by *arrow*) from drawings **1** through **4**. Within each drawing, the *x* axis represents the vector of wave propagation and the *y* axis represents wave amplitude. (After Hejnowicz 1980)

Where the phases of two waves are the same, the amplitude of cumulative change produced will be maximal. At sites at which these elements are in opposite phases, the cumulative changes produced will be minimal. In general, there will be a regular pattern of maximal and minimal amplitudes of cumulative change in the system.

Because the period of the waves propagated along the elements is the same, the pattern of amplitudes will be stationary, and would fit within an envelope (Fig. 12.5). Our hypothesis is that the envelope of such an amplitude pattern can provide positional information, or serve as a virtual "morphogenic map".

Because waves can be propagated through different elements at different velocities, waves of the same period need not have the same length. Wavelength may also be influenced by assemblage length, as the minimum or maximum of amplitude is likely to occur at the ends of the assemblage. Thus, one can imagine a basis for adjustment of the same morphogenic map to a range of sizes (assemblage lengths).

A morphogenic map based on superposed waves of this type would also have automatic temperature compensation. Even if periods changed with temperature, patterns of amplitudes (the envelope shape) would be unaffected. Such a system would have one other interesting property: at each point within the system, there would be a cyclic change of the superposed quantities, the summation of which determines the amplitude. If we interpret the pattern of amplitudes, the envelope shape, as a map, we should interpret the changes of phase (i.e., the change in the ratio of the amplitude of one wave to the amplitude of another) within the envelope as a clock. In this sense, the map and clock are inseparable. This feature has appeal when we seek to explain the integration of development in higher organisms in terms of virtual maps and clocks (but see Sec. 12.4.6).

The propagation of wave-like changes in the nonrandom, leftward (S) or rightward (Z) orientation of certain cellular events in the vascular cambia of trees is manifested in the grain pattern of the wood. Typically, the period of a grain orientation "wave", as recorded in the wood, is more than 10 years (Sect. 18.5.1). The orientations and spatial relations of wood cells are an "archival record" of morphogenic waves and events that have operated within and upon cambial initial cells at various hierarchical levels.

We propose that the set of interrelated phenomena underlying the waves that are finally manifested and recorded as changes in the cambium may collectively operate within a very broad range of time and space scales, as explained more fully in Section 18.5.5 (see also Zajączkowski et al. 1984). The same set of phenomena, with appropriate time and space scales, could thus be involved at several levels of the structural hierarchy.

The tremendous range of scale of morphogenic maps based on superposed behavioral waves gives this concept greater explanatory potential than the morphogen-gradient concept (Sect. 12.4.2.2), which has a rather narrow range of application. The morphogen-gradient concept cannot, for example, work at a scale in the range of the mean free path of diffusion — that is, in the micrometer range. This means that this concept cannot account for the formation of intracellular patterns such as those of secondary-wall thickening. A *fundamentally* different basis would be required for pattern formation on that small a scale.

In this discussion of morphogenic maps and waves, we have so far considered only the resultant amplitude pattern of superposed waves to be capable of transmitting positional information. It is also possible that such information is carried in the orientation of isophasic fronts of two- or three-dimensional waves. Wodzicki et al. (1979) found evidence of waves of changing concentration of diffusible auxin in the cambium of trees. They proposed that the spatial pattern of the vectors of wave propagation (which are normal to a constructed surface including all points in the same phase of the same oscillation) can form a morphogenic pattern map with near-stationary features.

For example, we would expect the propagated wave of auxin concentration to have different velocities in different layers of the cambial zone. The velocity is probably greater in the fusiform initial cells than in the differentiating xylem and phloem. Thus, the isophasic surface of the wave in typical basipetal auxin transport would be convex. The vector of propagation presumably is parallel to the longitudinal axes of the cells in the initial layer but makes an increasing angle with the longitudinal axes of the differentiating cells at progressively greater distances adaxial and abaxial to the initial layer. Zajączkowski and Wodzicki (1978) hypothesized that the angle between the isophasic front of the auxin wave and the longitudinal axes of cells provides positional information for cell differentiation.

This interpretation may significantly aid our understanding of the integrated development of large plants. Unlike much auxin research of the past several decades, which focused on cellular or molecular aspects, the auxin-wave theory of morphogenesis offers a broad conceptual framework. Obviously, individual cells and macromolecules have no "need" for phytohormones; organisms do. A directional (or polarizing) effect of auxin on cells within a large, complex organism is inherent in the auxin-wave theory (Zajączkowski et al. 1984).

12.4.4 Cell Lineages, Polyclonal Compartments, and Histogens

All the cells of a higher-plant body are derived ultimately from a zygote and penultimately from an embryo. The progeny of repeated cell divisions in the embryo and, later, in root and shoot meristems accumulate as assemblages of cell lineages or clones. Some assemblages are quite regular, as, for example, in the early embryo of many taxa and in root primordia of *Azolla* (Gunning et al. 1978). In these, there is a strict coincidence between the sequence of cell divisions, cell lineages, and formation of cell types.

The main method of studying cell-lineage patterns is clonal analysis (Poethig 1987) of heterozygotic plants having recessive mutations with visible, cell-autonomous phenotypes, involving, for example, pigment production. Loss of the dominant allele, and consequent disclosure of the recessive phenotype, may occur spontaneously or may be induced by irradiation or chemical treatment. A cell in which this disclosure has occurred can give rise to an identifiable cell lineage. Such studies provide information about the number of cells in a primordium at various stages of development, about the duration of growth phases, and about the rate and orientation of cell division.

Because it is probable that there are principles of morphogenesis common to all organisms, we will take an idea from research on differentiation of epithelium in insects (Garcia-Bellido 1975) and examine its applicability to plant development. This idea is that compartments composed of polyclones of cells are formed early in organismal development. In larvae of a strain of *Drosophila* heterozygous for some epithelial characters, Garcia-Bellido's group induced somatic recombinations by X-ray treatment.

These recombinations led to homozygotic cells, which then produced clones. In the adult insects, the positions and dimensions of the clones were studied in relation to the ages of the larvae when they were irradiated. We give an outline of Garcia-Bellido's concepts and evoke a hypothetical layer of cells as a model of the developing epithelium (Crick and Lawrence 1975).

Imagine that all cells of the hypothetical layer are at first heterozygous for a condition we will call "white". Then, by recombination, the homozygous state or mark that we call "black" appears in some cells. "Blackness", which has no effect on further cell development, is transmitted to all the cells that originate from a cell so marked. Thus, "black" clones appear in the layer. The earlier the cell is marked, the larger the clone that is eventually derived from it. The clones remain spatially discrete in spite of the innate ability of animal cells to creep over each other. The borders of early clones are not quite smooth or regular. The size of each clone tends to increase along the direction of fastest growth in the layer. Eventually, however, smooth lines develop, separating different regions of the layer, and those "black" clones that subsequently originate do not cross these lines, although the lines may pass through, and thus "partition", pre-existing clones. The position of these lines is the same in comparable layers of different individuals.

The regions delineated by these partitioning lines generally include cells of several clones and are therefore called **polyclonal compartments**. During further development, new partition lines may split a polyclonal compartment into two, then into four, etc. There is thus a generational hierarchy of compartments. A border between polyclonal compartments may or may not coincide with a morphological line in the resulting pattern.

Delineation of polyclonal compartments is more like the drawing of a grid of coordinate lines than it is like copying a sketch of a master pattern. In effect, a cell becomes subordinated to a set of coordinates. Its further development is guided by positional information that is interpreted in relation to compartmental borders. These borders resemble in function a civil engineer's reference lines, in relation to which streets are located and new buildings are sited, though the reference lines themselves are not obvious after the projects are finished.

Maturing primary tissues in stems and roots seem to be polyclonal compartments, in that each tissue is composed of the progeny of a number of apical initial cells, and crossing of compartment borders seems not to occur — or to occur only rarely. For example, a cell in the inner (adaxial) layer of the developing root cortex may divide periclinally, but the inner daughter cell is not added to the central cylinder. However, very close to the root apical initials, where the borders between the cortex and the central cylinder are not yet clearly delineated, the fate of some cells may not be readily predictable, and we cannot be certain that no crossing occurs. Clonal analysis (Poethig 1987) might yield useful information on this subject.

Hanstein's (1868) concept of **histogens** (tissue formers; Sect. 14.1.3.1), had it proven to be correct, would have supported the hypothesis that, in angiosperms, the histogens are clonal (not polyclonal) compartments. Hanstein regarded each histogen as consisting of specifically located initials or small groups of initials that produce layers or lineages of cells that give rise to particular tissues. However, studies of periclinal chimeras have shown that the initial-cell layers of shoot apices do not have fixed developmental fates (Sect. 14.1.5.2), and hence, rigorously, are not histogens in the sense of Hanstein. In certain roots there may be true histogens, but in many other roots no histogen initials can be identified.

Guttenberg's (1960) redefinition of the term "histogen" to mean a collection of meristematic cells that give rise to a defined part of an organ, though it may not have its own initial cell(s), seems more compatible with the concept of polyclonal compartments than does Hanstein's concept. Do the histogenic lineages that arise from Guttenberg's meristems (though not the meristematic initials themselves) constitute regions or compartments that have been delineated as "fields" in relation to positional coordinates of a morphogenic map — fields separated by borders that are not crossed?

We have thus far applied the term "compartment" to a part of a cell population committed to a developmental route that differs from the route taken by another part of the population. However, some botanists also use the term "developmental compartment" in a more vague sense, to mean a small group of cells, or even a single cell, the progeny of which form a patch or block of tissue within which the primordium of an organ will arise. The developmental compartment in this sense is that minimal set of cells that will contribute to the primordium of the organ. For example, by means of clonal analysis, the size of the primordial compartment for reproductive organs has been estimated as two cells in *Arabidopsis* and four cells in *Glycine* and in the primary spike of *Hordeum*. Clonal analysis of *Gossypium* revealed that when the embryo attains a globular stage with a distinct dermatogen, the surface layer already includes an eight-celled compartment for each cotyledon, a two-celled compartment for the first leaf, a one-celled compartment for the second leaf, and a three-celled compartment for the shoot apical initials (Christianson 1986). The related concepts of polyclonal compartments, developmental compartments, and primordial compartments need to be more rigorously defined before they can contribute much to general understanding of organismal development.

12.4.5 Differentiation in Relation to Existing Patterns

In a developing organism, individual cells are necessarily subordinated to "decisions" made and "programs" established at higher levels in the organizational hierarchy. Each cell seems to be entrained, limited, or inhibited by contiguous cells. The effect is especially strong in groups of cells, **meristemoids**, that are developing into cell types more specialized than the surrounding cells. Meristemoids are loci of residual meristematic activity in a region of somewhat older, differentiating cells. Examples are axillary bud primordia, developing provascular strands, and guard-cell mother cells. Early in their development, when their constituent cells are actively dividing, meristemoids seem to be especially effective in positively or negatively influencing nearby cells.

Some kinds of meristemoids, when first being delineated, inhibit similar meristemoids from differentiating nearby. Thus, a new meristemoid will not appear until a site is available at an adequate distance from existing meristemoids (Bünning 1965). This repulsion effect may be involved in the initiation of leaf primordia on shoot apices (Sect. 15.4), and in the siting of stomata in developing leaves (Sect. 16.3.4).

The mutual repulsion of meristemoids can explain how the number of elements in a pattern may depend on meristem size. For example, small embryos of *Pinus resinosa* may have only six cotyledons, whereas the largest embryos may have as many as 16. But, regardless of the number, the cotyledons are always arranged in a single cycle.

Some kinds of meristemoids may induce neighboring cells to differentiate along particular pathways. If the cells contiguous to those that are already differentiating follow the same developmental route as the meristemoid, we can speak of "contagious" differentiation and "positive induction". This kind of differentiation is especially prominent during the extension of provascular strands. Further, the original provascular strand that develops into a leaf trace is a template for subsidiary strands (Larson 1975); and the narrow vessels in the terminal part of an annual xylem increment can be templates for the much larger vessels that form in the initial part of the next annual increment (Kurczyńska 1986).

Differential contacts between cells may affect the patterns of inhibition and of propagation of induction, as illustrated by Bünning's (1951) model system based on root apices of members of the Brassicaceae. Trichoblasts in these roots occur in regular, longitudinal files separated by one, two, or three files of atrichoblasts, depending on the developmental stage. The trichoblast files characteristically are located above (abaxial to) the axial anticlinal walls of the cortical cells, whereas atrichoblast files are located above the main bodies of the cortical cells. Bünning concluded that the rhizodermal cells above the anticlinal walls in the cortex are more isolated physiologically than are the other rhizodermal cells, though he did not investigate the degree of plasmodesmatal connection of the two types of rhizodermal cells with underlying cells (Carr 1976).

The transfusion tissue in needles of *Pinus nigricans* (=*P. nigra*) shows a developmental pattern similar to that in the rhizodermis of Brassicaceae. The transfusion tracheids lie adjacent to the radial anticlinal walls between endodermal cells, whereas the transfusion parenchyma cells lie adjacent to the main bodies of the endodermal cells (Huber 1947). The siting of the first stomata in young leaves provides a further example of the importance of the degree of contact between cells in the propagation of induction. These stomata appear over relatively large intercellular spaces between the epidermis and mesophyll (Sect. 16.3.4).

12.4.6 The Role of Stresses and Strains in Growth and Development

Mechanical stresses have a wide range of effects on plant growth and development. In particular, stresses affect the rates and orientations of cell division, and cell differentiation. These effects are manifested structurally at the subcellular, cellular, tissue, and organ levels. Some of these effects, such as changes in the thickness of cell walls, are readily apparent in microscopic sections. Other effects are less obvious. It is possible that stresses are a continuing source of positional and morphological information that guides the development of an organ or plant in maintaining its adaptations to these stresses — mainly by cell divisions oriented so as to resist the stresses (Lintilhac 1984).

Before discussing the varied anatomical and morphological responses of plants to mechanical stresses, we briefly summarize the mathematical bases for quantifying stresses and strains in plants. The simplest case involves stresses that act in a single direction. For these, the stress, σ, at a point, P, is defined as the ratio of the force, F, to the elemental area, A, upon which the force acts. Thus, along any single direction:

$$\sigma = \lim_{A \to 0} \frac{F}{A} . \tag{12.1}$$

The value of σ at P, however, commonly is different in different directions, and, furthermore, stress is a function of the position of P in the plant body. If there is a vector field, $\underline{F}$, in the body (that is, if the orientation of the plane of A, or the direction of a unit vector normal to A, varies), then the above definition of σ leads to a tensor and σ is a tensorial quantity (Sect. 12.3.2). The way in which σ varies with the changing position of P depends strongly on body geometry.

The tensorial approach to studying stress distribution in the cell-wall meshwork and/or in the plasmalemma or cytoskeleton of a plant organ yields results similar to those obtained by using tensors to describe the principal directions of growth (Sect. 12.3.2). Their similarity lies in the inevitable linkage of stress with strain and with growth (Sect. 12.2.3.1).

After analyzing the distribution of stresses in growing organs, Lintilhac (1974b) concluded that, in a dividing cell, the new partition wall forms in the plane of minimal stress, and thus is normal to one of the principal trajectories of stress. Because the principal trajectories of stress generally coincide with the principal directions of growth, it is difficult to distinguish between them. An interesting premise arises from this work: the distribution of growth rates in a growing organ is such as to maintain a state of minimal shear stress in the middle lamellae during growth (see also Lintilhac and Vesecky 1980, 1981; Lintilhac 1984).

If the trajectories of the principal directions of growth (which can be derived though a tensorial analysis of growth) are to regulate organ shape, they must have a feedback relation with another factor that is a tensorial quantity of the same order and also dependent on organ geometry. Mechanical stress is a good candidate for such a factor. We hypothesize that growth and stress are inextricably interrelated and that stress distribution influences growth distribution in an organ. If the growth pattern of an organ changes, then the distribution of stresses within the tissue changes also. The new stress pattern alters the distribution of further growth. Thus, one can imagine a biophysical basis of morphogenesis, manifesting itself through symplastic and other growth.

There are many sources of mechanical stresses. Some, such as gravity, wind, and water, originate outside the plant. Others originate internally, generated by the growth and development of the tissues themselves. We consider externally derived forces first.

Though conflicting opinions are recorded in the early literature, externally applied mechanical forces are now generally thought to affect the anatomy and morphology of plant organs. Gravity is the most significant of these forces. It is omnipresent and nearly constant, and its effects have long been studied. Effects of wind are almost as ubiquitous as effects of gravity. Winds, though, are highly variable in time and space.

The effects of wind sway have been demonstrated by artificially holding stems in fixed positions. Rasdorsky (1925), for example, reported that, if bending stresses are prevented by providing mechanical supports and screens against winds, developing stems and petioles are weakened. In addition, Jacobs (1954) found that, if the lower half of a young *Pinus radiata* stem is mechanically restrained from swaying, a normal xylem increment is produced in the upper half, whereas in the lower half the increment is considerably reduced. However, after the restraints are removed, the lower stem thickens faster than the upper stem, until the normal taper is restored. Careful analysis of such experimental results reveals that the significant factor is the wind sway itself, not increased transpiration elicited by moving air.

Trees allowed to sway in the wind develop tapered stems of uniform bending resistance (Sect. 4.7.1), but similar trees prevented from swaying develop more nearly cylindrical stems (Jacobs 1939, 1954). Furthermore, woody stems that are swayed in a single plane become elliptical in cross section, with the longer axis in the plane of swaying (Rasdorsky 1925; Larson 1963).

According to Walker (1960), the genetic information concerning development of mechanically competent stems and petioles is not fully expressed unless there is mechanical stimulation during organogenesis. He found that, in *Datura* seedlings, the structural effects commonly ascribed to "etiolation" are, in fact, partly due to lack of swaying and bending. Mechanical shaking of seedlings growing in darkness reportedly counteracts some of the "etiolation" effects on structure. In general, mechanical shaking causes a decrease in stem length, and increases in the modulus of elasticity and the amount of collenchyma and sclerenchyma (Biddington 1986).

There is evidence that tension causes thick walls to develop in collenchyma and fibers of stems, petioles, hypocotyls, and roots. For example, in a small *Fagus sylvatica* root that was stretched by the thickening of a larger root beneath it, we observed that the cell walls were greatly thickened.

Stresses may be necessary for some tracheary tissues to differentiate. For example, in cambial explants of *Populus*, stresses corresponding to external unidirectional pressures of 5 kPa induced differentiation of tracheid-like elements, whereas without the pressure, parenchymatous cells proliferated. The differentiation of cambial derivatives in tangential flaps of attached, woody stem tissue also is strongly influenced by mechanical pressure: if no pressure is applied to the flaps, the derivatives do not differentiate normally (Brown and Sax 1962). Such results suggest that, in an intact stem, tracheids differentiate in regions of internal, radial compression.

For large, earlywood vessels in ring-porous wood to differentiate, stress conditions apparently must be just the reverse of those necessary for tracheids to differentiate in diffuse-porous, *Populus* wood. There is evidence that these vessels develop at sites that are temporarily free of radial and tangential compression, along the inner (adaxial) margin of the cambial zone (Sect. 19.2.1.3). It seems likely that the temporary lack of radial compressive stress, or even the existence of radial tension, in cambium that has started its seasonal growth is due to collapse of phloem that was functional the preceding fall. The lack of tangential compressive stress may be due to the tendency of the thin-walled cambial cells, which appear tabular in transection, to widen radially and narrow tangentially, under the characteristically high, early-season turgor pressures. In this developmental system, growth-generated stresses seem to contribute to propagating a pattern in which very wide vessels are formed each spring.

Studies of internally generated tissue stresses have revealed that, close to an organ surface, the pattern of principal stress trajectories is similar to the general orientational trends of partition walls. That is, in these regions, the lines of principal stress are either parallel or perpendicular to the surface, as are the divisions within the cells. For example, after elastooptical visualization of stresses in models, Lintilhac (1974a,b) proposed that, in an ovule of normal geometry, the principal stress trajectories are similar to the general orientational trends of partition walls in actual ovules. Interestingly, the embryo sac develops at a site of "pure" tension in the center of the ovule.

Morphogenesis may be partly guided by developmental "feedback" coming both from growth stresses, which change as the system changes, and from externally derived stresses. This could work as follows: for a young plant to grow and develop into a larger plant of different, but predictable, geometry, it presumably must always "know" what form it has already attained in relation to its overall developmental "plan". Stresses are suitable candidates for providing this information, because stresses depend strongly on form. Conceivably, a meristem may function as a virtual morphogenic engine, its output keeping the plant (or organ) structurally adapted to stresses generated by both internal and external forces. It may do this mainly by cell divisions oriented so as to resist the stresses (Lintilhac 1984). In this way, form could beget form automatically, with little or no reference to a "map" or "clock" (Sect. 12.4.3).

How could a cell-growth-regulating mechanism that involves wall stress work? Most likely, a protoplast is somehow able to detect and evaluate the stress, and then modulate its effects on the walls. It is difficult to conceive how a protoplast could measure wall stress directly. However, the protoplast might be able to respond to a variable that is a function of wall stress, such as elastic strain or piezoelectric effects in the crystalline components of cellulose walls.

The cytoskeleton, which has already been discussed in relation to polarity in cell development (Sect. 12.4.1), may also include filaments that are sensitive to stress. Thus, there could be a regulatory system in which growth-induced deformations operated via stress-sensitive cytoplasmic elements anchored in the plasmalemma. The stress sensors could conceivably be actin filaments, which are widely distributed in higher plants (Jackson 1982).

Some plants are extremely sensitive to mechanical stimulation as is exemplified by the existence of "touch induced" genes in *Arabidopsis*. As these genes are related to calmodulin (Braam and Davis 1990), they may be involved in transduction (via mechanical stress) of signals from the environment and thus in evocation of appropriate responses.

It is also possible that an intracellular sensor measures strains resulting from changes in the balance between internal turgor forces and externally applied forces. This system would ensure that the new partition wall is oriented such that it is minimally strained and is thus shear-free (Lintilhac 1984). Because the stresses that influence cell partitioning in the meristem are functions of the shape of that meristem, partitioning is also a function of meristem shape. But the mechanism by which the cell measures slight intracellular dimensional changes probably depends directly on strain and strain-relieving growth and only indirectly on stresses.

13 The Embryo

13.1 General Developmental Concepts

13.1.1 The Scope of the Subject

The process of sexual reproduction in seed plants is highly complex, in part because there is an alternation between asexual (diploid) and sexual (haploid) generations. In this chapter, we focus on the immediate structural precursors of the fusion nucleus or zygote, and the subsequent development of the fusion nucleus or zygote into an embryo.

Several terms have been used to refer to the study of plant embryos. Like Johansen (1950), we focus on **embryogeny**, the course of development of an embryo. We venture into the broader subject of **embryology**, which traditionally (see Maheshwari 1950) includes a series of asexual and sexual reproductive phenomena leading to syngamy, only as necessary to provide a background for developmental consideration of embryogeny. We use the term **proembryogeny** to refer to embryonal development from syngamy to the globular or subglobular stage, and **late embryogeny** for subsequent stages. The term "embryogenesis", while widely used (Raghavan 1986), lacks the precision we seek here.

Some recent literature follows the tradition of organizing embryological information taxonomically. However, some unifying principles and emerging morphogenic concepts that apply broadly to plant embryogeny are emerging, though comprehensive understanding at this level has not yet been attained. To the extent possible, we avoid classifying and listing in favor of generalizing and seeking principles.

13.1.2 The Milieu of Embryogeny: Megagametophytes, Egg Cells

The trees and other seed plants that are macroscopically visible are sporophytes and, technically, are diploid and asexual. Within the flowers of angiosperms and the strobili of gymnosperms, sporophytes typically bear megaspores and/or microspores that produce minute, parasitic, sexual plants of an alternate, haploid, gametophytic generation. Female gametophytes, or **megagametophytes**, are much larger than male gametophytes, and produce female gametes, or eggs. The megagametophytes and eggs are typically sequestered within ovules. In gymnosperms, the egg arises in an archegonium within the megagametophyte, whereas, in angiosperms, the megagametophyte produces the egg more directly. Male gametophytes are **microgametophytes**, familiar as pollen grains. They consist of notably fewer cells than do female gametophytes, and produce male gametes, or sperm.

13.1.2.1 Megagametophytes, Archegonia, and Eggs in Gymnosperms.

In gymnosperms generally, as in angiosperms, a megagametophyte develops from a megaspore. The megaspore, in turn, is produced by a meiotic process, megasporogenesis, within an ovule. Several megaspores are usually formed per ovule, but typically only one survives. It develops into a sac-like megagametophyte.

Megagametogenesis begins with a series of free-nuclear divisions in a megaspore. The resulting numerous, small nuclei become arranged in a peripheral layer of cytoplasm. Then, by an obscure process not immediately preceded by nuclear division, each nucleus becomes linked with six neighboring nuclei by sets of "spindle fibers". Phragmoplasts and anticlinal walls soon arise normal to the spindles, causing the surface of the megaspore, a nascent megagametophyte, to assume a honeycomb-like pattern. Then, driven by unknown forces, the nuclei, their connecting spindles, and the phragmoplasts begin migrating inward towards the center of the sac, forming the anticlinal walls of hexagonally columnar cells as they go. The cells, many of which do not form walls on the sides towards the center of the sac until late in their development, are the primary prothallial cells, or **alveoli** (Sokolowa 1890).

During the inward migration of nuclei and spindles, competition for space becomes severe. Some nuclei lag behind others. The regular hexagonal pattern of spindles and walls is broken, and walls form around lagging nuclei, enclosing them in cells before they reach the center of the sac. Only a few sets of columnar walls reach the central region and form end walls that abut on those of their counterparts from the opposite side. The megagametophyte then consists of radially arranged alveoli of varying lengths. As the megagametophyte matures, most of the alveoli become segmented into radial rows of cells, by a series of periclinal divisions. Those alveoli that remain unsegmented typically have reached the center of the sac before being walled off, and are near the micropylar end of the sac. They are larger than the others and function as archegonial initials.

Gnetum and *Welwitschia* do not have archegonia, but most gymnosperms do. Archegonia may occur singly, or in compact groups as archegonial complexes. In *Torreya*, there is one archegonium per megagametophyte; in *Ginkgo*, two archegonia are common; most conifers have four to seven; and Cupressaceae and Taxodiaceae have five to 50 or more. If there is a single archegonium, it is almost always at the micropylar end of the megagametophyte. Archegonial complexes also commonly occur at the micropylar end, but in some taxa they also occur laterally or even chalazally.

Spontaneously, or in response to stimuli emanating from approaching pollen tubes, the nucleus in each archegonial initial moves into the neck region, near the outer face of the megagametophyte, and divides. Subsequent, unequal cytokinesis forms a large, **central cell** and a small, neck-initial cell. The latter divides several times, forming a set of small neck cells. The central-cell nucleus enlarges, then divides unequally, into a large **egg nucleus** and a small ephemeral, ventral-canal-cell nucleus. As the egg nucleus matures, it enlarges and retreats from the neck region. Little is known about the forces driving these nuclear movements.

The egg cell, with its very large nucleus, is basically a later developmental stage of the central cell of the archegonium, after a small fraction of its nuclear material and cytoplasm have been lost to the ephemeral, ventral-canal-cell nucleus. There is considerable literature on the structure of the gymnosperm egg nucleus, which becomes surrounded by a dense cytoplasm rich in inclusions. Meanwhile, the small, derivative cells of neighboring, segmented alveoli divide and form a jacket around the archegonium, which itself develops a thick, pitted wall (Singh 1978).

13.1.2.2 Megagametophytes (Embryo Sacs) and Eggs in Angiosperms.

A typical megagametophyte of an angiosperm, like that of a gymnosperm, is derived from a haploid megaspore produced by megasporogenesis within an ovule. At the cellular level, though, megagametogenesis in angiosperms is very different from that in gymnosperms. Further, the cellular structure of the mature megagametophytes in the two groups is so different, in both pattern and scale, that homology between them is doubtful (Singh 1978).

Of the four megaspores produced by meiosis of an angiosperm megasporocyte, one, two, or all four may participate in formation of a single megagametophyte. Accordingly, angiosperm megagemetophytes can be categorized as monosporic, bisporic, or tetrasporic. Probably about 80% of angiosperms have monosporic megagametophytes.

In all types of angiosperm megagametogenesis, the megaspores that develop into megagametophytes tend to become very large and to produce thick cellulose walls lined with callose. In some taxa, this callose probably is a factor in symplasmically isolating the nascent megagametophyte from cells of the ambient, maternal sporophyte (Rodkiewicz 1970; Sect. 13.6). In other taxa, isolation may involve deposition of suberin.

The enlargement phase of the nascent megagametophyte is accompanied or followed by several free-nuclear divisions, resulting in formation of a coenocytic, sac-like, immature megagametophyte. In many megagametophytes, there are three cycles of free-nuclear divisions, producing eight haploid nuclei. These may be genetically identical, though rigorous proof is lacking. However, the environment of the coenocyte, or **embryo sac**, is highly polar, and the various nuclei, which are located in different regions of the polar field and possibly are not physically identical, do not behave identically.

Commonly, four nuclei gather near the chalazal end and four near the micropylar end of the sac. Soon, one nucleus from each group migrates toward the central region. These two are termed **polar nuclei** because they come from opposite poles. (Note that they migrate in opposite directions in the same polar field!) One of the three remaining micropylar nuclei then organizes itself and some surrounding cytoplasm into an elongated **egg cell**, delineated by a plasma membrane. The other two micropylar nuclei organize smaller, pear-shaped, **synergid cells**. Each synergid may have a filiform apparatus, a system of elaborately dissected, haustoria-like processes extending into the peripheral cytoplasm at the micropylar end of the sac. This structure may function as a transfer area (Schulz and Jensen 1977). The egg cell and synergids are usually incompletely walled and collectively are referred to as the **egg apparatus**. Functionally, this apparatus constitutes an archegonium, though no close homology with a typical gymnosperm archegonium is indicated.

Meanwhile, in most taxa, the three chalazal, or antipodal, nuclei organize three **antipodal cells**. These cells, like the synergids, may be only partially walled and may have haustorial processes or transfer areas. In other taxa, the chalazal nuclei generate a coenocyte, which may invade and digest neighboring tissues of the ovule (Johri and Agarwal 1965). In still other taxa, the three chalazal nuclei disintegrate as the mega-gametophyte matures.

The two polar nuclei, along with any cytoplasm not appropriated by the egg apparatus or by antipodal nuclei or cells, then constitute a large, residual, binucleate, **central cell**. Thus, a typical embryo sac is seven-celled and eight-nucleate, though variations are common. The central cell, because of its size, dominates the sac, though the egg cell may also be relatively large. The central cell of an angiosperm embryo sac *is not* closely analogous to the central cell in an immature archegonium of a gymnosperm megagametophyte. In an immature gymnosperm archegonium, the egg cell, functionally, *is* the central cell and dominates the archegonium.

13.1.3 Syngamy, Fusion Nuclei, and Zygotes

By pollination and pollen-tube growth, sperm nuclei arrive in the vicinity of an egg cell. **Syngamy** (union of gametes), or **fertilization** of the egg by the sperm, can then occur. The melding of the genetic material of the two gametic nuclei into a **fusion nucleus** restores the diploid condition. The fertilized egg, if it is a discrete cell, is a **zygote**. It is the first cell of the new, asexual, sporophytic generation. Its further development is the purview first of **proembryogeny**, then of **late embryogeny**. This development typically includes a globular or subglobular stage, but not, as is typical in animals, a hollow, blastula stage.

Rigorously speaking, gymnosperms do not generally form a zygote (as a discrete *cell* that contains a fusion nucleus and that soon divides), because the fusion nucleus undergoes several successive free-nuclear divisions. Thus, in these plants, pro-embryogeny begins before cell walls between nuclei are completed. One can speak of a **zygotic nucleus** in gymnosperms, but the term has limited usefulness.

13.1.3.1 In Gymnosperms. The means by which pollen-tube growth brings male gametes to an archegonium varies widely among taxa. The general result is that a male nucleus with some cytoplasm enters the archegonium and migrates to the vicinity of the nucleus of the egg cell, which is the central cell of the archegonium. This male nucleus, which is mobile, merges with the egg nucleus. A second male nucleus may also penetrate the egg cell in the archegonial neck region, or may remain just outside. Even if it enters the egg cell, it becomes immobile rather than migrating toward the nucleus. In *Ephedra*, as one of the male nuclei merges with the egg nucleus, the other may merge with a ventral-canal-cell nucleus, forming a diploid nucleus possibly capable of generating a few progeny cells (Khan 1940). Though vaguely suggestive of the "double fertilization" in angiosperms, the *Ephedra* process is not really comparable (see later).

As it migrates through the egg cytoplasm towards the egg nucleus, the mobile male nucleus brings with it a variable amount of male cytoplasm. Commonly, the two nuclei become surrounded by an increasingly dense cytoplasm that includes much of that accompanying the male nucleus but only a small fraction of the original egg cytoplasm.

Typically, both nuclei are filled with dense nucleoplasm. As they contact each other, their membranes fragment or develop pores through which their nucleoplasms fuse into a single mass, which initially is not well delineated from the similarly dense envelope of cytoplasm derived from the egg and the male gamete. This plasmic blend, which includes a large fraction of the egg and male nucleoplasm, and some of the male cytoplasm, constitutes the **neocytoplasm** (Camefort 1969; Singh 1978). As the new, diploid, fusion nucleus becomes organized, the neocytoplasm alone functions as the *cytoplasm* surrounding that nucleus and all subsequent, free-nuclear, proembryonic stages. The remaining plasmic fractions, including much of the original egg cytoplasm, do not participate in proembryogeny.

We emphasize that the "zygote" in most gymnosperms is not a discrete cell, but a fusion nucleus surrounded by neocytoplasm within an archegonial sac. There are several free-nuclear divisions, and the neocytoplasm becomes enriched with organelles and inclusions (Chesnoy and Thomas 1971; Willemse 1974).

13.1.3.2 In Angiosperms. A pollen tube, after arriving in the vicinity of a mega-gametophyte (called an embryo sac in angiosperms), typically enters the filiform apparatus of one of the synergid cells. This seems to trigger both degeneration of the penetrated cell (and sometimes of the other synergid also) and cessation of pollen-tube growth (Went and Willemse 1984). After entry, the pollen tube releases its cytoplasm, along with two sperm cells or nuclei and a vegetative nucleus. This may cause the synergid to burst. The cytoplasm, with sperm cells or nuclei and other inclusions, then has good access to the egg cell and the central cell, which are in part bounded only by plasma membranes. Though the process is not well understood, it seems plausible that the plasma membranes of a sperm cell and the egg cell meld, which then allows the sperm nucleus and possibly some male cytoplasm to meld with the contents of the egg cell. This presumably is followed by fusion (karyogamy) of the egg and sperm nuclei. A similar process probably occurs between the other sperm nucleus and the two polar nuclei in the central cell (Jensen 1972), thus achieving "double fertilization". The products of these fertilizations are a diploid, zygotic fusion nucleus and a triploid, endosperm fusion nucleus. Typically, an angiosperm zygote is briefly quiescent after karyogamy. The other fusion nucleus quickly begins to divide, forming a free-nuclear endosperm (Schulz and Jensen 1977), which later becomes cellular in most taxa.

13.1.4 Polarities and Symmetries

In both gymnosperms and angiosperms, the processes of megagametogenesis, oogenesis, and proembryogenesis occur in a polarized environment, along a micropylar-chalazal axis. This polarity is a factor in determining which megaspores develop into megagametophytes and in the positioning of the nuclei of these megaspores. The polarity is also shown in the movements and positions of the various cells and nuclei of archegonia in gymnosperm megagametophytes and egg apparati in embryo sacs, and in the distribution of vacuoles, organelles, and other inclusions in egg cells before fertilization (Raghavan 1976; Singh 1978).

Early polarity in gymnosperm oogenesis is manifested in the usually very unequal division of the central-cell nucleus to form a large egg nucleus and a small ventral-canal-cell nucleus. The egg nucleus moves polarly, away from the neck region (Sect. 13.1.2.1).

Early polarity is even more obvious in angiosperms. The several nuclei in an immature embryo sac migrate in a polar field as cells become organized around them. One cell, the egg cell, comes to be in contact with, or attached to, the embryo-sac wall at its micropylar end. Its wall is thickest in the contact region and may be very thin or lacking at its opposite, free, chalazal end. The micropylar half or more of the cell consists of a single vacuole and a thin, parietal layer of cytoplasm, whereas the chalazal part contains most of the cytoplasm, the nucleus, and some small vacuoles (Willemse and Went 1984). Syngamy occurs and proembryogeny begins in this polar environment.

During proembryogeny, symmetry is predominantly radial, around a polar axis. In some taxa, at some stages, the proembryo approaches spheroidal symmetry, though it fundamentally continues to be radially symmetrical and polar. During late embryogeny in both angiosperms and gymnosperms, bilateral symmetry tends to replace radial symmetry (Sects. 13.4.2; 13.5.2). The original polarity apparently remains unaltered.

13.2 Proembryogeny in Gymnosperms

In many gymnosperms, the fusion nucleus divides shortly after it has become well enough organized to form a bipolar spindle (Sect. 13.1.3.1). The two daughter nuclei, which may or may not be genetically identical and are embedded in neocytoplasm, migrate through the maternal cytoplasm toward the chalazal end of the archegonium. As they migrate, the nuclei typically divide again, forming a coenocytic quartet. The neocytoplasm associated with these nuclei becomes adpressed to the archegonial wall near the chalazal pole. Walls may begin to form between nuclei, either shortly after the quartet arrives in the chalazal region or after large numbers of additional nuclei have formed there.

Patterns of free-nuclear divisions and proembryogeny among gymnosperms can be divided into four types:

1. The cycad and *Ginkgo* type, with as many as 1024 free nuclei in *Dioon*, 512 in *Cycas*, and 256 in *Ginkgo*, though all of these nuclei do not become a part of a cellular proembryo
2. The conifer type (discussed further in Sect. 13.2.1), with 64 or fewer free nuclei, all of which participate in proembryogeny
3. The *Ephedra* and *Sequoia* type, in which walls are laid down between nuclei very early, forming four independent cellular units in *Sequoia* and eight in *Ephedra*, with each unit able to develop into a competing proembryo

4. The *Gnetum* and *Welwitschia* type, in which, in contrast to the preceding types, eggs are not borne in archegonia but differentiate from seemingly ordinary cells of the megagametophyte. In this type, several walled zygotes per megagametophyte may be formed directly from fusion nuclei. After some cell divisions, these extend branched and septate "suspensor tubes" (Sect. 13.2.2), some of which form subglobular proembryos at their tips

13.2.1 The Conifer Type: A "Basal Plan"

Of the four types of gymnosperm proembryogeny, the conifer type has been most studied and occurs in more living gymnosperm species than any other. Doyle (1963) discussed and interpreted the conifer type as a **basal plan** for gymnosperm proembryogeny, and Singh (1978) adopted his scheme. We use the term "basal plan" in the sense that they did.

In the conifer type of proembryogeny, all the free nuclei derived from the fusion nucleus come to lie in the chalazal region of the archegonium in a common body of dense neocytoplasm. *Athrotaxis* has four free nuclei, *Pinus* eight, *Cephalotaxis* 16, *Podocarpus* 32, and *Agathis* 64. On the basis of these and other differences, the conifer basal plan has four major variations: (1) Pinaceae, (2) *Pseudotsuga*, (3) *Athrotaxis*, and (4) *Callitris*. There are also several other, minor variations.

Research during recent decades has clarified structural relations and dynamics of proembryogeny in various conifers to such an extent that a precise new terminology can now be applied. The terms, defined next, may be understood in relation to Fig. 13.1. All references to orientation and tier levels assume the chalazal pole to be "lower" and the micropylar to be "upper".

Primary proembryo (pP): the proembryo at the stage of wall formation between nuclei (Dogra 1967)
Primary upper tier (pU): the upper tier of the pP, the cells of which are open on their upper sides (Dogra 1967; Doyle and Brennan 1971)
Primary embryonal tier or cells (pE): the lower tier or group of cells of the pP (Doyle 1963; Dogra 1967)
Internal division: a division of a cell of the pP, as distinguished from a free-nuclear division (Doyle 1963)
Upper tier (U): the upper tier derived from internal divisions of the pU tier (Doyle 1963)
Suspensor tier (S): the lower tier derived from internal divisions of the pU tier (Doyle 1963). Cells of this tier may elongate to form a suspensor, but are often dysfunctional.
Embryonal tier or cells (E): the tier or cells derived from the pE tier or cells after internal divisions (Doyle 1963)
Embryonal mass (EM): all E cells and their early progeny, which, if they undergo cleavage, may form separate embryonal masses or embryonal units (Singh 1978)

Singh (1978) demonstrated that rigorous use of this terminology allows one to write proembryogenic formulae for the various taxa of conifers.

All members of the Pinaceae follow the Pinaceae variation of the conifer basal plan, though *Pseudotsuga* shows a simplification thereof (Mehra and Dogra 1975). In the Pinaceae variation, the four free nuclei derived from the fusion nucleus, having arranged themselves in a tier at the chalazal end of the archegonial sac, divide, forming two tiers of four *nuclei* each. These tiers soon become cellular. The lower tier is equivalent to pE and the upper to pU (Fig. 13.1). Internal divisions in these two tiers then produce four tiers of four *cells* each. The lower two tiers (not always discrete) constitute the E group. The tier above is S and the uppermost tier is U. The U cells degenerate, and the S cells may be designated as a dysfunctional-suspensor (dS) tier (called the "rosette tier" in the older literature). Cells of dS may show some meristematic activity and have been thought capable of producing "rosette embryos", though this has not been confirmed (Berlyn 1972; Singh 1978). The upper four cells of the E group elongate to form a functional suspensor (called an embryonal suspensor; Sect. 13.2.2), and the lower four E cells form the embryonal mass (EM). Differential elongation of the four suspensor cells, which may have divided transversely to form strands, may cleave the EM into four units, each with a small EM at its tip. This is the basis of cleavage polyembryony (Sect. 13.2.3).

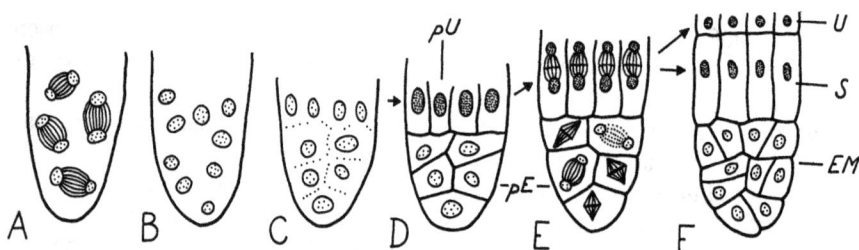

Fig. 13.1 A—F. Generalized schemata of "basal plan" of conifer proembryogeny. **A—D** Distal (chalazal) regions of nascent proembryo, representing synchronous divisions of free nuclei **A**, the progeny of which become arranged in upper and lower groups of nuclei **B** and **C**; wall formation between these then produces the cellular, primary proembryo, *pP*, represented by **D**, with a primary upper tier, *pU*, and a primary embryonal group, *pE*. **E—F** Later cellular proembryo, after internal divisions represented in **E** have produced upper tier, *U*, and suspensor tier, *S*, from *pU*, while *pE* cells have generated a larger embryonal group, which after further cell divisions becomes the embryonal mass, *EM*. See text for definitions and developmental details. (Modified after Doyle 1963)

Using the terminology and abbreviations just given, and the methods of Singh (1978), one can formulate proembryogeny in *Pinus* as:

$$4\,pU + 4\,pE$$
$$4\,U + 4\,S + 8\,E,$$

in which the first line represents the arrangement of nuclei as wall formation begins and the second line expresses the arrangement after a cycle of internal divisions in the pU and pE tiers. With this method, one can express any degree of complexity in proembryogenic processes in conifers, though one may need to use additional symbols and additional lines of formulation.

The utility of the terms and abbreviations introduced here becomes strikingly evident when applied to deviations or variations from the basal plan or from the Pinaceae variation thereof. For example, the *Pseudotsuga* deviation from the Pinaceae variation includes pU and pE tiers of four cells each, but internal divisions occur only in the pE tier. The eight E cells thus formed are arranged in two tiers of four. The upper four elongate into a functional suspensor; the lower four produce the EM; and the pU cells degenerate. No U or S tiers are formed (Allen 1946).

13.2.2 Suspensor Systems

Because of the different origins, morphologies, and developmental stages of structures that function as suspensors in various gymnosperms, we refer to them collectively as "suspensor systems" rather than "suspensors". If used rigorously, "suspensor" refers to a structure derived directly from the basal cells (which are located above the EM in conventionally oriented illustrations) of the early proembryo; that is, from the S tier (Fig. 13.1), which is usually dysfunctional in conifers. Such a suspensor, which is often lacking or ephemeral, is sometimes designated a "primary" suspensor, though that term has been used in other ways and can be misleading (Singh 1978). In most gymnosperms, especially conifers, the functional suspensor is "secondary", arising from E cells of the proembryo or from their progeny at the base of the EM rather than from S cells. Secondary suspensors in gymnosperms are of three types: embryonal suspensors, suspensor tubes, and embryonal tubes.

Embryonal suspensors (Es) are so named because they are derived from E rather than S cells. An Es may be a single elongating cell or a set (tier) of four cells elongating symplastically. The Es cell or cells are attached to S cells proximally and to E cells distally. Most conifers produce only one generation of Es cells, but, in *Pinus* and a few other genera, there may be several in succession, end-to-end (Singh 1978).

Suspensor tubes (st) are similar in origin to the Es type, but they do not, at first, have E cells at their tips. Instead, as the st elongates, the nucleus moves to the distal region of the tube and divides. Wall formation then delineates a small apical cell that functions as an E cell and can produce an embryo. This system is well developed in *Juniperus chinensis* (Tang 1948).

Embryonal tubes (et) are produced later in proembryogeny than Es and st types. They typically arise by elongation of proximal cells of the EM after the basal diameter of the mass has exceeded that of earlier suspensors. A suspensor system at the base of an EM tends to become multi-stranded as cells are added to it from the EM. During growth, the diameter of these systems becomes as large as or larger than that of the EM itself.

Generally, the first-formed cells of a gymnosperm suspensor system become very long and loosely coiled, convoluted, or folded. They then lose turgor and collapse as new cells continue to be added near the base of the EM. Because they elongate vigorously, the newer cells push against both the pad of collapsed suspensor cells and the base of the EM. This elongation of the functional suspensors late in proembryogeny pushes the growing EM through the chalazal archegonial wall and jacket cells, into the thallus of the megagametophyte. In all gymnosperms, this penetration is accompanied by development of a **corrosion cavity** in the central region of the megagametophyte. That cavity is the immediate environment of late embryogeny.

The other-than-mechanical functions of gymnosperm suspensor systems have not been as thoroughly studied as have those of their angiosperm counterparts, which often are endopolyploid and seem to produce and supply morphogens to the proembryo. Because all types of polyploidy are rare in gymnosperms, and because extensive endopolyploidy during gymnosperm embryogeny is not documented, the nonmechanical functions of the suspensors may be different in gymnosperms than in angiosperms (Sect. 13.3.2).

13.2.3 Polyembryony

A gymnosperm ovule generally is able to produce numerous eggs that can become proembryos. Further, in most taxa, several proembryos can develop from each fusion nucleus. Nevertheless, because of "competition", "selection", or "position effects", a typical mature gymnosperm seed contains only one surviving embryo. In conifers, fewer than 3% of seeds contain two or more embryos (Berlyn 1972).

Polyembryony is of two basically different types. **Simple polyembryony**, widespread among gymnosperms, results from fertilization of the egg in each of several to many archegonia of a megagametophyte. This type has also been called **archegonial polyembryony**, though this term is inappropriate for those few genera that lack archegonia.

Cleavage polyembryony is also common, especially in conifers. It arises from cleavage of the early EM, which is at the distal end of the suspensor system, into several (often four) smaller embryonal units. Each unit typically includes at least one functional pE cell or derivative. Cleavage seems to be caused by stresses arising from different elongation rates of the various strands of the suspensor system (Dogra 1967).

13.2.4 The Embryonal Mass

As mentioned in Section 13.2.1, gymnosperm embryonal cells, which are borne on the distal end of a suspensor system, constitute an **embryonal mass** (EM). In its earliest stage, an EM consists only of E cells, and later of several to dozens of generations of their progeny. As conventionally used, the term "embryonal mass" does not distinguish between (a) a mass of functional E cells and/or their derivatives borne by a multi-stranded suspensor system before cleavage, and (b) a smaller mass (or embryonal unit) resulting from cleavage of the original mass. Thus, an EM may arise from four or more E cells, or, after cleavage, from a single E cell borne on the end of a single-stranded, functional suspensor. In *Pinus*, there are usually four post-cleavage EMs, each of which is derived from a single E cell.

Early in proembryogeny, a pyramidal apical initial cell may be present in each EM in *Pinus* (Buchholz 1918), *Larix* (Schopf 1943), and probably other genera, though this initial is not distinguishable after 500 to 1000 cells have accumulated. Whether distinguishable as initials or not, cells near the distal end divide obliquely in various planes, establishing an ephemeral globular or subglobular EM. This mass is soon transformed into an elongate cylinder with a near-hemispherical distal end, which later becomes the shoot apical region (though this end is usually illustrated as pointed downward). Proximally, the EM is continuous with the suspensor system.

Whereas cells in the distal part of the EM divide in all planes, derivatives of E cells near the proximal boundary tend to divide transversely and may add some Es cells (Sect. 13.2.2) to the suspensor system. This restriction on orientation of cell division in the proximal region continues and may even become more pronounced as the EM becomes a well-organized proembryo and then an embryo. In most genera, lack of such restriction in the distal region means that even the surface layers continue to divide periclinally as well as anticlinally, preventing differentiation of a protoderm or tunica over the nascent shoot apex. Significantly, some taxa that later develop a tunica and corpus also develop a protoderm over the apical region of the EM. This occurs in *Podocarpus* (Konar and Oberoi 1969) and *Cephalotaxus* (Buchholz 1925).

Generally, after their earliest stages, gymnosperm EMs are radially symmetrical and show an apical-basal polarity, which becomes root-shoot polarity (Sect. 13.1.4). They have a superficial apical (distal) meristematic region and a deep-seated basal (proximal) meristematic region.

Although there is no sharp temporal boundary between proembryogeny and late embryogeny, it is convenient to assume that the transition occurs as the EM is thrust through the archegonial wall into a corrosion cavity in the megagametophyte, by vigorous elongation of the suspensor system.

13.3 Proembryogeny in Angiosperms

13.3.1 Orientations and Patterns of Early Cell Divisions

Immediately after syngamy, an angiosperm zygote has a wall around its micropylar end, but its chalazal end is bounded only by a plasma membrane. In all species studied, the wall is completed before or shortly after the fusion nucleus divides. With the puzzling exception of *Paeonia*, which forms a coenocytic proembryo somewhat like that of *Ginkgo* (Yakovlev 1969), cytokinesis follows nuclear division in angiosperm zygotes, at least in those species that have been studied. Thus, an angiosperm fusion nucleus behaves quite differently from a gymnosperm fusion nucleus, which, within its mass of neocytoplasm, undergoes several cycles of free-nuclear division.

Because a zygote forms in a polar environment, it is not surprising that its internal structure is also polar. Generally, the nucleus, which constitutes only a small fraction of zygote volume, is near the chalazal end of the cell, while a large vacuole occupies the micropylar end. It seems obvious that, if the zygote divides transversely to the micropylar-chalazal axis, the daughter cells will inherit notably different fractions of the cytoplasmic and vacuolar components. In most angiosperms that have been studied, division of the zygote is, indeed, transverse and asymmetric. This division typically produces a large, vacuolate, **basal cell** and a smaller, more densely cytoplasmic, **apical cell**. The latter extends into the body of the central cell, in which a mass of endosperm is proliferating from derivatives of the triploid fusion nucleus. The basal cell, at its end opposite the common wall with the apical cell, is attached to the micropylar wall of the embryo sac.

Proembryos of angiosperms, like those of gymnosperms, are conventionally illustrated with the "basal" cell uppermost and the "apical" cell pointing downward. With some exceptions, apical-cell progeny eventually form the embryo, and basal-cell progeny generally form a suspensor system and sometimes contribute to a root cap.

13.3.1.1 Segmentation Types and Classification Systems.

Transverse division of an angiosperm zygote into an apical and a basal cell is soon followed by division of each of these cells, forming a tetrad. The apical cell may divide longitudinally, obliquely, or transversely, while the basal cell may divide either longitudinally or transversely. Accordingly, six "segmentation types", based on orientation of divisions of the apical and basal cells, are theoretically possible among tetrad-stage proembryos. Variations of subsequent division patterns yield further segmentation types and subtypes.

Formal systems of classifying proembryogenic types have been constructed based on the orientations of the first three generations of divisions at the apical end, and on the putative contributions of the progeny of the original basal and apical cells to the embryo and suspensor (Maheshwari 1950; Johansen 1950). For decades, botanists attempted to refine these classification systems, but the results were disappointing. The formal systems of classifying segmentation in angiosperms do not distinguish between dicots and monocots, because these groups do not show divergent patterns of development until the proembryo has reached the early globular stage (Sect. 13.5.1). Further, although an entire species, genus, or family may have a particular segmentation type (Davis 1966), segmentation may be of the same type in taxonomically diverse groups (Natesh and Rau 1984). Also, a single family, genus, species, or even an individual, may exhibit several types. Evidently, not only genetic, but also epigenetic, factors determine the orientation of early cell divisions. Physical factors, such as mechanical stress, electrical potential gradients, and direction of the gravity vector, probably play a role.

In addition, several woody dicots in which proembryogeny has been studied deviate notably from the formal segmentation types (as discussed later). These deviations contribute to the emerging view that efforts to interpret proembryogeny of dicots in terms of formal segmentation types are probably futile.

A newer approach to classifying angiosperm proembryogenic types is based on criteria relating to formation of the first wholly internal cell (Periasamy 1977). This approach, which is more attuned to contemporary notions of morphogenic control than are the older, formal systems, can help refine thinking about controls of embryo morphogenesis. The broader utility of such a nonphylogenetic system, however, is not yet apparent.

13.3.1.2 Proembryogeny to the Globular Stage.

We will first summarize the well-known proembryogeny of the cosmopolitan weed, *Capsella bursa-pastoris*, which conforms to one of the formal proembryogenic segmentation types. Then, to illustrate the divergence of at least some woody dicots from formal segmentation patterns, we discuss proembryogeny in *Nerium oleander* and *Quercus gambelii*.

The *Capsella* zygote divides transversely, forming a large, vacuolate, basal cell and a small, densely cytoplasmic, apical cell (Schulz and Jensen 1968). Almost immediately, the basal cell again divides asymmetrically, forming a new, and still very large, basal cell and a much smaller, subapical cell. The subapical cell quickly undergoes several more transverse divisions, forming a filamentous suspensor of eight to ten cells. The cell at the base of the suspensor remains vastly larger than any other in the proembryo. This cell is seemingly unaltered throughout embryogeny (Schaffner 1906). Meanwhile, the apical daughter cell produced by the transverse division of the zygote divides further, forming first a quartet of cells (the **quadrant** stage) and then an octet (the **octant** stage). These cells form an early globular mass at the apical end of the proembryo. A significant morphogenic event is the periclinal division of cells of the octant stage, forming a 16-celled globular proembryo having eight protodermal and eight internal cells. Progeny of the internal cells produce all internal tissues of the embryo. Protodermal cells contribute little more than epidermis, because, due to unknown factors, they seldom divide periclinally.

Division of the zygote in *Nerium oleander*, as in *Capsella*, is transverse, but the apical and basal daughter cells do not differ greatly in size. The apical cell and its progeny undergo several transverse divisions, forming a three- to five-celled, filamentous proembryo. Meanwhile, the basal cell may divide transversely several times and then obliquely, which results in a typically uniseriate suspensor, with broadening in the region of attachment of the suspensor to the embryo-sac wall.

The orientation of cell divisions in the filamentous proembryo then changes from transverse to longitudinal or oblique, and each cell of the filament becomes a two-celled tier. This series of longitudinal divisions is soon followed by a second series, forming four-celled tiers and initiating three-dimensional growth. Further divisions in various planes among the cells of these tiers then form a globular proembryo. Periclinal divisions in the surface cells of all tiers soon produce a distinguishable protodermal layer, in which further divisions are mostly, but not entirely, anticlinal (Mahlberg 1960).

The zygote of *Quercus gambelii* is highly polarized. It usually has a large vacuole in the micropylar region, while most of the cytoplasm and the nucleus are in the chalazal region. However, it divides not transversely but obliquely, sometimes almost longitudinally. The "apical" daughter cell is much smaller than the "basal" one, which occupies all of the micropylar region and a flank of the chalazal region. The species thus does not conform to any of the formal segmentation types, nor do *Quercus* or the Fagaceae as a whole (Brown and Mogensen 1972). The large, basal cell soon divides obliquely again, partitioning off a small cell similar to the apical cell. The three-celled proembryo then becomes four-celled as the original apical cell divides longitudinally. Additional oblique and longitudinal divisions produce a small EM and a two-celled suspensor, which structurally does not develop further during embryogeny. When the EM of several dozen cells attains a distinct globular stage, it already has a well-defined protoderm (Brown and Mogensen 1972).

13.3.2 Suspensor Systems

Suspensor systems of angiosperm embryos vary greatly in size and gross morphology. Generally, these systems are likely to be derived rather directly from early progeny of the basal daughter cell produced by zygotic division, whereas in gymnosperms the origin of suspensor systems is more variable. Angiosperm suspensor systems also are less likely than gymnosperm systems to be multi-stranded in the region where they adjoin the embryo. These differences arise partly because angiosperms generally lack the secondary suspensors characteristic of many gymnosperms (Sect. 13.2.2).

Suspensor cells of angiosperms, like those of gymnosperms, are only ephemerally meristematic and are relatively short lived. Their cytoplasm usually begins to degenerate shortly after the embryo has attained the globular stage. However, in some genera, including *Capsella*, the basal cell may maintain its integrity long after other suspensor cells have collapsed (Raghavan 1976).

In angiosperms, as in gymnosperms, suspensors tend to consist of stabilized populations of relatively large, highly vacuolate cells. Proembryos, in contrast, consist of small, densely cytoplasmic, highly meristematic cells. In angiosperms, these differences begin with the unequal apportionment of cytoplasmic and vacuolar components to the cells produced by the typically transverse division of the zygote, which is structurally polar. Early progeny of the basal cell tend to grow rapidly in volume, whereas the total volume occupied by the first several generations of progeny of the apical cell may be less than, or barely exceed, the volume of the original apical cell. Thus, suspensor volume usually greatly exceeds proembryo volume during early proembryogeny. Suspensor growth then rapidly declines and ceases as the proembryo becomes established.

Some angiosperms, including *Penaea*, *Tilia*, and many genera of the Fabaceae, lack suspensors. However, in those members of the Fabaceae that have them, suspensors vary widely in size and structure — from long filaments to stocky masses and berry-like clusters much larger than the proembryos to which they are attached (Lersten 1983).

Angiosperm suspensor systems, and their gymnosperm counterparts, have several possible functions, varying in relative importance with the taxon: (1) mechanical action in pushing the embryo through the wall of an archegonium or embryo sac; (2) absorption of metabolites and nutrients from parental tissues and translocation of them to the growing embryo (Yeung 1980); and (3) elaboration of growth-regulating substances and nutrients and translocation of them to the embryo (Yeung and Sussex 1979; Natesh and Rau 1984). However, suspensor systems are absent or vestigial in some taxa, indicating that they are not universally essential in higher-plant embryogeny.

Though reliable data are few, plasmodesmata apparently interconnect some angiosperm suspensor cells (Simoncioli 1974) and may also connect the suspensor with the embryo. A symplasmic connection between embryo and suspensor would be consistent with evidence that the suspensor performs a nutritional function for the embryo. In several angiosperm families, suspensors produce haustorial extensions that may absorb solutes from other tissues of the ovule (Natesh and Rau 1984).

The common occurrence of extreme endopolyploidy in angiosperm suspensor cells may be associated with the physiological relationship between the suspensor and embryo. It is possible that the large amounts of suspensor-cell DNA lead to augmented synthesis of proteins or growth regulators, which are then available to the growing embryo (Picciarelli et al. 1984).

In some species, the large suspensor cells with giant nuclei have wall ingrowths characteristic of transfer cells. These ingrowths may be related to transport from suspensor to embryo (see references in Natesh and Rau 1984).

13.3.3 Polyembryony

Several types of polyembryony occur in angiosperms, as in gymnosperms, although in general polyembryony is rare in angiosperms. Broadly defined, the term "poly-embryony" refers to embryos produced not only as a result of syngamy, but also by various types of apogamy or apomixis (Lakshmanan and Ambegaokar 1984), processes that are common in angiosperms but practically unknown in gymnosperms. However, we follow Johansen (1950) and include in the concept of polyembryony only those multiple proembryos arising from a zygote or zygotic nucleus.

The "simple polyembryony" of gymnosperms, resulting from fertilization of the egg in each of several archegonia in a megagametophyte, has a near-counterpart in angiosperms, in the fertilization of the egg in more than one megagametophyte (embryo sac) in an ovule. This type of polyembryony occurs in *Casuarina* and *Citrus* (Bhojwani and Bhatnagar 1978).

Cleavage polyembryony, widespread in gymnosperms, is less so in angiosperms. In gymnosperms, it is commonly associated with differential elongation of various parts of a multi-stranded suspensor system (Sects. 13.2.2; 13.2.3), whereas in angiosperms it is likely to be initiated by branching or budding of an EM derived from the apical daughter cell of a zygote. It is common in orchids, but seems to be rare among arborescent angiosperms. Nonetheless it reportedly occurs in *Cocos nucifera* (Whitehead and Chapman 1962) and probably also in *Hamamelis virginiana* (Mathew 1980).

13.4 Late Embryogeny in Gymnosperms

Embryogeny is a continuous process, not naturally divisible into distinct stages. Yet, as mentioned in Section 13.2.4, the term "late embryogeny" in gymnosperms has come to refer rather loosely to the stages that occur after the elongating suspensor components have pushed the EMs through the archegonial wall. More specifically, it refers to the stages following development of a three-dimensional nascent embryo, with numerous wholly internal cells and a layer of surface cells that function as a protoderm, though their divisions may not be exclusively anticlinal. Late embryogeny is a time of histogenesis and organogenesis. Early in this stage, the root and shoot apical meristems are delineated and the plant axis is established (Allen 1946; Spurr 1949; Guttenberg 1961).

By convention, the frame of reference for illustrating structures and events of late embryogeny is usually inverted compared with that of proembryogeny. Thus, whereas proembryos are illustrated with the shoot apical end lowermost, late embryos are shown with that end uppermost.

13.4.1 Embryonic Meristems, Cytohistological Zonation, and Histogenesis

As it enters late embryogeny, a typical gymnosperm embryo is a subglobular or elongate-cylindrical mass of small, densely cytoplasmic cells, with a hemispherical distal end. Proximally, it is continuous with a suspensor system consisting of large, vacuolate cells. The cells of the EM usually are derived from a single post-cleavage embryonal cell and presumably are genetically identical. Before they begin to differentiate, the cells in the interior of the mass may be considered simply as constituting an **embryonic meristem**.

The cells begin to show regional differences as populations become delineated among them. These differences are a response to cell position relative to the surface and relative to the suspensor cells, and also are a response to stresses arising from the geometry of three-dimensional, symplastic growth. As in the proembryo (Sect. 13.2.4), there is a proximal region, in which cell divisions tend to be transverse, and a distal region, in which divisions occur in all planes.

A boundary between proximal and distal regions gradually becomes evident as a discontinuity in the pattern of cell-wall orientation. The core of this zone of discontinuity consists of a small group of embryonic-meristem cells, which, possibly in response to positional information or signals, become the **root-organization center**.

Delineation of a root-organization center in a gymnosperm embryo marks the beginning of internal cell differentiation, which is soon followed by tissue differentiation (Fig. 13.2). Establishment of this center delineates a hypocotyl/shoot axis in the distal region of the subglobular embryo. The root-organization center is a deep-seated region of cell origin, typically consisting of a transverse, plate-like group of meristematic cells. This region lies within the core of the proximal region of the post-cleavage EM, near the region of its merger with the suspensor system.

The cells of the root-organization center at first differ overtly from their neighbors only in that the orientation of their planes of division is mostly transverse rather than random. As these cells change in this way, their rate of division tends to decline. However, by divisions of proximal and distal progeny cells, the root-organization center indirectly contributes to a large fraction of the total mass of the embryo. These progeny may divide rapidly and begin to function as tiers or groups of clone-forming mother cells; the progeny of these, in turn, begin to construct the central column or "columella" of the root cap and the central cylinder of the root-hypocotyl axis. Much of the EM is derived from this "intercalary" growth (Schopf 1943; Allen 1946; Spurr 1949; Singh 1978). Those cells of the original organization center that do not migrate away are the precursors of the root-apical-initial cells in the mature embryo and seedling (Sect. 17.1.3).

The shoot apex — specifically, the apical meristem of the epicotyl — originates quite differently. It begins as the dome-like, distal part of the globular EM and thus at first consists of undifferentiated, embryonic-meristem cells. We will not try to settle the mostly semantic question of whether the cotyledons are the first appendages produced by the shoot-apical-meristem, or whether the shoot-apical-meristem arises within the part of the original dome-like structure remaining after the cotyledons are initiated. We focus on the meristem of the nascent epicotyl, which is present after the cotyledons have been initiated.

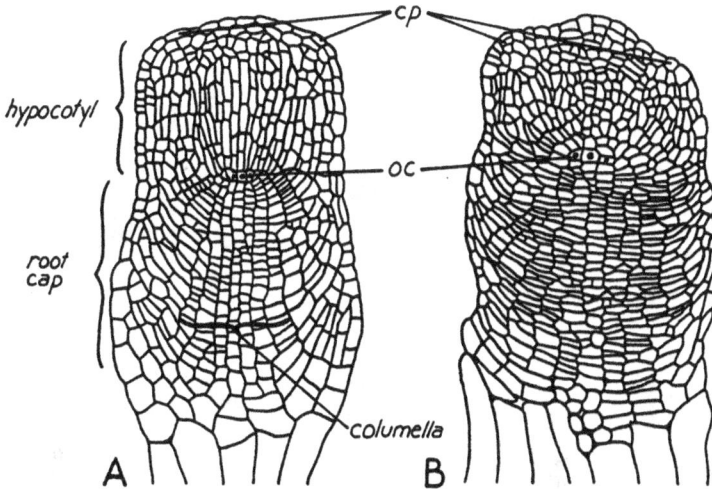

Fig. 13.2 A,B. Cell patterns during early stages of histogenesis in two conifer embryos, drawn with shoot end uppermost and suspensor system at bottom. The region of the deep-seated root-organization center (*oc*) is marked by cells with nuclei drawn in. **A** *Thuja*, with a columella of root cap cells becoming evident below the *oc*. At this stage, the nascent root cap region occupies half or more of embryo volume. The hypocotyl constitutes most of the remainder, and the epicotyl consists only of the superficial cells between the nascent cotyledonary primordia (*cp*). **B** *Cedrus* at a comparable stage. The root-cap region is more voluminous than in *Thuja*, though the columella is less distinct. A superficial shoot-meristem mound, and incipient cotyledonary primordia, are evident. (**A**, redrawn in part after Strasburger 1872; **B**, redrawn in part after Buchholz and Old 1933)

Unlike the root-organization center, the embryonic epicotyl meristem is relatively superficial. Its few cells are larger, more highly vacuolated, and less active than are the embryonic-meristem cells. These cells may be regarded as the progenitors of the central-mother cells (Sect. 14.1.3.1). Orientations of divisions among these cells are not notably restricted. Though they are an ultimate source of cells for the construction of the epicotyl later, these putative shoot-apical-initial cells contribute less to constructing a mature gymnosperm embryo than does the root-organization center (Allen 1946; Spurr 1949; Singh 1978).

A nascent pith becomes recognizable because of numerous transverse divisions among the progeny of the subapical initials and of residual embryonic-meristem cells, though files of cells as regular as those in the columella in the embryonic root cap are rare. Then, in response to unknown factors, the cells beneath the subapical region (but still within the nascent hypocotyl/shoot axis), immediately abaxial to the embryonal pith, begin to divide axially rather than transversely. Because the entire embryonic axis is elongating rapidly, these cells become long and narrow. They form a "ring" of provascular tissues or provascular bundles. Partition walls formed by cell divisions in the first provascular meristem and tissue, unlike the partitions formed earlier in proembryogeny, do not have minimal areas (Sect. 12.3.1).

Residual embryonic-meristem cells between the protoderm and the provascular ring continue to divide in seemingly random orientations, and develop into the cortex. This region may or may not be continuous with the peripheral zone of the nascent root cap.

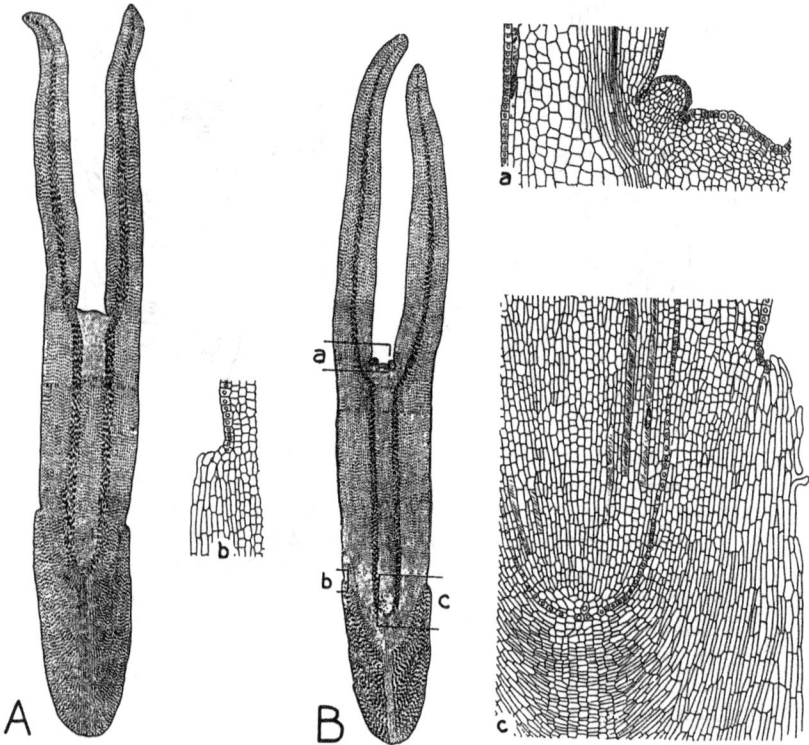

Fig. 13.3 A,B. Drawings of median longitudinal sections of mature *Cedrus* embryos. Each *stroke of shading* in **A** and **B** represents the orientation of the major axis of a single cell. Only two of the eight to ten cotyledons are represented in these sections. **A** *C. libanotica* Link (= *C. libanitica, C. libani). Length*: 1.10 mm. Note large root cap with columella, and provascular or vascular strands extending through radicle, hypocotyl, and cotyledons, but not into epicotyl. **B**. *C. atlantica. Length*: 1.07 mm. The root cap is smaller than in *C. libanotica*, but 2the columella is more distinct. Provascular or vascular strands, though well defined, do not extend toward or into the foliar primordia, which are present in the axils of the cotyledons. Drawings *a, b,* and *c* (all x 100) are based on sections of the embryo in **B**, as indicated; protodermal and endodermal cells are drawn with nuclei. In *b* and *c*, note that the protoderm ends abruptly in a juncture zone with the root cap. Suspensor cells, which have collapsed and been partly resorbed, are not shown. Protoxylem elements that have differentiated beyond the provascular stage are *shaded*. (Buchholz and Old 1933)

In some gymnosperms, protoderm covers the entire surface of the embryo and suspensor system; in others, including the Pinaceae, protoderm covers the distal end (the shoot/hypocotyl region) and ends abruptly at a juncture zone between the root cap and hypocotyl (Singh 1978; Fig. 13.3). The developmental controls that prevent protoderm from differentiating beyond the juncture zone, and determine the location of this zone, are obscure. Protoderm typically is present, at least distally, before internal histogenesis begins. One to several hypodermal layers may underlie it.

As internal tissues (and cotyledons) differentiate in the hypocotyl/shoot axis, internal tissues also differentiate in the root axis, as a basal continuation of the hypocotyl. These tissues include some provascular tissue and a cortex, but often no distinct pith.

13.4.2 Organogenesis: Cotyledons and Root Cap

As pith, provascular tissue, and cortex differentiate along the axis of a gymnosperm embryo, broad shoulders develop on the flanks of the distal region. Elevation of buttresses on these shoulders, sometimes resulting from activity of subapical meristem cells, marks the initiation of cotyledon primordia. These primordia are further elevated by periclinal divisions in the hypodermal layers and sometimes in the surface layer as well. As the cotyledon primordia elongate, provascular tissue advances acropetally into them, as a continuation of that already present in the hypocotyl.

Embryos of most gymnosperms have only two cotyledons (Butts and Buchholz 1940) and thus are bilaterally symmetrical. However, many conifers have more than two, and in the Pinaceae the number is quite variable among, and even within, genera and species. Thus, in some gymnosperms the number may not be determined solely by genetic factors, but, rather, may be affected by the conditions under which the embryo develops.

In gymnosperm embryos, the root cap may become quite massive but shows limited cytodifferentiation. Generally its tissues are not even potentially permanent. Its cells vary in size and shape due to regional differences in prevailing orientation of cell division (Fig. 13.3). These differences may be partly ascribable to stresses related to symplastic growth. Embryonal root caps may be covered by or embedded in remnants of suspensor systems.

13.5 Late Embryogeny in Angiosperms

In general, we consider that the globular stage marks the transition from proembryogeny to late embryogeny in angiosperms. During late embryogeny, a globular mass of at most a few dozen cells, which have differentiated only into internal, embryonic-meristem cells and superficial, protodermal cells, gradually becomes a mature embryo, composed of thousands of cells organized into well-defined tissues and organs. In the embryological literature, late embryogeny in angiosperms has often been treated as a progression through several imaginatively named stages — heart-shaped, torpedo-shaped, and walking-stick-shaped. These terms seem appropriate for *Capsella*, but less so for some other taxa.

Before we discuss the formation of internal tissues in angiosperm embryos, we will look at the divergence between monocot and dicot embryogeny. This divergence occurs in the globular stage.

13.5.1 Monocot-Dicot Divergence

The broad, general features of early proembryogeny in monocots and dicots are similar. Then, in the globular stage, the progeny of the four-celled (quadrant-stage), proembryogenic, apical tier show quadrant-based differences in development between the two groups. In some, or perhaps most, dicots, the progeny of two diametrically *opposite* quadrant cells produce the two cotyledon primordia (as in *Capsella*), while the progeny of the other two quadrant cells are involved mainly in initiating the epicotyl. In monocots, cells of two *adjacent* quadrants typically are the progenitors of the cells of the single cotyledon, while the progeny of the two remaining adjacent quadrant cells construct the epicotyl (Lakshmanan 1972). Thus, the epicotyl and the single cotyledon in monocots seem to arise from the apical quadrants as regularly as do the epicotyl and two cotyledons in many dicots (Swamy 1979).

In monocots, the single cotyledon, which arises on one flank of the apical region, grows very rapidly, while the primordial-epicotyl apical meristem, which arises on another flank, languishes. As a result, the epicotyl meristem seems to be subverted to a lateral position, but it is not clear whether this meristem is truly lateral or not (Natesh and Rau 1984).

13.5.2 Embryonic Meristems, Cytohistological Zonation, and Histogenesis

We refer to the collective mass of undifferentiated cells in the interior of the globular embryo of angiosperms as "embryonic meristem", because we have difficulty reconciling the various definitions and usages of terms such as fundamental meristem, ground meristem (except when referring to a specific embryonic region), and promeristem. We have also used the term "embryonic meristem" for the cells within the subglobular to early cylindrical embryo of gymnosperms (Sect. 13.4.1).

During late embryogeny in angiosperms, there is cytodifferentiation, coupled with active cytogenesis. As a result, the embryonic meristem becomes differentiated into ground meristems named for the tissues they produce — specifically, cortical ground meristem and pith ground meristem. Delineation of these ground meristems typically leaves residual areas of embryonic meristem, which give rise to shoot- and root-apical meristems and possibly a provascular meristem.

In some angiosperm species, cytodifferentiation has been interpreted as beginning among the progenitors of the radicle cortex. This region is abaxial to a region that is analogous to the root-organization center in gymnosperm embryos (Fig. 13.2). The cells of the nascent cortical ground meristem become increasingly vacuolated. As the globular embryo begins to elongate, differentiation of these lightly staining cells is propagated into the presumptive hypocotyl region. In *Malus* (Meyer 1958) and *Juglans* (Nast 1941), but not in *Downingia* (Kaplan 1969), differentiation of outer, or cortical, ground meristem is soon followed by differentiation of an inner ground meristem in the presumptive pith region.

As seen in transection, an annulus of less vacuolated, more densely staining cells remains between the lightly staining, inner and outer ground-meristem regions. This annulus, more obvious in angiosperms than in gymnosperms, is variously referred to as residual meristem, meristem ring, prodesmogen, procambium, and provascular tissue. The term of choice depends partly on one's concept of the sequence of differentiation. In a population of uniform, uncommitted, embryonic-meristem cells, does differentiation first set apart provascular tissue between an inner and outer region of undifferentiated cells, or do inner and outer ground-meristem regions differentiate first, leaving a "residual" meristem between them? This question is perhaps of little consequence because a residual meristem would be ephemeral in any case, soon differentiating into provascular tissue. Recognizing this situation, Meicenheimer (1986) proposed the term "residual meristem/procambium" (RMPC). Mindful of experimental studies on vascular differentiation (Sect. 14.3.1), we use the term "**provascular tissue**", but in the sense of RMPC. As in gymnosperms (Sect. 13.4.1), provascular cells in angiosperms tend to divide by partitions that are near maximal rather than minimal in area.

Cytohistological zonation and histogenesis in *Nerium oleander*, as described and interpreted by Mahlberg (1960), differ somewhat from the patterns just discussed. In *Nerium*, distinctive tissue zones become detectable in the embryonic meristem when the globular embryo is about 100 μm in diameter. These include a highly meristematic protodermal layer. Most divisions in this layer are anticlinal, but some are oblique and periclinal, especially near the suspensor.

In contrast to the angiosperm species already described, internal histogenesis in *Nerium* begins with differentiation of a pith ground meristem of enlarged, vacuolate cells, just beneath a group of embryonic-meristem cells that are the progenitors of the shoot-apical initials. Differentiation of pith ground meristem in *Nerium* progresses downward, by vacuolation and cell growth, until it is within a few cells of the progenitors of the root-apical initials. No pith differentiates in the root. As the pith ground meristem is differentiating, a cortical ground meristem differentiates by a similar process in the peripheral region of the embryonic meristem. Differentiation of this meristem progresses downward to the margins of the suspensor.

A thin layer of undifferentiated embryonic meristem remains between the pith and cortical ground meristems. By cell enlargement and division, the most abaxial part of this layer soon becomes recognizable as a **lateral expansion meristem** of the cortex. Activity of this meristem is nonuniform, both axially and circumferentially. Axially, it is most active in the upper hypocotyl and least active near the suspensor. Circumferentially, it is most active in two opposite quadrants, and ultimately gives rise to cotyledon primordia (Sect. 13.5.3). This nonuniformly distributed meristematic activity, combined with generally higher growth rates in the distal half of the embryo, leads to a heart-shaped stage. The heart-shaped embryo is bilaterally symmetrical in its distal region but remains radially symmetrical in its proximal region, especially near the suspensor.

The lateral expansion meristem and the cortical ground meristem produce a cortex of variable width. The cortex abuts adaxially on a thin residual "ring" of embryonic-meristem cells, which differentiate into a provascular meristem. This differentiation begins within the lobes of the "heart" and progresses basipetally. In striking contrast to neighboring cells derived from the lateral expansion meristem, provascular-meristem cells undergo little lateral expansion. Instead, they become long and slender, as do the provascular cells of other angiosperms and of gymnosperms (Sect. 13.4.1).

In the *Nerium* embryo, both root- and shoot-apical meristems can be interpreted developmentally as residual groups of embryonic-meristem cells present in the late globular stage, not as new structures differentiating after pith, cortical, and provascular regions have been delineated (Mahlberg 1960). The shoot apex is delineated gradually, beginning early in the globular stage and continuing through the heart-shaped stage. Throughout the globular stage, cell divisions are more frequent in the distal than in the proximal region of the shoot, and, functionally, the nascent shoot apex extends across the whole dome-like, distal part of the embryo. The protoderm covering the apex constitutes an embryonic tunica. The several irregular layers of cells in the apical corpus region are a residuum of embryonic-meristem cells, which, though hardly differentiated, constitute the central cell zone of the nascent shoot apex.

The *Nerium* root apex, like the shoot apex, becomes discernible as a residuum of embryonic-meristem cells, after pith, cortical, and provascular meristems have been delineated. The root apex, however, is not subprotodermal as is the shoot apex. No continuous protoderm, in a rigorous sense, forms between the root end of the embryo and the suspensor system. Instead, the root apex arises at this interface, occupying a gap in the protoderm. Its progeny, in addition to constructing tissues of the primary root axis, also form the central part of the root cap (Sect. 13.5.3).

13.5.3 Organogenesis: Cotyledons and Root Cap

The origin and early development of cotyledons have been studied in relatively few angiosperms, so that it is difficult to generalize about broad groups of these plants. However, in *Nerium*, each of the two cotyledonary primordia originates through localized periclinal divisions in the subprotodermal layers at the periphery of the dome-like, distal part of the globular proembryo. They arise in those regions that were most broadened by activity of the lateral expansion meristem (Sect. 13.5.2).

The primordia at first are two buttresses of cells. Within each buttress, a subdermal, cotyledonary meristem becomes organized. The cells of these meristems divide in various planes, and their progeny become the internal cells of the cotyledons. The cotyledons, as they elongate, remain covered by protodermal cells, which divide almost exclusively anticlinally. Simultaneously, differentiation of cortical and provas-cular tissue, but not pith, is propagated into the cotyledon primordia from the hypo-cotyl/shoot axis. The subdermal, cotyledonary meristems remain active until the cotyledons are several hundred micrometers long (Mahlberg 1960).

Not all taxonomic "dicots" are genetically determined to produce two cotyledons. For example, as many as 12% of mature *Malus* embryos (of some cultivars) have three cotyledons (Meyer 1958). Further, 87% of embryos of the tropical tree, *Degeneria*, regularly have three cotyledons, and the remaining 13% have four (Swamy 1949). Some taxonomic "dicots", including *Claytonia virginica* and *Anemone apennina* (Haccius and Fischer 1959), have monocotyledonous embryos.

The embryonic angiosperm root cap does not, as a whole, derive from any single group of progenitor cells. In *Nerium*, it begins by periclinal divisions of early proto-dermal cells around the suspensor attachment site. Contributions are also made by capward derivatives of the root-apical-meristem initials, which are located in the gap in the protoderm of the root apex. Cells from these two sources gradually form a continuous layer across the boundary zone between the embryo and suspensor. The root cap of the mature embryo arises from periclinal divisions of cells in this layer (Mahlberg 1960).

The relative size of the root-apical gap in the protoderm and of any rib-meristem-like columella along the axis of the root cap are variable. However, tissue differentiation within the embryonic root cap is always very limited, and the outer regions of the cap are sometimes not sharply delineated from suspensor tissue.

13.6 Developmental Controls During Embryogeny

Control of embryo development is so complex that we can here try to address only a few central questions. First, there is a question fundamental to all of embryogeny: how does the embryo, a new individual and generation, develop as an organism separate from, but not totally independent of, the tissues of surrounding, older generations? That is, how can the zygote or zygotic nucleus, a new diploid generation, pursue a developmental pathway different from that of the megagametophyte, which is haploid? How, in turn, can the megagametophyte be developmentally independent of the older, diploid-generation cells around it?

Two levels of hierarchical control seem to be involved in the development of new generations. At one level, there is control of activation of segments of genetic information. At another level, there is control of rates of growth and development.

Control of activation of genetic material seems to reside in the new organism itself. This independence from surrounding organisms is manifested in the severing of symplasmic connections between the generations. As a result, the various generations may constitute separate symplasms. For example, in at least some taxa, callose is deposited on the walls of the megasporocytes before these cells give rise to the megaspores, which in turn give rise to the megagametophytes. Because callose deposits plug the plasmodesmata during microsporogenesis (Górska-Brylass 1968), it seems likely that callose has a similar effect during megasporogenesis, isolating the gameto-phytic generation from the older, sporophytic generation (Rodkiewicz 1970). Although plasmodesmata reportedly connect the functional megaspore of *Glycine max* to the nucellar cells (Kennell and Horner 1985), these connections appear to be broken at a later stage, and, in angiosperms generally, there may be no functional plasmodesmatal connections between a mature embryo sac and its parent sporophyte. Similarly, in gymnosperms, the megaspore membrane, which surrounds the megagametophyte, probably is symplasmically discontinuous with the parent sporophyte (Singh 1978).

Within an embryo sac or megagametophyte, there may be some plasmodesmatal connections, but only at an early developmental stage. For example, in gymnosperms, the cytoplasm of the egg cell is in symplasmic contact with the cytoplasm of other cells of the megagametophyte (Maheshwari and Singh 1967). In *Capsella* (Schulz and Jensen 1968) and perhaps other angiosperms, there reportedly are some plasmodesmatal connections of the egg cell with the synergids and the central cell, or "polar" nuclei. In the *Capsella* zygote, these connections persist, but disappear during an early phase of embryo development. No plasmodesmata seem to occur in the wall of the zygote in *Quercus gambelii* (Singh and Mogensen 1975) or in *Hordeum vulgare* (Norstog 1972). In *Ledum* and *Rhododendron*, a callose wall is deposited around the zygote shortly after fertilization (Williams et al. 1984). In general, embryos maintain their symplasmic independence during development.

Although it seems that there must, at certain times and locations, be some barriers to information transmission to the ovule and zygote from other cells and tissues, this isolation raises the question of how the temporal aspects of development, such as rate of embryo growth and onset of dehydration and dormancy, can be integrated across boundaries (diploid-haploid-diploid) between the generations. Classical embryology mostly did not address this problem.

Information may be transmitted as a result of the apoplasmic continuity between the generations, which is maintained despite the symplasmic discontinuity. The older generations may also modulate the rate of embryonic development via mechanical forces. For example, mechanical forces can greatly influence embryo morphogenesis in pteridophytes (Ward and Wetmore 1954; Steeves and Sussex 1989), and probably also in embryos of seed plants (Lintilhac 1974a,b). Although the embryo of a seed plant is less accessible to developmental study than is that of a pteridophyte, it seems evident that the embryo-sac fluid, which is the cytoplasmic and vacuolar contents of the central cell, must exert an osmotic effect on the embryo. Because this fluid is enclosed by a plasmalemma, it may also exert turgor stress on the embryo. The endosperm, which typically develops within the central cell, may exert further mechanical stress.

Other physical factors may also play a role in embryogeny. For example, in certain physical systems and in zygotes and early proembryos, the tendency towards a minimum of free surface energy leads to the formation of partition walls that have a minimal area and that initially meet existing walls at right angles (Sect. 12.3.1). In later cell divisions in the embryo, physical regulation of orientation is increasingly modulated by intra-organismal developmental control. This may be why some taxonomically distant groups have the same early division pattern, whereas their later division patterns differ — sometimes enough to allow phylogenetic inferences to be made.

14 The Primary Shoot: Apex and Caulis

14.1 The Primary Shoot Meristem

14.1.1 General Characteristics

In higher plants, the meristematic cells at the tip of a main shoot axis, or a branch thereof, constitute a **primary shoot meristem**, or **shoot-apical meristem**. All the tissues of each main or branch axis originate from the cells of the primary shoot meristem at the end of that axis. Of course, in an ultimate sense, the tissues of the entire shoot originate from the primary shoot meristem of the embryo.

A primary shoot meristem produces two classes of components: **cauline** (stem-like, or constituting a caulis) and **phylloid** (leaf-like, or foliar). The foliar organs begin as primordia on the lower flanks of the meristem. If the leaves and leaf primordia are removed from the shoot, only the caulis remains. Several additional terms pertain to the primary shoot meristems of higher plants:

The **shoot apex** (narrow sense), at the distal end of the caulis, is that part of the primary shoot meristem that is not yet differentiated into cauline and foliar components. Rigorously, the shoot apex is not part of the caulis, but in general discussions it is deemed to be so because it is physically integral with the caulis and because the profile of the shoot apex complements and completes the profile of the caulis.

The **subapical meristem** includes all the primordial cauline nodes and inter-nodes delineated by initiation of foliar primordia. It includes neither the shoot apex (narrow sense) nor the foliar primordia themselves. Basally, the subapical meristem melds with the region of stem elongation, in which coordinated cell extension growth and cell division result in final delineation of nodes and inter-nodes.

Intercalary meristems are persistent cauline meristems separated from the apical and subapical meristematic regions by segments of more mature stem tissues. Intercalary meristems are commonly coin-shaped regions, oriented transversely to the axis at the bases of internodes.

Although these terms are adequate for discussing primary shoot meristems, much of the literature does not defer to such restricted definitions. Many authors have used "shoot apex" in a broader sense than defined here, and many authors refer to "apical domes".

The **shoot apex** (broad sense) includes the shoot apex in the just-defined narrow sense, as well as the subapical meristem, as defined here, and those foliar primordia that do not overtop the vertex of the meristem, though this last is an arbitrary criterion. We use the term in its broad sense primarily in discussing the work of others who so use it.

In many gymnosperms and dicots, the shoot apex (narrow sense) takes the form of a beautifully regular **apical dome**. This dome is that part of the shoot apex (broad sense) that is distal to the youngest leaf primordia, and therefore is "bare". Apical domes typically are paraboloidal or hyperboloidal and sometimes nearly hemispherical. Other apices may have the form of a terrace between a descending flank on one side and an ascending flank on the other. Still others are saddle-shaped, located between two opposite, distichous primordia. Sometimes, the upper surface is even cup- or bowl-shaped.

The range of shapes of the external surface of the shoot apex is matched by a range of shapes of the internal boundary between the apex and the subapical meristem beneath. For example, that rare apex that is a true hemispherical dome is delineated from the subapical meristem by the uppermost (primordial) nodal plane, which is the most recently formed in a series of transverse nodal planes geometrically defined by the axils of the successive leaf primordia. In theory, this plane is orthogonal to the surface at the base of the dome and to the shoot axis. This basal-plane concept is also applicable to apices that, though not hemispherical, have high dome-like profiles. However, it is clearly inapplicable to apices that have low dome-like, flat, or concave surface profiles; on the basis of studies of cell arrangement and cell lineage, it seems obvious that such an apex should have a cup-like basal (interior), boundary, orthogonal to the surface of the shoot apex in the axil of the youngest leaf primordium and to the cauline axis.

Thus, the base of a shoot apex may vary in shape from transverse to cup-shaped. If the shoot apex is a hemispherical or other high dome, the basal boundary is a transverse plane. In lower domes, the basal boundary is more deeply cupped, becoming hemispheric in a flat dome. In those rare apices that have a concave upper surface, the lower boundary in theory approximates the boundary of a sector of a sphere larger than a hemisphere.

According to this concept, the basal boundaries of many shoot apices tend to be somewhat dish- or cup-shaped. Nevertheless, it is convenient, and has become a convention in discussing and analyzing growth of dome-shaped apices, to imagine their basal boundaries as being approximated by transverse planes normal to the shoot axis. Likewise, subapical meristems are conventionally represented as including a set of transverse planes, each passing through one primordial-leaf axil (less commonly two or more axils). The part of the subapical meristem between two successive transverse planes is a **frustum** (a solid segment of a cone delineated by two parallel planes intersecting the cone). Each cauline frustum supports at least one foliar primordium on its periphery. The combination of the frustum and the primordium it bears is a **frustum-primordium unit**.

Geometrically, the subapical-meristem region can be imagined as a stack of frusta (Fig. 14.1). However, in some apices, especially in distichous shoots (Sect. 14.2.1), the region resembles not so much a stack of coin-like frusta, delineated by parallel transverse nodal planes, as it does a stack of transversely oriented, wedge-shaped sections, each tapering from maximal thickness at the base of the nascent primordium to zero thickness 180° opposite. It can be shown mathematically, though, that these wedge-shaped sections have the same thickness at their centers, and the same volumes, as their coin-like counterparts (Romberger and Gregory, unpublished calculations). Thick-stemmed monocots, however, may have a structure requiring a still different geometric model (Sect. 14.2.2).

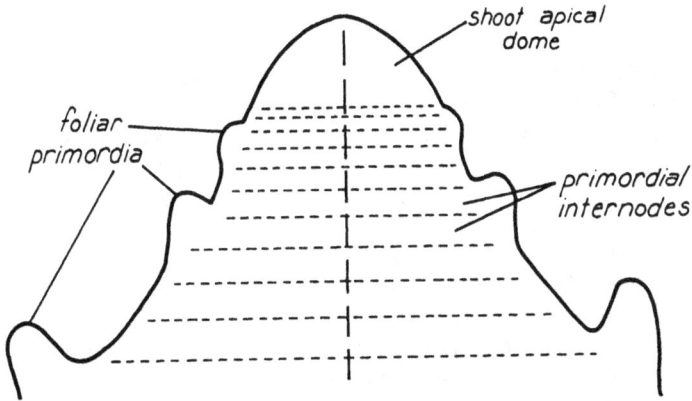

Fig. 14.1. Schematic of median axial section of a shoot tip with the subapical region of the caulis represented as a stack of frusta delineated by imaginary, primordial nodal planes (-----) normal to the axis. Such a diagram seems simplified because only a few of the many foliar primordia that may be present around the circumference of the shoot tip are intersected by any one median axial plane. The upper (younger) frusta are shown as axially shorter than the lower ones, as they would be in a bud. The basal regions of the frusta typically elongate into internodes during shoot extension. Foliar primordia are represented in several sequential developmental stages: protrusion, buttress, and phyllopodium (Sect. 16.2)

The **initial cells,** in the upper part of the shoot apex (narrow sense), are the ultimate source of all cells of the epicotyl (Sect. 14.1.2). This region is a permanent meristem (though the initial function may devolve to a succession of cells; Sect. 14.1.3.2). This is a region of continuing functional embryogeny. The lower part of the shoot apex, when active, is characterized by cytogenesis and the beginning of differentiation into stem tissues. Its peripheral region, the organogenic zone around the lower flanks of the dome, is the site of foliar initiation.

Initiation of a new foliar primordium (or set of primordia) adds a frustum to the subapical caulis and thus delineates a higher basal boundary of the shoot apex. If the volume growth of the apex per plastochron (see later) is larger than the volume of cells "exported" from the apex to the new frustum-primordium unit, the excess is a "capital investment" that increases the volume and growth potential of the "working" apical meristem. The investment relative to the total growth has been calculated in some meristems and expressed as an investment ratio. After being as high as 24% in 10-day-old *Picea abies* seedlings, it declined to zero as the apical dome attained a steady-state volume, in 130-day-old seedlings (Romberger and Gregory 1977). A decline in apical volume, indicating a negative investment ratio, accompanies development of male strobili in some conifers (Romberger and Gregory 1974).

Meristematic growth typically occurs throughout the primary shoot meristem, though at varying rates. This growth has several components: (1) periodic initiation of foliar primordia and delineation of cauline frusta, (2) volume growth of existing cauline frusta, and (3) growth of foliar primordia and the meristematic parts of leaves.

Because growth occurs throughout the apical meristem, the successive leaf-frustum units in the meristematic shoot at any instant form an axial developmental series. Each frustum-primordium unit is one morphogenic time unit older than the one immediately above it. The morphogenic time between the initiation of the foliar primordia borne on two successive frusta is a **plastochron**. Plastochrons have no constant relation to chronological time (Sect. 16.5.2). Morphogenic time relations among successive leaf-frustum units are useful in calculating various relative growth rates from anatomical measurements.

The volume of a shoot apex is more dynamic than its shape, because its height and diameter may change appreciably on a time scale of hours or days, even if the overall shape of apex and caulis is nearly constant. Volume of the apex increases during a plastochron, then suddenly decreases when the initiation of a leaf primordium (or set of primordia in other than spiral monojugate systems) delineates a new frustum-primordium unit at the base of the apex. Shoot apical-dome volume and surface area are minimal just after a new primordium is initiated.

Foliar primordia, which by their position delineate cauline frusta, vary in sectorial size, both with ontogenic stage of the apex and with the taxon. A primordium may encompass nearly the whole circumference of a frustum-primordium unit, or only a small fraction of it. The spatial aspects of initiation of primordia and arrangement of leaves on the caulis are discussed in Chapter 15.

Because the leaf primordia are crowded in the bud or shoot tip, their growth imposes mechanical stresses in the shoot-apical meristem (Williams 1975). Any deviation from symplastic growth of the various primordia and frusta increases mechanical stress. Feedback from this stress can, we suggest, evoke corrective growth (Sect. 12.4.6). Mechanical stresses are probably a factor in regulating symplastic growth in various parts of the primary shoot meristem, and possibly also in regulating organogenesis on the flanks of the apex (Lintilhac 1984).

14.1.2 Initial Cells: Their Location and Slow Growth

Meristematic cells must take one of two developmental courses. One course is to divide repeatedly. This maintenance of the meristematic function provides potential for continued rejuvenescence and postponement of mortality. The other course is eventually to differentiate according to position within the organism. That is the way to structural specialization, to functional maturation, and — except for rare, post-maturation rejuvenation — to aging, senescence, and death.

Relatively few cells can be at the focus of meristematic activity and thus be, for a time, the apparent source of the cells of the shoot. Those many cells that are displaced from the meristematic focus become part of an outward and downward flow of differentiating progeny. The few cells that remain at the focus for many mitotic cycles can be thought of as the **initial cells** and as being repeatedly rejuvenated by divisions.

In this section, we focus on the developmental concept of initial cells, describe ways in which they can be identified, and compare growth rates of these cells (and their immediate derivatives) with growth rates of other cells in the primary shoot meristem. We defer until Section 14.1.3.2 our treatment of another important aspect of initial cells — their relative permanence. In many taxa of higher plants, the long-term composition of the group of cells harboring the initial function is probably more dynamic than static.

The position of an initial cell, in the distal region of the shoot apex, is precarious. Figuratively, if it maintains itself as an initial, it is doing a successful balancing act atop a pedestal. A slight inclination to either side is apt to lead to a "fall" from the initial position into differentiation. Once a cell is displaced, even slightly, from the meristematic focus, it becomes susceptible to further displacement as the cell or cells between it and the focus grow and divide.

We can appreciate the implications of this short-term stability in the face of long-term instability by conducting a "mental experiment". Let us imagine that we have available a technique by which we can at some instant mark each cell in the distal part of a dome-like shoot apex with a different color, and that the progeny of each cell will retain this color marking. That is, suppose that we assign one of the 2000 or so distinguishable colors to each of as many cells and then allow the apex to grow normally. With this imaginary technique, we can "see" the progeny or clones of cells derived from the marked cells as ontogeny proceeds. What will happen to the distribution and variety of colors within the apex during one or more days of growth?

As the variously colored cells divide, forming clones, they seem to move downward relative to the vertex. It soon becomes obvious that the number of colors in the apical dome is decreasing. Those colors that were given to cells low down in the dome are first to disappear. In their places, we see small clones of cells with colors that were originally given to cells farther up. We also see that the lower borders of clones move downward faster than do their upper borders. This is because the cells within clones are growing and dividing. A clone that does not show different rates of downward movement between upper and lower borders is not growing axially. The migration of the clones of cells constituting the shoot apex has been called ontogenic (or diplontic) drift (Balkema 1971).

Eventually, the apex attains a steady state, its cells being either of a single color or of only a few colors. Clones of all other colors will have "fallen" into the regions of differentiation and maturation. If the apex has several persisting colors, they are apt to be arranged in one or several discrete layers and a core. A layer may consist of two or three sectors of different colors, but never of a mosaic of colors. Those few cells having colors that persisted in the distal part of the apex through the entire experiment were the functional initials during that period, though their permanence over weeks or months is not proven or implied.

Regrettably, this experiment cannot be done, because no such color-marking technique exists. One can, nevertheless, be quite confident of the predicted results because genetic mutations or ploidy changes that have been studied in chimeral apices (Sect. 14.1.5) yield similar results.

It follows from this thinking that the concept of shoot-apical initial cells in higher plants is not basically a morphological-anatomical one. The initial cell is better defined developmentally than morphologically. In contrast, in all bryophytes and many pteridophytes, the developmental concept conforms with a morphological one — that is, there is a single initial cell that is distinctive in appearance as well as being developmentally unique.

A seemingly anatomical-morphological method of apical-growth analysis sometimes allows initial cells of seed plants to be identified, but the criteria used in this method actually are fundamentally developmental, in that they depend on the recognition and mapping of the borders of cell lineages (clones) and noting their movement with time. Lineages having upper borders that do not drift downward are originating from initial cells. The uppermost cells of these lineages are always located at or near the vertex of the apical dome and at the apical extremity of the layer or sector of the layer that they perpetuate. The initial cells of such layers or sectors must contact the axis and, with few exceptions, must divide only anticlinally. The fact that these criteria for identifying initial cells are long-term developmental rather than anatomical-morphological is not always appreciated.

Inferences about growth-rate distribution in the shoot apex have been drawn using a variety of techniques, including comparisons of mitotic indices and incorporation of DNA precursors (Lyndon 1976). These studies have shown that growth rates in the distal region, where the initial cells are located, are lower than rates in more proximal regions. In all annuals studied, frequency of division in the distal regions was only half or less of that in the proximal part. For example, in *Datura*, the mitotic index of the distal region was about half that of the proximal region (Corson 1969). In *Oryza*, the difference was about eightfold (Rolinson 1976). In *Coleus*, the mitotic indices in the distal and proximal regions were 0.5% and 3.1%, respectively, while the corresponding percentages of nuclei that incorporated ^{3}H thymidine were 1.6 and 5.7 (Saint-Côme 1966).

In trees, for which fewer data such as these are available, mitotic indices of less than 0.5% in the distal zone and 3% or more in the proximal parts have been found. Our unpublished studies of the arborescent genera *Alnus*, *Betula*, *Fagus*, *Juglans*, *Magnolia*, *Platanus*, and *Quercus* indicate that, in contrast to herbs, there are obvious cytohistological differences (Sect. 14.1.3.1) between the distal zone (the initial-cell groups and their immediate derivatives, plus the so-called central mother cells) and the more proximal regions.

In some plants, growth rates in the various apical regions can also be measured indirectly, as we have done for *Chamaecyparis pisifera* (Fig. 14.2). In the central regions of such apices, there are easily distinguishable strands of axially elongated cells that are rich in resin and that no longer divide. If we assume that the cellular pattern is steady, then individual cells "flow" through the apex, successively assuming the positions **A**, **B**, and **C**, shown in the figure. From the lengths of successive, non-dividing cells, we can infer that the growth rate in the distal part of the *Chamaecyparis* apex is low compared with that in the proximal part.

In our opinion, however, there are no significant data from either herbaceous or arborescent plants to indicate that the distal zone contributes no cells to the vegetative shoot. As Wardlaw (1957) noted, the initial cells would need to divide only very infrequently to provide continuous, if slow, replenishment of the cells proximally. Even if each cell of the distal zone divided only twice a year, the region corresponding to the *"anneau initial"* (Sect. 14.1.3.1) would during a year be repeatedly replaced by progeny of the initial cells. This is an inevitable consequence of the typical dome shape of most shoot apices, the relative numbers of cells in the two zones, and their relative growth rates.

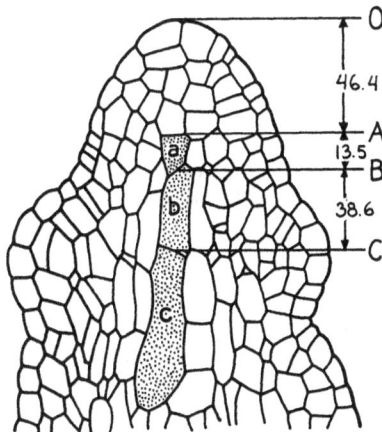

Fig. 14.2. Cell size and arrangement in a median longitudinal section of a typical *Chamaecyparis pisifera* shoot apex. The dimensions here labelled *OA*, *AB*, and *BC* are mean values of comparable measurements in ten such sections. The relative lengths of the resin-filled cells (*stippled*) constituting the strand *a-b-c* can be used to estimate the normative linear growth rate at comparable levels in the mean axis. See text for further interpretation. (Redrawn after Hejnowicz 1973)

Nevertheless, some botanists have considered the "initial function" to be associated with high rates of cell division. Accordingly, because the most distal cells of the shoot apex usually grow and divide slowly, these workers have sought the functional initial cells more proximally. The concept of an *"anneau initial"* in the peripheral, organogenic region of the shoot tip expressed this tendency (Plantefol 1947; Camefort 1956).

In botanical science, there often is lack of agreement on rigorous, authoritative definitions, and authors appeal to "common" understanding or usage. We believe that our definition of the term "initial cells" as the *ultimate source* of all cells of the shoot agrees with common usage as applied in a developmental rather than a histological context. According to our usage, the initial cells *are not* centers of rapid cell generation. They *are* the ultimate sources of all cells, though they *are not* the immediate functional sources of most cells.

In the short term, the direct contribution of an apical initial cell to production of shoot tissues may be imperceptible. In the long term, however, its contribution, as measured by the size of the clone (lineage) derived from it, tends to be major.

14.1.3 Concepts and Terminologies of Shoot Apical Structure and Function

14.1.3.1 Established Perspectives, Based on Origin, Histogenic Function, and Cytohistological Zonation. The structure and function of shoot apical meristems can be discussed from four perspectives. Three are well established; the fourth is emergent. The three established perspectives emphasize, respectively: (1) **origin** of parts (layers, zones), (2) **histogenic function** of the parts, and (3) **cytohistological differences** between parts. The fourth, the emerging perspective, focuses on the **relative permanence of initial cells**, as revealed by loss or retention of the mutations manifested by chimeras. In this section, we review the terminology pertaining to the three established perspectives. We consider the fourth perspective in Section 14.1.3.2.

From the first of the three established perspectives, we see one or several **germ elements**, each consisting of a relatively few cells, including its own initial cell or cells. From the second perspective, we see histogenic regions, or **formative elements**, each consisting of dozens or hundreds of cells that are rapidly dividing. The third perspective reveals **structural elements** by emphasizing cytohistological zonation. Unfortunately, in describing apical-meristem structure from these three perspectives, different authors have used the same terms with different intentions. We will refer to these established perspectives by the key terms (already introduced) of (1) "origin", (2) "histogenic function", and (3) "cytohistological zonation". The expression "structured versus stochastic" describes the emergent, fourth perspective, which focuses on the relative permanence or impermanence of individual initial cells (Sect. 14.1.3.2).

Consideration of the "origin" perspective (No. 1) to the exclusion of others leads to such terms as "apical layers" (Stewart et al. 1974). However, cell arrangements in the so-called "layers" are not always discrete. For example, cells of the innermost of the several layers (and sometime those of the outer layers) may divide periclinally as well as anticlinally and hence contribute to the central region of the shoot apex as well as to the more peripheral regions.

The term **histogen** has been applied to layers or zones of the shoot apex in two somewhat different ways. The priority of usage goes to Hanstein (1868), who incorporated both "origin" and "histogenic function" perspectives (Nos. 1 and 2) in his definition. He introduced the term to designate those "germ elements" (as collections of cells) of the root or shoot apex that have their own initial cells *and* that predictably give rise to specific tissue systems of the mature organ. In the shoot apex, Hanstein distinguished three histogens — dermatogen, periblem, and plerome — which he thought to generate the epidermis, the cortex, and the central cylinder, respectively. We now know that three such distinguishable germ elements, each with its own initial cells, do not occur generally in higher-plant shoot apices. Even in those shoot apices in which they occur, the histogenic function of the germ elements does not necessarily conform to Hanstein's proposals. Therefore, Hanstein's histogen terminology is now little used to describe the structure of shoot apices, though it still has some currency relative to root apices (Sect. 17.1.1).

The second meaning of "histogen" comes from Guttenberg (1960), who used the term only to refer to "histogenic function". As a consequence of his work, "histogen" is now more commonly used in the sense of "formative element" than of "germ element". Guttenberg's rationale was that cell differentiation and histogenesis begin not at the vertex of the shoot apex, but some distance below it. Thus, the "histogens" usually can be distinguished as layers or strands of cells in the lower flanks of the shoot apex, after their role in tissue formation has been manifested by beginning cytodifferentiation. Guttenberg's histogens may or may not have their own initial cells, depending on the cellular organization at the vertex, and probably are usually polyclonal compartments (Sect. 12.4.4), though in extreme cases they may be monoclonal.

Other sets of terms that have been used to describe internal structure and putative function within shoot apices generally apply to the "cytohistological zonation" perspective (No. 3) or to cytohistological differences combined with "origin" or with presumed "histogenic function". Differences in a wide range of cell properties and activities have been used as criteria to delineate cytohistological zones, including: orientation and frequency of cell divisions; relative growth rates; synthesis of various cytoplasmic, nuclear, and cell-wall constituents; and development of organelles. Such differences may be detectable singly or in combination. For example, different combinations of rates of volume growth and rates of cell division can produce cells of differing sizes in adjacent regions. Differences in growth rates in various directions can interact with changes in orientation of cell divisions to effect obvious differences in cell shape.

The diverse cytohistological terminology applied to shoot apices reflects actual diversity in zonation less than it reflects superposition of pairs of zones — each pair being based on "more-less" judgments of a single attribute. To illustrate: assume that a certain cytohistological feature is obviously at a "more" level in a certain zone of the shoot apex relative to the remainder of the apex, which is thus at a "less" level. An apex, though, is likely to show differences in more than one attribute, and thus more than one pair of "more-less" zones is usually distinguishable. These pairs of zones may be variously superposed, forming a pattern of three or four or more subzones based on several attributes each. Examples of paired zones based on single attributes are: tunica and corpus, distal and proximal regions, more densely staining and less densely staining. Separate discussions of such subzones, however, would be difficult and of doubtful utility. Therefore, in what follows we confine ourselves to discussing paired zones based on single attributes.

Tunica/corpus terminology, introduced by Schmidt (1924), is derived from studies combining the "origin" and the "cytohistological differences" viewpoints. Schmidt defined the tunica as the peripheral part of the apex, usually consisting of one cell layer or several layers, within which divisions are exclusively anti-clinal. The corpus is the core of the apex, within which the orientation of cell divisions is unrestricted. The corpus and each layer of tunica have their own separate initials or small sets of initials. In apices in which the outer cells of the corpus show a predominance of anticlinal divisions, the border between tunica and corpus is not sharp.

Apices of *Araucaria* and some related gymnosperms have tunica/corpus zonation, but in most Pinales, Ginkgoales, and Cycadales, some periclinal divisions occur in the surface cells, even near the vertex. Thus, the surface layers are not truly discrete and persistent. Some angiosperms have four or even five tunica layers, but most have no more than two. Shoot apices having one or several discrete tunica layers overlying a corpus are sometimes referred to as "stratified".

Though zonation based on **proximal** versus **distal** regions is simple in theory, the border between the two is poorly delineated. The distal zone includes the vertex and the apical initial cells. Its lower boundary is orthogonal both to the shoot axis and to the surface of the apical meristem, and thus is somewhat cup-shaped in a dome-like apex. The rest of the apex is proximal. Cytohistological differences between these zones are due mostly to slower growth in the distal zone, resulting in, for example, differences in stainability and in numbers of organelles.

Differences in intensity of histological staining are derived mostly from differences in nucleic-acid content of nuclei and cytoplasm. Most stainable are generally the small, densely cytoplasmic, rapidly dividing cells around the lower periphery of the apical dome, just below the proximal/distal boundary. This intensely stainable zone is the "eumeristem" of some authors (see later).

In addition to "more-less" schemes of zonation, there is a scheme focused on an "initial ring", and another on a "central-mother-cell zone".

Plantefol (1947, 1948) proposed the term *anneau initial* for the region in which each foliar "helix" (somewhat suggestive of a contact parastichy; Sect. 15.2.2) terminates in a leaf-generating center. Collectively, these centers form a ring around the lower flanks of the shoot apex (broad sense) and constitute the highly meristematic "eumeristem" or "organogenic region" of some authors. Distal to the *"anneau initial"* is the *zone apicale* in gymnosperms (Camefort 1956) and the *méristème d'attente* in angiosperms (Buvat 1952a,b). The *"zone apicale"/"méristème d'attente"* is regarded as the least active zone in the vegetative shoot apical meristem. In angiosperms, if an apex is evoked to floral development, the *"méristème d'attente"* is quickly transformed into the most active zone. The interior part of the basal region of the apex, which is encircled by the *"anneau initial"*, is the *"méristème médullaire"*, which produces cells that mature into pith.

This terminology is not widely used outside the French language community, partly because many workers do not regard the true initials as being in the organogenic zone. A broadly related terminology, focused on a "central-mother-cell zone", has historical priority and wider continuing use.

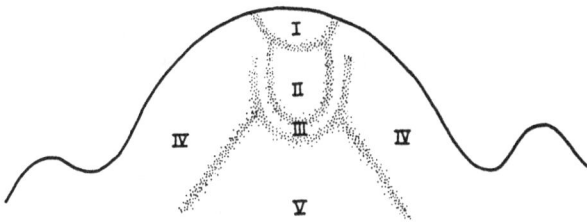

Fig. 14.3. A widely accepted scheme of cytohistological zonation within the shoot apex, based on Foster's (1938) interpretations of the pattern in median longitudinal sections of *Ginkgo biloba* apices. **I** apical-initial zone; **II** central-mother-cell zone; **III** transition zone; **IV** peripheral zone; **V** rib meristem. Some zonal boundaries may be distinct, others poorly defined

In *Ginkgo* (Foster 1938), the **central-mother-cell zone** occupies the upper (but not uppermost) central region of the shoot apex (Fig. 14.3). Foster called the constituent cells "central mother cells" because he believed that their progeny give rise to most of the internal tissues of the subapical region. Because cells of the central-mother-cell zone divide slowly, and are also replaced slowly by progeny of the apical initial cells, which are located above them, Foster considered the central mother cells neither as ultimate sources nor as prolific generators of cells. When cells inevitably drift downward through this zone, they change in cytological character. The central mother cells are the largest cells of the shoot apex and contribute heavily to the pith.

Foster's (1938) terminology also included an **apical-initial zone** (or group) and a **peripheral zone** (Fig. 14.3). The apical-initial zone, distal to the central-mother-cell zone, consists of all the cells near the vertex, including the initial cells and their most immediate derivatives. Except for their smaller size, these cells are superficially similar to the central mother cells. Being located close to the vertex and dividing infrequently, they drift downward slowly. The rate of cell division seems to be lower in the apical-initial zone and central-mother-cell zone than elsewhere in the apex (Jacobs and Morrow 1961; Gifford et al. 1963; Saint-Côme 1966; Steeves et al. 1969; Mauseth 1976).

The peripheral zone, which is partially analogous to the *"anneau initial"*, is the entire peripheral part of the proximal region, including a large fraction of the volume and cell population of the shoot apex. The cells of the peripheral zone are very actively meristematic and are intensely stainable, because they are small and densely cytoplasmic and tend to have large nuclei with a 4C complement of DNA.

Another apical region that can be regarded as a cytohistological zone is the **rib meristem**, or **central meristem**. As the former term has priority, we use it, in the sense proposed by Schüepp (1926) and adopted by Foster (1938; Fig. 14.3). Schüepp noted that, because of almost exclusively transverse division, the meristematic cells in the central part of the proximal region are arranged in regular lineage files that are generally parallel to the axis. In low, broad shoot apices, the rib meristem may seem to be in the upper part of the subapical meristem, whereas in tall, dome-like apices, it may occupy much of the lower central core region of the dome (Fig. 14.3). Rib-meristem cells, which are derivatives of the central mother cells distal to them, in turn produce the cells that develop into pith. Rib-meristem cells divide more frequently than do central mother cells.

14.1.3.2 A Newer Perspective: Structured Versus Stochastic Apices.

In the emergent, fourth perspective on shoot apical structure and function, some apices are seen as having initial cells that are permanent stewards of the initial function, whereas the initials in other apices are just temporary incumbents, soon to be replaced. **Structured apices** are considered to have a small, stable set of relative-ly permanent initials (Stewart 1978). Devolution of the initial function is determ-inistic. **Stochastic apices**, in contrast, are seen as having a pool of uncommitted meristematic cells within which, due to a combination of "fitness" criteria and fortuitous positional effects, the initial function devolves, always temporarily, to some of the cells. The process of devolution in these apices is essentially stochas-tic or probabilistic and not deterministic.

Most shoot meristems are probably neither strictly structured nor wholly stochastic, but somewhere in between. The lower vascular plants tend toward strict structure and the more advanced forms seem to be more stochastic (Klekow-ski 1988). There may, however, be many exceptions to this broad generalization (Rogers and Bonnett 1989).

Central to the concept of stochastic apices is the tenet that, of the two daught-ers of an initial, one, both, or neither may carry on the initial function. This is because new initials are selected not by simple mother-to-daughter devolution, but by dynamic processes, including both random and nonrandom elements. Nonran-dom processes include relative vigor or fitness, which determines the survival times of the various cells in the pool of potential initials. The phylogenetic ascendancy of advanced gymnosperms and angiosperms, especially those having highly stratified tunica-corpus apices, may be in part due to the efficiency of their apices in deleting deleterious mutations and in fixing advantageous ones (Klekow-ski 1988). The advent of stochastic apices thus might have increased the rate of evolution. This subject is approached again in relation to stability of chimeras (Sect. 14.1.5.1) and to the prevalence of shoot-tip abortion (Sect. 14.2.5).

The old term, "promeristem", may be redefined in terms of stochastic proper-ties of apices. This term was originally histological, designating that part of a shoot or root apex within which there is no indication of differentiation. But that definition is of little value, as recognition of early stages of cytodifferentiation depends on the techniques employed. In contrast, concepts of stochastic properties of cells at positions appropriate for initial cells, and of selection for the initial function, allow the term **promeristem** to be defined precisely — as including the current functional initials and those cells from among which replacement initial cells are selected (Sussex and Steeves 1967).

14.1.4 The Functional Germ Line and Its Protection by Low Mitotic Rates

As has long been known (Babcock and Clausen 1927), there is no direct coun-terpart in higher plants to the rigorously distinct germ line present in the gonads of higher animals. As a consequence, the somatic cell lines of plants, which eventually produce sporogenous tissues and meiocytes, can accumulate mutations that may be transmitted to gametes and zygotes of later generations. This is especially significant in long-lived plants such as trees because many cell generations may intervene between zygote and meiocyte (Klekowski 1988). The evolutionary and developmental consequences of the lack of a rigorously defined germ line in higher plants are only beginning to be appreciated (Walbot 1985; Klekowski 1988).

Plants, of course, do have a continuous line of undifferentiated cells from zygote to meiocyte. By default, this line functions as a germ line, even if it is not rigorously defined and closed. There is wide agreement that, in dicots, the functional germ line tends to be carried by the second layer (L-2) of the shoot apex. In monocots, the first layer (L-1), as well as L-2, may contribute (Stewart and Dermen 1979), whereas in gymnosperms the situation is unclear. To emphasize the difference between the zoological and botanical connotations of the term "germ line", we use the term "functional germ line" for plants.

It is widely accepted that repeated DNA replication and mitosis increase the probability that genetic information will be lost or degraded. Conversely, inhibition of DNA synthesis prolongs the cell cycle and reduces that probability (Haber 1972). The evolutionary tendency toward reduced growth rates in the distal regions of the apex can be interpreted as providing a reservoir of cells that have undergone relatively few mitotic cycles. In plants not having strictly structured apices, replacement initial cells can be selected from this reservoir by partly stochastic processes. Though the subject is controversial, the replacement of initials may indirectly protect the germ line by promoting loss of defective mutant genotypes (Klekowski et al. 1985). Protection is especially important in trees, because of their long period of vegetative growth between syngamy and production of spores of the next generation. This need for protection may help explain why the shoot apices of trees seem to have more pronounced cytohistological zonation than do those of annuals (Sect. 14.1.2).

In angiosperms, the tunica commonly is two-layered, and the functional germ line usually is included in the second layer. In a steady-state shoot apex, the tunica cells grow in only two dimensions. Thus, the axial growth rate near the vertex is nil for the depth of the tunica. Because of the restriction against periclinal division in tunica cells, the number of somatic cell generations in the germ line in angiosperms is particularly low, somewhat lower than would be expected on the basis of the generally low rate of cell division in the interior distal regions.

The further long-term significance of a stratified, tunica-corpus pattern of shoot-apical organization for germ-line stability is not yet clear, due to the complexity of this pattern as it relates to the retention or the loss of mutations. Such meristems consist of several component meristems (the individual tunica layers, and a corpus), each having its own initials. Each tunica layer is rather structured, whereas the corpus is more stochastic. However, even the tunica layers have some stochastic properties, resulting from selection of replacement initials within a layer and from displacements of initials between layers. Via displacements, those components of the meristem that are outside of the functional germ line may contribute to the germ line to some extent (Sect. 14.1.5.1). This means that selection of functional initials, and consequently the deletion or fixation of mutations, is apt to be highly complex. One attribute of stratified meristems is clear: they accumulate somatic mutations, as is evidenced by the existence of persistent chimeras. This accumulation may give increased adaptive value to ontogenic features that promote loss of disadvantageous mutations (Klekowski et al. 1985; Klekowski 1988).

The location of the germ line in gymnosperms is still controversial. Archesporial cells are almost invariably subepidermal *in position*, sometimes deeply so, but have often been interpreted as being more superficial, even protodermal, *in origin* (Singh 1978). Developmental knowledge is still too fragmentary to justify speculation on differences in protection of the functional germ line between gymnosperms having a tunica and those lacking one (but see Klekowski 1988).

If, during evolution, growth in the distal part of the apex has tended to be reduced as a means of protecting the germ line, why has this region not reached the extreme of no growth at all? One must consider that elimination of weak or abnormal cells through competition for space via diplontic selection protects against disturbances of genetic information. Growth activity, even if weak, promotes selection between cells, whereas absence of growth does not. We know, for example, that, due to very intense competition among initial cells in the cambium, many of the less vigorous cells are eliminated as initials, whereupon they differentiate (Sect. 18.4.4). To achieve similar diplontic selection of initials in stochastic shoot apices, there must be some growth and division in the distal region (Sect. 14.1.3.2).

We mentioned in Section 14.1.2 that the mitotic index in the distal regions of shoot apices of those few tree species that have been investigated is lower than 0.5%. In these trees, what would be the duration of the mitotic cycle in L-2, which, in angiosperms at least, includes the functional germ line? In those shoot apices in which mitosis lasts about five hours (Edgar 1961; see also Stewart and Dermen 1970), a mitotic index of less than 0.5% would be equivalent to only one mitosis/cell/1000 h; hence, the cell cycle would exceed 40 days. If such an apex were to grow for 2 months a year, an initial cell in the distal region would be expected to divide only once or twice a year. Thus, in a 50-year-old tree, the number of cell generations in the functional germ line might still be little more than 100.

14.1.5 Shoot Apical Chimeras and Development

14.1.5.1 Attributes, Definitions, Origins. The term "**chimera**" is applied to plant organs derived from genotypically nonuniform shoot-apical meristems, and also to whole plants in which the shoots are chimeral. Let us assume that a mutation in at least one cell causes a shoot-apical promeristem (Sect. 14.1.3.2) to become genotypically nonuniform. Generally, the more distal the mutated cell is within the promeristem, the longer and wider is the sector of its progeny cells. If the mutated cell is near the lower border of the promeristem, the cell and its progeny will drift away from the center of meristematic activity and soon differentiate and no longer divide; there will be only a short, narrow sector of cells of the mutant type, and the chimera will be ephemeral. In contrast, if the mutated cell is in the most distal region of the promeristem, its progeny may eventually displace all other cells in a particular sector of a meristematic layer. Such a chimera is relatively persistent. If two or three cells at the vertex of a layer are bearers of the initial function, a mutation in one of them may result in a persistent mutant sector having an angular width of 180 to 120°.

Chimeras are of several major types. If the genotypically different cells are confined to, and dominate in, only one of several apical layers of initial cells, we speak of a **periclinal** chimera. If there is only a single discrete apical layer and the sectors of that layer are genotypically different, the chimera is **sectorial**. If there are two or more apical layers, each with its own initial or initials, and one layer has a sector that is genotypically different, the chimera is **mericlinal**. Periclinal chimeras can occur only in plants in which each of at least two apical layers has a set of relatively persistent initial cells. If there is only one set of apical initials, sectorial but not periclinal chimeras can arise (Neilson-Jones 1969).

The general characteristics of periclinal chimeras are conventionally denoted by three or more letters, each representing a readily recognizable attribute of a layer, considering the outermost first. For example, a WGG periclinal chimera has three apical layers: the outermost (L-1) has attribute W (white) whereas the next two (L-2, L-3) have attribute G (green) instead.

If, after a stable phase, a chimeral plant begins to form shoots consisting entirely of one genotypic component, some type of chimeral **dissociation** has occurred, commonly as a consequence of **displacement** of mutant or other initials in the apical meristem. For example, an atypical, periclinal division in an initial cell may result in local **duplication** of a layer, which leads to inward displacement of the layers of cells located beneath it. Thus, a WWG chimera may become a WWW type (which could also be designated a WWWG type). Another kind of displacement is **perforation**, in which, for example, a cell of the second or a deeper layer intrudes into the next more distal layer, or vice versa (Fig. 14.4). Thus, a WGG type can become a GGG type. Either duplication or perforation may lead to a mericlinal chimera in the layer to which a cell of a variant genotype was displaced.

Some chimeras have easily recognizable phenotypes. For example, in **cytochimeras**, the components differ in ploidy. In **plastogenic chimeras**, the components differ in type of plastids. There are various types of chlorophyll-deficient plastids, which form the basis of plastogenic chimeras in which some layers are "white" and others "green". Chimeral phenotypes can often be identified by "variegata" or "marginata" types of foliage. However, some instances of leaf variegation are ascribable to patterned distribution of subepidermal gas spaces rather than to chimeras (Hara 1957a).

Chimeras can be elicited experimentally by treating shoot apices with agents that induce mutations or changes in ploidy. For example, in *Prunus, Castanea, Vaccinium*, and some other genera, diploid-tetraploid and related chimeras have been produced by applying colchicine, which temporarily inhibits normal spindle formation in dividing cells and thereby induces polyploidy (Dermen 1947, 1953; Dermen and Diller 1962). Carefully modulated ionizing radiation also has been used to induce somatic mutations and various types of chimeras, which then can be studied by clonal analysis (Poethig 1987).

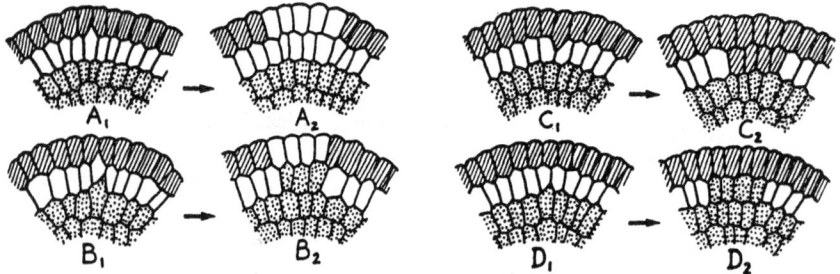

Fig. 14.4 A—D. Schematics of displacements of the "perforation" type among initial-cell layers of a chimeral tunica having cells of three morphologically distinguishable types. In each pair of drawings, the *left* one represents an earlier and the *right* one a later developmental stage. A_1, A_2 Perforation of L-1 by cells derived from L-2 initials. B_1, B_2 Perforation of L-1 by cells from L-2, and of L-2 by cells from L-3. C_1, C_2 Perforation of L-2 by cells from L-1. D_1, D_2 perforation of L-2 by cells from L-3, with no disturbance of L-1. (Based, in part, on work of Bergann and Bergann 1962)

Very rarely, chimeras arise spontaneously as a result of grafting. Adventitious buds differentiating in the callus at the graft union derive some of their initial cells from the stock and some from the scion. The several components of these graft chimeras may then belong to different species, or even different genera. Among trees and shrubs, two well-known intergeneric graft chimeras include cells from *Mespilus germanica* and *Crataegus monogyna* (*Crataego-Mespilus asnieresii* and *C.-M. dardari*). Another chimera has components from *Laburnum* and *Cytisus* (*Laburno-Cytisus adamii*). Chimeras have also been produced by experimental interspecific and intergeneric grafting, notably intergenerically between species of *Solanum* and *Lycopersicum* (Dermen 1969; Neilson-Jones 1969).

Both *Crataego-Mespilus asnieresii* and *C.-M. dardari* have persisted for about a century (Neilson-Jones 1969). *C.-M. asnieresii*, in which the outermost apical layer came from *Mespilus germanica* and the next two layers from *Crataegus monogyna*, could be designated an MCC type of chimera. Its shoots superficially resemble those of *Crataegus*, but the epidermal hairs are like those of *Mespilus*. Nonchimeral shoots sometimes appear on *C.-M. asnieresii* and *C.-M. dardari*. Characteristically, these shoots are either of a pure *Crataegus* or *Mespilus* type, with no sign of transmission of genetic characters from one component to the other despite many years of close contact. This negates the once prevalent idea that the entities we have called "graft chimeras" might be graft hybrids instead (Neilson-Jones 1969; Dermen 1969). We can infer that the system of positional-information referencing (Sect. 12.4) within the apical meristem is general enough that cells of the different taxa in a chimera can interpret it harmoniously.

14.1.5.2 What Studies of Chimeras Have Revealed About Ontogeny.

Chimeras usually show normal patterns of histogenesis (Stewart et al. 1972). Though sometimes a mutation in a specific apical layer may change the relative histogenic contribution of that layer, there is still harmonious interaction among the layers (Stewart et al. 1974). Some inferences about the development of the shoot apex and caulis can be drawn from studies of chimeras:

1. In species having a tunica, the outermost initial layer, L-1, always gives rise to the epidermis and normally contributes little or nothing to other tissues. The rare periclinal divisions in L-1 lead to instabilities of the displacement type (Stewart and Burk 1970).

2. The amount of tissue derived from the second layer, L-2, varies, even among different sectors at the same level of the stem or in positionally equivalent leaves (such as leaves at the same node). The major tissues derived from L-2 are usually the hypodermis and any thickened parts of the cortex. Despite the variability of the L-2 contribution, however, reproductive cells in angiosperms usually are indirectly derived from this layer; that is, the functional germ line is likely to be made up of L-2 progeny (but see Sect. 14.1.4). "Displacements" of the duplication or perforation type into L-2 from L-1 or L-3 are rare (Stewart and Burk 1970).

3. Layers deeper than L-2 make a variable contribution to histogenesis, as is exemplified by peach and apple (Dermen 1953; Stewart et al. 1974). In these plants, in which the shoot apices have three- to five-layered tunicae, periclinal chimeras of the types ABB, BAB, and AAB are quite stable, but they sometimes give rise to the much less stable chimeras AAAB, AAAAB, BAAB, and BAAAB. Though only the outer two layers are highly stable, the inner layers are stable enough to allow study of their contribution to internal tissues in ploidal chimeras of the types 2n-2n-2n-2n-4n and 2n-2n-2n-4n-4n (Dermen 1953; Dermen and Darrow 1960). By using simple microscopic techniques which allow 4n cells to be recognized (Fig. 14.5), Dermen (1953) was able to analyze young stems having 4n cells in their inner apical layers and to determine the contributions of the different apical layers to tissues in different sectors of the stem. In the 2n-2n-2n-2n-4n type, the vasculature was commonly derived from L-4 or L-5. In the transverse section shown in Fig. 14.5C, the border between diploid cells (derived from L-4) and tetraploid cells (derived from L-5) crosses the vascular tissue several times radially, but does not run tangentially within that tissue. Similarly, our own unpublished observations of sectorially chimeral stems of *Juniperus sabina variegata* (an "ever-sporting" periclinal chimera) reveal that the border between green and white components also may cross the vascular tissue radially but not obliquely, and does not run tangentially within it. Does this indicate that the vascular tissue is a polyclonal compartment (Sect. 12.4.4) that is determined relatively late in histogenesis of the stem? The question is open.

4. The ontogenic time of fixation (determination) of the attributes of the epidermis can be studied with chimeras. For example, one can ask: when in the development of the epidermis is it determined that most of its cells will be nonchlorophyllous, despite their genetic greenness? If there are synchronized periclinal divisions in the protoderm (which functionally is a layer of epidermal

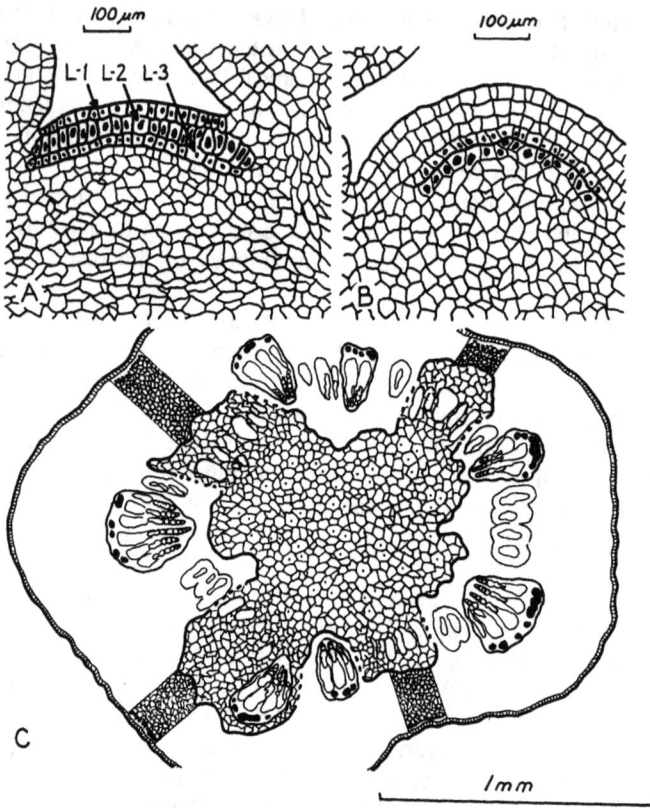

Fig. 14.5 A—C. Schematics drawn from photomicrographs of sections of diploid-tetraploid cytochimeras of **A** *Pyrus malus* (apple) and **B,C** *Prunus persica* (peach), with irrelevant anatomical detail omitted. The much larger nuclei of tetraploid cells allow them to be distinguished from neighboring diploid cells; thus, the several chimeral layers and the tissues derived from them can be distinguished. **A** Axial section of shoot tip in which the cells of *L-2*, between the *heavily inked lines*, are tetraploid, while those of *L-1*, *L-3*, and more internally located cells are diploid. **B** The cells of all four tunica layers are diploid and the corpus cells are tetraploid. **C** A cross section of the twig produced by the meristem in B. The *heavy line* encloses tissue consisting of tetraploid cells, and shows the irregularity (with constraints) of the border between the two cytotypes of a cytochimera. (**A** drawn from a photomicrograph by Dermen 1955, **B** and **C** drawn from photomicrographs by Dermen 1953)

mother cells), resulting in a multilayered nonchlorophyllous epidermis and perpetuation of the nonchlorophyllous phenotype, one can conclude that there was prior fixation of the nonchlorophyllous condition.

5. Studies of mericlinal ploidal chimeras in *Vaccinium macrocarpon* (Dermen 1945) and several other species (Stewart and Dermen 1970) have shown that a sector derived from one initial of a layer occupies 1/3 or 1/2 of the compass of the shoot and may maintain that angular width during the initiation of many

leaves. But the compass may then change: for example, from 1/3 to 0, or to 1/2 or 2/3; or from 2/3 to 1/3 or to the whole. It thus follows that a mutation or displacement has occurred and that it typically involves only one of two or three cells located at the vertex of a particular apical layer (these cells functioning as the temporary initials of the corresponding sectors). Typically, the initial cell of one sector of the layer has been replaced by a derivative of the initial cell of a neighboring sector of the layer.

6. Initial layers composed of some genotypes may contribute less to mature tissues than do layers composed of other genotypes. That is, genotypical differences between layers can lead to competition between them, though normal gross morphology of all plant organs is nevertheless maintained (Stewart et al. 1974). The extent of competition between genotypes may vary with the environment. For example, in *Buxus sempervirens* leaves, the potentially "green" L-1 produces only epidermis in outdoor plants, but in plants grown in a shaded and warm glasshouse, L-1 contributes to the outer layers of mesophyll also (Pohlheim 1969).

7. In the initial layers of dicots, dissociations of the displacement type are rather rare, and perforations are very rare. In some gymnosperms, however, with their commonly less stable outer cell layers, displacements are frequent (Fig. 14.4). This results in **ever-sporting** periclinal chimeras, as occur in Cupressaceae, especially in *Juniperus*. In unstable, periclinal, "white-green" chimeras of this type, the shoot apices (except, sometimes, cells near the vertex) consist of a genetically "white" L-1 and a "green" L-2. That is, on the flanks of the apex and in the subapical region, the cells of L-1 divide only anticlinally, and hence the new lateral shoots, with their internal tissues derived from L-2, are green. However, near the vertex of the apex, some periclinal divisions may occur in L-1. This eventually leads to displacement of the green, L-2 tissue by white tissues derived from L-1, and the older shoots thus may become all white. Such displacements are a normal part of development in taxa in which there are relatively frequent periclinal divisions in cells of the outer apical layer.

These latter chimeras are called "ever sporting" because, though the chimeral status is ephemeral in any single shoot apex, it reappears in others. Typically, a green chimeral shoot on such a plant produces many green leaves and buds, and then, due to displacement, becomes first sectorially white-green and ultimately uniformly white. The white parts, being dependent on the green parts for assimilates, are short lived. The plant produces a series of generations of chimeral branches (Pohlheim 1971a,b). A model system based on stochastic processes and diplontic selection of replacement initials was devised to explain these chimeras in *Juniperus davurica* (Ruth et al. 1985).

14.1.6 Theoretical Considerations of Growth Rates and Orientation of Cell Division in the Shoot Apex

The shoot apex is a more complex structure than is the root apex (Sect. 17.1). The pattern of growth of a shoot-apical meristem is less steady than is that of a root-apical meristem because the periodic formation of leaf primordia produces an obvious cyclic or pulsating component. The growth of a shoot-apical meristem can be considered as consisting of two components: a steady growth of the axial caulis, and a superposed cyclical initiation of foliar primordia. In this section, we will concentrate on the steady component — that is, the growth of the caulis.

Microscopic studies indicate that shoot apices grow symplastically, and that newly formed partition walls have either an anticlinal or periclinal orientation. The walls thus intersect each other in an orthogonal manner. This orthogonality is preserved during growth, and the walls, though displaced with respect to the vertex, retain a recognizable anticlinal or periclinal orientation. A consequence of this stability is that the cellular pattern in the apex remains steady as the individual cells move down through it. The situation is similar to that in a well-designed water fountain, in which the display pattern is steady because the individual water jets or droplets follow the same routes as did their predecessors. However, maintaining anticlinal or periclinal orientations of cell walls demands a special distribution of growth rates (Sect. 12.3.2).

Sachs (1874 and earlier) already recognized that the periclinal/anticlinal orientation of cell walls is a consequence of an equivalent orientation of the cell partition walls during cytokinesis and the distribution of subsequent growth such that this orientation is maintained. However, only recently has there been development of approaches to understanding how growth rates are distributed such that this can be accomplished (Sect. 12.3.2).

Application of tensor analysis to growth reveals that orthogonal orientation can be maintained during growth only in those elements that are oriented along the principal directions of growth — i.e., in the directions along which linear growth rates at a point are maximal and minimal (except where growth is isotropic, as by definition all directions are principal directions there). If elements are orthogonal, but not aligned along the principal directions of growth, the angle between them will change during growth. The periclines and anticlines represent trajectories of the principal directions of growth. Such trajectories are present in every shoot apex, though they are not always easily recognizable. Close to the surface periclines may be seen as lines tangent to the surface, and anticlines as normal to it. Inside the apex, the periclines and anticlines may be less readily discernible; the more anisotropic the growth, the more nearly are newly formed cell partitions normal to the principle directions of growth and the more obvious are the periclines and anticlines.

The two natural coordinate systems (Sect. 12.3.2) most likely to occur in shoot-apical domes are confocal and fan-like (Fig. 14.6)). Within each of these types, especially the confocal, there may be variants, depending on the actual shape of the periclines and anticlines (which, theoretically being orthogonal, cannot be independent). These variants may be parabolic, hyperbolic, or some other shape (Fig. 12.4). Confocal systems, which are the most common, can be recognized by their tendency to form a tunica. If there is no anticlinal component of the vector of displacement velocity, $\underline{V}$, the tunica is strictly limited to a depth equal to that from the vertex to the focus (and there is no anticlinal growth in that layer). If there is an anticlinal component of $\underline{V}$, though weak, the tunica may not be strictly delineated, and we can speak of a **potential tunica** (Foster 1939).

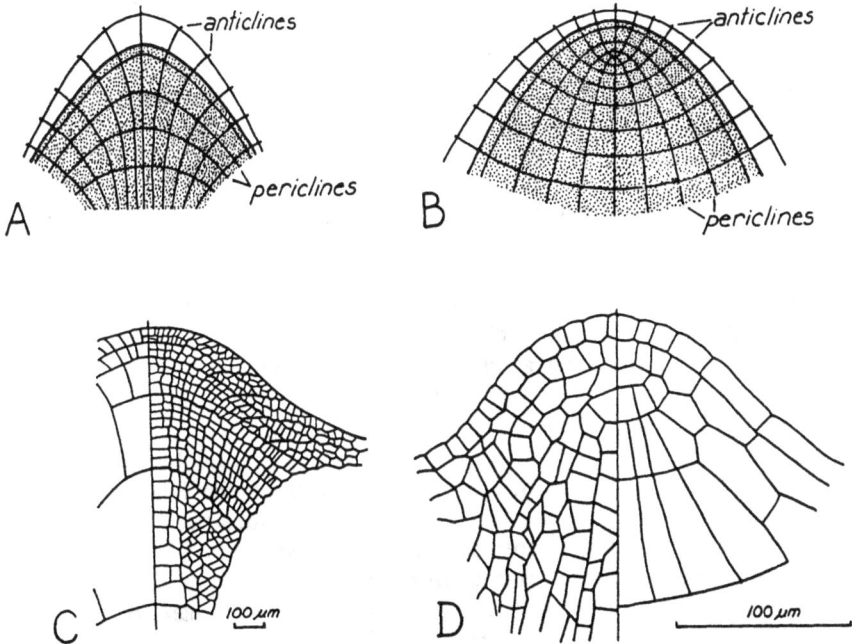

Fig. 14.6 A,B. Graphic approximations of theoretical patterns of fan-like or fountain-like **A** and confocal **B** orthogonal sets of anticlines and periclines of natural coordinate systems that may occur in median longitudinal sections of superficially paraboloidal shoot-apical meristems. **C,D** Examples of patterns of cell arrangement in shoot apices of **C** *Microcycas* and **D** *Sequoia*, redrawn from photomicrographs of median longitudinal sections. **C** is an example of the fan- or fountain-type of pattern, whereas **D** approaches the theoretical confocal type. Packets of cells of the same lineage can be recognized in both apices. The cells in the *left half* of **C** and the *right half* of **D** are drawn only as packets. The orientations of some individual anticlinal partitions do not necessarily conform with orientations of packet borders. (**C** based on photomicrographs by Foster 1943 and interpretations by Schüepp 1966; **D** based on photomicrographs by Cross 1943)

If the system of anticlinal and periclinal cell walls in the shoot apex is fan-like, there is no tendency towards tunica formation (Fig. 14.6 A,C). This type is characteristic of the cycads. Periclines and anticlines, if recognizable at all, may be less apparent than borders of cell clones (or "packets"). One must keep in mind that the latter are not necessarily aligned along continuous periclines or anticlines. For example, a packet border at one place may be an anticlinal wall; at a nearby place it may also be an anticlinal wall, but one that is not in the same anticlinal plane as the first one, and so on for many cell lengths. The result is that the whole packet border may be curved or inclined with respect to any single anticline. Such a border may be confused with an anticline (Fig. 14.6 D).

How can we obtain empirical data on growth-rate distributions within shoot apices in order to evaluate mathematical models such as the ones just presented? The most useful empirical data would be direct measurements of displacement velocities along a meridian on an apical-dome surface, but such measurements are not yet available. We have estimated these velocities indirectly in shoot apices of *Chamaecyparis pisifera* (see Fig. 14.2). Inferences about growth distribution have also been drawn on the basis of mitotic indices and/or incorporation of DNA precursors (see Sect. 14.1.2).

14.1.7 Deductions About Apical Structure Based on Surgical and Excision Experiments

The literature detailing decades of in situ microsurgical experimentation and of in vitro cultural studies with excised shoot apices testifies that the shoot-apical meristem has a remarkable ability to maintain or recover its cytogenerative function after extensive wounding, and even after excision from the plant. This is strikingly illustrated by the ability of a small part of the apex to regenerate a whole apex after the original initial cells have been killed by microsurgical incisions. In such experiments, it is obvious that the initial function has been assumed by other cells and that the original apex was not strictly structured.

For example, after a shoot apex of *Vicia faba* is bisected by a median longitudinal incision, each of the halves organizes itself into a new apex. This reorganization begins with delineation of two new cytogenerative centers on the upper flanks of the original apex, at sites removed by many cell diameters from the incision and from the remains of the original initial cells (Pilkington 1929). Also, if *Lupinus* apices are quartered longitudinally, each quarter generates a new apex. But, if cut into six sectors, some sectors generate whole apices and others die; and if cut into eight sectors, none regenerate (Ball 1952a,b).

Other evidence of the tenacity of the regenerative function is the ability of just a few cells near the vertex to construct a new shoot apex after microsurgical removal of most flanking tissues. For example, in *Solanum tuberosum*, after 95% of the flanking tissues (relative to the basal area of the apex) are removed by cuts parallel to the axis, cells near the surface of the remaining central panel can regenerate a complete apex. The central panel may include as few as 12 surface cells, which have symplasmic continuity with the remainder of the meristem only through a column of central mother cells and differentiating pith cells. The panel enlarges considerably before a functional group of initial cells is again evident (Sussex 1952).

14.2 Development of the Primary Shoot Axis

14.2.1 Formation of Internodes

In a frustum-primordium unit (Sect. 14.1.1), the nascent node, which is the region that bears the primordium (or primordia), is delineated from the nascent internode by the restriction of elongation growth to the internode. The cells in this elongating region grow longitudinally and divide transversely many times. They commonly form longitudinal files, or "ribs", each rib originating from a single "mother" cell. These files thus constitute a rib meristem (Sect. 14.1.3.1). Cells in the nodes are more randomly arranged (Smith and Lew 1970). Boundaries between nodes and internodes are more imaginary than real, however, especially in the subapical region (Eames 1961).

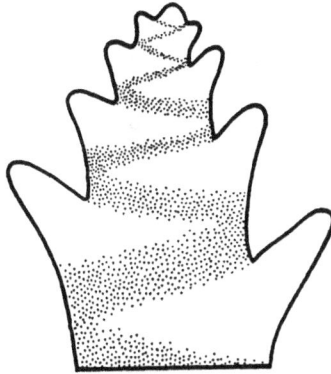

Fig. 14.7. Diagram of a median longitudinal section through a distichous shoot of *Celtis occidentalis*, based on a *photomicrograph*. The shoot is represented as a stack of frustum-primordium units. *Shading* indicates the highly meristematic, lower regions of the "frusta", which by their meristematic activity produce cells that expand and differentiate into internodal tissues. The *unshaded, wedge-shaped regions*, bearing the leaf primordia, produce the nodes. (Redrawn after Hejnowicz 1980)

It is usually not obvious how internodes develop between the densely packed leaf primordia in the upper part of an embryonic shoot. We will consider this process first in the relatively simple, distichous caulis of *Celtis*, which can be diagrammatically represented in only two dimensions (Fig. 14.7). In each frustum-primordium unit, the side opposite the leaf primordium is thinner than the side bearing the primordium. This gives the unit a wedge shape. Each unit consists of two regions. The lower, plate-like region (the "frustum") is relatively fast growing and gives rise to the internode. The upper region, which bears the primordium and is distinctly wedge-shaped, grows more slowly and becomes the node. Separation of the plate-like, primordial internodes by the wedge-shaped, primordial nodes, and the faster growth of the internodes, cause a young distichous caulis to assume a zig-zag shape. This shape may persist in mature twigs, as, for example, in *Robinia*, though in many taxa the caulis straightens during late stages of extension growth.

In shoots with spiral phyllotaxy, the principles of organization of frustum-primordium units and of internodal elongation are similar to those in distichous shoots, but are more difficult to illustrate in two dimensions. This is because, in spiral systems, each frustum-primordium unit is, in a sense, "rotated" relative to its nearest neighbors. In addition, the three-dimensional shape of a frustum-primordium unit depends on the shape of the young caulis, which typically is somewhat conical, and on the angular width of the primordium. If the primordium almost encircles the caulis, the frustum-primordium unit is nearly plate-like but still somewhat thinner opposite the primordium. If the primordium is narrow, the frustum-primordium unit is an obviously wedge-shaped conic section. Three-dimensional models based on stacks of such conic sections allow one to understand how a young caulis (as the core of an "embryonic shoot") develops into a long-shoot with obvious nodes and internodes.

The ultimate length of internodes depends on both the number of transverse cell divisions and final length of cells. Taxa vary considerably in the contributions of these factors to internode elongation. For example, in *Helianthus*, during a 60-fold increase in length of the first internode, cell length increased about 12-fold and cell number about 5-fold. In contrast, in *Syringa*, cell length increased only 3-fold, but cell number increased about 65-fold (Wetmore and Garrison 1966).

In *Liquidambar* seedlings grown under a range of light conditions, the final length of cells in the cortex and pith was relatively invariable (Lam and Brown 1974). It seems probable that final length of these cells is genetically determined and depends only slightly on external factors, unless they are so extreme as to cause etiolation. In some other plants also, external factors reportedly strongly affect the number of cell divisions within internodes and hence internode length, though cell length is only moderately affected (Sachs 1965).

14.2.2 Primary Thickening of the Stem

Due to "**primary thickening growth**" below the base of a steady-state shoot apex, a mature primary stem is always larger in diameter than is the shoot apex from which it originated. In gymnosperms and dicots, primary thickening growth typically occurs in episodes correlated with histogenic processes within the caulis — in particular, the radial component of early volume growth in the precursor cells of the pith and cortex. For example, in continuously growing apices of long-shoots of 90-day-old *Picea abies* seedlings, the diameter of a newly formed frustum-primordium unit increases little for about 15 plastochrons (down to about 50 μm below the base of the apical dome); then, for the next 15 to 20 plasto-chrons, caulis diameter increases so rapidly that 400 μm below the base of the apical dome the diameter is double that in the first 15 plastochrons (Romberger and Gregory 1977). Another episode of primary thickening may follow this early episode, well before secondary thickening growth becomes dominant.

In contrast, in stems of bulbous and many arborescent monocots, the shoot tip tends to have a pronounced, radially symmetrical, head-and-shoulders profile, with the head (the apical dome) on a short neck (the youngest part of the caulis), subtended or surrounded by enormous shoulders (the subapical caulis). Common-ly, the apical dome sits in a shallow pit, the shoulders extending above the vertex of the dome. The extensive primary thickening growth of these monocots is due mainly to the activity of a diffuse, **primary thickening meristem**, arranged as in Fig. 14.8. Activity of this meristem results in pronounced anticlinal growth in the newly formed primordial internodes. The anticlinal files of cells that result (as seen in median longitudinal sections) are steeply inclined in the upper primordial internodes, but become less steep, then nearly transverse, lower in the caulis. As a result, the internodes, if visualized in three dimensions, are nested like cups at the top, then bowls, and then plates, at progressively lower levels (Fig. 14.8). If viewed in transection, the inclined files of cells in the upper internodes appear as concentric circles (periclinal files).

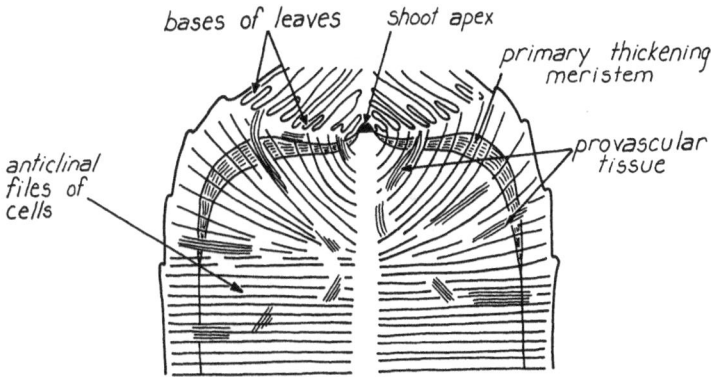

Fig. 14.8. Schematic illustrating shoot apical region of typical thick-stemmed monocot having well-developed primary thickening meristem but lacking secondary thickening. (Redrawn after DeMason 1983)

The reorientation of files of cells with increasing plastochronic age results from symplastic growth of the caulis, in which, in the vicinity of the primary thickening meristem, the principal growth rate in the anticlinal direction prevails over that in the periclinal (meridional, or axial) direction. Relative elemental growth rates are not necessarily uniform along the anticlinal files. If they are higher in the future pith region, medullary primary thickening predominates; if higher in the peripheral regions, cortical primary thickening predominates.

In thick-stemmed monocots, the primary thickening meristem, which is diffuse rather than uniseriate, is often a broad subapical region underlying the shoulders of the caulis (as indicated in Fig. 14.8), but sometimes it extends far down the flanks and grades into a secondary thickening meristem, which is not closely analogous to a vascular cambium. This arrangement occurs in many arborescent monocots (Tomlinson and Zimmermann 1969; DeMason 1983), in which it results in some types of so-called anomalous secondary thickening (Stevenson and Fisher 1980; Yarrow and Popham 1981).

In some tropical monocotyledonous trees, the primary thickening meristem produces a primary stem that appears obconical (i.e., inverted-cone shaped) in median axial section (Fig. 14.9). That is, the subapical meristem becomes broader with the passage of time and thus with height in the stem. Eventually, as the obconical, primary axis becomes reinforced by increments of secondary tissue that are thick basally and progressively thinner upward, the stem attains a more uniform diameter (Tomlinson and Esler 1973).

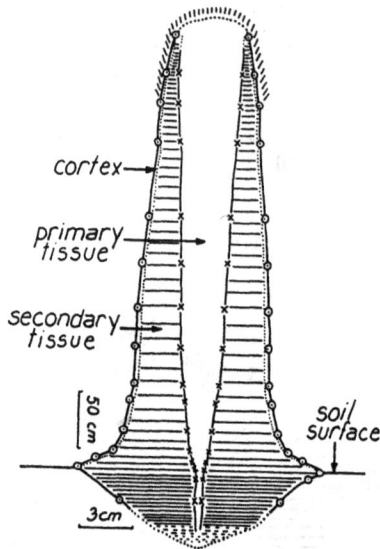

Fig. 14.9. Diagram of median axial section of sapling of *Cordyline australis*, indicating relative distribution of primary (*white*) and secondary (*lined*) tissue; based on measurements of distances between sets of points *o* and *x* at various levels in a specimen in which the distance from the soil surface to the base of the leafy crown was 4 m. The vertical axis is foreshortened 16-fold relative to the horizontal axis. Dimensions and limits shown for the cortex are arbitrary. The primary axis is obconical. Accretion of secondary tissue in the lower stem allows the primary meristem to become wider with age. (Redrawn after Tomlinson and Esler 1973)

14.2.3 Monopodial Versus Sympodial Growth

Among temperate-zone trees, except in some scale-leaved conifers, there are two modes of shoot extension growth. In the **monopodial** mode, terminal buds, which are located at the tips of the main axis and its branches, expand into shoots. After a shoot has expanded, a new terminal bud forms at its tip. Then, after a period of dormancy, the shoot apex enclosed in the bud undergoes further extension growth, typically during the following growing season. In this mode of growth, a shoot-apical meristem is active for several to many years, though interrupted by dormant periods. This mode typically leads to a strongly excurrent growth habit.

In many other woody plants, the apical meristems do not form dormant terminal buds at the end of a season's growth activity. The mode of shoot extension growth in these taxa is **sympodial**. Secondary (lateral, or axillary) buds, rather than true terminal buds, expand into shoots. An axillary bud that functions as a terminal bud is a **pseudoterminal bud**. At the end of a season's growth, the shoot tip aborts or senesces (Sect. 14.2.5). The following spring, another axillary bud, usually the most distal one surviving on the twig, assumes the role of a pseudoterminal bud.

14.2.4 Formation of Buds

A bud is an unextended, partly developed shoot having at its tip the apical meristem that produced it. In a fully formed, mature bud, the apical meristem is usually hidden and protected by primordial leaves and scales that it had initiated earlier. Once a bud is beyond the primordial or formative stage, the potentially active subapical meristem within the caulis is relatively inactive. In such a bud, commonly considered dormant, there is little elongation of primordial internodes, and the nodes are not well delineated.

There are three major types of buds: terminal; lateral, or axillary; and adventitious. These originate quite differently, though when mature they may seem similar. Lateral buds begin as bud primordia lateral to a shoot apex, typically in the axils of young leaf primordia. Adventitious buds originate when relatively uncommitted parenchymatous cells, such as occur in wound and other callus tissue, become organized into new shoot meristems. Terminal buds, in contrast, are formed from an existing shoot apex and its subapical meristem, as the foliar primordia change their developmental patterns.

14.2.4.1 Terminal Buds. Most nonadventitious buds on a tree originated in leaf axils. With few exceptions, even the meristems within true terminal buds, which characterize taxa having monopodial growth (Sect. 14.2.3), are ultimately of axillary origin. But, after the axillary buds of monopodial taxa have opened and the shoots within have expanded, the shoot tips do not abscise or senesce as they do in trees that grow sympodially. Rather, at the end of a growing season, the shoot-apical meristem and any leaf primordia that have formed on a terminal leader or branch tip become enclosed in bud scales. Thus, the apical meristem within a true **terminal bud** goes through cyclic episodes of formation of bud scales (cataphylls) and leaf primordia (Sect. 16.7), interrupted by periods of dormancy. In contrast, the shoot tips of trees that grow sympodially (Sect. 14.2.3) have a noncyclic pattern of development of foliar organs (Romberger 1963).

Because the apical meristems within true terminal buds generally originate as ordinary axillary buds, we will discuss, in the following section, general aspects of origin and early development of axillary buds, and defer to Section 16.7 a more detailed discussion of the cyclic versus noncyclic patterns of development of foliar organs.

14.2.4.2 Lateral, or Axillary, Buds. Because leaf primordia on a typical meristematic caulis are crowded, space for lateral-bud primordia is very limited. Geometrically, the most feasible site for a bud primordium is in a leaf axil, where the meridional distance to the leaf above is maximal. A bud developing at an axillary site is *developmentally* associated with the subtending leaf, is usually located below the level of the next higher frustum, and therefore is not displaced away from the axil during the internodal elongation of that frustum. Thus, the bud typically remains axillary in location, though the final position of the bud atypically may be either cauline, or — as in *Hibiscus cannabinus* — foliar (Kundu and Rao 1955).

Axillary buds, which are lateral to the main axis of the stem, are the primordia of lateral branches. After their subtending leaves have abscised, axillary buds are, strictly speaking, no longer "axillary". They are then more appropriately called "lateral" buds. We use the term "axillary" to specify origin and early development of buds, and "lateral" to refer to a bud's position. Most lateral buds are axillary in origin.

A primordial axillary bud does not protrude from the surface of the subapical-meristem region of the caulis until the subtending leaf primordium is in its second, third, or even later, plastochron. Nevertheless, the orientation of cell divisions in the axil of a leaf primordium still in its first plastochron may already differ from that in neighboring cells. Divisions of tunica cells, which ordinarily are exclusively anticlinal, are not fully anticlinal at this site (not orthogonal to the caulis surface), as can be seen in median axial sections. It is as if, in this region, a new natural coordinate system were being superposed on that of the caulis. The tunica cells are beginning to divide anticlinally relative to this new coordinate system.

Because cell divisions in at least the two outer tunica layers of the parent caulis in dicots are anticlinal either to the caulis surface or to the surface of the newly forming bud primordium, cells from the tunica layers of the parent apex become the tunica layers of the axillary apex. Thus, there is direct continuity of the second layer, L-2, through which the functional germ line is transmitted (Sect. 14.1.4).

Shoot apices of many conifers lack a distinct tunica-corpus organization. But, even in these, the surface cells at the loci where axillary bud apices are formed divide only anticlinally, though surface cells at loci of leaf-primordium formation often divide periclinally. This restriction on periclinal divisions in young axillary bud apices enables a succession of axillary apices to continue the periclinal chimeral state, even if during later development there are periclinal divisions in the outer cell layer at the vertex of the axillary bud (Sect. 14.1.5.2).

When an axillary bud apex is still in the early stages of protrusion, it initiates the primordia of its first leaves, the prophylls. In dicots, there are usually two prophylls, but sometimes, as in *Cuminum cyminum* (Shah and Unnikrishnan 1969), there is only one. After initiating prophylls, an axillary bud follows one of two courses of development: **syllepsis**, which is continuing development with no episode of rest or inhibition; or **prolepsis**, in which development is interrupted by such episodes.

The axillary shoots of temperate-zone trees commonly, but not universally, are proleptic, whereas in tropical trees, there are various combinations of syllepsis and prolepsis (Hallé et al. 1978). In sylleptic axillary shoots, there is typically a long basal hypopodium (analogous to an internode) below the first node, and the prophylls are similar to foliage leaves. In contrast, proleptic axillary shoots have short basal internodes separated by nodes bearing bud scales. However, the distinction between syllepsis and prolepsis is often not sharp.

In temperate zones, proleptic buds are likely to become dormant during the summer and remain dormant through the winter, then expand into shoots in the spring. During the summer in which they become dormant, the buds may form second-order bud primordia in the axils of their leaf primordia. When the first-order buds develop into lateral shoots in the spring, the second-order buds may develop into branches of these shoots, or, as is very common in woody dicots, may remain dormant as winter buds that will not expand until the second season after their inception.

Buds are not initiated in the axils of all leaves. In orthotropic shoots of the Cupressaceae, for example, many leaves have neither axillary buds nor even recognizable residual axillary meristems, whereas other leaves already have axillary buds by their third plastochron. Similarly, in plagiotropic shoots of *Thuja*, *Chamaecyparis*, and related genera of Cupressaceae, buds seem to develop only in some axils; they are arranged in a pattern that results in a characteristic symmetry of the branches. However, persistent "detached" meristems in apparently "empty" axils may produce axillary buds even after several years of quiescence. This may occur in apparently normal shoots, but is more common, and produces larger buds, after frost or other damage to existing buds. Such late-formed axillary buds are derived mostly from superficial cell layers (Fink 1984).

14.2.4.3 Adventitious Buds. Shoot buds can arise from cells that are somewhat differentiated and not obviously meristematic. Such buds are called "adventitious" and can be viewed as evidence of the totipotency of cells. The cells of such a bud have no direct developmental relation to the apical meristem.

Adventitious shoot buds may form on practically any part of an intact plant or on plant fragments. They often appear in wound-callus tissue. Adventitious buds that appear on stem or root fragments always develop on the shoot-apical ends, especially on any callus present there. In leaves and in stem segments having living epidermal cells, adventitious buds, or even adventitious embryos in some taxa, may be derived from a single epidermal cell or from a group of epidermal or subepidermal cells.

On roots, adventitious shoot buds may arise with or without prior wounding (Sect. 17.6). It is noteworthy that, on periclinally chimeral plants, shoots from adventitious buds of root origin are often unlike the "parent" shoot because practically all root tissues are endogenous, arising from derivatives of initial layers beneath those that perpetuate the attributes of the chimera (Neilson-Jones 1969).

Adventitious buds usually develop directly into shoots, without a dormant period. When long-dormant lateral buds of axillary origin finally become active, they are easily mistaken for adventitious buds (Priestley and Swingle 1929; Romberger 1963). Thus, adventitious buds are less common than may be supposed.

14.2.5 Abortion and Senescence of the Shoot Tip

Many temperate-zone woody plants, particularly angiosperms, have a sympodial mode of branching. In these, growth of the main axis is continued not by a terminal bud, but by one or more upper lateral buds, usually of axillary origin (Sect. 14.2.3). Depending on the taxon, the fate of a shoot tip of such a plant is usually either: (1) "abortion" with active abscission, or (2) simple cessation of meristematic function, followed by parenchymatization, with eventual withering of the tip and possible shedding of dead tissue, but with no active abscission. Apices of short-shoots of *Pinus* are of the second type, but can for a time revert to the meristematic condition if the long-shoot which bears them is decapitated.

The two types of shoot-tip fate have not always been well differentiated in the literature, and the term "shoot-tip abortion" (or "apical abortion") has been applied to both types. We use the term inclusively, to refer to either type of fate.

In addition, in many taxa vegetative apices of some post-juvenile shoots under certain conditions undergo another kind of fate — transformation into determinate, reproductive apices. This process typically "uses up" the vegetative apices of leading shoots, and often of other shoots as well.

During seasonal dormant periods, trees and shrubs in which shoot-tip abortion occurs have obvious lateral buds covered by scales. In some of these plants, there is active abscission, and, in others, passive withering or shedding. Plants having shoot-tip abortion include the genera *Ailanthus*, *Albizia*, *Betula*, *Bumelia*, *Catalpa*, *Carpinus*, *Castanea*, *Celtis*, *Cercis*, *Cercidiphyllum*, *Citrus*, *Corylus*, *Diospyros*, *Fagus*, *Gleditsia*, *Gymnocladus*, *Maclura*, *Morus*, *Platanus*, *Paulownia*, *Rhamnus*, *Robinia*, *Salix*, *Sambucus*, *Sapinda*, *Staphylea*, *Syringa*, *Tilia*, *Ulmus*, *Viburnum*, and others (Romberger 1963).

In temperate-zone trees, shoot-tip abortion usually occurs during spring or summer, when conditions for vegetative growth are good. Shoot-tip abortion is controlled by internal mechanisms, which may be triggered by signals of internal or of indirect, external origin. The effects of photoperiod are well documented. For example, in *Robinia* and *Catalpa*, short photoperiods can greatly hasten shoot-tip abortion, and long photoperiods can delay it. *Gleditsia*, too, is quite responsive in this regard (Neville 1969). The timing of shoot-tip abortion also varies with plant age and vigor (Millington and Chaney 1973).

In vitro experiments strongly indicate that the apical meristem of *Syringa vulgaris* remains programmed to senesce and die, even if removed from the influence of subjacent tissues (Garrison and Wetmore 1961), though the factors controlling the shoot-tip abortion are not well understood. In contrast, in *Castanea sativa*, abortion can be delayed for up to two years by in vitro cultural stratagems, indicating that influences from older tissues may promote the abortion (Codaccioni 1963). However, even when abortion was long delayed, the shoots did not produce dormant terminal buds with scales.

The process of shoot-tip abortion with abscission usually begins with a yellowing of the tissue distal to the site where abscission will occur. This site in turn is just distal to the node bearing the lateral bud that will become the "pseudo-terminal" bud after the more distal parts of the tip abort. The yellowing rapidly progresses acropetally in the stem and in each young leaf thereon. In *Ulmus*, the yellowing process runs its course in about two days. In some taxa, appreciable tissue is aborted: loss of 80 to 100 mg (fresh weight) of tissue per shoot tip is not unusual. In *Tilia*, the aborted tissue may include several partially expanded internodes and leaves, along with several well-developed axillary buds (Romberger 1963).

The symrodial branching habit, which is characteristic of species having shoot-tip abortion, transfers the function of continuing development of the shoot axis to an apex that is proximal to, and thus older than, the terminal apex. Because it is older, this more proximal apex has had time for a greater number of cell divisions to occur during the process of selection of functional initial cells (Sect. 14.1.3.2). This might increase the efficiency of deletion of disadvantageous mutations (Klekowski 1988).

14.3 Vascularization of the Primary Shoot

14.3.1 Development of Provascular Strands, Leaf Traces, and Vascular Sympodia

A perennial land plant that is to become large and self-supporting must have a means of thickening and strengthening its stem as it grows. It must also have a way to supply increasing amounts of water and solutes to the various parts of its increasingly large body. Some evolutionary lines of plants have met both these requirements by developing first provascular tissues, and, later, a vascular cambium (Chap. 18), which ensheathes all except the very youngest root and stem segments.

Provascular tissue (sometimes also called "procambium"; but see Sect. 13.5.2) arises as strands of cells that are more narrow and elongate than neighboring cells. This shape results from local increases in the ratio of longitudinal to transverse cell divisions. Strands of provascular cells, as seen in transections of the upper part of the subapical meristem, appear as local groups of cells within a ring of densely staining cells that generally do not differ in shape from neighboring cells. This ring has been called a provascular, or residual, meristem (Sect. 13.5.2). It seems likely that determination of which of the cells will become provascular and subsequently vascular is a function of the positions of the meristematic cells within the ring, in relation to signals derived from leaf primordia above (see also Sect. 15.5).

In pteridophytes, some differentiation of provascular cells into primary vascular elements may occur in the absence of leaf primordia, though the presence of primordia stimulates vascular differentiation below them. In seed plants, in contrast, if leaf primordia are absent, differentiation does not proceed beyond the provascular stage unless the influence of the leaves is artificially replaced, by, for example, indoleacetic acid and sucrose. Under the influence of leaf primordia, strands of provascular tissue develop into leaf traces. These traces become separated from each other by more highly vacuolated cells, which develop into interfascicular ray parenchyma on either side, and into leaf-gap parenchyma above the trace divergence (Esau 1965b; Meicenheimer 1986).

In recent decades, the spatiotemporal pattern of vascular differentiation has come to be better understood, partly through use of the "optical shuttle" system for making simulated motion pictures from series of transections of plant axes

(Zimmermann and Tomlinson 1966). This technique allows one to view the development of provascular strands and vascular bundles as accelerated in time. As a result, expressions such as "the strand leads to (or toward)" become dynamic. Among the most detailed of such studies are those of Larson (1975, 1976, 1977, 1979, 1980) on *Populus*.

Two principles seem to govern the earliest stages of provascular-strand development in shoots of many seed plants:

1. The first-formed, so-called "original", provascular strands develop acropetally, as branches of "parent" leaf traces. These early, acropetally developing strands are the first components of the nascent leaf trace. Several such traces and, therefore, several "original" provascular strands, may lead to a single primordium.

2. Additional provascular strands, often called "subsidiary" strands, typically begin differentiating at the base of the leaf primordium. In most dicots, they at first use an "original" strand as a template, developing along it both basipetally and acropetally, as in *Fraxinus pennsylvanica* (Larson 1985) and *Gleditsia triacanthos* (Larson 1984). (In some monocots, the subsidiary strands use the original strand more as a guide than as a template.) Then, in both dicots and monocots, the subsidiary strands may, within a leaf blade, diverge from the original strands and become the provascular precursors of lateral veins.

The number of subsidiary strands accompanying an original provascular strand varies. In *Populus*, the original strand is complemented by six subsidiary strands. In *Fraxinus excelsior*, we have observed that there are as many subsidiary strands as leaflets (7 to 11), one strand leading to each leaflet.

The expression "gives rise", as used in discussions of vascularization, usually means "initiates a branch". The branching of an original provascular strand occurs at a specific developmental stage of the leaf primordium (N) to which the original strand leads. The branch strand, which represents a new original provascular strand, differentiates acropetally, and may appear before protrusion of the leaf primordium that it will serve (Esau 1965b; Larson 1975). In shoot apices of the *Populus deltoides* seedlings studied by Larson, provascular strands were already developing acropetally for the next six primordia that would appear. This means that the original strands that ended in six existing leaf primordia, and were their central traces, were giving rise to "offspring" original strands for six primordia that did not yet exist. Each younger primordium would emerge above — and in the same sector of the apex as — the existing leaf from which it was deriving its provascular strand. That is, the original strand ending in leaf primordium N was initiating a branch that would become the central trace of primordium $N + n$, in the same sector of the stem. (Here n is one of the characterizing numbers of the corresponding phyllotactic series; Sect. 15.2.2).

Development of a new central trace accelerates rapidly once the trace has entered a leaf primordium. The trace increases notably in thickness, both by cell growth and division within the strand, and by the acquisition of subsidiary strands differentiating from the surrounding provascular tissue.

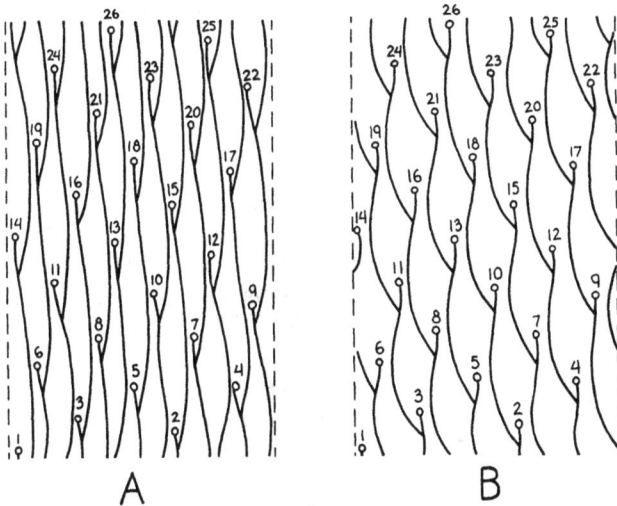

Fig. 14.10 A,B. Schematic illustration of the concept of vascular sympodia in primary stems. *Lines* represent central leaf traces; *small circles* represent leaves, which are *numbered serially* in order of initiation. The system of leaf traces is arranged as if on the surface of a tube, represented here as if cut open (*dashed lines*) and "unrolled" onto a plane. Every *nth* leaf is served by the same sympodium, and in the steady state there are n sympodia present in transections of the primary stem. In **A**, $n = 13$. In **B**, $n = 8$. The number n is always one of the numbers describing the phyllotaxy (Sect. 15.1) of the shoot. (Redrawn after Hejnowicz 1980)

An anastomosing continuum of sequentially appearing central leaf traces is called a **vascular sympodium** (Fig. 14.10). The number of vascular sympodia present in a stem is related to the phyllotaxy. If a discrete sympodium serves every nth leaf, then, during steady-state growth, n vascular sympodia must be present at any level in the stem. The number n, therefore, is one of the character-izing numbers of the phyllotactic series (Sect. 15.2.2).

Some facts about sympodia that help explain the number present and the relationship between sympodia and spiral phyllotaxy include:

1. Vascular sympodia are arranged more or less parallel to the stem axis and never cross each other, although their branches in a topographic sense may seem to interconnect laterally (Esau 1965b).

2. A vascular sympodium, in branching sympodially, produces a new original strand, which becomes a trace of the primordium that is next to be initiated in that sector of the caulis.

3. Branching of a vascular sympodium occurs at a definite developmental stage of the youngest leaf in its sector and usually precedes initiation of the prim-ordium that the branch will serve.

4. The original (branch) strand that first enters a newly formed primordium becomes its central trace.

5. An original strand may bifurcate before it becomes a trace. The "extra" strand that arises in this way typically becomes a lateral leaf trace. In *Ulmus*, for example, there are two lateral traces for each central trace (Smithson 1954). Alternatively, the extra strand may become a new original strand, thus initiating a new sympodium.

Formation of new sympodia is especially common if the rate of formation of leaf primordia accelerates relative to the rate of leaf development, so that the number of recently formed primordia in the bud increases. The new branch strands enter the "extra" primordia formed during this phyllotactic change. The pattern of the increased branching of the sympodia is very orderly and is closely coordinated with the phyllotactic changes, which follow precise geometric rules (Sect. 15.3.3).

According to some workers, genetic programs cause phyllotactic transitions to occur at specific stages of development of the provascular systems, as described by Larson (1980) in *Populus* seedlings. Because differentiation of the original provascular strand (which later becomes the central trace of the leaf) occurs before differentiation of the leaf primordium to which it will lead, Larson favored the idea that precocious leaf traces determine the sites of leaf initiation, and that the control of siting of primordia should be sought in the developing vascular system rather than in the shoot apical dome.

However, there is still a lack of knowledge about controls over patterns of either vascularization or phyllotaxy — and the interactions between them. We find it hard to imagine how an increase or decrease in the number of primordia and of sympodia, according to a precise geometric pattern, can be coordinated unless one pattern is clearly subordinated to the other, or unless primary vascularization and phyllotaxy are both controlled by some higher-level system. The latter possibility has strong appeal (Sect. 15.5).

14.3.2 Differentiation and Maturation of Primary Phloem and Primary Xylem

The development of functional primary phloem and primary xylem from provascular strands involves differentiation, from precursor cells, of the characteristic cell types that perform the early translocating function — that is, the **protophloem** sieve elements and the **protoxylem** tracheary elements. Later metaphloem and metaxylem differentiate. We can describe the differentiation of these cell types either from the viewpoint of the changes that an individual cell undergoes as it becomes specialized for functioning in transport (cytodifferentiation, as discussed in Sect. 12.2.2), or from the viewpoint of the temporal and spatial patterns of differentiation that a series of such cells undergoes within the provascular strands. Because cytodifferentiation of the conducting cells is somewhat similar between primary and secondary tissues, and because of the paramount importance of secondary tissues in arborescent plants, we defer the major part of our discussion of vascular cytodifferentiation to the chapters on secondary vascular tissues (Chaps. 19, 20). Here, we briefly describe the aspect of cytodifferentiation that differs most between primary and secondary elements — that is, the wall-thickening pattern of the tracheary elements. Then, in subsections that follow, we focus on the temporal and spatial patterns of differentiation that a series of cells undergo within a provascular strand.

In protoxylem tracheary elements, wall thickenings are deposited in the form of annular or helical bars that allow the elements to stretch as internodes are elongating. The bands of secondary wall and underlying primary wall become liberally lignified, but the primary-wall areas between the secondary bands are not susceptible to lignification. Apparently, some cytoplasmic constituents, such as microtubules (Sect. 19.2.1.1), act not only in localizing cell-wall thickening according to a predictable pattern, but also influence the selective deposition of lignin. Areas of primary wall unprotected by secondary wall and by lignification are then partially hydrolyzed, increasing wall extensibility and aiding lateral water transport.

Unlike the protoxylem tracheary elements, which are adapted to extension during maturation, the metaxylem elements, which have ladder-like and other patterns of thickenings, are only minimally extensible. Not surprisingly, metaxylem elements generally appear in axial stem segments only after cessation of extension growth.

14.3.2.1 Protoxylem and Protophloem.

Protoxylem and protophloem elements differentiate directly from cells on the adaxial and abaxial margins, respectively, of the provascular strands. The first cells to mature in the vascular bundles that develop from provascular strands are the protophloem sieve elements, which appear as files of cells continuous with the protophloem in the parent bundle. Being located in an elongating region of the stem, these elements undergo symplastic elongation as they differentiate. Once mature, they usually no longer divide. Early diagnostic features of differentiating sieve elements are: (1) wall thickening without lignification, (2) deposition of callose pads on the developing sieve areas, and (3) in angiosperms, synthesis of so-called P-protein (Sect. 20.3.1.2).

In *Populus* stems, the first sieve element typically appears near the base of an original provascular strand three plastochrons before protrusion of the primordium of the leaf that the strand will serve. As the primordium protrudes, a sieve-element file rapidly differentiates acropetally, but it does not enter the primordium until the latter is about 500 μm long and four to five plastochrons old (Larson 1975).

In fast-growing stems of *Actinidia* (which has 21 vascular sympodia, compared with only 13 in *Populus*), the first mature protophloem appears near the base of a leaf trace belonging to a leaf seven plastochrons old, in continuity with the protophloem of the parent trace, when the parent leaf has a plastochronic age of 28 (the "parent" leaf being the next older, and lower, leaf served by the same sympodium). That file of young protophloem cells enters the base of the young leaf three plastochrons later. Protophloem appears in the provascular strand of a young leaf trace at about the same time as protophloem matures in lateral veins of the parent leaf (Puławska 1965).

Fibers commonly develop in the protophloic, or at least the extraxylary, parts of the vascular bundles (Blyth 1958). Secondary walls are deposited when the internode in which the fibers are located stops elongating. Before they mature, those young fiber cells that were located between the precociously obliterated protophloem sieve elements grow, not just symplastically with neighboring cells, but also intrusively, pushing their tips between neighboring cells. This increases the number of cells that appear in cross sections through the bundle. In *Pisum*, at least, those factors that stimulate the fibers to grow intrusively, and to mature, originate in the leaves and migrate basipetally (Sachs 1972; Aloni 1976).

Maturing protoxylem elements are relatively easy to study, largely because their wall thickenings, which contain microcrystalline cellulose deposits, become brilliantly visible under a polarizing microscope. Histological clearing and staining can also make these wall thickenings visible.

The first tracheary elements typically arise at the base of a leaf primordium. From there, differentiation proceeds both acropetally into the leaf primordium and basipetally into the stem. Simultaneously, protoxylem files differentiate acropetally in the stem, in continuity with existing xylem strands below. In *Populus* stems having 13 sympodia, protoxylem appears first in the central leaf trace during the fifth plastochron of the leaf primordium served by that trace. This protoxylem matures basipetally, and usually merges with a metaxylem file belonging to the parent trace. These metaxylem files are basipetally continuous with files of tracheary elements of the secondary xylem, so that there is a long continuum of tracheary elements (Larson 1976, 1979, 1980).

The first-formed protophloem and protoxylem elements function only a short time before internodal elongation tears and obliterates them. They are functionally replaced first by late-formed protophloem and protoxylem and then by metaphloem and metaxylem.

14.3.2.2 Metaphloem and Metaxylem.

In the middle part of a provascular strand, between the protoxylem differentiating on the adaxial side and the protophloem differentiating on the abaxial side, are some residual meristematic cells. These cells mostly divide tangentially. The resulting files of radially aligned, cambium-like cells constitute a meristematic tissue that Larson (1976) proposed to call "metacambium" to distinguish it from the vascular cambium (a secondary meristem), which produces the secondary plant body. The inner and outer derivatives of the metacambium differentiate into metaxylem and metaphloem, respectively. The middle layer remains meristematic, but gradually develops into the vascular cambium as its cells assume the shapes characteristic of fusiform and ray initials (Sect. 18.6).

Typically, metaxylem elements begin to differentiate in internodes that are in their final stages of elongation, and mature after the internodes have stopped elongating. In *Populus*, though, all metaxylem reportedly matures *prior to* the cessation of internode elongation (Larson 1976). Thus, the criterion that various authors have used to distinguish "protoxylem" from "metaxylem" — i.e., that protoxylem but not metaxylem matures before internodes cease to elongate — is not precise. The transition between proto- and meta-vascular tissues is gradual and therefore the two can be distinguished only somewhat arbitrarily (Esau 1965b, 1977).

Protoxylem is formed all along those vascular bundles or parts of bundles that develop from original provascular strands, though subsequent elongation and development may obliterate it. However, in those bundles or bundle parts that develop from subsidiary strands, the basipetally developing protoxylem may not reach some levels in the stem before metaxylem begins to differentiate at these levels. There, metaxylem rather than protoxylem may occur adjacent to the pith, while in the same bundle at a higher level, protoxylem may be present, especially close to the base of the leaf that the bundle serves. Even near the leaf base, however, some bundles may have no detectable protoxylem.

14.3.3 Vascularization of Axillary Buds

Quite detailed information is available on vascularization of the axillary buds of *Populus*, in which the vascular system originates from provascular strands that either are branches of the central trace of the axillant leaf or are branches of other strands that delineate the central leaf gap at that node. Two provascular branches of the central trace of the axillant leaf primordium develop acropetally before the axillary bud apex begins to protrude. As protrusion begins, these provascular branches advance acropetally into it and later become the central traces of the two prophylls (Larson and Pizzolato 1977; Pizzolato and Larson 1977).

We may draw some interesting inferences from studies of differentiation in the provascular strands that connect axillary apices with the vascular system of the parent stem in *Coleus* and *Phaseolus*. Protophloem, here as elsewhere, always matures acropetally, in continuity with existing protophloem. Protoxylem, in contrast, matures both acropetally from existing xylem toward the axillary bud base, and basipetally from leaf primordia within the bud. Discontinuity between basipetally and acropetally advancing protoxylem is an important factor in regulating bud growth. If the bud is senescent rather than merely dormant, protophloem does not enter it, and the discontinuity between the basipetally and acropetally advancing protoxylem persists (Sachs 1970).

In axillary buds of *Populus deltoides*, both protoxylem and metaxylem are initiated near the base of each primordium (whether prophyll, scale, or leaf) and differentiate basipetally within the bud. These early traces collectively are "bud traces". The xylem derivatives of the first and second pairs of bud traces are directed downward and join the main stem vasculature, but those of the third pair are directed upward into the main stem. The primary vessels of the bud traces differentiate along the margin of the branch gap (also called "bud gap": a parenchymatous region in the vascular cylinder of the stem, usually confluent with the similar, and subtending, leaf gap). They form radial files of vessels that are adjacent to the primary xylem of leaf traces in the stem. The plant initiates secondary xylem in the stem before it does in the branch. Consequently, the last-formed metaxylem vessels of the bud traces are continuous with secondary vessels in the stem (Larson and Fisher 1983). It is noteworthy that the polarity of the cambium in the axils of axillary buds may be deviant. Circular vessels (indicating circular polarity) often appear there (Sect. 12.4.2.1).

14.3.4 Hormonal Regulation of Shoot Vascularization

The acropetal progression of protophloem differentiation suggests that physiological processes in the antecedent leaf or leaves influence its differentiation. In contrast, the primordium into which the provascular strand is developing, rather than an antecedent one, seems to influence the differentiation of protoxylem. Jacobs and Morrow (1957), working with *Coleus*, found a correlation between the differentiation of xylem in provascular strands and auxin production by the leaf associated with those provascular strands. Later, Wangermann (1967) demonstrated that exogenous auxin can replace the leaf in exerting this effect, and also showed that exogenous radioactive auxin is transported basipetally in provascular strands. The polar transport of inducing substances, especially of auxin, almost certainly has a central role in controlling vascular differentiation (Sect. 12.4.2.1).

Sachs (1969) induced vascular strands in root and stem cortex in *Pisum* by applying IAA in lanolin to cut surfaces at various sites. He found, not surprisingly, that the induced strands developed basipetally — that is, in the direction of auxin transport. Existing vascular bundles in decapitated *Pisum* epicotyls attracted vascular strands induced by auxin applied to the upper ends of tongues of stem cortical tissue. But, if auxin was supplied to the apical end of the vascular cylinder also, then the strand induced in the tongue might or might not merge with the preexisting bundles, depending on the relative concentrations of auxin applied at the two sites. Increasing the concentration in the existing bundle decreased the probability of merger, but increasing the concentration in the induced bundle increased that probability. We see this as closely analogous to the induction of a connecting vascular strand between an axillary bud and the vascular cylinder in the stem. A strand from the bud cannot merge with a vascular bundle in the stem if an actively growing terminal shoot apex is present and presumably producing and exporting auxin; consequently, further development of the axillary bud is inhibited.

Further studies (Sachs 1975, 1984) revealed that, during the early stages of vascular differentiation, there is positive feedback between the ability of cells to transport the inducing factor polarly and the transport of that factor (Sect. 12.4.2.1). As the factor being transported, auxin, itself induces further vascular differentiation, any route of transport of the factor, however inefficient it may be initially, soon becomes efficient. Thus, polar transport of auxin from the leaves toward the roots may gradually induce a vascular system joining them.

15 The Primary Shoot: Phyllotaxy

15.1 An Approach to Phyllotaxy

The terms "phyllotaxy" and "phyllotactic" refer to the pattern of arrangement of leaves or other lateral appendages on a cauline axis. The same terms are also applied to studies of the principles underlying the origin and maintenance of these patterns.

More than a century of research on phyllotactic patterns has generated a voluminous literature, emphasizing geometric properties and formal mathematical description and classification. We will limit mathematical discussion primarily to geometric dynamics, and will discuss these dynamics only as necessary to give some coherence to our ideas on how phyllotaxy might be controlled and on its developmental interpretation.

To begin, we advance several simple assumptions applicable to practically all spiral phyllotactic patterns and to most others. We point out, however, that the assumptions we make and the approaches we discuss here were arrived at mainly through study of gymnosperms and dicots. We discuss phyllotaxy in monocots only peripherally.

Our first assumption is that we introduce no gross errors when, in the early discussions in this chapter, we disregard axial dimensions in the organogenic, subapical regions of the shoot tip and emphasize radial dimensions and azimuths. This widely accepted assumption is useful in that it allows phyllotaxy, which is a patterned attribute of an entire three-dimensional shoot tip (Fig. 15.1A), to be represented as a composite projection of a set of transverse sections through the axils of all primordia onto a two-dimensional polar-coordinate system (Fig. 15.1B). This method permits the whole circumference of a shoot tip to be represented on a single drawing; accordingly, the azimuthal positions of all foliar elements can be shown. If the dynamism of a living, growing apex were to be represented using this polar system, each primordium (indicated as a point or an outline) would move gradually outward along a radius, as it and the frustum bearing it grew. New primordia would appear near the center of the system at the ends of short radii having quite predictable azimuthal positions relative to some reference radius.

Our second assumption is that each frustum-primordium unit grows symplastically with its neighbors, never rotating appreciably relative to them. It then follows that the azimuths of the radii passing through the centers of the various primordia (and thus the angles between the radii) remain unchanged as dimensions increase with growth. This is the geometric basis of the principle of situational similarity. We invoke this principle later, as it applies to developing primordia and the spatial relations between them.

Our third assumption is that initiation of a new primordium is a relatively sudden, local event, marking the end of one plastochron and the beginning of the next. The new primordium, by its axial position, also delineates a new frustum-primordium unit, thus moving the base of a dome-like apex (the most common shape among shoot apices; Sect. 14.1.1) upward by the axial thickness of such a unit.

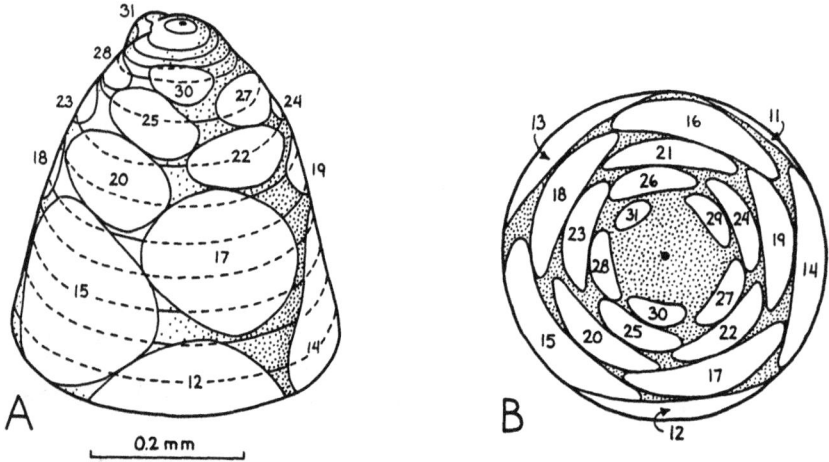

Fig. 15.1 A,B. Drawings based on a model of the shoot tip of a 15-day-old *Linum usitatissimum* seedling. A *black dot* represents the theoretical vertex in both drawings. A Perspective drawing. The primordia (except for number 31) are represented as having been excised at the caulis surface. Primordium bases are *numbered* in order of initiation. *Dashed lines* represent the edges of serial cross sections of the shoot tip from which measurements were made to construct the model. **B** Transverse projection diagram of the same modeled shoot tip. For more detailed interpretation see Section 15.2.2. (Modified after Williams 1975)

These three assumptions provide a geometric and temporal framework within which phyllotaxy can be considered as a patterned attribute of a developing shoot tip. Note that, whereas the initiation of a single primordium is a short-term, local event, the development and maintenance of a phyllotactic pattern is longer term, requiring action of control systems at the level of the organism or organ, as well as the tissue and cell (Sect. 15.5).

In a set of radii terminating in primordia, projected onto a polar-coordinate system, as described, the lengths of the radii to the various primordia necessarily constitute a graded series. The longest radii lead to the oldest primordia, which are borne by the lowest, oldest frusta. The shortest lead to the youngest primordia, borne on the most recently delineated frusta. If we direct our attention first to the oldest, lowest primordium, then to the next oldest, and so on to the youngest, our gaze takes the path of a spiral passing through all primordia in the chronological order of their initiation (Fig. 15.1B). If we admit a third dimension, this spiral rises as it progresses inward towards the youngest primordium. Its direction of rotation may be either clockwise or anticlockwise. Thomas and Cannell (1980) treated in detail the mathematical properties of this **generative spiral** and its role in formal descriptions and theories of phyllotaxy.

Usually, the generative spiral is not obvious, and almost certainly it has no objective reality. Yet, we can assign initiation numbers (serial numbers) to primordia along this spiral, and then use those numbers to characterize the more obvious spirals, usually termed **parastichies**, which compose the most common phyllotactic patterns. It has become accepted to number primordia either in chronological order of initiation, from oldest to youngest (1, 2, 3, ...n, n+1, ...), or in the reverse order, from youngest to the oldest (n, n-1, n-2, ...). We use both methods, as seems appropriate to the situation.

Though it has no physiological reality, a generative spiral can be envisioned as the path of a primordium-initiating signal moving around the base of the apex in a saltatory manner. In the most common phyllotactic systems, the leading end of the spiral moves through an arc of more than 120° but less than 180° between pauses during which primordia are initiated. This is the **divergence angle** (Sect. 15.2.3). The leading end of the spiral always appears to move upward and inward relative to its earlier path. Yet, during vegetative growth, the spiral never arrives at the origin of the polar system (the vertex of the apical meristem) because every point through which the spiral passes is continually increasing its radial and axial distance from the vertex, as the shoot tip grows.

We apply the concepts of generative spirals and of certain kinds of parastichies to spiral phyllotactic systems, in order to discern some plausible geometric, yet biologically based, rules governing the siting of new primordia in relation to older ones. Such rules promote efforts to define the general problem of the biological control of phyllotaxy.

15.2 Some Attributes of Phyllotactic Patterns

15.2.1 Decussate, Distichous, and Spiral Patterns; Generative Spirals

The relatively simple phyllotactic patterns are not the most common ones. The most common are based on obvious spiral elements, whereas the simplest are decussate (from Latin: *decussis*, the number 10, or X; hence, X-shaped) and distichous (from Greek: *di-* plus *stichos*, row; hence, in two rows). Neither decussate nor distichous patterns have obvious spiral components (though both patterns have sometimes been interpreted in terms of spiral constructs). In decussate patterns, each frustum bears two primordia, 180° apart. Each pair is perpendicular to the pairs borne by the frusta above and below. In distichous patterns, leaves appear as two axial rows, 180° apart along the caulis, with each frustum bearing either one or two leaves; thus, leaves may be either opposite or alternate. In a basal projection onto a polar-coordinate system, the radii leading to the leaves on a distichous shoot constitute just two opposing sets.

In dicots, the first pair of plumular leaf primordia usually is initiated perpendicular to the axial plane of the two cotyledons, and thus the early phyllotactic pattern tends to be decussate. In monocots, the first plumular leaf is sited 180° from the single cotyledon, and a distichous pattern is likely to follow.

Another relatively uncommon type of phyllotaxy has elements of both simplicity and complexity. This is the **verticillate** (or **whorled**) type. In this type, three or more primordia are borne on a single frustum or on a set of frusta that fail to grow axially. Each such set is separated from the sets above and below by internodes that elongate extensively. In *Casuarina*, phyllotaxy seems to be of a purely verticillate type. However, in *Acacia verticillata*, among others, although the primary pattern is verticillate, vestiges of spirality are evident within the whorls (Dormer 1972).

The most common, as distinct from the simplest, phyllotactic patterns are based on spirals. The generative spiral (the least steep of all the spirals) may appear in the patterns, but is usually not obvious. Various steeper spirals are more obvious. In a simplistic physiological sense, these appear in the patterns because the siting of each new primordium is most strongly influenced by two nearby existing primordia, one of which is older and lower than the other. The physiological asymmetry in this siting is almost certain to result in geometric asymmetry as well. In fact, the triangle delineated on the curved surface of the caulis by lines connecting the centers of a new and two older primordia tends to be skewed. (Such triangles in three-dimensional space cannot be well represented in two-dimensional projection drawings using a polar-coordinate system.) Further, as a result of the principle of situational similarity (Sect. 15.3), the same asymmetry and skewness apply to all primordia and all plastochrons during steady-state growth. Packing of the skewed triangular units on the surface of the caulis generates patterns with spiral elements.

Unlike decussate, distichous, and purely verticillate phyllotactic patterns, spiral patterns exhibit **chirality**; that is, they are either right- or left-handed. In describing and analyzing a spiral phyllotactic pattern, its chirality is taken to be that of its generative spiral(s).

Most spiral phyllotactic patterns can be analyzed by assuming a single generative spiral. However, in some relatively large apices, the spacing and timing of primordia initiation can be systematized and ordered only by assuming that two generative spirals, 180° apart, wind around the organogenic region in the same direction. Thus, two new primordia rather than one primordium are initiated during each plastochron, and each cauline frustum bears two primordia rather than one primordium. In this system, one spiral passes through, for example, primordia numbered ...10, 11, 12, ..., in order of initiation, while the other passes through ...10' 11', 12',..., also numbered in the order of initiation. Primordia, say, 10 and 10', are located at the ends of radii 180° apart and are not developmentally distinguishable. A pattern requiring the assumption of two congruent generative spirals is **bijugate** (from Latin: *bi-*, twice or doubly, plus *jugum*, yoke, as a yoke of oxen; hence, two operating in unison, or existing in pairs). Very large apices may even have tri-, tetra-, or k-jugate systems, requiring assumption of three, four, or k generative spirals, respectively. In the terminology used by some authors, mono-, bi-, tri-, or k-jugate systems are said to have 1, 2, 3, or k-fold **symmetry**, respectively (Richards 1948).

Among the plumules in populations of seedlings of higher plants, there are roughly equal numbers of right- and left-handed generative spirals. Imbalances, such as those reported in *Picea abies* seedlings (Gregory and Romberger 1972), tend to be small.

However, in the post-seedling stages of many woody species, the direction of generative spirals in branches is commonly opposite (antidromous) rather than the same (homodromous) as that in the parent axis or branch order (Sect. 12.4). In *Salix* and *Populus*, homodromous and antidromous shoots occur in about equal numbers, but in *Crataegus monogyna* and *Sarothamnus scoparius*, among other species, as many as 95% of branches may be antidromous (Raunkiaer 1919, Dormer 1972).

The phyllotactic pattern of a plant may change during development. Patterns in seedling plumules, juvenile shoots, and mature shoots may all be different. Likewise, shoots arising from terminal buds may differ in phyllotaxy from those originating from axillary buds. Thus, it seems probable that although phyllotactic patterns may be specified in general detail by genetically driven developmental programs, some of their specific details may result from the interaction of many variables (Sect. 15.5), including various types of shoot asymmetry.

15.2.2 Parastichies: Contact and Noncontact

In the buds or shoot tips of many plants having spiral phyllotaxy, all but the very youngest leaf primordia are so densely arranged on the surface of the caulis that they contact each other, often on four sides (Fig. 15.1A). Because of the regularity of their arrangement, the primordia that contact each other constitute visible arrays, or rows, typically in two directions, and sometimes in three. Such rows are segments of spirals, varying in steepness. Some of these ascend to the right, some to the left. A linear array, or row, of foliar elements that are in mutual contact is a **contact parastichy** (from Greek: *para-*, beside, or closely resembling, plus *stichos*, in a row; thus, a row-like arrangement). Those geometrically similar contact parastichies that wind around a shoot tip, parallel to each other, constitute a **set** of contact parastichies. There are definable geometric and temporal relations between the primordia within a contact parastichy. In addition, these relations between primordia are related to the number of contact parastichies within a set.

One may also discern rows and sets of spirally arranged foliar elements that are not in contact, but that nonetheless have regular geometric and temporal relations. These are **noncontact parastichies**. When a bud-borne embryonic shoot expands, the contacts that existed between the foliar elements in the bud usually are lost, and contact parastichies become noncontact parastichies. The latter commonly are not obvious on casual inspection, but can be identified by careful analysis.

Because of their easily quantifiable features, contact parastichies allow phyllotactic patterns in buds and shoot tips to be described and classified. In addition, contact parastichies are more meaningful biologically than are noncontact parastichies because the primordia that are in contact exert physical forces on each other. These forces are significant in growth dynamics (Williams 1975).

Now, let us assign numbers to all the primordia in the subapical region of a caulis in the order of their initiation along an assumed generative spiral, as in Fig. 15.1A,B. We can then analyze the relationships among the initiation numbers. We will limit much of this analysis to the common, monojugate phyllotactic patterns, in which a single generative spiral passes through all leaf primordia on a shoot tip. However, the same principles, after attunement, can be applied to bi-, tri-, or k-jugate systems.

The initiation numbers assigned to the primordia in a monojugate system show two distinctive relationships. First, the difference between initiation numbers of the successive primordia in any one contact parastichy is constant and is the same in all parastichies of the set to which that parastichy belongs. Second, this difference between initiation numbers equals the number of parastichies in the set to which that parastichy belongs. Thus, that single number (the difference between initiation numbers of successive primordia in a contact parastichy) uniquely characterizes an entire set of contact parastichies in an apex.

For example, in Fig. 15.1A,B, the steeper set of contact parastichies, rising to the right, is characterized by 5. The less steep set, rising to the left, is characterized by 3. This means that, in the steeper set, the initiation numbers of the primordia differ by 5 and that there are 5 parastichies in the set. In the less steep set, the initiation numbers differ by 3 and there are 3 parastichies in the set. Such spiral phyllotactic systems are specified by the characterizing numbers of the two most obvious sets of contact parastichies. Sometimes, a third, less obvious, set is also present. The example in Fig. 15.1A,B is a 3:5 phyllotactic system.

If we consider the sets of contact parastichies of a shoot tip in the order of their increasing steepness, we see a regular alternation of direction: left, right, left, etc. Further, the higher the characterizing number, the steeper is the set of parastichies. Thus, if there are two sets of contact parastichies, one is steeper than the other and they slope in opposite directions. If there are three, the steepest one ascends in the same direction as the least steep. A parastichy so steep as to be parallel to the axis is an **orthostichy**. In most shoot tips, the least steep parastichy is the generative spiral, which is a noncontact parastichy that passes through all foliar elements in the order of their initiation. If two shoots differ only in the direction of their generative spirals, then their equivalent contact parastichies (of the same steepness) appear related as mirror images.

Models of phyllotaxy based only on visible sets of intersecting contact parastichies take slight cognizance of the time sequence of initiation of primordia (Meicenheimer 1979). The generative spiral, in contrast, has a closer conceptual relation to the actual events in the organogenic region. Angles between radii to successively initiated leaves along a generative spiral can be thought of as representing time.

The geometric attributes of noncontact parastichies are not determined by the size and shape of the foliar primordia relative to the size of the caulis. For example, in the transverse projection diagrams of Fig. 15.2, each primordium is represented by a point, and, in part (A) of this figure, the points are joined by several possible sets of parastichy lines (1, 2, 3, 5, 8, and 13); the shapes of the primordia impose no limitations.

When the apex and subapical caulis are maintaining a steady-state size and shape, the size and shape of the primordia will determine which of the parastichies possible for mere points can become contact parastichies of real primordia. Thus, as shown in Fig. 15.2B, the characterizing numbers of the contact parastichies that can occur simultaneously are determined by the major directions of growth of the primordia.

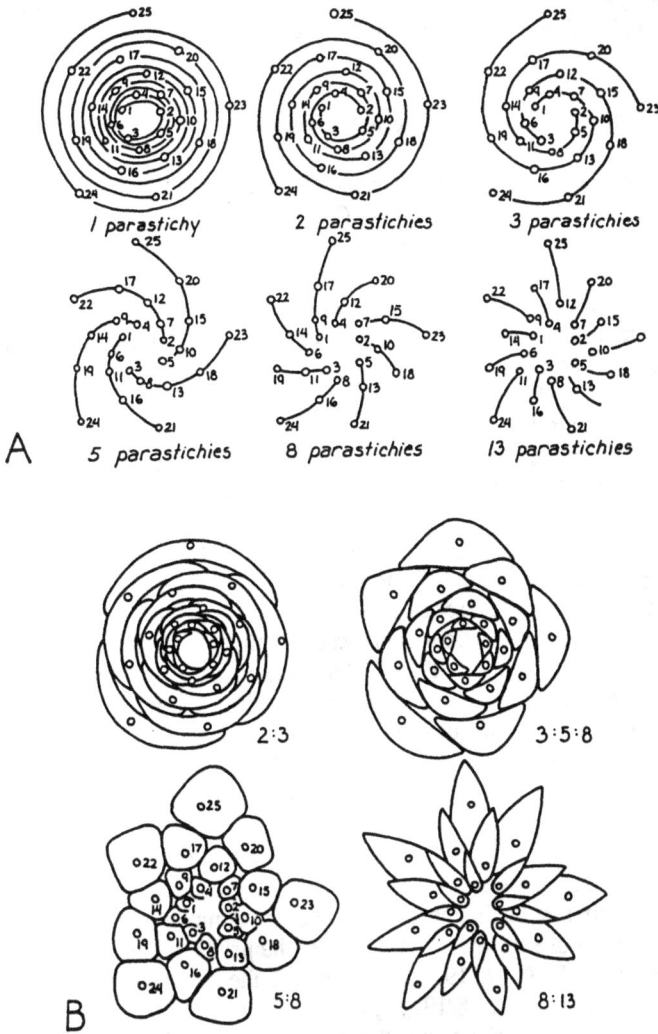

Fig. 15.2 A,B. Transverse projection diagrams of various parastichies that can develop from different sets of primordia with initially identical space relations. Centers (*small circles*) of comparable primordia have the same relative geometric locations in all diagrams in **A** and **B**. Primordia appearing in the same relative locations have the same initiation numbers in all diagrams, whether or not numbers are shown. **A** Connecting the centers with lines drawn in various ways reveals six different, geometrically possible, contact parastichies. The *1* parastichy is identical with the generative spiral. **B** The shapes of the primordia determine which contact parastichies among the possibilities depicted in **A** actually develop. Extreme tangential growth (widening) of primordia results only in *2* and *3* parastichies; strong widening and moderate radial growth (thickening) result in *3, 5,* and *8,* but not *2* parastichies; moderate radial and moderate tangential growth favors *5* and *8* parastichies, and extreme radial growth produces *8* and *13,* but not *5,* parastichies. (Redrawn after Hejnowicz 1973)

Most spiral phyllotactic systems have two, or perhaps three, sets of contact parastichies, and each foliar element may belong to noncontact parastichies as well.

15.2.3 Fibonacci and Other Series: Divergence Angles

In the phyllotactic systems that are most common among those dicots and gymnosperms that have been investigated (Fig. 15.2B), the numbers characterizing the parastichies (2, 3, 5, 8, and 13) are members of a well-known mathematical series: 1, 2, 3, 5, 8, 13, ... In this series, each number, with the exception of the first two numbers, is the sum of the two preceding numbers. This is the **primary Fibonacci series**. (Fibonacci series are named after the thirteenth century Italian mathematician, Leonardo Fibonacci).

An intriguing feature of the spiral phyllotactic patterns corresponding to the Fibonacci series is that, if L, M, and N are successive numbers in the series, and if N is the number of leaves that the generative spiral passes through in going from a chosen first leaf to the next leaf positioned very nearly above the first, then L is the number of turns that the generative spiral makes between those two leaves. Thus, necessarily, the average angle of divergence between two successive leaves or primordia along the generative spiral is approximated by L/N x 360°. The numbers of the Fibonacci series provide a series of L/N fractions: 1/3, 2/5, 3/8, 5/13, 8/21, ... The decimal value of these fractions approaches a limit of 0.38197, which, as a fraction of a circle, specifies a limit of 137.509° for the divergence angle. Plants having phyllotactic patterns categorized by the contact parastichies 3:5, 5:8, 8:13, ... are in fact characterized by a mean angular divergence angle of about 137.5° between successively initiated primordia. This is the **primary Fibonacci angle**, which characterizes the most common of the phyllotactic patterns among those dicots and gymnosperms that have been studied (Figs. 15.2, 15.3A).

In less common spiral phyllotactic patterns, the numbers characterizing the parastichies are members of series analogous to the Fibonacci series. These **analogous series** have different beginning numbers, but the numbers after the first two have the same relation to one another as do those in the primary Fibonacci series. The analogous series: 2, 5, 7, 12, 19, 31, ... provides the denominators of a series of fractions: 1/2, 2/5, 3/7, 5/12, 8/19, ..., analogous to the L/N fractions of the Fibonacci series. (For mathematical reasons, the numerators are still from the Fibonacci series.) This analogous series of fractions approaches a limiting value of 0.4198 and thus indicates a limit of about 151° for the divergence angle. Another analogous series: 1, 3, 4, 7, 11, 18, 29, ... provides the denominators of the fractions: 1/4, 2/7, 3/11, 5/18, 8/29, ..., which indicate a divergence angle of about 99°. Phyllotaxy of this type can occur in *Picea abies* (Fig. 15.3B). Series that imply divergence angles of 78° and 64° also are known.

If a monojugate system becomes bijugate, as may happen if an apex increases in size, the single generative spiral is replaced by two congruent spirals winding around the apex 180° apart. Thereafter, two primordia are initiated per plastochron and each newly delineated frustum bears two primordia rather than one primordium (for example, n and n'). As a consequence, the number of congruent contact parastichies

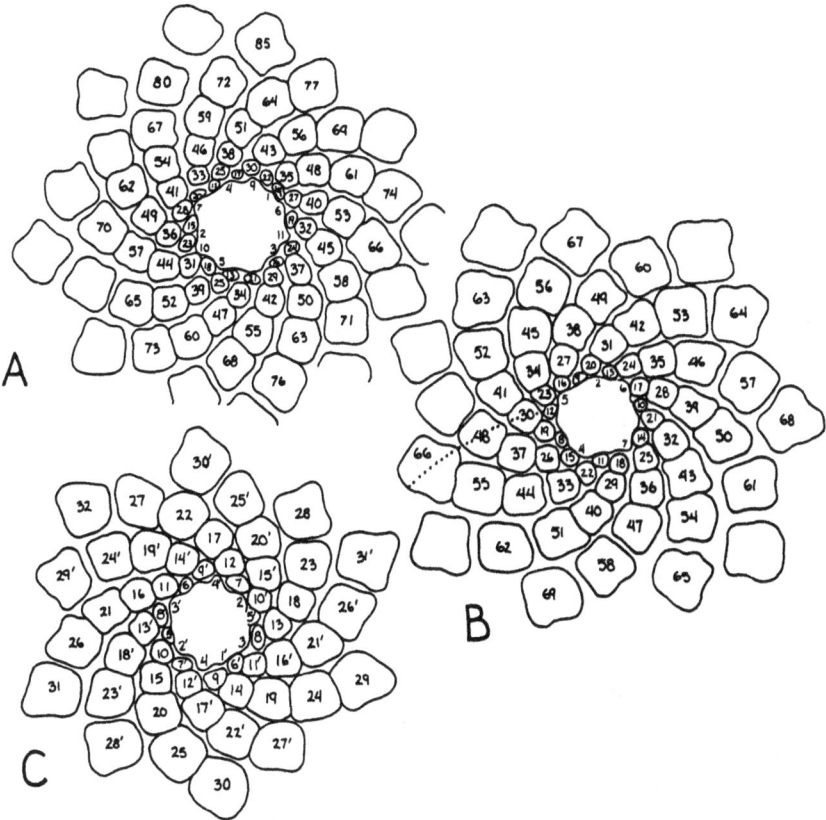

Fig. 15.3 A—C. Phyllotactic patterns found in individual *Picea abies* seedlings from the same population. **A** An obvious 8:13 primary Fibonacci pattern, in which the mean azimuthal divergence angle between successively initiated primordia is about 137.5°. **B** A 7:11 pattern of an analogous series in which an 18 parastichy is also discernible (*dotted line*). The divergence angle is about 99°. **C** A bijugate variant of the primary Fibonacci pattern. Parastichies *3*, *3'*, *5*, *5'*,*8*, and *8'* are present. The divergence angle is about 68.7° (half of the primary Fibonacci angle). (From original drawings by R. A. Gregory)

characterized by a particular number, such as 3 or 5, is doubled, because, for example, three 3 parastichies + three 3' parastichies are counted as a total of six parastichies. Thus, in such systems, the number of parastichies in a set is equal to twice the difference between the initiation numbers of the successive primordia in any one contact parastichy. In effect, two Fibonacci (or analogous) systems are operating in harmony (Fig. 15.3C). The divergence angle between successively initiated primordia in a bijugate Fibonacci system (when all primordia are considered) is 68.75°, or one-half the primary Fibonacci angle of 137.5°. Similar arguments hold for k-jugate systems.

15.3   Developmental Dynamics of Phyllotactic Patterns: Situational Similarity

There is a flavor of the occult in the number series characterizing the types of phyllo-taxy that can occur in nature. Sets of numbers from the primary Fibonacci and certain analogous series characterize commonly occurring phyllotactic systems, whereas other analogous series, equally easy to write, characterize systems that can have no reality. Yet, perhaps there is no more mystery in these series than there is, for example, in the numbers characterizing the possible and impossible degrees of symmetry in crystals. In a crystal, each lattice node is similar (disregarding edge effects), and the principles of physics underlying crystallization are everywhere the same. In growing shoot tips that have attained at least a short-term developmental steady-state, each primordium, at a specific plastochronic stage, fits a set of descriptors pertaining to spatial relations with neighboring primordia, to physiological status, and to developmental attributes similar to those that fitted its predecessor primordium one plastochron earlier.

Though the primordia around an apex are not as nearly identical as are the mole-cules in a crystal, we propose that the siting and development of primordia initiated in sequence on a steady-state apex can be understood in the context of a **principle of situational similarity**. The manifold similarity between sequentially initiated primordia on a shoot apex leads to both developmental and geometric regularities, some of which are manifested in series of parastichy numbers that characterize spiral phyllotactic systems.

This situational similarity is not to be confused with the **principle of similitude**, which is a basis of the branch of mathematics known as dimensional analysis, and is also of special relevance to biology (see Thompson 1942). The principle of situational similarity is based on the geometric concepts of congruence and similarity and on the general predictability of outputs of developing systems given similar inputs and operating conditions during similar developmental stages. Dormer (1972), in his discussion of "succession of parts", briefly mentioned a "principle of similarity" as the basis of a geometric consideration of phyllotaxy. According to his principle of simil-arity, angles remain unchanged as linear dimensions increase during growth. We have superposed on this principle considerations of implied biophysical and biochemical control systems, to arrive at a principle of situational similarity. This principle can be applied to each primordium and also to the triad consisting of the most recently initiated primordium, n, and its nearest neighbors, usually n-2 and n-3. Similar triads follow each other repetitively on the surface of the caulis.

Thus, situational similarity posits the existence of numerical and spatial similarities among whole series of sequentially initiated primordia. This becomes obvious when one notes that, if numbers assigned to primordia are based on the most recently initiated being designated as n and the next youngest as n-1 (usually the most conven-ient and sometimes the only feasible scheme), initiation numbers are only *temporary* labels. With each passing plastochron, the numerical designation of a primordium changes by one. Thus, in two plastochrons, primordium n becomes n-2, and n-3 becomes n-5; in three plastochrons, n-3 becomes n-6, and n-2 becomes n-5, etc. Further, as primordia generally do not migrate except as dictated by symplastic growth, their azimuthal positions relative to one another within the bud are stable over time, though absolute distances between their centers increase with growth.

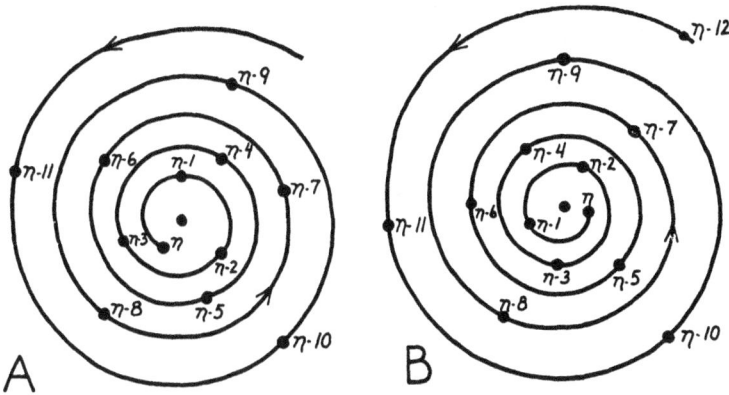

Fig. 15.4 A,B. Diagrams such as these are helpful in visualizing the consequences of certain postulated rules of phyllotaxy. The perspective is that of a shoot apical dome viewed from above. The inwardly spiraling lines are two-dimensional basal projections of three-dimensional generative spirals winding up the slopes of the dome. The vertex is at v; n is the most recently initiated primordium; and $n-1$, $n-2$, ... are successively older primordia. In both A and B, there are more than 2, but fewer than 3, primordia per $360°$ along the generative spiral, which is *anticlockwise* in direction. In A it is assumed that a physiological system controls the siting of n to be closer to $n-3$ than to $n-2$. A rationale explained in the text predicts that the phyllotaxy thus generated will have characterizing numbers from the primary Fibonacci series, 2, 3, 5, 8, ... In B control of the siting of n is such that n is closer to $n-2$ than to $n-3$. The rationale referred to in A then predicts that the phyllotaxy generated in B will have characterizing numbers from the series 2, 5, 7, 12, ..., a series analogous to the primary Fibonacci series

What are the consequences of this similarity and regularity for spiral phyllotactic systems? According to the literature and in our experience, the simplest overtly spiral system (that is, a system having onefold symmetry) typically has more than two, but fewer than three, primordia per turn of the generative spiral. For this system, the divergence angle must be in the range of $120°$ to $180°$. If the angle is close to $180°$, as in most grasses, the phyllotactic pattern can be described as distichous (often with slight torsion). If the divergence angle is appreciably less than $180°$, yet still obviously more than $120°$, then the most recently initiated primordium, n, is located in the sector between n-3 and n-2, whereas n-1, being at least $120°$ away, is not a possible nearest neighbor of n (Fig. 15.4).

If we apply the similarity principle to systems with twofold symmetry (having two generative spirals), we find that there are more than four, but fewer than six, primordia per cycle of the pair of generative spirals. The angle of divergence is thus $< 90°$ but $> 60°$. In the following discussions, we generally apply our arguments to systems having only one generative spiral and thus onefold symmetry (i.e., monojugate systems), but in principle the arguments also apply to systems having k generative spirals and k-fold symmetry.

In common monojugate systems, n is closer either to the older (n-3) or the younger (n-2) of its two nearest neighbors. The concept of situational similarity allows us to examine the consequences of either case. As we develop the implications of this concept, we urge the reader to consult the diagrams in Fig. 15.4 to help visualize possible sitings of the youngest primordium, n, relative to the successively older primordia n-1, n-2, n-3, etc. The reader may wish to draw additional diagrams to help visualize other phyllotactic patterns described later.

15.3.1 Which Parastichies Can Occur?

To prove that a certain contact parastichy is *possible*, it is enough to show that initiation numbers of primordia that may come into contact differ by a whole number that belongs to a particular series (Fibonacci or analogous). However, although a certain contact parastichy may be geometrically possible, its realization depends on the directions of growth of the young primordia (Sect. 15.2.2).

Which primordia are apt to come into contact? Only those located in the same sector of the periphery of the organogenic region; that is, those having vicinal radii on a transverse projection. On the basis of very simple assumptions, one can predict which primordia *can possibly have* vicinal radii. For example, assume that each primordium, when initiated, confronts situational similarity in the form of a physiologically mediated rule ensuring that n is located closer to n-3 than to n-2 (Fig. 15.4A). From situational similarity, it follows that n-3 is between n-6 and n-5 (because n-3 is *the same* primordium that was n when n-6 was n-3 and when n-5 was n-2). By similar reasoning, n-6 is between n-9 and n-8, etc. Further, if n is to the right of n-3, then n-3 is to the right of n-6. Thus, n-6, n-3, and n make a 3 parastichy that ascends in the anticlockwise direction when viewed laterally (that is, from the oldest to the youngest primordium; Fig. 15.4A).

Similarly, it follows that if n-2 is between n-5 and n-4, and if n is to the left of n-2, then n-2 is to the left of n-4. Thus, n-4, n-2, and n establish a 2 parastichy (Fig. 15.4A), which ascends in the opposite direction from the 3 parastichy. Accordingly, the contact parastichy pair 2:3 is possible and can be realized if the size and shape of the growing primordia are such (Fig. 15.2B) that no other, older primordia in the same sector separate n from n-2 or from n-3.

Which primordia might these be? Assuming again that n-2 lies between n-5 and n-4, we can also say (as we already stated) that n-3 lies between n-6 and n-5. Therefore, n-5 must be between n-3 and n-2, as is n (Fig. 15.4A). Consequently, n-5 is in the same sector as n. There are, in theory (if we temporarily overlook our initial assumption that n is closer to n-3 than to n-2), three possibilities of positioning of n with respect to n-5: (**a**) n is on the same radius as n-5, which implies that the 5 parastichy is an orthostichy, and the divergence angle, α, is 144° (that is, there are exactly five primordia produced in just two cycles of the generative spiral; $720°/5 = 144°$); (**b**) n is between n-3 and n-5, and angle $\alpha < 144°$; and (**c**) n is between n-2 and n-5 and angle $\alpha > 144°$.

In possibility (**a**), n-5 would not intrude between n and n-2 or between n and n-3, and the contact parastichy pair 2:3 could theoretically be realized. But, from the statement that n is closer to n-3 than to n-2, which sets limits on the divergence angle, it follows mathematically that $\alpha < 144°$. (That is, $360°$ divided by 2.5 turns of the generative spiral yields an angle $\alpha = 144°$; more than 2.5 turns would yield an angle $\alpha < 144°$.) This limitation on the value of α excludes cases (a) and (c), leaving case (b). Thus, n is between n-3 and n-5 and $\alpha < 144°$ (Fig. 15.4A). Accordingly, there may be a contact parastichy pair 3:5 instead of 2:5, because n-5 can intrude between n and n-2. The 3:5 pair is realized if there are no other, older primordia in the same sector that could (depending on the growth characteristics of the primordia) separate n from n-5 and/or from n-3. What primordia, if any, might these be?

Again, situational similarity predicts that n-3 is between n-6 and n-8 — because n-3 is the same primordium that was n when n-6 was n-3 and n-8 was n-5, as in (b). By similar logic, n-5 is between n-8 and n-10. Consequently, n-8 is in the same sector as n; that is, n-8 is between n-3 and n-5 (Fig. 15.4A). Again, there are three possibilities of siting n: (aa) n is on the same radius as n-8, implying that the 8 parastichy is an orthostichy, and $\alpha = 135°$ (that is, eight primordia are initiated in three cycles of the generative spiral; $1080°/8 = 135°$); (bb) n is between n-5 and n-8, and $\alpha > 135°$; or (cc) n is between n-3 and n-8, and $\alpha < 135°$.

In possibility (aa), n-8 would not intrude between n and n-3 or between n and n-5, and the contact parastichy pair 3:5 could theoretically be realized. However, the assumption that n is closer to its older neighbor, n-3 (and therefore $\alpha < 144°$, as discussed above), narrowly excludes (aa) and (cc), leaving (bb). In (bb), there is a possibility of a contact parastichy pair 5:8 instead of 3:5, because n-8 intrudes between n and n-3. This possibility will be realized if there is no older primordium between n and n-8 or between n and n-5. But it can be shown that the older primordium, n-13, can intrude between n and n-5; thus, the contact parastichy pair 8:13, as well as the 5:8 pair, is possible. The growth characteristics of the primordia would determine whether either pair (or both pairs) were realized.

Summarizing now for cases in which n is closer to its older neighbor: contact parastichies characterized by 2, 3, 5, 8, 13... (numbers of the Fibonacci series), are *possible*. Of course, if the contact parastichies are the pair 3:5, then the noncontact parastichies that can coexist in the pattern are 2, 8, 13,.... Note that the possible contact parastichy sets were identified without specification of a divergence angle, α, except in a broad range: $\alpha < 144°$ for the parastichy pair 3:5; and $\alpha > 135°$ for the 5:8 parastichy pair. This range, $135°$ to $144°$, includes the theoretically limiting value of α ($137.5°$) that can be derived mathematically from the series of L/N fractions from the Fibonacci series (Sect. 15.2.3).

But there is a second possibility waiting to be examined: under the hegemony of a different physiologically mediated siting rule, n could be initiated closer to n-2 than to n-3 (Fig. 15.4B). It can be shown that if n is sited closer to n-2, then angle α will be $> 144°$. This excludes cases (a) and (b), leaving case (c), in which n is between n-2 and n-5 (Fig. 15.4B). Thus, a possible contact parastichy pair includes 2:5 but not 2:3. Is another contact parastichy possible? Yes, because, by situational similarity, n-2 is between n-4 and n-7, and n-5 is between n-7 and n-10. Thus, n-7 is in the sector between n-2 and n-5, which also includes n. The third possible contact parastichy therefore is 7. The numbers characterizing the possible contact parastichies, then, belong to the series 2, 5, 7, 12,... (analogous to the Fibonacci series, 2, 3, 5, 8,...., which cannot occur under this particular siting rule).

To review: if there are more than 2, but fewer than 3, primordia per turn of the generative spiral, the youngest primordium (n) will be closer either to the older (n-3) or to the younger (n-2) of its lateral neighbors. If n is closer to its older neighbor, the phyllotaxy will be of the common type based on contact parastichies characterized by 2, 3, 5, 8,... (the Fibonacci series; Fig. 15.4A). If n is closer to its younger neighbor (n-2), phyllotaxy will be of a much less common type, based on contact parastichies characterized by 2, 5, 7, 12,... (Fig. 15.4B). This type has been found in some cones of *Pinus pinaster*, in vegetative shoots of *Abies alba*, and in some *Betula* and *Magnolia* flowers.

Under yet a different siting rule, there are more than 3, but fewer than 4, primordia along each turn of the generative spiral. The divergence angle then is < 120° but > 90°. Necessarily, the youngest primordium, n, is located between n-4 and n-3. If n is closer to n-4 than to n-3, then the possible contact parastichies are characterized by numbers from the "analogous" series, 3, 4, 7, 11,... (as are the parastichies in Fig. 15.3B). This can be proven by the rationale and logic already invoked, without appeal to a narrowly limited divergence angle, as can be shown by drawing diagrams analogous to those in Fig. 15.4. Likewise, it can be shown that, in a system in which n is sited closer to n-3 than to n-4 but other factors are unchanged, the possible parastichies include 4, 5, 9,..., members of another "analogous" series.

Contact parastichies characterized by numbers of the Fibonacci series (2, 3, 5, 8....) constitute the most common type of phyllotaxy among those dicots and gymnosperms that have been investigated. Typically, in dicots, the shoot tip begins its ontogeny with two cotyledons (in the embryo) or with two prophylls (in lateral buds). These are usually as far apart as possible — that is, 180°. In some species, this opposite pattern continues indefinitely. In most species, however, a generative spiral develops. It typically begins to produce more than 2, but fewer than 3, primordia per turn, and the incipient primordium, n, is usually located closer to n-3 than to n-2 (Fig. 15.4A). We can now ask why this phyllotaxy, characterized by members of the Fibonacci series, is so common. Does it confer a selective advantage on the plant? Possibly, yes. Leaves arranged in this way may intercept light more efficiently (Leigh 1972) and may also have advantages related to packing under physical constraints in the bud (Williams 1975).

15.3.2 Divergence Angles and Control of the Phyllotactic Pattern

Elaborate mathematical systems (many based on the divergence angle) have been proposed to explain phyllotactic patterns. Williams (1975), who believed that such systems were artificial and unnecessarily complex, quoted from the classic book *On Growth and Form* by D'Arcy W. Thompson (1942, p. 920): "... I, for my part, see no subtle mystery in the matter, other than what lies in the steady production of similar growing parts, similarly situated, at similar successive intervals of time." Thompson thought that the elaborate mathematical interpretations of phyllotaxy offered by Church

(1920) were essentially irrelevant. However, neither Thompson nor Williams went on to show which aspects of phyllotaxy arise solely from similarity principles.

In the preceding section, we indicated that to construct a spiral phyllotactic system in which the divergence angle approaches a predicted value, it is sufficient to invoke the principle of situational similarity and then to make two simple assumptions: (1) the number of primordia initiated in one turn of the generative spiral is somewhere between two small whole numbers (such as 2 and 3, or 3 and 4); and (2) the most recently initiated primordium, n, is closer to either (a) the older, or (b) the younger of the two between which it is located. (Assumptions 2a and b may be expected to specify different phyllotactic systems.)

If we accept the validity of the principle of situational similarity and these simple assumptions, we can consider the role of the divergence angle in a new light. Whereas some workers have assumed that the divergence angle controls the phyllotactic pattern, we propose that the phyllotactic pattern is determined largely by whether a new primordium is initiated closer to the older or to the younger of its lateral (sectorial) neighbors, and that the divergence angle, rather than controlling the phyllotactic pattern, is a consequence of that pattern. From this line of reasoning, it follows that the shoot apical meristem need have no system for measuring distance along the generative spiral and for emanating signals to induce initiation of primordia at regular intervals, equivalent to a divergence angle. We think it unlikely that the plant uses such a strategy. (Indeed, there is no compelling need for physical reality of the generative spiral, either.) We know, in fact, that divergence angles measured on plants vary widely from their theoretical limiting values (Williams 1975). We conclude, as did Schwabe (1984), that the divergence angle is the result of mutual interaction of primordia; that is, it is a consequence of growth and changing geometry and is not genetically specified.

15.3.3 Changes in Order and Changes in Type

In a very young shoot tip, the primordia are large compared with the apex. Each primordium is under the influence of all the others around the lower margin of the apical dome, because all distances are short and all groups of cells can receive signals or influences from all other groups of cells. As the apical dome increases in circumference during ontogeny, however, the widths of the primordia do not necessarily increase proportionately. Space becomes available on the caulis for an increasing number of primordia. Eventually, many primordia surround the base of the apical dome. Each primordium is then reduced to the status of "one of many". Interactions between n and the next youngest primordia, n-2 and n-3, in that sector of the caulis are probably weakened by the increased distances between them. In such apices, cells at a site competent to initiate a new primordium, n, may interact more strongly with a pair of primordia older than n-2 and n-3, in the same sector as n. These older primordia may be, for example, n-3 and n-5, or n-5 and n-8. They are neighbors of n, based on continuity of ontogenesis under the same rules of siting of primordia and situational similarity as before. In most apices, n is located closer to the older member of the pair of its nearest neighbor primordia.

Shoot tips in which many primordia surround a large apical dome have phyllotactic patterns that are of a higher **order** than are those of young shoot tips, in which domes are relatively small and primordia relatively large and few. The term "order" refers to the position, within the Fibonacci or an analogous series, of the numbers characterizing opposing parastichy pairs. A change in order arises from a uniform increase or decrease in the circumference of the base of the apical dome relative to the size of the primordia (Williams 1975; Romberger and Gregory 1977). Within the Fibonacci series, an 8:13 system is of higher order than a 5:8 system. Changes in order can sometimes be induced experimentally (Maksymowych and Erickson 1977).

Changes in order of phyllotaxy may also occur during the further growth of the frustum-primordium units of the young shoot. For example, due to primary thickening growth, the radii of frustum-primordia units increase with age and thus with distance below the shoot apex. The primordia consequently diverge, each along its own radius from the axis, and, as a result, they become farther apart in absolute distance. As this occurs, contacts between primordia may be lost and contact parastichies dissipated. New contact parastichies may be established. Generally, these are of a higher order than are the earlier ones. Thus, there often are more parastichies per set at the base of a bud-borne shoot than near the apex, and more parastichies near the periphery of a "head"-type inflorescence (as in *Helianthus*) than near the center.

Phyllotactic systems that are characterized by numbers belonging to different series differ in *type* rather than in order. For example, a 7:12 system differs in type from an 8:13 system. The literature includes many reports of phyllotactic changes occurring during natural or experimentally controlled development, often without distinguishing between changes in order and changes in type. Changes in type appear to be much less frequent than changes in order. In a change of order, there is a change in scalar relations, but not in the topology of the phyllotaxy. Mathematically, changes in order are homologous with continuous transformations, whereas changes in type, being topologically based, are homologous with discontinuous transformations (see Harris and Erickson 1980).

15.4 Advances Toward a Theory of Phyllotaxy

For a century, most of the extensive literature about the origin of phyllotactic patterns has focused on one of two areas, to the exclusion of the other. These areas are: (1) mathematical and spatial descriptions (or analyses) of patterns found in nature or of models intended to emulate natural patterns; and (2) efforts to discover the dynamic biological control systems that regulate the initiation and siting of primordia. The need to combine these approaches is now widely recognized. Progress has been made by using computer-aided simulation methods and mathematical models to test the adequacy of hypothetical physiological and biophysical systems to generate patterns similar to those observed in nature.

We cannot here attempt to evaluate the various theories and hypotheses that have arisen from mathematically inspired or supported experimental work (see compilations by Schwabe 1984; Jean 1984). However, guided by mathematical and biophysical studies (Veen and Lindenmayer 1977; Young 1978; Schwabe and Clewer 1984, among others), we choose to emphasize a body of ideas that collectively can be called a "diffusion theory of phyllotaxy", though this theory is not wholly distinct from various "field" theories that may also have merit. Ideas on involvement of diffusible inhibitors in the siting of new primordia, as expressed by Schoute (1913) and amplified by Richards (1951), have gradually gained credence because of supporting experimental and mathematical work, and because of the weakness of other theories.

It is unclear why primordia can appear at the base of a dome-like apex, but not on the dome itself. Functionally, the distal regions exert a kind of "dominance" over the regions nearer the base of the apex. In the dermal and subdermal layers, this influence may take the form of constraints on orientation of cell divisions. New primordia may arise when and where small sets of cells are released from constraints and undertake several cycles of divisions in planes that are periclinal rather than anticlinal, and then grow in the anticlinal direction. A basic physiological question underlying phyllotaxy is: what determines where these local areas of independent development arise?

More than three decades of work by M. and R. Snow (1931, 1947) and R. Snow (1965) provided evidence that siting of new leaf primordia is strongly affected by existing primordia, which seem to inhibit initiation of new primordia nearby. For example, if a recently initiated primordium is mechanically destroyed, the next one to be initiated will be closer to the site of the destroyed one than it otherwise would have been. Likewise, destruction of one member of a newly initiated decussate pair evokes a transition to a spiral pattern, at least temporarily. The position of a new primordium is not necessarily changed by microsurgical undercutting of the presumptive site of the primordium, severing direct vascular continuity. Thus, vascular influences alone may not determine the positions of primordia. However, if growth regulators that interfere with auxin transport are applied through the vascular system of *Chrysanthemum*, thus causing elongation of the youngest internodes, the phyllotaxy changes from spiral to distichous, and the divergence angle increases from 137.5° to 180° (Schwabe 1971).

The hypothesis that young primordia repel each other provides a basis for interpreting this change from spiral to distichous phyllotaxy. As the axial distance between young primordia is increased, the repulsive effect of n-2 (and n-1) on the siting of n is reduced so that n can arise directly above n-2 (as is typical in a distichous pattern). Experimental data from several taxa have confirmed that if the youngest and next youngest internodes elongate either naturally, as in the onset of flowering, or after experimental treatment with growth regulators, the divergence angle increases, often to 180°. Conversely, if elongation is restrained, divergence decreases and phyllotaxy becomes spiral (Meicenheimer 1981, 1982; Schwabe 1984).

Various interpretations offered for these and similar experimental results suggest that an inhibitory morphogen moves only a short distance from its site of synthesis in young primordia before it is inactivated or otherwise removed from the organogenic region. Some investigators have postulated the joint operation of two morphogens, one an inhibitor diffusing away from primordia, and the other an inactivator of the inhibitor, derived from other regions of the caulis (Sect. 12.4.2.2). However, it is simpler to assume a single morphogen, moving mainly by basipetal polar transport in the symplasm (Schwabe and Clewer 1984). If this morphogen is presumed to be auxin, which is known to move basipetally in the symplasm by active polar transport, the total prohibition on initiation of new primordia below older ones is readily explainable (Schwabe 1984). In addition, a small fraction of the morphogen may diffuse away apoplasmically from sites of production in all directions, including acropetally. Thus, there would be a finite concentration of inhibitory morphogen throughout the organogenic region beneath the apical dome. As long as the concentration of morphogen exceeded a threshold, no new primordia would be initiated. But as growth proceeded, dome dimensions would increase and the morphogen would be subjected to various attenuations, until its concentration dropped below a critical threshold in some locality. The cells in this locality would then be released from constraints on orientation of cell divisions and begin initiating a new primordium.

Determining whether the morphogen postulated by a diffusion theory of phyllotaxy is or is not auxin will be difficult. However, a diffusion theory involving auxin as an inhibitor of primordial initiation seems plausible enough to justify experimental work because: the apoplasmic fraction of the cross-sectional area of the apical meristem is in the range of 1 to 5% of the total; passive diffusion rates of auxin are known; the acropetal distances involved are short; and plastochrons are long enough for diffusion to occur (see Meicenheimer 1979, 1981, 1982).

Concepts based on diffusion of morphogens and reactions to them do not explain how small sets of cells at the base of an apical dome are released from prior constraints and undertake periclinal divisions. Such concepts pertain to a pre-pattern for the phyllotaxy, whereas the actual cellular processes by which a leaf primordium is initiated are not considered. But there is another possibility: siting of leaf primordia and the actual process of leaf initiation could be controlled by physical processes in growing tissues. The concept of a reinforcement-field theory of phyllotaxy (Green 1985) is based on such a perspective. This theory is compatible with the finding that a new leaf primordium develops a "hoop" of reinforcing cellulose on its surface. This reinforcement extends towards the apical dome and modifies the reinforcement patterns extending from older primordia. The site of most abrupt angular change in the reinforcement pattern on the dome's surface becomes a new center of a new hoop reinforcement and this center is the site of the next primordium. This process occurs cyclically and, if during each cycle similar conditions prevail, a regular phyllotactic pattern is produced, as can be understood in the context of the principle of similitude (Sect. 15.3).

15.5 Hierarchical Control of Phyllotaxy?

A phyllotactic pattern is an attribute of an entire shoot tip. In steady-state systems, patterned regularities involving triads of primordia such as n, n-2, and n-3, or their successor triads (n-1, n-3, and n-4; n-2, n-4, and n-5; and so on, after one, two, or more plastochrons, respectively) may be expected to produce regular phyllotactic patterns merely on the basis of their situational similarity. However, rigorously considered, a plant progressing through its normal vegetative ontogeny is not a long-term, steady-state system, and its phyllotaxy typically changes.

For example, an increase in circumference of a shoot apex due to growth may, especially if that growth is asymmetrically distributed, provide space for a new primordium that is the first member of a new parastichy. Such changes, which are not necessarily rare, result in nonconformities in the crystal-like regularity of the pattern of the triads of primordia arranged around the apex. Further, changes in rates of radial and axial growth of the maturing frustum-primordium units (as they become internode-node units) may further distort the patterns of triads, transforming them into patterns having different mathematical characteristics (Zagórska-Marek 1987). In this sense, transitions of phyllotactic patterns (which are beautifully recorded in the arrangement of leaf bases on twigs of some conifers) are evoked by changing growth rates in the primary shoot as various parts of it pass through successive phases of non-steady-state conditions. Thus, the control of the several different apical growth rates is part of the control of phyllotaxy.

Further, if the rate of primordium initiation (and plastochron duration) changes relative to the rates of radial and axial growth of the frustum-primordium units, the morphology of the shoot tip also is affected. The relations between these various rates determine the extent to which primordia accumulate as part of a compact embryonic shoot within a bud, and also determine the shape of the bud-borne shoot — such as whether it becomes a miniature strobilus-like structure, as in dormant *Picea* buds, or a disc-like head, as in the floral buds of composites.

Complex changes in phyllotactic patterns occur during shoot development, and complex systems generally consist of interrelated subsystems. The control of the subsystems typically operates in a hierarchical manner (Jean 1982). Although, in the preceding section, we abstracted some ideas invoking the combined involvement of symplasmic polar transport and nonpolar apoplasmic diffusion of a morphogen, and presented them as an advance toward a plausible theory of phyllotaxy, we now hasten to express doubts that any easily comprehensible theory can in itself explain either all attributes of observed phyllotactic patterns or their ultimate origin.

An unavoidable consequence of accepting a diffusion theory of phyllotaxy as sketched here is that phyllotaxy then becomes subject to modulation by a great number of factors, including, among others: plastochron duration; absolute size of the apical dome; absolute and relative size of newly initiated primordia with respect to the apical dome; lengths of the youngest internodes, along with their rates of increase; rates of apoplasmic diffusion of primordium-derived morphogens laterally and acropetally; rates of metabolic and nonmetabolic inactivation of morphogens; rates of symplasmic basipetal transport of primordium-derived morphogens; and relative proportions of symplasmic and apoplasmic space in the organogenic region. Of course, the various elements in this open set of rates and dimensions are themselves variables influenced by still other factors.

In our view, it is unrealistic to seek a simple system that regulates phyllotaxy separately from other aspects of primary shoot growth, such as vascularization. A close coordination between patterns of phyllotaxy and of vascularization is demonstrated by the orderly increase in the branching of vascular sympodia that occurs if there is an increase in the rate of leaf initiation by a shoot apex. The new branch strands enter the "extra" primordia (Sect. 14.3.1). Commonly, in an enlarging apex, the number of sympodia increases until it attains the next higher number in the series characterizing the prevailing phyllotactic pattern, with no indication of stable intermediate values. For example, if, in an enlarging apex, the established parastichies are characterized by 5 and 8, the number of sympodia may increase to 13, which is the next higher number in the Fibonacci series, 5, 8, 13,.... Conversely, if the rate of primordium initiation decreases, the branching of sympodia decreases, and the number of sympodia may decrease to a lower number in the same series.

Such close relations between phyllotaxy and vascularization indicate that a complex regulating system is operating. Present knowledge is too limited even to arrange the elements of such a complex system in a plausible ascending or descending order within a hierarchy. Jean (1982) addressed the problems of comprehending hierarchical control of phyllotaxy along with control of patterns of primary vascularization, but did not propose that one derives from the other.

Perhaps both phyllotaxy and vascularization respond to parts of the same hierarchical control system. In this case, arguments that one is subservient to the other are irrelevant. Indeed, if the morphogen postulated in some variants of the diffusion theory of phyllotaxy is auxin, one can imagine that there is a link — perhaps an unavoidable one — between vascular and phyllotactic patterns. For example, the acropetally advancing provascular traces could, by functioning as routes of basipetal polar transport, influence the local concentrations of auxin near their apical termini. Locally low concentrations of auxin in the symplasm could reduce the amount of auxin "leaking" into the apoplasm and thus reduce the amount passively diffusing acropetally into the organogenic region.

16 The Primary Shoot: Leaf Development

16.1 Introduction: The Leaf Concept

Consider the problem of writing an adequate and botanically accurate definition of "leaf". The following definition, partly facetious, is synthesized from several dictionary entries. Leaf: a lateral outgrowth from a stem, constituting one discrete element of the foliage of a plant and usually functioning in photosynthesis, being one of such outgrowths as arise in regular succession from a shoot apex, consisting typically of a flattened blade attached to the stem by a petiole, and which in transection has an adaxial and abaxial epidermis, one or both of which are perforated by stomata, and between these dermal layers has layers or masses of palisade and spongy mesophyll cells containing chloroplasts, this mesophyll tissue being permeated by labyrinthine gas spaces and by veins of vascular tissue which may be accompanied by mechanical tissue; the whole structure being distinguishable from a leaflet, cladophyll, or phylloclad, by presence of a shoot bud in the axil between its base and the subjacent stem.

Obviously, even such a cumbersome definition fails to encompass all organs that *function* as, and are readily recognized as, leaves. It does not, for example, accurately define the mature foliage "needles" of *Pinus*, nor does it allow for the frequent to general lack of buds in the axils of many gymnosperm organs that function as leaves, such as those of *Picea*. It is evident that, whatever definition one might construct, there will be organs exceptional to it or seeming to contradict it (Goebel 1880; Ryder 1954). Rather than trying to refine our definition of leaf further, we will begin a consideration of those features of morphogenesis common to leaves of a wide range of taxa.

16.2 Initiation and Early Development of Leaf Primordia

A leaf begins as a localized protrusion, which gradually becomes a leaf primordium, on the flanks of a shoot apical meristem. In pteridophytes, each primordium consists of progeny cells of a single, persistent leaf apical cell that strongly resembles the pyramidal apical cell of the pteridophyte shoot. In seed plants, however, no such single apical initial cell is discernible, either at the shoot apex or at the site of the initiation and development of a leaf; rather, leaf initiation, protrusion, and development are integrated activities of progeny of groups of cells. Such groups may include proto-dermal cells on the flanks of the shoot apex, and generally include subsurface and more deeply lying cells. None of these are identifiable as initials, but their progeny may make major contributions to the internal tissues of the leaf.

The relative roles, in various taxa, of protodermal, hypodermal, and more deeply seated cells in initiation and early development of leaf primordia are, in effect, a continuing influence of the structure of the shoot apex (Sect. 14.1.3) on histogenesis. The relative contributions by cell lineages derived from apical layers L-1, L-2, etc., probably are largely controlled by systems related to those that determine whether an apex has a tunica/corpus structure. The contributions may also be influenced by the wall reinforcement pattern in the surface cells of the apical dome and by space and volume relations. Depending on the taxon, these systems allow or prohibit periclinal divisions in surface cells.

The common opinion that angiosperms generally have "stratified", tunica/corpus-type shoot apices, whereas gymnosperms generally do not (Sect. 14.1.3), is inaccurate. Actually, among gymnosperms, there are three distinct types of apical structure (one of which is the same as that prevailing in angiosperms):

1. In *Ginkgo, Zamia,* and most members of Pinaceae, there is no restriction on periclinal divisions in surface cells of the apex or its flanks; there is no apical stratification (no tunica or true protoderm); and the progeny of surface cells can contribute to internal tissues.
2. In Taxodiaceae, Cupressaceae, and Taxaceae, there also is no stratification in the region of the apical initials, but there is a distinct protodermal layer (within which there are no periclinal divisions) on the lower flanks of the apex, including the organogenic region, in which foliar primordia arise.
3. In Araucariaceae, possibly in some members of Podocarpaceae, in Gnetaceae at some ontogenic stages, and in angiosperms generally, tunica/corpus stratification prevails throughout the apex (Strasburger 1872; Cross 1942; Johnson 1943; Sacher 1954, Rom-berger 1963; Napp-Zinn 1966).

Thus, in *Taxodium, Cryptomeria, Cunninghamia,* and some other gymnosperms, and in angiosperms generally — wherever there is a true protoderm in the organogenic region of the apex — the L-1 cells contribute primarily to the superficial, dermal tissues of the new leaf and hardly at all to internal tissues (Cross 1940, 1942). In contrast, in *Pinus, Abies, Picea,* and some other genera of gymnosperms, there is no discrete protoderm in the organogenic region, and periclinal divisions are abundant in surface cells during leaf initiation (Strasburger 1872; Heimerdinger 1951); consequent-ly, surface cells contribute substantially to internal tissues of the leaf primordium.

The early development of leaf primordia is influenced by spatial and volume relations with neighboring primordia and with the shoot apical meristem (Sect. 15.2.2). Nonetheless, leaf form is mostly a result of controlled, patterned cytogenesis and cell volume growth within the primordium itself. Cytogenesis is primal in early develop-ment but is gradually supplanted by cell expansion within the leaf as the major agent of morphogenesis (Dale 1988; see also Green 1985; Sect. 15.4).

In angiosperms generally, the early protrusion (or "elevation") of leaf primordia is closely related to divisions of hypodermal and deeper layers of cells (Napp-Zinn 1973). The orientation of these divisions has been called "periclinal", but one can question the accuracy of describing the orientation of cell divisions occurring during early protrusion in terms of periclines and anticlines based on a cauline coordinate system. There is evidence that a new spatial reference system is established early in leaf-primordium initiation. An orientation that may seem obliquely periclinal in the cauline system is actually anticlinal in the leaf-primordium system. Cells involved in early stages of protrusion may already have changed their axiality and polarity of growth (Green 1980).

Leaf initiation is generally confined to the lower, "organogenic" flanks of the shoot apical meristem, but it is not necessarily a narrowly local phenomenon. Relative to the periphery of the parent axis, the angular width of the base of a young leaf primordium, being partly "determined" by phyllotaxy and subjected to the spatial constraints of contact parastichies (Sect. 15.2.2), is quite variable among taxa. In some woody plants, the base of a typical foliar primordium may span 120° or more. In *Liriodendron* and *Magnolia*, for example, it nearly encircles the apex (Postek and Tucker 1982). In *Picea*, though, it typically ranges between 10° and 20°.

After initial protrusion, a leaf primordium grows by cell division and expansion into a mound-like foliar buttress. After the buttress stage and before extensive internal differentiation and histogenesis, the primordial leaf axis is called the **phyllopodium** (or proaxis). In most taxa, cytogeneration in the phyllopodium is briefly localized in the apex. It soon becomes generally distributed, and then basally localized.

As the base of the phyllopodium grows wider and takes more space on the surface of the caulis, according to the phyllotactic pattern (Sect. 15.2), the distal part of the phyllopodium, by apical and intercalary growth, becomes a finger-like appendage. This appendage often curves toward the shoot apex.

In ferns, apical growth of leaf primordia and leaves may continue for most of a growing season or may even be indeterminate, whereas in seed plants apical growth stops early. In many conifers, it ends before the primordia are 400 μm long (as in *Taxus*, *Picea*, *Pinus*, and *Taxodium*). In some gymnosperms having large leaves, including *Agathis* and *Podocarpus*, it lasts longer. Among angiosperms, also, apical growth varies in extent, usually being greater in species having large mature leaves (Sonntag 1887). In *Drimys winteri*, this length is about 1.2 mm (Gifford 1951); in *Viburnum rufidulum*, about 1.6 mm (Cross 1937). An extreme case is the branch-like, compound, semi-indeterminate leaves of *Guarea* spp. (tropical forest trees of the Meliaceae). The tips of these leaves bear a bud from which new pinnae expand (Steingraeber and Fisher 1986).

Generally, after a phyllopodium ceases to grow apically, it extends mainly in its basal parts. In many taxa, extension growth continues to be greatest in basal regions during later developmental stages. Extended basal growth is especially pronounced in long-needled *Pinus* species, in which it may persist for most of the first season, and sometimes part of the second. In many monocots, intercalary growth near the leaf base is so persistent and well defined that this region constitutes a **basal intercalary meristem**. Because of such meristems, animals can graze in pastures without destroying the grass.

Early cessation of apical growth of leaf primordia allows early differentiation and functional maturation of cells in the leaf tips. This can be viewed as enabling photosynthesis in the apical regions to contribute to overall leaf carbohydrate economy while vascular tissues that pass through the meristematic and extending basal regions still consist partly, or largely, of functionally deficient, immature elements subject to constriction and disruption by growth stresses.

16.3 General Aspects of Development of Leaf Primordia into Leaves

If enough information were available, the development of many forms of leaves could be discussed as variations of an orderly series of overlapping temporal phases: (1) protrusion of a foliar buttress from the profile of the parent shoot apex; (2) formation of a near-microscopic, conical or cylindrical, primordial axis, the phyllopodium; (3) organization of marginal, submarginal, and plate meristems, along with early expansion of a broad lamina (common in angiosperms, less so in gymnosperms); (4) differentiation, within the phyllopodium and expanding lamina, if present, of the rudiments of a foliar venation system via longitudinal divisions, symplastic growth, and possibly intrusive growth; (5) further expansion of the lamina, with opening of intercellular gas spaces, differentiation of stomatal apparati, and maturation of vascular and accessory tissue as the leaf becomes functional.

The available information, however, is inadequate to allow connected discussion of all aspects of leaf development using this ideal, temporal-phasic approach. Therefore, we have supplemented that approach with others as appropriate.

16.3.1 Regional Meristems — Marginal, Submarginal, Plate, Abaxial, Adaxial, and Intercalary

After the phyllopodial stage, most of leaf morphogenesis typically can be ascribed to the differential meristematic activities of various subregions. It is convenient and customary to assign to these regions the term "meristem" modified by a descriptive adjective — i.e., apical, intercalary, marginal, submarginal, and plate. We use these terms for convenience in discussion, but emphasize that such "meristems", as cell groups, are typically not discrete and seldom consist of sets of identifiable initials comparable to those of shoot apices. Whatever we can briefly say about the activities of various regional meristems is inadequate and inaccurate, because growth includes both meristematic and expansionary components. Further, the activities of these meristems, though separately considered here, are well integrated, forming structural patterns beautifully adapted to leaf function.

As apical growth is typically of short duration and accounts for only a small fraction of the volume of the phyllopodium, we do not discuss it with the regional meristems. In many gymnosperms and some angiosperms, the apical cells, which often are protodermal derivatives, produce only a spine-like tip (Cross 1940; Napp-Zinn 1973).

16.3.1.1 Marginal and Submarginal Meristems.
In many taxa, while the phyllopodium is still very small (0.1 to 1.0 mm long), a patterned distribution of cytogeneration and cell enlargement produces two parallel **marginal meristems** along its flanks. Often ridge-like, these meristems are rarely diametrically opposed along the flanks of the phyllopodium, except in some relatively narrow leaves. Rather, the marginal-meristem ridges usually are adaxially displaced, and the original adaxial surface of the phyllopodium becomes the bottom of a groove between them. The adaxially located, marginal meristems then begin forming the opposite halves of a primordial lamina that is convolute, or folded "upward" along the midrib. In some taxa, the marginal meri-

stems are abaxially displaced, producing lamina halves that are revolute, or folded "downward" along the midrib (Hara 1957b).

In some gymnosperms, marginal-meristem activity, if any, is diffuse, and no lamina as such is produced. The phyllopodium then develops into a narrowly columnar leaf having a round, triangular, square, rhomboid, or similar cross section, as is typical in *Pinus*. In other gymnosperms, including *Taxodium* (Cross 1940), marginal meristems account for all lateral growth by cell division. Because this activity is not supplemented by plate-meristem activity, the leaf remains quite narrow.

The marginal initials and their immediate progeny may be regularly arranged, as was demonstrated in *Xanthium* by Maksymowych (1973). In developing foliage leaves, the marginal initials, due to their position, are commonly considered to be a single layer of cells that divide almost exclusively anticlinally to the surface of the marginal ridge. The immediate progeny (still protodermal) cells of the putative marginal initials also divide mostly anticlinally and differ from the initials only in position.

In taxa having apices with a tunica, or having true protodermal layers at least in the organogenic region, the lineages of the marginal initials generally are continuous with those of the tunica, or protoderm, of the apical meristem. In taxa lacking a true protoderm, lineages of marginal initials are continuous with cells of a **de facto protoderm**. Such a layer, though occupying an equivalent position, differs from a true protoderm in that its cells divide periclinally as well as anticlinally. Accordingly, progeny of a de facto protoderm, unlike progeny of a true protoderm, can become component cells of internal tissues. Whatever the lineage of the marginal initials, their progeny collectively always include a protoderm or proepidermis — which may give rise to additional layers before it matures into an epidermis (Sect. 16.3.3).

The submarginal "initials", constituting a putative **submarginal meristem**, are located immediately proximal to the marginal initials. Submarginal initials are mostly derived from subdermal cells of a shoot apex via lineages that are separate from those of the marginal initials. Sometimes, however, the submarginal initials are supplanted by derivatives of marginal initials (Sect. 16.3.2). The progeny of submarginal initials may organize themselves into four or more layers of cells that, after further divisions, differentiate into some of the internal tissues of the lamina. These tissues, however, do not necessarily include the middle layers with reference to laminal thickness, nor much of the internal tissue in the mid-laminar regions with reference to area.

Marginal- and submarginal-meristem activity is not necessarily uniform along the entire phyllopodial axis, and thus the ridge-like rudiments of the laminar halves are not necessarily continuous along the axis. If meristematic activity is uniform to the base, the lamina will be entire and sessile; if it does not reach the base, the leaf will be petiolate; if it is fractioned into segments, the leaf will be deeply lobed or compound (Sect. 16.4). Stipules, which may be discontinuous with the main ridge, may develop from localized, opposite, marginal meristems at the very base of the phyllopodium. If the ridges gradually diminish apically and then end near the phyllopodial vertex, an acutely tipped leaf will be formed. If the ridges, as they approach the phyllopodial vertex, continue to grow vigorously, they may merge, forming a slender, inverted-V configuration. Such merged marginal meristems give rise to various types of hooded, reniform, and peltate leaf laminae.

Some authors consider the lateral edges of an expanding lamina as marginal meristems and the remaining parts as submarginal, "intercalary", or "plate" meristems.

The term "intercalary meristem", however, is more often applied to regions of relatively persistent meristematic activity at the base of the whole phyllopodium or growing leaf. We therefore use the term "plate meristem" (see later) for meristematic regions that are not marginal or basal and that contribute to area growth of a lamina by cell division. Functionally, though, there may be no clear distinction between marginal, submarginal, and intercalary growth, as growth rates in neighboring areas commonly are similar.

16.3.1.2 Plate Meristems.
In an expanding lamina of a broad-leaved angiosperm or gymnosperm, growth arising directly from marginal and submarginal meristems is usually soon complemented and partly superceded by less localized production and expansion of cells in nonmarginal regions. The meristematic cells in these regions function as **plate meristems**. This diffuse expansion typically does not occur in narrow leaves with prominent longitudinal parallel venation. The expansion resulting from plate-meristem activity has also been called "dilatation" growth, but we prefer to restrict use of that term to proliferation of phloem-ray tissues (Sect. 20.5.4).

In plate meristems, divisions anticlinal to the laminar surfaces predominate. Progeny cells are arranged in semi-regular layers parallel to those surfaces. Divisions in these meristems contribute to construction of broad laminae, though not to peripheral regions or surface layers. Plate meristems generally have no identifiable initial cells and usually no definite borders, especially where they meld with any remaining ground meristem around the nascent midvein region in the core of the phyllopodium.

Mesophyll and vascular tissue differentiate within the plate meristem and its recent derivatives (Sects. 16.3.5; 16.3.6). In later stages, the mesophyll differentiates into palisade and spongy types. The palisade mesophyll usually differentiates from the adaxial subepidermal layers, and the spongy mesophyll is derived from the remaining layers. In some leaves, palisade mesophyll differentiates from abaxial layers also. Apparently, provascular tissue sometimes originates in a specific mid-layer (Sect. 16.3.6.2). The lineages of plate-meristem cells, as derived from marginal and submarginal meristems (which, in turn, are traceable to specific cell layers of the shoot apex; Sect. 16.3.2), are not necessarily of prime importance to morphogenesis. Again, the position of a cell seems to be more important than its lineage.

16.3.1.3 Abaxial and Adaxial Meristems.
In a phyllopodium, development of lamina halves from marginal meristems is often accompanied by extensive abaxial cell proliferation, commonly leading to formation of a fleshy, abaxially protruding midrib. Meanwhile, subdermal cells within the groove between the marginal-meristem ridges along the adaxial side of the phyllopodium may divide regularly enough to justify the term "adaxial meristem". This meristem may also contribute to midrib tissues.

In some taxa of both angiosperms and gymnosperms, the lamina-building function of the marginal meristems is usurped by an adaxial meristematic ridge (or an adaxial-abaxial pair of ridges), which then constructs a lamina oriented in an adaxial-abaxial plane. In some taxa, such as *Acacia melanoxylon* (Boke 1940), the mature "leaves" can be interpreted as phyllodes (flattened petioles) derived from pronounced adaxial-meristem activity in the basal region; the lamina proper remains relatively undeveloped. Among gymnosperms, abaxial-adaxial meristems occur in juvenile leaves of

Podocarpus dacrydioides, and in nonjuvenile branches of *Acmopyle pancheri* and *Dacrydium* (Lee 1952).

16.3.1.4 Intercalary Meristems. Small, local, intercalary meristems may occur wherever vascular strands pass through rapidly expanding tissues, as in petioles, midribs, and larger veins. Much more significant and persistent than these meristems, however, are the basal intercalary meristems of phyllopodia and primordial leaves in taxa having strongly linear or needle-like foliar organs. These intercalary meristems are similar to rib meristems of stems and roots, in that they and their recent progeny divide almost exclusively transversely. They produce longitudinal, clonal files of cells, which, except near the tip, may give rise to most of the internal tissue of the leaf.

Does a basal intercalary meristem consist of a disk-like set of initials, with each initial heading an axial file of progeny cells on its apical side? Among derivatives of a basal intercalary meristem, all cells within an axial file or set of files typically differentiate and mature into similar tissues, though neighboring sets of files can differ. Is this because the initials heading the files have undergone quantal divisions (Sect. 12.2.2.1), and differentiation of their progeny cells is therefore predetermined? Or, are the progeny cells responding more to positional information? These questions cannot yet be answered.

Leaves in which most tissues are derived from basal intercalary meristems tend to have strongly parallel venation. Because the younger cells are near the base of the leaf, maturation of vascular tissues appears to proceed basipetally. However, provascular strands already present in the phyllopodium before a basal intercalary meristem becomes established may mature into veins acropetally.

16.3.2 Insight from Chimeras and Clonal Analysis

As discussed in Section 14.1.5, periclinal chimeras effectively "label" cells derived from discrete layers of the shoot apex. Therefore, studying the distribution of chimeral cell layers can yield information on the contributions of progeny cells of specific apical-meristem layers to regional meristems in the phyllopodium and leaf. Most relevant are studies of chimeras in which the several apical layers (L-1, L-2, L-3, and sometimes L-4) differ in chlorophyll or other pigment content, or in ploidal level. Most studied have been chlorophyll-variant chimeras, such as W-G-G, G-W-G, and W-W-G-G, and ploidal chimeras, such as 4-2-4, 2-2-4, and 2-4-2. The letters and numerals in these designations refer to successive layers of the shoot apex (L-1 being outermost), where W represents the "white" phenotype, G represents "green," and 2 and 4 represent "diploid" and "tetraploid", respectively. Extensive studies of these periclinal chimeras indicate that cells derived from L-3, and not just from L-1 and L-2, are involved in leaf-blade development, though the relative contributions of these layers may be highly variable.

The white-margin, green-center form of *Pelargonium x hortorum*, a G-W-G plant, illustrates what can be learned from chimeras (Stewart et al. 1974). In this plant, as in most angiosperms, cells derived from L-1 are typically confined to the surface of, successively, the primordial buttress, the phyllopodium, the marginal meristem, the protoderm, and finally the epidermis — where they do not manifest their genetic ability to synthesize chlorophyll, except in stomatal guard cells. Thus, the epidermal cells, though genetically "green", do not mask the fact that the mesophyll in the marginal parts of the leaf is "white", having originated from the genetically "white" L-2, via the submarginal meristem.

Studies of chimeral leaves indicate that contributions of marginal and submarginal meristems (i.e., L-1 and L-2) to laminae are easily overestimated. For example, in the nonmarginal part of the leaf blade of *Pelargonium x hortorum*, the genetically "white" L-2 contributes a single layer subjacent to upper and lower epidermal layers (as determined microscopically); to the naked eye the white layers are masked by the numerous layers of green mesophyll derived from L-3. In addition, our own studies of a *Fraxinus excelsior* periclinal chimera of the G-W-G type, which has white-marginate leaflets, revealed that cells of L-3 lineage contribute significantly to internal tissues of the leaflets. Our interpretations of similar analyses of chimeral leaves of *Ligustrum ovalifolium* and other woody species (Dermen and Darrow 1960; Stewart and Dermen 1975) lead to similar conclusions. Induced mutations and clonal analysis have produced further evidence that laminae in at least some taxa are derived from cells of three lineage layers (Dulieu 1970; Poethig 1987). The hypothesis that the submarginal initials are the primary source of the mesophyll is not supported.

Information gained from study of chimeras and clonal analysis is consonant with thinking that, though distribution of lineages of apical layers may, for example, determine leaf pigmentation patterns and hence some types of function, leaf form is mostly unaffected and its control is to be sought elsewhere.

16.3.3 Protodermal, Epidermal, and Hypodermal Tissues

A major determinant of the contribution to internal tissues by progeny cells of early de facto protodermal layers is the extent to which periclinal as well as anticlinal divisions occur in the outermost cell layer (L-1) in the organogenic region of the shoot apex. Periclinal divisions are rare in primordia arising on apices with discrete tunica layers, but are frequent in the Pinaceae and other gymnosperms lacking a tunica and/or lacking a discrete, persistent protoderm in the organogenic region (Sect. 16.2). In these taxa, surface cells, constituting a de facto, and possibly stochastic (Sect. 14.1.3.2), protoderm have the potential of contributing substantially to internal tissues. For example, in *Pinus nigra*, the leaf buttress at first consists of a de facto protoderm of large cuboidal cells and a similar, underlying layer of promesophyll cells. Then, the surface (de facto protodermal) cells, by anticlinal *and* periclinal divisions, produce proepidermal, prohypodermal, or additional promesophyll cells. Some of the promeso-phyll cells, in turn, by further divisions produce proendodermal cells. Even much of the transfusion tissue (which is located toward the leaf axis from the endodermis) probably ultimately has a protodermal origin (Huber 1947; Heimerdinger 1951).

However, the final extent of the contribution of dermal layers to the mature leaf is not always directly and immediately determined by the relative frequency of early periclinal divisions in L-1 and L-2 of the leaf primordium. This is because the progeny of these layers may make a continuing contribution to internal tissues. We mentioned previously that there is an early shift of cytogeneration from apical regions of the phyllopodium to the midregions and then, in many taxa, to a basal intercalary meristem. If cells derived from periclinal division of the protoderm in the early primordium eventually become a part of the set of initials of the intercalary meristem in the young phyllopodium, then these cells can continue to give rise to progeny that develop into internal tissues. For example, the lineage of an axial file of cells in the midregion of, say, an immature pine needle or a grass leaf might lead back to an initial cell in the basal meristem; in turn, the lineage of this initial might lead back to a progenitor dermal or subdermal cell (Huber 1947; Heimerdinger 1951).

Regardless of the lineage of protodermal and hypodermal cells, and whether or not their progeny contribute to internal tissues, they have a special position and acquire special functional attributes that are well known (Sects. 3.2.2; 3.3), especially in broad, deciduous leaves of the common angiosperms. Less familiar are the protoderm, epidermis, and hypodermis of the xerophytic, "needle" leaves of Pinaceae and some other gymnosperms. Though inclusive surveys have yet to be made, the surface layers in some of these taxa are quite different from those of angiosperms. In *Pinus resinosa*, for example, a mature secondary leaf is enclosed in a uniseriate layer of elongated epidermal cells having such thick walls that the lumina are mere slits (Gambles and Dengler 1982a). These epidermal cells, which are dead at maturity, are underlain by a layer of thick-walled, almost fiber-like, hypodermal cells, which retain some attributes of living cells at maturity.

The epidermis, as it matures from protoderm, may serve a *developmental* function for the leaf, in addition to its final physical role in providing either armor-like protection and structural rigidity, as in *Pinus* needles, or a resilient, hydraulically supported envelope within which a fragile aerenchyma is suspended. The expansion of protodermal cells as they differentiate into mature, often tabular, epidermal cells may have a role in physically driving the primordial and expanding leaf toward functional maturity. This cell-expansion growth, which reaches a maximum after the rate of cell division in the leaf has declined, may provide the force to move apart the cells in the underlying mesophyll (though lysigenous processes may also be involved; Sect. 16.3.5) and thus to form the gas-space system. At this stage, the dermal layers, being under periclinal compression while the inner tissues are under tension, give hydraulic stiffness and support to the expanding leaf. After the gas-space system has been generated and overall leaf expansion declines, epidermal and hypodermal cells may undergo pronounced wall thickening and perform a mechanical support function.

Protodermal/epidermal cells adjacent to the midvein and other large veins grow symplastically with components of bundle-sheath extensions, bundle sheaths, and vascular tissues (Sect. 16.3.6.2), resulting in only limited intercellular gas-space formation in these locations. In contrast, the formation of gas spaces in the mesophyll involves locally **nonsymplastic growth** relations between the protoderm/epidermis and the underlying or overlying promesophyll. Too little is known about details of this aspect of leaf development to justify speculation about its controls. However, the correlation between stomatal distribution and distribution of large, subdermal gas spaces suggests that the distribution of nonsymplastic growth and of stomata may be under related control systems.

In many species, the anticlinal walls of leaf epidermal cells are sinuous. In some entire genera, including *Picea* (Marco 1939), this sinuosity is so extreme that the cells seem to be interlocked with "dovetailed" joints. In addition, projections from the outer walls extend outward into the cuticle. Though common in both angiosperms and gymnosperms, sinuous epidermal cell walls are not equally apparent at all developmental stages (Campbell 1972). They tend to be more evident on abaxial than adaxial surfaces. In humid tropical forests, they are more likely to occur in understory than in overstory species (Pyykkö 1979). Their significance and mode of formation are not well understood.

16.3.4 Stomata

Stomata have been studied extensively for more than a century, as is documented by a vast literature. Although detailed morphological descriptions of mature stomata of many taxa are available, developmental information is quite fragmented. Most studies of stomata, even those intended to be developmental, have focused on local protodermal/epidermal cell groups. However, study of stomatal development at the organ level is also essential.

Protodermal cells cover the surface of a phyllopodium or young leaf primordium. Most of these, especially in angiosperms, are derived from the activity of the marginal meristems. A marginal meristem consists of a narrow axial strip of protodermal initial cells undergoing repeated axial anticlinal divisions. The cells that it produces typically undergo further divisions, but most soon lose their meristematic capability. They become highly vacuolate, expand, and begin differentiating into epidermal cells.

Some protodermal cells, however, lag behind their neighbors in differentiation. These are **residual protoderm cells**, a kind of **meristemoid**. If these cells directly, or after further divisions, produce stomatal complexes, they are **S-residual protoderm cells** (Patel 1978). These cells have also been termed "stomatal meristemoids" (Fryns-Claessens and Cotthem 1973; Stevens and Martin 1978). Study of sequences of cell divisions among progeny of S-residual protodermal cells, especially in gymnosperms and monocots, in which active basal intercalary meristems result in stomatal complexes that are arranged in orderly rows, with an obvious age gradient, led to various systems of classifying stomatal ontogenic types. Among dicots, in which ontogenic sequences of stomatal complexes do not usually occur in rows, stomata have often been classified on the basis of their morphological features at maturity, along with *conjecture* about their development.

The systems of stomatal classification proposed by Fryns-Claessens and Cotthem (1973) and by Stevens and Martin (1978) apply to both angiosperms and gymnosperms. These systems accommodate observations that different ontogenic subtypes can yield stomatal complexes that are structurally indistinguishable at maturity. By some criteria, these systems seem to have bridged the gap between morphological and ontogenic approaches. In abstract, these systems, like some earlier ones, recognize three major ontogenic types of stomata:

1. **Mesogenous** (formerly **syndetocheilic** in gymnosperms): a guard-cell mother cell *and* all subsidiary cells, or a single ring-like subsidiary cell, are derived by a series of divisions from one S-residual protoderm cell (a meristemoid).
2. **Perigenous** (formerly **haplocheilic** in gymnosperms): all subsidiary cells are derived from protodermal cells other than the guard-cell mother cell, which itself divides only once (and that equally), to form two guard cells.
3. **Mesoperigenous**: the two major subsidiary cells of the guard cells differ in origin, one being a sister of the guard-cell mother cell (thus mesogenous) and the other originating from a different protodermal cell (thus perigenous).

These basic type designations indicate the origin of the subsidiary cells, while ignoring the presence and origin of "neighboring" protodermal cells that may also be in contact with the guard cells but that are not demonstrably "subsidiary" — i.e., not functionally distinguishable from ordinary protodermal cells. Stevens and Martin (1978) remedied this by subdividing each type according to whether the guard cells are surrounded completely by subsidiary cells or are surrounded by a mixture of subsidiary and neighboring cells. Thus, the three basic types became six. In addition, Stevens and Martin (1978) added a seventh — the **agenous** type — to accommodate those stomatal complexes in which there are no obvious subsidiary cells; instead, the guard cells are directly surrounded by protodermal cells of other lineages. Mature stomata of many arborescent taxa seem to be agenous.

These seven types have been further divided into subtypes, on the basis of number, orientation, and equality or inequality of divisions of the protodermal cells involved. As research progresses, categories may be redefined, or whole systems may be supplanted by others.

Practically all extant gymnosperms have perigenous stomata, in which subsidiary cells are not of the same lineage as the guard cells. In *Torreya, Pseudotsuga, Pseudolarix, Larix, Pinus, Sciadopitys,* and *Thujopsis,* "neighboring" protodermal cells differentiate into subsidiary cells without first dividing, and stomata are thus of a "monocyclic" perigenous subtype. In contrast, in almost all other extant gymnosperms, including *Ginkgo* (Kausik 1974), the "neighboring" cells divide longitudinally, producing inner subsidiary cells and a "wreath" of outer cells. They are thus of an "amphicyclic" perigenous subtype (Ziegler 1987). The contributions of such formal categorizing to phylogeny or to understanding of the controls of stomatal ontogeny or of leaf development are not yet clear.

Unfortunately, all systems of stomatal classification are based on sets of terms lacking rigorous, universally accepted definitions. For example, "stomatal meristemoid", "stomatal-residual protoderm cell", "guard-cell initial", "guard-cell mother cell", and "stoma mother cell", though addressing closely related concepts, do not have the same meanings to all authors. Likewise, the terms "subsidiary", "accessory", "contact", and "neighboring" have been used in different ways (Tomlinson 1974; Patel 1978). Even the terms "mesogenous", "perigenous", and "mesoperigenous", as we have defined them, are not accepted by all (Patel 1978). These terminological problems impede progress in understanding stomatal development.

Stomatal classification systems based on ontogenic features generally have only tacitly addressed the question of whether a stomatal complex arises from a single "determined" protodermal cell or from several such cells. The question of when meristemoids, via one or more quantal cycles (Sect. 12.2.2.1), might be determined to develop into stomata has received little attention. Another neglected question concerns the extent of involvement of subprotodermal cells in stomatal siting and determination, particularly in formation of substomatal gas space. Why are stomata usually sited over large intercellular gas spaces? Is limited symplasmic contact with subepidermal cells a critical factor in the siting, insuring that stomata develop only over large enough spaces? What is the stepwise order of development of these spaces, the guard cells, the accessory cells, and the stomatal pore itself? Though these questions cannot yet be answered, recognizing them broadens the relevance of research on stomatal development.

Some commonly occurring features of mesogenous or mesoperigenous stomatal development can be outlined, though they may not all occur in any one species and though there are many variations. First, in response to positional information, perhaps including proximity to gas spaces in underlying tissues, an S-residual protodermal cell becomes densely cytoplasmic at one end, then divides unequally. The larger daughter, by further division, may contribute variously to the subsidiary cells. The smaller daughter, a guard-cell mother cell, divides unequally, producing a pair of guard-cell precursors. As these mature into guard cells, a stomatal pore develops between them as an intercellular space along their common middle lamella. In all major types of stomatal development, some of the precursors of subsidiary cells may also divide unequally.

Except in some simple perigenous types, unequal cell divisions are cardinal events in the genesis of stomata (Willmer 1983; Sack 1987). The daughter cells of these divisions usually differ grossly in size and in the disposition of their contacts with other cells. Usually the smaller of the unequal daughters is above a large intercellular gas space between subepidermal mesophyll cells.

As guard cells mature and attain the shape that characterizes the taxon, there is deposition of patterned secondary wall thickenings, opening of the pore along the middle lamella between the cell pair, and splitting of the cuticle over that opening, followed by formation of new cuticle on the exposed walls of the pore (Willmer 1983).

The secondary thickening of guard-cell walls is nonuniformly distributed. The pattern of thickening and the pattern of cellulose microfibril arrangement are characteristic for the species — or at least for the leaves in a particular phase of ontogeny of the plant — but they are variable between taxa. The highly ordered pattern of microfibrils embedded in an amorphous matrix usually allows turgor changes to produce changes in cell curvature along the arched axis, without changing the distance between the poles of the stomatal axis. This allows stomatal pores to open and close. These microfibrillar patterns are probably controlled by arrays of microtubules (Palevitz 1982).

Plasmodesmata may connect young guard cells with subsidiary or other neighboring epidermal cells, but then reportedly disappear during maturation (Carr 1976). Wall ingrowths similar to those of transfer cells may also be present. The proplastids of guard cells develop into functional chloroplasts, whereas the proplastids of other epidermal cells typically do not. In numerous grasses (Brown and Johnson 1962) and some orchids (Ziegler 1987), however, the ordinary epidermal cells do have functional chloroplasts. Generally, no plasmodesmata interconnect mature guard cells, but guard cells in grasses may be connected by "canals" remaining after incomplete cytokinesis in the guard-cell mother cell.

Distribution patterns of stomata have been considered to arise from mutual "repulsion" among guard-cell initials. Repulsion, however, does not seem to characterize all taxa. In some, there may be apparent repulsion, which actually has a purely spatial rather than a physiological basis. For example, no inhibitor need be evoked to explain the minimal distances between stomata in *Crinum americanum*; these distances are a geometric consequence of the derivation of the guard cells, and subsidiary and possibly other surrounding cells, from the same S-residual protodermal cell (Sachs 1974). This argument can be applied to all mesogenous stomata, but seems inappropriate for perigenous types. Interestingly, in *Ginkgo*, which has perigenous (haploceilic) amphicyclic stomata, there commonly are "twin" stomata, of both the side-by-side and the end-to-end types (Kausik 1974). Likewise, in *Tsuga canadensis* (also perigenous and amphicyclic), in which stomata are arranged in orderly rows on the abaxial epidermis, subsidiary cells of successive stomata in a row may be in contact or be separated by only a single cell (Gambles and Dengler 1974). There is no evidence of repulsion.

16.3.5 Mesophyll

In broad-leaved taxa, as cells in the various regional meristems divide, the internal tissues of a typical primordial lamina come to consist of four to six or more layers of cuboidal, mostly still meristematic cells, with almost no intercellular spaces (Smith 1934; Pray 1955; Dengler and Mackay 1975; Franck 1979). This condition persists only briefly in epicotyls of growing seedlings. However, within buds, it may continue for months until the buds open and the leaves expand. Most of the internal cells then rapidly enlarge, differentiate, and mature into mesophyll cells as a pervasive system of intercellular spaces opens among them. In a primordial leaf, mesophyll precursor cells constitute a cellular matrix that embeds the precursors of other internal tissues, which differentiate with no development of intercellular spaces.

Commonly, but with many exceptions and variants, there are two basic kinds of mesophyll, palisade and spongy. Both kinds are chlorenchymatous, though palisade cells may have much larger complements of chloroplasts than do spongy cells. **Palisade mesophyll** cells typically are columnar, with their major axes oriented normal to the epidermis. They maintain contact at their ends with other cells, usually including epidermal or hypodermal cells at one end, but become separated from their palisade neighbors by narrow gas spaces. **Spongy mesophyll** cells, in contrast, have complex shapes variously described as lobed, armed, or stellate. Each cell, with its arms or lobes, typically has several axes. These have a range of orientations, though in laminar leaves the major axes tend to be paradermal. The shapes of mature mesophyll cells depend on the rate and orientation of cell divisions, and on how much the cells grow in various directions.

A primordial lamina can be considered as a layered population of precursor cells, in which differentiation seems to be influenced more by cell position and by the course of development in nearby layers or regions than by clonal lineage. It is as though each layer of cells could separately interpret and respond to an organ-level "map and clock" appropriate to its position and its temporal stage (Sect. 12.4). To illustrate: in many leaves, cells in the promesophyll layer immediately subtending the upper epidermis continue to divide anticlinally after their neighbors above and below have ceased doing so. For example, during early leaf expansion in *Catalpa bignonioides*, the number of cells per unit paradermal area is the same in the epidermis and first promesophyll layer, but the promesophyll cells undergo two cycles of divisions after the proepidermal cells have ceased to divide. Thus, the ratio of cell populations in these adjacent layers becomes 4:1. Rates and extent of enlargement in the paradermal plane are also different. The extent of enlargement totals 11-fold in the nascent epidermis, but only eightfold in the nascent mesophyll (Mounts 1932). Gas space in the mesophyll accounts for the difference. The promesophyll cells then elongate rapidly in a direction normal to the epidermis and become columnar palisade cells. Meanwhile, their abaxially subtending neighbors extend themselves laterally, becoming spongy-mesophyll cells.

It is possible to think of growth and differentiation within the mesophyll as being controlled by a complex, biophysical-cybernetic system, involving cycles of growth- and turgor-induced stresses that evoke strains, which potentiate further growth, etc. Such control, however, would require elaborate timing and rate-modulating systems, as well as polarity-sensitive systems to specify principal directions of growth in different parts of a cell (Sect. 12.4.6).

Mesophyll growth involves, along with cell divisions, striking examples of differential expansion and modes of maturation of neighboring cells. If cells are to follow such divergent courses of differentiation, or if stresses are to be locally altered by patterned failure, in space and time, of cell cohesion and the opening of gas spaces (see later), then the constraints of tissue-level symplastic growth characteristic of meristematic regions must be lifted. Growth in the mesophyll, in fact, is formidably complex compared with the symplastic growth of a dome-shaped shoot-apical meristem (Sect. 14.1.6). Methods of growth analysis useful in studying steady-state growth in apical meristems are of limited relevance to studying mesophyll growth.

The gas spaces that open due to a patterned failure of cohesion along the middle lamellae between mesophyll cells are continuous with substomatal cavities and stomatal pores. By facilitating gas exchange, these gas spaces allow the leaf to become photosynthetically functional.

We do not yet know how the pattern of separation and cohesion between mesophyll cells is specified. Also incompletely understood are the lytic processes in the middle lamella preparatory to separation; the mechanics of the separation (but see Sect. 16.3.3); and the means by which capillary water is excluded from intercellular spaces as they open. Though bounded by elements of the water-saturated, cell-wall apoplasm, the spaces do not usually become water filled. Apparently, the surrounding surfaces are nonwettable from the moment the spaces open.

A proposal regarding intercellular gas-space formation was offered by Roland (1978b). It may apply to mesophyll and other tissues and is supported by histochemical and ultrastructural evidence from *Phaseolus aureus* hypocotyls. This proposal is that inception of a gas-space system is preceded by patterned differentiation of a nonpolysaccharide, "splitting layer" within or locally replacing the pectic middle lamella. In response to slight tension, this specialized layer is split into two "mirror-image" films, which provide a hydrophobic lining for the cell walls bordering the opening cavity.

Roland's (1978b) proposal is relevant to opening of gas space at T wall configurations, in which a partition wall between sister cells meets the maternal cell wall

(Jeffree et al. 1986). At these "corners", walls tend to pull apart passively, after lytic processes, as cells "round out" due to turgor-derived wall stress. However, gas spaces in *Glycine max* leaves are also initiated in *intercalary* positions — that is, along cell edges remote from T configurations (Weston and Cass 1973). In angiosperms, intercalary gas spaces seem positionally related to beginnings of so-called "invaginations" of walls in the early development of "armed" cells of the spongy or palisade type. Development of these "invaginations" is related to active localized wall growth in the absence of any clearly definable local stress-strain/growth relations. In various gymnosperm leaves, there is developmentally quite different wall "invagination" involving locally heavy deposition of new wall material and opening of gas spaces during development of mesophyll cells (see later).

As a leaf expands and a patterned cohesion failure between cells leads to opening of an intercellular gas-space system, most neighboring cells cohere tenaciously to one another along small local areas of wall. These remaining areas of contact between neighboring cells typically occur at the ends of lobes or arms of spongy-mesophyll cells, and at the ends of columnar palisade cells. A spongy-mesophyll cell may have lobes extending from a central body and contacting similar lobes or main bodies of six or more neighboring cells. These cells have large surfaces exposed to gas space and small common surfaces available for transport between cells.

Within a genus, there may be great variation in the number of layers and relative volume of palisade mesophyll. Further, this tissue may occur subjacent to the adaxial epidermis only, or also in a similar relation to the abaxial epidermis. For example, a study of 47 species of *Ficus* revealed that all had one to several layers of adaxial palisade mesophyll overlying some spongy mesophyll, and that 27 had some abaxial palisade tissue also (Philpott 1953). Some species had only palisade and no spongy mesophyll (see also Powers 1967).

Within a single species, or within an individual, the number of mesophyll layers in young "shade" leaves tends to be less than in "sun" leaves. In *Fagus sylvatica* this aspect of leaf differentiation is controlled by light conditions during leaf development rather than during initiation of leaf primordia (Eschrich et al. 1989).

A striking feature of mesophyll cells of various species of *Pinus*, in which segregation into palisade and spongy types is not well defined, is extensive development of a pattern of wall ingrowths, which are called "invaginations", as already mentioned. As a result of these ingrowths, most mesophyll cells, which are initially barrel-shaped, are partially subdivided into six, eight, or more seemingly adpressed, crenate lobes or "arms" (Harris 1971; Campbell 1972; Gambles and Dengler 1982a). Some authors refer to these lobes as "arms". This may be confusing because these crenate lobes are different from the "arms" of spongy mesophyll cells. Many crenate lobes have no contact areas with lobes of neighboring cells and thus are not "conjugal" arms. In addition, they are formed differently from arms that arise via a pattern of local persistent cohesion of walls of neighboring cells, with cohesion failure elsewhere.

In *Pinus strobus*, mesophyll wall ingrowths appear after there has already been local axial growth of the primordial leaf and separation of mesophyll precursor cells into thin, irregular, porous, transversely oriented plates alternating with similarly dimensioned gas spaces. Then, local thickening and hardening of cell walls in the epidermis and hypodermis curtail axial growth in that segment of the leaf, and further radial growth in that region is also slight.

Although an ingrowth of a mesophyll cell wall is sometimes called an "infold", it is, rigorously, an *ingrowth*, in that it does not begin as a passive invagination of an existing wall by active external forces but rather as a narrow hoop, or partial hoop, of local thickening. This thickening then extends itself into the cell lumen in the manner of a slowly closing, irregular iris diaphragm, and divides the cell into lobes. The inwardly advancing front of the ingrowth, resembling in section a wall folded back on itself, is blunt to flat. The thickening has the structure of a double wall with a common middle lamella. After the ingrowth is well advanced, the middle lamella weakens and gas space opens between the lobes of the cell (Harris 1971).

Similar ingrowths probably also occur in *Pinus nigra* (Campbell 1972), *P. resinosa* (Gambles and Dengler 1982a), and *P. sylvestris* (Tetley 1936), but are not evident in *Abies balsamea* (Chabot and Chabot 1975) or *Tsuga canadensis* (Fischer and Dengler 1977). A related kind of development may occur in *Sambucus* and other angiosperms (Tetley 1936).

After extensive studies in dicots, Wylie (1939, 1946) proposed that, in general, leaves having a large proportion of palisade mesophyll have densely reticulate venation patterns, with small interveinal distances. This presumably is a consequence of the poor lateral-transport capabilities of palisade cells. Philpott's (1953) study of the diverse tissue arrangements in leaves of 47 species of *Ficus* generally supports Wylie's (1939) proposal. A corollary of this proposal is that extensive palisade tissue *requires* a finely ramified venation system. There are, however, exceptions to this.

Leaves of some taxa, such as *Liriodendron tulipifera*, have palisade cells with numerous lateral contact areas. This "arm palisade" is somewhat intermediate between the typical forms of palisade and spongy mesophyll. "Arm-palisade" cells have short lobes extending to the overlying epidermal and subtending spongy mesophyll cells, but have more and longer arms contacting their lateral neighbors. In certain paradermal sections, these cells show a distinct stellate shape (Pray 1954). In this respect, they resemble **paraveinal mesophyll** cells.

In *Glycine max*, the paraveinal mesophyll is typically a monolayer of specialized spongy mesophyll (Weston and Cass 1973). It is derived from a specific layer of ground-meristem cells already distinguishable in the primordial lamina. After the lamina expands, these cells abut on and cohere to the "lower" ends of palisade cells and also have contact areas with conjugal arms of subtending spongy mesophyll cells. Mostly, however, they are a paraveinal network of cell bodies and conjugal arms seemingly aiding translocation between veinlets and the mesophyll cells in areolae. In *Populus deltoides*, the paraveinal mesophyll is sometimes two cell layers thick (Russin and Evert 1984).

In some leaves lacking a discrete paraveinal mesophyll, the cell walls of ordinary mesophyll form an anastomosing system that is voluminous enough to play a major role in apoplasmic distribution of water and solutes. For example, in *Fagus grandifolia*, mesophyll-cell-wall volume is about 11% of total leaf volume (Dengler and Mackay 1975). In contrast, this volume is only about 2.5% in *Tsuga canadensis*, a cold-climate xerophyte having much lower transpiration (Gambles and Dengler 1974).

Detailed information from a small number of species indicates an early developmental tendency toward opening of gas space between many, or most, mesophyll precursor cells. The internal gas-space system develops further as the leaf expands and as stomata differentiate. Possibly because the process begins early, it typically occurs without extreme schizogeny or programmed cell mortality. Exceptions to this growth program occur in, for example, *Avicennia marina* (a mangrove), in which continuous gas-space systems, along with substomatal cavities, are not clearly defined until leaf expansion is about half completed. Most of the smaller spaces then seem to form through some kind of "wall splitting" and mild schizogeny, as in other taxa, but substomatal cavities and other large spaces seem to result from localized lysigeny and autophagic cell mortality (Chalain and Berjak 1979). Apparently, gas-space formation begins so late in leaf expansion that the residual potential for local cell expansion is inadequate to generate large spaces by growth processes alone.

When palisade cells, due to failure of cohesion along the middle lamellae of their common walls, move apart and become mostly surrounded by gas space, they may remain interconnected by a web of microstrands of seemingly elastic pectic substance (Carr et al. 1980b; Jeffree et al. 1986). The strands may be analogous to those that appear when one pulls apart paper sheets stuck together with rubber cement. The strands, somewhat like a set of guy wires, may brace the palisade cells, which in many species lack arms conjugal with neighboring cells. It is conceivable also that the strands serve in translocation or signal transmission. Similar strands interconnect some maturing guard cells as the stomatal pore opens, and connect guard cells with subsidiary cells (Carr et al. 1980).

16.3.6 Foliar Venation: Developmental Controls

Venation patterns are as varied as leaf shape. Even within a species, the various foliar organs (cotyledons, primary leaves, bud scales) usually differ as notably in vasculature as in morphology. Some correlations between venation patterns and modes of leaf development are evident.

Columnar, acicular, or linear leaves, which have cells derived mostly from basal intercalary meristems, tend to have veins aligned parallel to the major leaf axis, in both angiosperms (as in *Zea*, *Poa*, and *Lilium*) and gymnosperms (as in *Abies*, *Pinus*, and *Picea*). In some broad-leaved taxa, there is a less common kind of "parallel" venation, in which the lateral veins within each lamina half are parallel to one another, but all lateral veins are nearly orthogonal to the midvein (as in *Canna*, *Musa*, and *Ravenala*). This type seems characteristic of leaves having regular and moderately persistent marginal-meristem activity, along with axial elongation growth distributed throughout the leaf.

Such correlations between venation and leaf growth have been interpreted as indicating that venation patterns are *consequences* of the mode of growth. For example, Prantl (1883), proposed that there are three major types of leaf growth, each with its resulting venation pattern. These types of growth, in modern terms, are:

1. **Basiplastic** (basally formative), in which meristematic activity is transitory at the tip and becomes basally localized early in development (becoming a basal intercalary meristem).

The major region of cell expansion migrates from the subapical region to a region some-what distal to the basal intercalary meristem. Marginal-meristem activity is minimal. Venation is primarily parallel in an axial orientation, as in many gymnosperms and monocots.

2. **Pleuroplastic** (laterally formative), in which a provascular midvein differentiates within the early phyllopodium while the apical meristem is active; then, as the phyllopodium elon-gates, its "marginal" meristems produce much of the lamina (Sect. 16.3.1.1), while the tip region gradually becomes inactive. The rates and principal directions of expansion growth of laminar tissues relative to those of the midvein subsequently determine the angle between secondary veins and midvein. The resulting patterns include those common in dicots.

3. **Eocladus** (early branching), characterized by early delineation of "branches" as lobes or pinnae, each with its own "midvein". This pattern may result from patterned segmentation (or "fractionation"; see Sect. 16.4) of marginal meristems while the phyllopodium is still entirely meristematic. The lobes or pinnae themselves may then develop basiplastically, pleuroplastically, or eocladusly. The resulting leaves are compound or doubly compound.

Prantl's (1883) leaf developmental types were widely accepted, sometimes in modified form, as plausible and useful. Histogenic data supporting them, however, were scant (Hagemann 1970; but see Cusset 1986).

Some interpretations of leaf growth, such as Prantl's, predate knowledge of phytohormones or morphogens that might influence vascularization. They also predate general awareness of patterns of regional-meristem activity in leaf primordia; of the relations between orientations of cell divisions, principal directions of growth, intercell-ular gas-space formation, and provascular-strand definition; and of vascular-element differentiation. Even so, the principles behind some of the ideas put forward long ago are still relevant.

For example, primordial dicot leaves often show patterns of cytogeneration by meristems localized in certain regions (Sect. 16.3.1), and growth by cell expansion in other regions. Active meristematic regions are sinks for various fixed carbon com-pounds, especially sugars, and are sources of auxin and probably of other morphogens. Cells in regions of expansion are sinks for morphogens and water, and, to a lesser extent, for fixed carbon compounds. In simplified terms, these traits of the separate regions of cell generation and cell expansion evoke a flow of "sugar" towards the regional meristems and a flow of "auxin" (with water and solutes) towards the regions of cell expansion. We have assembled the following skeletal working hypothesis integrating this new knowledge with the older ideas about vein patterns and leaf growth patterns.

While the phyllopodium is still wholly meristematic, with cell division especially active in its apical region, the flux of morphogens to and from that region evokes provascular determination in cells aligned along the axis. As a result, a provascular strand propagates itself acropetally within the phyllopodium. In this process, the provascular cells elongate axially and divide mostly axially, while the ground-meristem cells flanking them divide transversely or randomly. This provascular strand becomes the central provascular trace and is the precursor of the midvein of the leaf (Sect. 14.3.1). A subsequent shift of meristematic activity to the middle and basal regions of the phyllopodium is accompanied by changing patterns of gradients of morphogens. Subsidiary strands may then follow the original strand as a template, but be propagated basipetally rather than acropetally.

By similar logic, the activity of marginal meristems in generating sheets of pro-mesophyll "ground-meristem" cells would evoke other patterns of gradients of mor-phogens. For example, a discrete marginal meristem would approximate a line source of some morphogen(s), such as auxin, and a line sink for "sugar". Any existing, functional vascular strand could serve as a source of some morphogens and a sink for others. A four- to six-layered sheet of meristematic and expanding cells would serve as a field across which morphogens could move from sources to sinks. Such movement of morphogens would establish a polarity and — figuratively, like water flowing across an erodible plain — induce channelization of flow (Sect. 12.4.2.1), which would then influence orientation of planes of cell partitions and direction of cell growth (see Sachs 1984). Thus, second-order vascular strands would be determined as pleuroplastic leaf growth is established. By considering plate rather than marginal meristems, and local cell expansion, such a system can be invoked for each order of veins.

Many aspects of this hypothetical system are unclear. For example, primary vascular elements differentiate not directly from ground tissue, but within provascular strands. The cells in these strands already differ morphologically from the cells of the promesophyll matrix (ground meristem), mostly in having narrow, elongated shapes. How do induction, determination, or differentiation of provascular strands differ from determination or differentiation of the primary xylem or phloem elements themselves?

From a different perspective, it is evident that the earliest stages of vascular development in leaves occur in a context of symplastic growth. It is nonetheless likely that a pattern specifying future areas of cohesion and separation between adjacent cells is determined very early. According to this pattern, cells determined to become provascular continue to grow symplastically and to cohere, with little or no gas-space formation. Promesophyll cells, in contrast, undergo patterned separations and nonsym-plastic growth as they become palisade or spongy mesophyll (Sect. 16.3.5).

16.3.6.1 Vascular and Associated Tissues in Gymnosperm Leaves.

Information available on early development of gymnosperm leaves indicates that apical growth is brief and that most cells of a mature leaf are derived from a basal intercalary meristem, as, for example, in *Taxodium* and *Cunninghamia* (Cross 1940, 1942). The elongation of these basally derived cells, mostly in the proximal third of the leaf, accounts for most foliar extension growth. The shift of the locus of major meristematic activity from tip to base occurs after some provascular strands have already been determined. There must be accompanying changes in the dynamics of morphogens, even reversals of gradients and ratios, at sites of histogenesis. Such changes, if their spatial and temporal aspects were known, might explain shifts from acropetal differentiation of early provascular strands (when the activity of the apex is great) to basipetal differenti-ation of later strands (when basal intercalary meristematic activity becomes great). Likewise, intense intercalary meristem activity could be correlated with differentiation or maturation of certain vascular elements progressing in both directions from the sites of such activity (Sect. 14.3.2.1).

In general, **transfusion tissue** completely or partially surrounds the vascular bundles of gymnosperm leaves. This tissue differentiates in the region interior to the endodermis (if present) and exterior to the parenchymatous bundle sheath. It is derived mostly from ground-meristem or promesophyll cells, and typically consists of transfusion tracheids, transfusion parenchyma, and Strasburger cells (Ghouse and Yunus 1974; Kausik 1976; Hu and Yao 1981). Transfusion parenchyma, though always spatially associated with vascular tissue and having tracheary elements with bordered pits and variously patterned secondary-wall thickenings (Campbell 1972; Walles et al. 1973), also has some cell types intermediate between vascular parenchyma and mesophyll. This is especially characteristic of leaves lacking a discrete endodermal sheath. Thus, transfusion tissue is difficult to delineate and define rigorously (Worsdell 1897). Its histogenic origin also is not clear. In *Pinus nigricans* (= *P. nigra*), transfusion tissue may be derived partly from promesophyll cells, which in turn may have had protodermal origins (Heimerdinger 1951).

The proportion and spatial arrangement of the cell types in transfusion tissue vary with the genus, and with the extent to which basiplastic growth is supplemented by pleuroplastic growth during leaf development. As marginal-meristem growth increases, more of the transfusion tracheids elongate in an orientation transverse to the major leaf axis, though many tracheids remain elongated along the leaf axis or irregularly cuboidal. The transfusion tissue also tends to form lateral, wing-like extensions of bundles rather than merely a sheath. For example, in *Pinus nigra* needles, which have little or no marginal-meristem activity, transfusion tissue surrounds the vein like a sheath (Huber 1947). In contrast, in *Podocarpus macrophyllus* leaves, which have extensive marginal-meristem growth and approach a broad-leaved shape, transfusion tissue differentiates on either side of, rather than ensheathing, the veins; there being no endodermis, it extends a short distance into the mesophyll (Griffith 1957). Transfusion tracheids in this species may be elongated orthogonally to the vein.

Leaves of a few gymnosperms, though they develop basiplastically, also have extensive marginal-meristem activity (pleuroplastic growth) and approach "broadleaved" shapes. *Agathis dammara* leaves, for example, have numerous parallel veins that traverse the lamina axially, alternating with resin ducts. Each vein has some transfusion tissue on its flanks (Kausik 1976). Thus, a broad lamina is served by a simple multiveined system, with most veins laterally supplemented by transfusion tissue.

In a few other gymnosperms, leaves are moderately broad and not multiveined, indicating that they have a diffuse lateral translocation system. One system of this kind is based on **accessory transfusion tissues.** These tissues, in near continuity with ordinary transfusion tissue, develop outside the endodermis (if present) and bundle sheath, across the mesophyll, toward the leaf margins. These lateral extensions of transfusion tissues, located between adaxial and abaxial mesophyll layers, and having large intercellular gas spaces, are somewhat analogous to the paraveinal mesophyll of some angiosperm leaves (Sect. 16.3.5). Accessory transfusion tissues are well developed in some species of *Podocarpus* (Griffith 1957; Schoonraad and Schijff 1974), *Acmopyle* (Kausik and Bhattacharya 1977), and *Dacrydium* (Lee 1952). They may occur in less well-defined form in numerous other genera (Hu and Yao 1981). Accessory transfusion tissue is very slow to mature and may offer material for study of control of differentiation of mesophyll cells by gradients of morphogens and positional effects.

A question arises from knowledge of the long duration of basal-intercalary-meristem activity during gymnosperm leaf development: how is the vascular tissue that is fully functional in the apical regions maintained through this basal zone while new, immature cells are intercalated between mature cells above and below? This has hardly been studied.

16.3.6.2 Vascular and Associated Tissues in Angiosperm Leaves. Development of leaf venation in angiosperms, especially dicots, is generally more complex than in gymnosperms. Vascular supply to broadly laminar angiosperm leaves involves not transfusion tissues, but rather a hierarchical system of vascular bundles. These seem to merge or branch, forming several orders of discrete veins and veinlets. These ultimately divide the laminar tissue into areolae — small compartments in which no mesophyll cell is far from a vascular supply.

Gymnosperm leaves typically have a basiplastic mode of growth, though sometimes they also have a component of pleuroplastic growth. In contrast, all three major modes of growth occur among angiosperm leaves — basiplastic, pleuroplastic, and eocladus (Sect. 16.3.6; but see also Jeune 1982, Cusset 1986). Basiplastic growth is the common mode in monocots, with a few groups also exhibiting a complementary pleuroplastic mode. In dicots, leaf-growth modes are less clear, because regional meristems are not sharply distinguishable. Predominantly basiplastic leaf growth is uncommon. Pleuroplastic growth mediated by discrete or diffuse marginal-meristem activity, complemented by plate-meristem activity, is nearly general. In addition, phyllopodia of many dicots undergo eocladus growth, in which marginal-meristem regions are subdivided into lobes or segments. These then can undergo modified pleuroplastic growth, forming variously lobed simple leaves or compound leaves.

Leaf venation begins with differentiation of provascular strands. These strands, at first consisting only of immediate derivatives of an aligned group of presumptive provascular mother cells (in some of which a quantal division may have occurred), are quickly enlarged by further axially oriented divisions, and also sometimes by incorporation of laterally adjacent, symplastically growing, cells. These provascular cells form the "original" provascular strand, or central leaf trace, in the phyllopodium. The original strand typically is the major pattern element or template around which the midvein (first-order or primary vein), with its core of vascular elements, is organized. Along the original strand, various lateral and subsidiary vascular bundles and strands may differentiate — often basipetally. But the differentiation of, say, protoxylem from provascular cells is not necessarily continuous with existing protoxylem (Ashworth 1963). Vein development is usually being completed as the lamina attains about 80% of its final area (Dale 1982).

There seems to be no general "stem-to-branch" relation between the primary vein (midvein) and secondary veins. For example, in some leaves, specific vascular bundles in the lower midvein and petiole are identifiable as the vascular connections to specific regions, lobes, or leaflets, as Yin (1941) demonstrated for *Carica papaya* (Sect. 10.2.2.4). Further, after leaf abscission, discrete "bundle scars" commonly are visible within the leaf scars on twigs. Further, in *Populus deltoides*, the midvein is an aggregation within which the identity of particular vascular bundles is maintained (Isebrands and Larson 1980); functionally, these bundles remain dedicated to specific areas.

Although details vary greatly among taxa, the primordial laminae that are organized by marginal and plate meristems consist of three to eight layers of meristematic cells; there are typically five layers during the time of provascular differentiation (Smith

1934; Pray 1955; Dengler et al. 1975). The adaxial and abaxial protodermal layers are slightly differentiated by virtue of position. The three internal, promesophyll layers are undifferentiated, but differentiation of the provascular strands typically begins in the positionally unique, median layer (Pray 1955; Franck 1979). This has consequences in adjacent layers.

For example, progressive, midvein-to-margin differentiation of provascular precursors of secondary veins in *Ostrya virginiana* is accompanied by divisions in the subprotodermal cells just abaxial to the provascular cells (Franck 1979). These cells expand, forming abaxially protruding ridges that delineate intercostal panels. Cells adaxial to the nascent veins remain small. These abaxial/adaxial growth differences result in laminar folding. Then, by further differential cell division and expansion, the young, already conduplicate leaf becomes plicated, before it is 1 mm long. Most of the intercostal panels are still only five cell layers thick. Seven or eight secondary veins are delineated on each side of the midvein, usually in acropetal sequence. Then, simultaneously throughout the leaf, tertiary provascular strands are differentiated at regular intervals across the intercostal panels between secondary veins. Again, early differentiation involves mostly the midlayer of cells. Fourth- and fifth-order veins, which are first delineated as provascular strands, also originate via division and elongation of cells in the midlayer. These are formed in part as the bud opens and the leaf expands. Veinlets of the sixth or seventh order intrude into the ultimate areolae as the leaf matures; the end of each of these veinlets is unbranched (Franck 1979).

The venation process in *Liriodendron tulipifera*, though the pattern is less regular, is similar to that in *Ostrya virginiana*, except that two to four terminating and branching vein endings intrude into most areolae. These final branchings include both xylem and phloem, though the phloem is hardly distinguishable from parenchyma. The ultimate tip is always a tracheary element (Pray 1954).

Areolae lacking vein endings are rare in *Liriodendron tulipifera* leaves (Pray 1963), whereas leaves of *Quercus boissieri* and *Q. calliprinos* have many areolae with few or no vein endings (Fahn 1974). Intrusive growth may be involved in the differentiation of vascular tissue in vein endings of the *Hedera* leaf (Magendans 1983, 1985).

Veins in many common tree species arise from provascular beginnings in the middle layer of five layers. The other two internal layers typically become involved at later stages of differentiation (Franck 1979). Cells from these other layers may contribute a little to the vascular tissue proper, but contribute mostly to the "bundle sheath" and "bundle-sheath extensions". Significantly, as laminar expansion continues, cells in the developing veins, bundle sheaths, and any extensions grow symplastically, allowing no gas space to open. During this growth, the protoxylem elements may be stretched. There is, however, no good evidence that high-order veinlets are broken or that discontinuities or free endings are formed in this way (Pray 1963).

In the complex meshwork of veins in dicot leaves, there are many segments to which no "root-end" versus "shoot-end" polarity can be assigned. These segments are mostly anastomoses between neighboring veinlets. How might such seemingly nonpolar segments be formed? Sachs (1975) induced differentiation of nonpolar segments of vascular strands by repeated alternate application of auxin at two sites. The induced segments connected the strands leading basipetally from the two sites. Similar processes may be involved in some aspects of morphogenesis in leaves with net-like venation (Sachs 1981b).

In some species of *Euphorbia*, veins that develop into or within areolae may be discontinuous longitudinally, even in mature leaves (Herbst 1972). This is little known in other genera. More widespread and better known is radially incomplete differentiation of provascular cells of a strand into xylem or phloem elements. In these strands, the outermost cells generally differentiate into a layer of **bundle-sheath parenchyma** (also called border parenchyma, or mestome sheath in older literature). The bundle sheath increases the effective volume and surface of the vein complex by several to many fold. This is significant because most, or all, water and solutes passing between vascular and other tissues of the leaf must pass through this sheath (Armacost 1944).

Although bundle-sheath cells cohere strongly to their vascular neighbors and to each other, patterned gas spaces open along their interfaces with mesophyll cells. Some sheath cells have a companion-cell relation with adjacent phloem elements; some have transfer areas where they adjoin mesophyll cells (Pate and Gunning 1969); and seemingly all may sometimes serve as storage depots. Many woody plants, being of the C3 type, have bundle-sheath parenchyma cells that lack functional chloroplasts, though the analogous cells in C4 plants may be intensely green (Sects. 5.3.1; 5.3.2). However, some C3 plants, including *Fagus grandifolia* (Dengler and Mackay 1975), have green bundle-sheath cells. These are elongated transversely to the vein and extend into the mesophyll gas space.

Many of the veins in thin, broad, deciduous leaves, such as those of many common temperate-zone forest trees, develop **bundle-sheath extensions** (Wylie 1952). These plates of parenchymatous, colorless tissue, differentiating from putative promesophyll cells adaxial and abaxial to the bundle sheaths of most veins, grow symplastically with the vascular tissue. Thus, gas spaces do not open in those regions. Typically, the extensions connect the sheaths of small or intermediate veins with the dermal layers. In C3 plants, the extensions are unpigmented and thus not photosynthetically functional. The extensions probably are nonvascular conductive routes, which, via the epidermis, serve the mesophyll tissues.

Perennial, xerophytic leaves, with thick, multiple dermal layers and thick mesophylls, tend not to have bundle-sheath extensions. In these leaves, influences from the sheathed vascular strands do not prevent opening of gas spaces in the more distant, subdermal mesophyll layers abaxial and adaxial to the strands.

16.4 Development of Compound, Dissected, and Other Multilaminate Leaves

Many kinds of compound leaves occur among angiosperms, especially dicots. Among extant gymnosperms, overtly compound leaves are common only in cycads. For example, the cycad *Bowenia* has bipinnate leaves (Chamberlain 1919). Leaves of *Ginkgo biloba*, which lack discrete midveins, may be multilobed, but never compound;

some of these leaves have deep clefts, defining five or more lobes, though bilobed or entire fan-shaped leaves are more common.

Like a simple leaf, a compound leaf begins as a single buttress that becomes a phyllopodium, which then develops two axially oriented, parallel ridges. These are the visible manifestations of incipient marginal meristems. They are usually referred to as being lateral to the phyllopodium, but commonly are adaxially displaced.

In many angiosperms, marginal-meristem ridges on a primary phyllopodium can, early in their development, extend themselves axially by incorporating neighboring tissues toward the leaf apex or base. The ridges can also undergo **fractionation**, becoming divided into local segments separated by gaps within which there is minimal additional marginal-meristem activity. During fractionation, there is strict control of timing and rate of elongation of the various axial phyllopodial segments (Hagemann 1970, 1984). Typically, segments bearing marginal meristems elongate little, while those lacking them elongate notably, forming either a petiole or interleaflet segments of a rachis. Meanwhile, the "fractions" of the marginal meristem organize structures that function as leaflet primordia, and then as secondary phyllopodia, with secondary marginal meristems, secondary midveins, etc. If the marginal meristems of the leaflets also undergo fractionation, the leaf becomes doubly compound. Variations in timing and in relative rates of marginal-meristem activity, and in timing and rates of axial elongation of phyllopodial segments delineated by fractionation, can yield a variety of compound-leaf forms, even in the same genus.

The flexibility of this system of compound-leaf development is illustrated by the leaves of three species of *Sorbus* (Merrill 1979). All have the same basic venation pattern. *S. decora* leaves, however, typically are compound with six or more pairs of opposite leaflets and a single terminal leaflet; leaves of *S. alnifolia* are simple, with six or more pairs of opposite secondary veins; and leaves of *S. hybrida* are half-compound, having several basal pairs of leaflets and a basally lobed, but otherwise entire, terminal lamina. The same principles are also evident in the pinnately, ternately, and palmately compound leaves of *Decaisnea fargesii*, *Akebia trifolia*, and *Stauntonia hexaphylla*, respectively, of the Lardizabalaceae (Sugiyama and Hara 1988).

Compound leaves of some monocots develop by mechanisms not involving marginal-meristem fractionation. For example, in many palms, notably *Raphis excelsa*, submarginal plate-meristem regions initiate the primordial lamina halves as continuous sheets of tissue. Early in its development, each primordial sheet becomes plicated, thus coming to resemble the wall of a closed camera bellows. However, the margin itself, a narrow band of submarginal tissue, and the incipient rachis remain unplicated. Once initiated, laminal plications may be augmented by intercalary growth. The marginal strip of tissue is then abscised or fragmented. Leaflets finally become delineated by abscission or cleavage of either the abaxial or adaxial edges of the plications or by cleavage in intercostal regions. (Kaplan et al. 1982; Kaplan 1984).

Leaves of some other monocots, notably those of many species of *Monstera*, develop perforations. These apparently result from programmed necrosis in localized regions (Kaplan 1984), but full developmental details are not available. At first, narrow strips of fragile tissue intervene between some perforations and the leaf margins. As these strips fragment or are abscised, the holes become sinuses that effectively divide the blade into leaflets. The range of adult leaf types is wide. For example, leaves of

Monstera adansonii, though perforated, have entire margins, whereas leaves of *M. subpinnata* and *M. tenuis* are distinctly pinnate and seemingly compound.

16.5 Leaf Development from an Organ-Level Perspective

Hitherto, we focused on functional sets or regional populations of meristematic cells, and on the early generations of their progeny. The potential for meristematic activity of these cells declines with increasing generation numbers. Eventually, seemingly according to their positions within the primordium, the cells come under the influence of various programs of directed differentiation. As a result, most cells become constituents of tissues or of multicellular organelles, rather than idioblasts or meristemoids.

16.5.1 Time, Cell Number, Cell Size, and Organ Size

A young leaf on an embryonic shoot within a bud is small in relation to its final size. Nevertheless, the major outlines and structural features of this "embryonic" leaf may already be present. As this leaf, which consists of a large number of small cells, emerges and expands to full size, it undergoes a great increase in cell number and in mean cell size. Therefore, the total increase in tissue volume is enormous.

Most of the cell divisions and cell volume growth that contribute to development of a typical leaf occur after it emerges from the bud, even though a larger number of cell generations occur before than after emergence. For example, in a *Lupinus* seedling, when the fifth leaf emerges from the bud, it has about 9×10^5 cells, compared with about 900 cells in the leaf primordium during the mid-protrusion stage. If we assume that all cells of the primordium divide at equal rates, there are about ten cell generations between protrusion and the beginning of leaf emergence. During leaf expansion — from emergence to maturity — the cell number increases to about 14×10^6, or about 16-fold. This, however, requires only four more cell generations. While cell number increases 16-fold, mean cell volume increases by about tenfold. Thus, the total volume growth of the leaf after emergence is about 160-fold, though on average only four more generations of cells are produced (Sunderland 1960).

Detailed ontogenic studies have been done for relatively few of the vast numbers of woody taxa. However, based on available information, three generalizations can be made: (1) an embryonic leaf assumes the shape of a mature leaf very early; for example, a *Tilia americana* leaf has developed its oblique cordate base, and an *Acer saccharum* leaf has delineated its characteristic three lobes, before either attains a length of 1 mm; (2) histology of embryonic laminar foliage leaves seems to be relatively uniform among taxa; and (3) increase in laminar area during early leaf development is due almost entirely to increase in cell number, whereas cell enlargement becomes the dominant mode of growth during the later phases of leaf expansion (Smith 1934).

Comparative studies of leaf development are difficult, partly because coupling between chronological time and morphological manifestations of developmental time is often indefinite and weak. This coupling is especially tenuous during periods of dormancy or inhibition of buds. Yet, a quantitative recounting of leaf development requires an appropriate morphological time scale. Strictly chronological time is not suitable because chronological ages of leaves at the same morphological-developmental stage, but growing under different local environmental conditions, may differ greatly. But, if ages of primordia or leaves are specified in plastochron units, they become comparable in a morphological-developmental sense. Such thinking led to formulation of an "index" specifically pertaining to leaf development, as is detailed next.

16.5.2 The "Plastochron Index" and "Leaf Plastochron Index"

Obviously, early developmental stages of a leaf on an epicotyl can be closely specified by referring to the length of the leaf as a fraction of its final length. However, the length in conventional, linear units is not a good index because differences in length are usually not linearly related to a timed sequence of developmental stages. Under some conditions, the logarithm of leaf length can be used as a quantitative index of development of an epicotyl and its leaves. Using this approach with *Xanthium*, Erickson and Michelini (1957) formulated a **plastochron index.**

The validity of a developmental index for epicotyls based on the logarithm of leaf length depends on the validity of three assumptions about the growing system:

1. The time interval between initiation of successive foliar primordia along a single generative spiral by the same shoot apex is nearly constant; that is, plastochron duration in clock time is nearly constant.
2. The early growth in length of an expanding leaf is exponential, and a plot of the logarithm of length against time is linear. ·
3. The relative rates of elongation of young leaves on the same axis are similar.

These assumptions are valid for the early growth of epicotyls of seedlings of the many angiosperms (especially dicots) and some gymnosperms that have potentially indeterminate growth. They tend not to be valid for growth of leaves on shoots that are initiated within buds, then become subject to correlative inhibition or dormancy, and expand only during a later season, as the buds open. Thus, the plastochron index is useful in expressing the developmental age of first-season epicotyls and the leaves they bear, but less applicable to twigs and leaves of older plants. The index is not directly applicable to epicotyls having opposite, whorled, or bi- to multijugate phyllotactic systems. When these assumptions hold, the plastochron index (*PI*) is expressed as:

$$PI = n + \frac{\ln L_n - \ln ref}{\ln L_n - \ln L_{n+1}} , \qquad (16.1)$$

where n is the epicotylary "serial number" of the youngest (smallest) leaf having a length greater than the reference length (ref); L_n and L_{n+1} are the lengths of leaves n and $n+1$. In practice, ref is often set at 10 or 20 mm. Leaf lengths are also expressed in mm. For example, if, on a seedling epicotyl, the sixth and seventh leaves are 11.5 mm and 7.5 mm long, respectively, with ref fixed at 10 mm, then

$$PI = 6 + \frac{\ln 11.5 - \ln 10}{\ln 11.5 - \ln 7.5} = 6.325 \ . \qquad (16.2)$$

If derived and used in this way, PI is an expression of the developmental age of the epicotyl, not of the whole organism. From the PI of an epicotyl, one can obtain the developmental age of any leaf thereon — that is, the **leaf plastochron index** (LPI) — from the relation

$$LPI_k = PI - k \ , \qquad (16.3)$$

where k is the serial number of the leaf of interest. Thus, in the example in Eq. 16.2, the LPI for leaf 5 is 1.325; for leaf 6, it is 0.325; and for leaf 7, it is -0.675. Leaves longer than ref have positive LPI values; those shorter than ref have negative values.

Though PI and LPI are widely referred to as "plastochron" indices, they are not based on the plastochron, the time interval between initiation of successive leaves. Rather, they are based on the interval between attainment of an arbitrary reference length by successive leaves. Within their limitations, both indices are useful in studying leaf development (e.g., Erickson and Michelini 1957; Maksymowych 1973; Lamoreaux et al. 1978), including primary vascularization of leaves on seedling epicotyls (Larson 1975). These indices help provide an integrated view of development at the organ level, allowing comparison of the stages of development of the various constituent tissues.

16.5.3 Leaf Development and the Plastochron Index

Maksymowych (1973) made a detailed study of the leaf plastochron index (LPI) in Xanthium, an herbaceous plant. However, because detailed information is also available for Populus deltoides (e.g., Isebrands and Larson 1973; Larson 1975), which is woody (and evolutionarily advanced), we use this plant to illustrate application of LPI.

The information abstracted here is derived from studies of series of Populus deltoides leaves taken from shoots when the 16th leaf (counting from the base) just attained the reference length, 20 mm, and thus had an LPI of 0.0. On this scale, leaves at LPI = 6.0 are just attaining maturity.

There are many leaves shorter than 20 mm and thus having negative LPIs. In these leaves — even those still folded within the bud — the laminae are well delineated. Especially clearly defined are the adaxial epidermis, two layers of palisade mesophyll, a central layer that will contribute the vascular and closely related tissues, three layers of spongy mesophyll, and an abaxial epidermis.

When LPI is about 1.0, the spongy-mesophyll cells cease paradermal growth and begin passively to separate. The mean spacing between these cells becomes about twice the paradermal dimensions of the cells. Separation in the palisade mesophyll begins two plastochrons later and continues until the spaces between cells approximately equal mean cell width. Though the spongy mesophyll begins to separate two plastochrons before the palisade, the cells of both tissues have a mean paradermal width of about 12.5 μm when they begin to separate.

Cell maturation progresses basipetally in the two epidermal layers and the palisade and spongy mesophyll. Development of intercellular spaces in the mesophyll also advances basipetally, as in *Xanthium* (Maksymowych 1973). Unlike the *Xanthium* leaf, however, the *Populus deltoides* leaf is amphistomatous and, near the tip, some substomatal intercellular spaces are already present in both the lower, spongy mesophyll and the upper, palisade mesophyll when the *LPI* is only about 1.0. In addition, mature stomata are already present on both surfaces near the tip when *LPI* = 0.0. Elsewhere on the leaf, the stage of stomatal development differs on the two surfaces. On the abaxial surface, the maximum density of mature stomata is attained at about *LPI* = 4.0 (3.0 near the tip), whereas on the adaxial surface, the maximum density is not attained until *LPI* = 5.0 (except near the tip). After *LPI* = 5.0, further laminar expansion results in a decrease in stomatal density, due to "dilution" effects. In a mature leaf, the distribution is nearly uniform throughout.

Maturation of the midvein and major secondary veins is acropetal, whereas maturation of minor veins is basipetal and lateral. In the midvein, at *LPI* = -1.0, protoxylem and protophloem are mature all the way to the leaf tip, whereas expanded areolae with mature vein endings are present only at the very tip. At *LPI* = 0.0, areolae at the lamina base are still incompletely delineated by xylem strands. At *LPI* = 3.0, all veins in the distal half of the lamina have mature xylem, while those in the proximal half mostly have single files of protoxylem elements and those at the very base are still discontinuous. Development of bundle sheaths and their extensions also progresses basipetally. These are already developed in the lamina tip at *LPI* = -1.0.

Because it matures precociously, the tip of an expanding *Populus* leaf is capable of photosynthesis while the lower parts are still immature. Early functional maturity at the tip is physiologically correlated with early acropetal maturation of vascular tissue in the midvein and major laterals. Major lateral veins are thus available to import assimilates into the growing leaf, to supply water to drive growth, and later to export assimilates from the tip toward the base and beyond.

Photosynthesis per unit leaf area approaches its maximal rate as *LPI* approaches 4.0 (Dickmann 1971). The whole developmental program is, of course, well integrated, and by the time intercellular spaces and stomata have developed throughout the lamina, mature vein endings are present in expanded areolae, and the leaf can function.

16.6 Development of Scale Leaves Versus Foliage Leaves

Technically, a scale leaf is a **cataphyll**. Although the early development of cataphylls often is overtly the same as that of foliage leaves, mature cataphylls differ from true foliage leaves in having few or no stomata and poorly developed mesophyll, with little gas space. Adaxial and abaxial meristems contribute little. Thus, cataphylls typically lack thickening along the midvein. In addition, cataphylls have poorly developed venation. If there is intercalary-meristem activity, it is narrowly localized in the basal region. In contrast, marginal-meristem activity is intense and strongly concentrated along the leading edges. Accordingly, cataphylls are thin and broad. In the thin, membranous scales of many conifers, cells at the margins may even proliferate as monolayers. In contrast, the needle-leaf primordia in these trees have almost no marginal-meristem activity.

The first evidence that a foliar primordium will develop into a scale rather than a foliage leaf is rapid expansion, both laterally and axially, in the proximal part, and inhibition of growth in the distal part. Early growth in the proximal part is often followed by sclerification or other adaptation to protective functions, rather than by differentiation into palisade and spongy mesophyll. Development of scale leaves commonly is a first phase of the formation of a terminal bud (Sect. 14.2.4.1).

16.7 Sequential Formation of Different Foliar Types by a Shoot Apex

In most perennials, including practically all woody taxa except perhaps the scale-leaved conifers, a shoot apical meristem can initiate foliar primordia and direct their early development according to one of two modes.

In one mode, associated with sympodial branching, the apex initiates a single succession of primordia, the first of which develop into prophylls, the next into cataphylls (scales), and the last into foliage leaves. After the last leaf primordia have been initiated, which may be more than a year after the first prophylls were formed, the apical meristem senesces and withers away, or is actively abscised while still alive and succulent (Sect. 14.2.5). In this mode of growth, the apical meristem does not revert to initiating primordia that can develop into cataphylls, or scales, of a bud.

The other common mode of activity is associated with monopodial branching, which is the general growth pattern of all woody species that do not exhibit apical abortion. In monopodially branching species, the apical meristem produces a potentially indeterminate succession of primordia. A series of leaves is followed by a series of scales, which in turn is followed by a series of leaves, etc. Extension growth of a shoot tip tends to be episodic or seasonal, because a period of dormancy, during which the apical meristem is enclosed in a scale-clad bud, commonly intervenes between episodes of visible shoot extension.

In both sympodially and monopodially branching taxa, a new shoot apex, arising in the axil of a leaf primordium, usually begins activity by initiating a pair of prophylls. Then, it initiates a series of primordia that become cataphylls, serving as bud scales. Next, it initiates primordia that will become the lower foliage leaves on the shoot to be expanded from the bud during the following growing season. In temperate zones, the bud, with its partially formed "embryonic shoot" within, goes from summer dormancy into winter dormancy. Slow development may continue during winter dormancy (Romberger 1963).

In the spring, meristematic activity and volume growth resume, the bud opens, and the embryonic shoot with its preformed leaves expands. The apical meristem may then initiate additional primordia, which immediately begin to develop into leaves. Typically, they develop quite uninterruptedly from initiation to maturity in a few weeks or months, without a dormant period. Thus, the leaves of these shoots are of two developmental types: "early" leaves that overwintered in the dormant bud, and "late" leaves that were not initiated until spring. Vigorous shoots may have many late leaves. Further, in species exhibiting sympodial growth and having shoots of two distinctly different classes based on internode length, one class (short-shoots) tends to have only early leaves, whereas the other (long-shoots) usually has late leaves also. The early and late leaves often differ in morphology, as in *Populus* (Critchfield 1960). Members of the Pinaceae, which grow monopodially, have only early leaves, whereas the apices of long-shoots of many deciduous trees with monopodial branching can produce late leaves also. This allows them to respond quickly to favorable growing conditions by producing additional leaves.

We briefly discuss apical development in four of the many species in which the apex alternates between producing leaves and scales.

In the first of these, *Pseudotsuga menziesii*, the shoot apical dome is small and only slightly active when the bud opens and the embryonic shoot begins extension growth in the spring. As the shoot expands, the apical dome becomes more active, increases its volume, and initiates a series of primordia that develop into bud scales. These scales soon ensheathe the dome and all its younger primordia. The volume of tissue given up to these primordia, however, is smaller than that of the new cells generated by the apical dome, and the dome thus enlarges further. The dome produces a further series of primordia, which will become the leaves of the shoot to be expanded the following spring. Thus, the typical leaf primordia are almost a year old when they expand. These primordia accumulate around the base of the apical dome. Gradually, a conical "embryonic shoot" forms. It may include a hundred or more primordia arranged in beautifully systematic spirals (mostly contact parastichies). Towards the end of this phase of leaf initiation, the apical dome is giving up to new primordia more tissue volume than it generates, and is becoming smaller. It gradually returns to the small volume it had at the season's beginning, and becomes almost inactive (Allen 1947).

The dormant terminal buds of *Pinus lambertiana* and *P. ponderosa* also contain all the primordia for the leaves and cataphylls that will appear in the following season. The axils of many of the primordial cataphylls already contain primordial short-shoots with small apical meristems. Growth activity in the spring begins with elongation of the main axis of the embryonic shoot. The axillary short-shoots (often called "dwarf shoots") also renew their development. The apical meristem of the main shoot remains quite inactive while its axis elongates. Then, after the new shoot has made a good part of its extension growth, and the needles of the short-shoots have burst through their sheath of scales, the main apical meristem is reactivated.

The first of the new primordia that the apical meristem initiates develop into cataphylls that are "sterile — that is, lacking dwarf-shoot primordia in their axils. The internodes between these cataphylls will not elongate until the next season. Slow production of sterile cataphylls continues throughout the period of rapid shoot elongation. Then, as shoot elongation slows, the apical meristem becomes more active, rapidly initiating a second series of cataphyll primordia. These differ from those of the first series in that they bear dwarf-shoot primordia in their axils, and thus are "fertile". During the late summer, long-shoot bud primordia arise in the axils of a few cataphylls. Activity of the apical meristem then declines, as it produces another series of sterile cataphylls. These become the scales of the bud that will become visible in the following season and will expand the season after that (Sacher 1954, 1955a,b). Thus, each season, the apical meristems of long-shoots produce three series (sterile, fertile, sterile) of cataphyll primordia. The foliage leaves are produced by short-shoot meristems initiated in the axils of the fertile cataphylls, not by the long-shoot apex.

Terminal buds of *Liriodendron tulipifera*, unlike those of the already mentioned *Pinus* species, do not necessarily contain primordia of all the leaves that will be expanded. These buds contain about eight primordia, whereas 10 to 20 leaves commonly are expanded per season. The bud scales begin to open in mid-March in New Jersey (Millington and Gunckel 1950). By late March, the preformed leaf primordia are visibly enlarged and the apical meristem has resumed activity. By mid-April, young leaves are expanding, internodes between them are elongating, and new leaf primordia are being initiated. These new primordia and the internodes between them expand later in the same season. During late April, the new shoot typically develops two to five short lateral branches, from newly initiated lateral-branch primordia (Millington and Gunckel 1950). Additional leaf primordia are initiated during July and August. In early July, however, there is a change in developmental pattern, marked by inhibition of elongation of internodes between the new primordia. Several of these new primordia develop into pairs of scales. These may be interpreted as pairs of stipules belonging to phyllopodia in which laminar development was inhibited. Elongation of internodes between the pairs of scales is permanently restricted. After a brief period of scale development, the pattern reverts to production of primordial foliage leaves, which will not be expanded until the next spring. Thus, a new bud is formed. By early September, the bud typically contains eight partly developed leaves. Mitotic activity in the apex slows and stops in early October (Millington and Gunckel 1950).

17 The Root

17.1 The Root Apex

The root apex is not clearly demarcated from more basal regions of the root. In contrast, the shoot apex, because it initiates foliar primordia, is relatively easily delineated from more basal regions of the shoot. The shoot apex, as narrowly defined (Sect. 14.1.1), is that part of the primary shoot meristem that is not yet differentiated into cauline and foliar components. Here, we arbitrarily designate as "root apex" the region in which cells originate and elongate. In addition, in some roots, early stages of cytodifferentiation occur in regions of continuing cytogeneration and thus must be considered as occurring in the apex, though the later stages of differentiation occur subapically. Because the distinction between "apex" and "subapical regions" cannot be sharp, our use of these terms is not rigorous.

The root has no morphogenic cycles analogous to plastochrons, because the surface of the root apex does not typically initiate primordia of leaves or buds. Rather, branch roots arise from deep-seated, mature tissues some distance behind the apex.

Though the structure of root apices varies greatly, an almost universal feature in seed plants is the root cap, which somewhat resembles a cap pulled down over a skull. The presence of this cap means that root apical initial cells, unlike shoot apical initial cells, are deep-seated. The root apical meristem is neither entirely in the "skull" nor in the "cap". The cells of the proximal part of the cap are meristematic, as are the cells of the distal part of the root body proper.

17.1.1 Terminology

A rigorous, broadly applicable definition of the root apical meristem is difficult to provide. This is partly because of overlapping terminologies. Another complicating factor is that there commonly is a center of minimal meristematic activity in the very location where the most active initial cells formerly were thought to be. Though the existence and necessity of this "quiescent center" have been confirmed by mathematical-biophysical reasoning and modeling, and by physiological-biochemical investigation, the developmental significance of this region is not yet fully integrated into a rigorous terminology of root meristems.

Each of the various sets of terms that have been used to explain primary growth and development in roots has some relevance and continues to be used. A brief overview and glossary follows, with elaboration later as necessary:

Histogens (Hanstein 1868): Hanstein applied his **histogen theory** to apices of roots as well as shoots (Sect. 14.1.3.1). He recognized three axially superposed tiers of (topographic) initial cells and their early meristematic derivatives. These "tissue formers" are the **dermatogen, periblem,** and **plerome.** In the root, derivatives of the dermatogen were presumed to form the rhizodermis; those of the periblem and plerome to form the cortex and stele, respectively. The root cap was seen to have a variable origin. Later, the term **calyptrogen** was assigned to a fourth histogen, which purportedly produced the root cap in those monocots having a discrete cap (Janczewski 1874).

Although some roots, at some developmental stages, conform to histogenic concepts, these concepts and the related terminologies are not broadly applicable. For example, roots of members of Juglandaceae, Tiliaceae, and some other taxa typically have two rather than three or four discernible tiers of topographic initials that might be called histogens. Further, the sets of initial cells in roots of some other angiosperms and of most gymnosperms cannot be considered as grouped into tiers or histogens at all. Guttenberg (1960) modified the histogen terminology in adapting it to his theory of initial centers and central cells (see below).

Körper and **Kappe** (Schüepp 1926): The Körper-Kappe theory of root apical structure, though not conflicting with the histogen theory, addresses different questions. By categorizing the orientations of cell divisions that cause splitting of clonal-lineage files of cells (Sect. 17.1.4; 17.1.5), it allows the origins of these files to be traced to cells on the margins of a cytogenic zone, and the delineation of an outer, Kappe, region from an inner, Körper, region.

This method is similar to tracing a radial file of fusiform xylem and phloem cells back to its origin in a specific fusiform initial in the vascular cambium (Sect. 18.4.1). But, although a Körper-Kappe boundary is distinct and predictably located in roots of some taxa at some developmental stages, it is indistinct or variable in roots of others. In some species, including *Fagus sylvatica*, the boundary in large root tips falls well within the cortex, making the cortex partly Kappe and partly Körper; in small roots only the cap and rhizodermis are Kappe (Clowes 1950). Thus, the boundary may change as root tips enlarge. In those angiosperms that have distinguishable root-cap initials, as do many monocots, only the cap is Kappe; all else is Körper (Clowes 1961). The significance of a Körper-Kappe delineation is not simple to interpret developmentally (Sect. 17.1.5).

Central cells and **initial centers** (Guttenberg 1960): Guttenberg considered typical seed-plant root apices to be constructed by the progeny of a small number of apparent initials, or **central cells,** located near the pole of the stele. According to him, divisions of these cells are not regularly oriented and the cells are not especially active meristematically. Guttenberg considered these cells to be the ultimate progenitor cells rather than the immediate sources of myriads of tissue-building cells. That is, central-cell progeny rather than central cells themselves generate most cells of the root.

Quiescent center (QC) (Clowes 1956, 1958) and **promeristem** (Clowes 1950, Sussex and Steeves 1967): Guttenberg's (1960) realization that cells in the central region of the meristem may divide less frequently than those at the periphery was compatible with the emerging concept of a virtual quiescent center, encompassing the initial center and some of its progeny cells (Clowes 1954a). Guttenberg (1968), however, did not fully embrace the QC concept.

The term QC denotes a hemispherical or lens-shaped region of relatively inactive cells at the foci of root apices (Clowes 1954a, 1958). Discovery of the meristematic lassitude of this region, exaggerated in early work, caused confusion about appropriate use of the term "promeristem". Should this term refer only to the active, functional initials on the QC margins, or should it also include the reservoir of less active, but ultimate-progenitor, initial cells within the QC? We have chosen to use "promeristem" in the same sense in roots as in shoots (Sect. 14.1.3.2). That is, the promeristem includes both the active, immediate initial cells and those uncommitted, potential initial cells to which the active initial function can devolve as conditions change.

Open and **closed** apices (Guttenberg 1960; Clowes 1981a): According to Guttenberg, if the cells of the initial center are arranged in distinct, histogen-like tiers, and each tier generates discrete clones of progeny constituting developmental compartments, with no cells crossing between them, the apex has a closed structure. If no distinct tiers are discernible among the initial cells, the apex is open. Guttenberg realized that there could be no distinct "histogens" in open types.

Gradually, as a more dynamic concept of the root apex, including a QC, became widely accepted, connotations of the terms "open" and "closed" changed, in recognition that these need not be permanent states. The newer concept of the root apex transfers open-closed characteristics from the realm of genetic determination to that of epigenetically controlled morphogenic processes that change during development. Now, a meristem is called "closed" if the cap is discrete and is derived from discrete initials, and "open" if a cap, cortex, and rhizodermis, or all root tissues, are derived from a common group of initials. In both types, the functional initials are assumed to be located on the periphery of a QC.

Although we cannot yet abandon even the oldest of the above terminologies, we perceive histogen concepts to be of decreasing relevance. We expect terminologies based on the QC, and on cell files or clones having progenitors within the QC, to be of increasing relevance.

17.1.2 The Quiescent Center (QC)

The physiological anatomy of QC's is well documented for root apices of many angiosperms (Clowes 1961, 1984; Feldman 1984) and some gymnosperms (Chouinard 1959; Wilcox 1962; Peterson and Vermeer 1980). "Quiescent", as used in the root literature, does not mean "dormant" or "ameristematic", but does imply meristematic activity lower than in surrounding regions, and maintenance of the diploid condition.

The QC is better defined with respect to the mitotic cell cycle than by particular cell-wall arrangements or characteristics of cytoplasm and organelles. In the QC, cell cycling is strictly controlled (Clowes 1975): Some QC cells divide only once every few days or weeks; some do not cycle at all; and all remain strictly diploid. In *Zea*, the mean duration of the cell cycle is 230 hours within the QC, 10 hours in the nearby cap meristem and 25 hours in the central cylinder, beyond the opposite border of the QC. Although studies have shown that DNA, RNA, and proteins are synthesized relatively slowly in the QC and that titers of these substances are generally low (Clowes 1975), some enzymes may assay at higher levels in the QC than elsewhere.

QC cells are mostly in the G_1 phase of the cell cycle. Under ordinary conditions, their proliferation may be inhibited by metabolic products, signal substances, or unknown factors emanating from the active cells surrounding the QC, as is indicated by experimental treatments that disturb the normal functioning of the surrounding cells (Clowes 1970; Webster and Langenauer 1973).

The QC is commonly shaped like a thick lens, with a diameter half that, or less, of the whole root tip. It may include only a few dozen cells, or as many as several hundred. A large, actively growing apex is more likely to have a detectable QC than is a small, less active one. A small QC may remain undetected, however. Further, there is evidence that in the *Capsella* embryo the root QC may have a very early origin in a single cell (Raghavan 1990). Detection of such early and small QC's, however, is difficult.

Clowes (1954a) realized that the requirements of symplastic growth make a QC a geometric necessity for a meristem in which the planes of cell division are related as in root apices. The basis of this geometrical necessity can be visualized by referring to the natural-coordinate system and the vector field of the displacement velocity (Sect. 12.3.2; Hejnowicz 1989.

In a root in steady-state growth, cell division in the QC does not augment the QC cell population, because cells gradually migrate toward the boundary and leave the QC at a rate compensating for mitoses within. In this way, the very slowly cycling, diploid cells of the QC slowly but inevitably replace the functional initials on the QC boundary. The concept of the QC as a source of replacements for temporarily functional initials weakens earlier concepts of histogen-like initial groups, central-initial-cell groups, or other permanent initial cells occupying the same locale. Because the QC is a reserve of totipotent, diploid cells having constrained meristematic activity (Clowes 1981a; Gahan and Rana 1985), analogies may be drawn between QC cells and stem-cells of some animal systems (see Barlow 1987).

QC cells can be considered as the founders or progenitors of the functional initial cells, which, in turn, generate files of progeny cells. We follow Barlow (1976) in calling the potential initials within the QC **founder cells**. The active, functional initials on the periphery of the QC we call simply **initials**. The cells of the QC and the functional cells on its periphery — that is, the entire population of undifferentiated, diploid, meristematic or potentially meristematic cells at or near the focus of the root apex — constitute a promeristem.

If the files of cells contributing to the main body of the root regularly arise from discernible functional initials on the proximal and peripheral margins of the QC, and the files or groups contributing to the cap (or Kappe, which may include some rhizo-dermal and cortical tissue) regularly arise from a separate group of initials adjacent to the distal face of the QC, then the apex is "closed". If there is no such separate group or tier of Kappe initials, the apex is "open". In an open meristem, the initials distal to the stelar pole may function as initials of the cap alone, or of the cap, rhizodermis, and cortex. Some dozens of these cells may be episodically active, dividing mostly transversely (Clowes 1981a) and contributing cells to the periphery of the cap as well as to the columella. That is, the QC boundary may shift during root development, sometimes including and sometimes excluding these cells. Some apices are closed in the embryonic stage, become open in the young seedling, and then revert to the closed state (Byrne and Heimsch 1970a,b; Armstrong and Heimsch 1976).

The cellular dynamics and processes of devolution of the initial function within the root promeristem, which includes the QC, are probably less ordered and more stochastic (Sect. 14.1.3.2) than was formerly supposed. This has been the case with the shoot promeristem.

17.1.3 Functional Initial Cells

A cell that is displaced from the QC to the region of the functional initials becomes susceptible to epigenetic and cytological changes that leave it capable of only a limited number of further mitotic divisions. In general, the proliferative capacity of meristematic cells decreases with increasing distance from the QC (Barlow 1987). Whereas orientations of divisions within the QC may be either random or ordered, divisions of functional initials are mostly oriented in such a way that a discrete file of progeny cells is established.

The cellular changes accompanying displacement from the QC are sometimes considered "quantal" (Barlow 1976), though we reserve that term for cell cycles that result in "determination" of clonal files of progeny cells (Sect. 12.2.2.1). Quantal cycles lead to files of cells that are "determined" to differentiate into specific tissues. This "determination", though, is not necessarily irreversible, in contrast to earlier concepts of determination.

Divisions that are not quantal may be exclusively proliferative or may be formative as well as proliferative. In the context of cell files, **proliferative** divisions are transverse and contribute only to cell number; **formative** divisions are longitudinal and contribute to the form of the organ as well as to cell number. Via proliferative and formative divisions, the initials and their progeny lay down the primary tissues of the root. The progeny may eventually become endopolyploid and unable to complete normal mitosis (Evans and Van't Hof 1975).

All active initials are similar in that they are progeny of undifferentiated founder cells, but the remaining potential for mitotic division of these cells and their progeny seems to vary regularly with position in the root apex (González-Fernández et al. 1968). This suggests a cycle-regulating system, possibly based on gradients of two substances (Sect. 12.4.2) that separately or jointly regulate nuclear DNA synthesis and mitosis with cytokinesis (Barlow 1976). The requirements of such a system might be satisfied by an auxin and a cytokinin, but other phytohormones also might be involved.

17.1.4 Körper and Kappe

A root tip consists of a coherent bundle of cell files radiating from a topographic cytogenic center or meristematic focus. In some pteridophytes, this focus or center is a single apical cell. In seed plants, it is a multicellular promeristem that usually encompasses a QC. In median longitudinal sections, one can commonly follow files of cells that curve away from the meristematic focus and become parallel in the zones of elongation and maturation. In the curved part of their courses, single files often become two files, as a result of longitudinal (formative) division (Sect. 17.1.3). Each new file may bifurcate again. Where two files butt against one, the common walls of three cells (one "sister" and two "daughters" of the other "sister") form a T (or Y) configuration, with the stem being the common wall between the new files, and the capital being the last transverse wall of the old single file (Fig. 17.1).

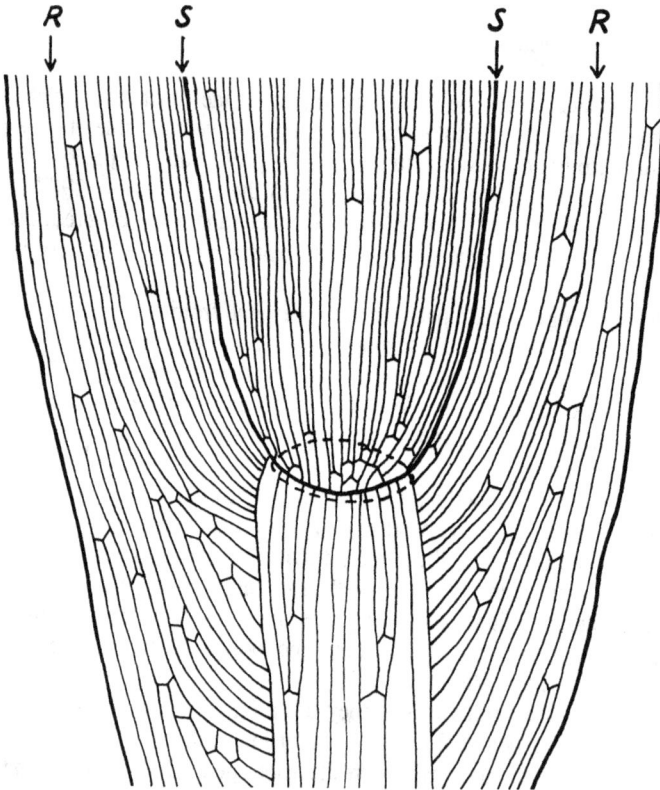

Fig. 17.1. Cell-file pattern and Körper-Kappe delineation on a median section of a *Fagus sylvatica* root apex. Walls resulting from transverse proliferative divisions within files are not shown. In the Körper, the stems of the Ts marking formative divisions point toward the base of the root. In the Kappe, the Ts are oppositely directed. The boundary between Körper and Kappe usually is not sharp. There may be a transition region in which few or no T wall configurations occur. A few Ts do not conform to the pattern. The analysis was based on a cell pattern similar to that shown in Fig. 17.3. The area enclosed by the *dashed line* represents the location of the QC, though its borders here are conjectural. S marks the limit of the vascular cylinder or stele. R marks the files that will become rhizodermis after the overlying cap tissues disintegrate. (Redrawn after Clowes 1961)

The orientations of the T configurations of cell walls can provide information about distribution of growth and cell division within the root tip. The region of the tip in which the number of cell files increases with distance from the meristematic focus, and in which the stems of the T's point toward the root base, is the Körper. The region in which the number of cell files decreases with distance from the meristematic focus (the latter including the cells along the edges of the columella), and in which the stems of the T's point toward the columella, is the Kappe. (The columella often cannot be assigned to either Körper or Kappe.)

Körper and Kappe regions are neither cytohistological compartments nor histogens. The boundary between them may lie in different histological regions in different (or even in the same) species, and even in the same root at different developmental stages.

Körper-Kappe analysis reveals that root construction is regulated by control of the plane of cell division, especially among the functional initials near the QC. For example, in simplified schematic outline, consider two functional initial cells adjacent to the lateral boundary of the columella where it abuts on the stelar pole (Fig. 17.2A). Both initials undergo transverse divisions followed by a longitudinal (formative) division in one of the daughters (Fig. 17.2B). However, because the two initials receive different positional information, the formative division in the upper packet occurs in the daughter distal to the columella, and the initial function devolves to the proximal daughter, whereas, in the lower packet, the formative division occurs in the daughter proximal to the columella, and both of the daughter cells produced by the formative division assume the initial function. This pattern is repeated, resulting in larger cell packets differing in the direction of T configurations (Fig. 17.2C). The boundary between the two packets is the Körper-Kappe boundary (Clowes 1950).

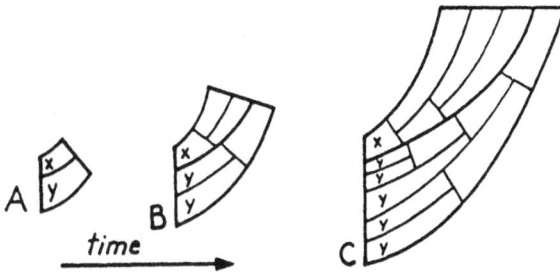

Fig. 17.2. Schematic drawings of Körper and Kappe root-apical-initial cells before and after formative divisions. T wall patterns form where one cell file becomes two. Transverse walls resulting from ordinary proliferative divisions are not shown. Derivatives of *initial x* divide in a pattern characterizing the Körper, whereas *initial y* and its derivatives divide according to a Kappe pattern. All drawings represent groups of cells just to the *right* of the columella where it approaches the stelar pole. For further details, see text. (After Clowes 1950)

In this example, the increased number of Kappe cell files along the margin of the columella allows the peripheral cap tissue to grow symplastically with the extending columella, which in this analysis behaves as Körper tissue. At the same time, the increase in number of Körper cell files with increasing basal distance from the stelar pole accommodates the primary radial growth of the root. In the Kappe region, in contrast, formative divisions, which add cell files, are most prevalent in the functional initial cells at the distal border of the QC and in the region immediately adjacent to the columella, and decrease with distance from the QC. Shifting of the Körper-Kappe boundary with root development can be interpreted as indicating that the occurrence and position of formative divisions respond to changing positional information.

17.1.5 The Root Cap

Although superficially similar, nearly all root apical meristems that have been studied appear to be either "open" or "closed", if examined histologically. They may, however, change from one state to the other during development. In an open meristem, founder cells in the distal part of the QC can contribute initials and progeny to the cap. In contrast, in a closed meristem, a thick, polylamellate wall may accrete between the root proper and the cap. The cap then has no reservoir of noncycling cells that can

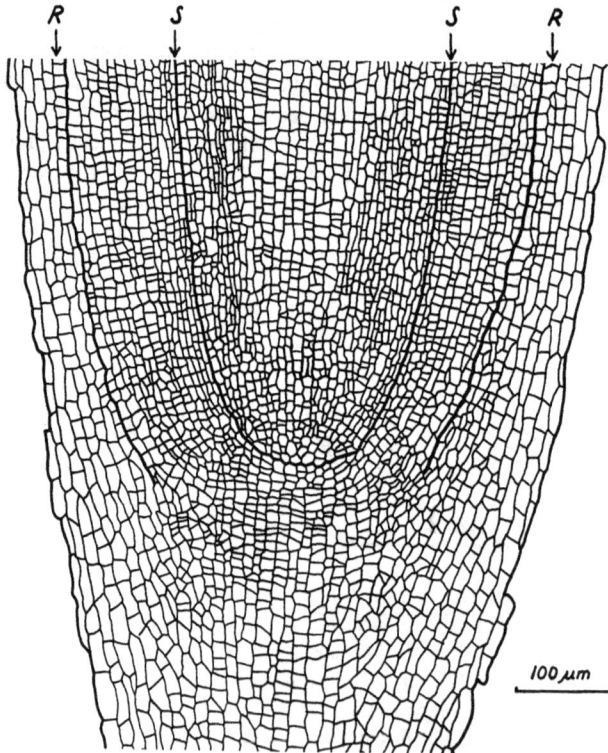

Fig. 17.3. Cell-wall pattern visible in median longitudinal section of root apex of *Fagus sylvatica* similar to that upon which Fig. 17.1 was based. *S* and *R* have the same meaning as in Fig. 17.1. Note that, before cell-file-pattern analysis and Körper-Kappe delineation, the columella is not prominent and other root-cap tissues are not sharply separated from the cortex and rhizodermis. (Redrawn after Clowes 1961)

replace, and thus rejuvenate, its set of initials. Lack of such a reservoir could lead, successively, to cessation of mitosis, endopolyploidy, and senescence of the cap, followed by early demise of the primary-root tip, to be functionally replaced by secondary roots. This would lead to a fibrous root system. Indeed, some Poaceae and other monocots, which generally have closed root apices, do have discrete caps and fibrous root systems (Haberlandt 1914). Woody plants, in contrast, tend to have open meristems, nondiscrete caps, and long-lived, woody components in their root systems (Sharma and Sharma 1987). In fact, insofar as we know, arborescent plants beyond the germination stage — even arborescent monocots, such as many genera of palms (Drabble 1903) and some species of *Yucca* (Arnott 1962) — generally have open root apices and nondiscrete caps. Some species of *Bambusa* are a possible exception.

We briefly review the developmental pattern of the root cap in *Fagus sylvatica*, which is probably not atypical of woody plants (Clowes 1950). This cap is not well demarcated from the more permanent part of the root tip, which pushes the cap through the soil. This poor delineation is evident from the cell-wall pattern seen in median longitudinal sections (Fig. 17.3). Cell-file delineation and Körper-Kappe analysis of this pattern (Sect. 17.1.4), as shown in Fig. 17.1, can provide information about the distribution of cell division and growth within the cap and adjacent, permanent tissues. Nevertheless, assigning files of cells to Körper, Kappe, or an indeterminate region does not reliably predict their differentiation or involvement in histogenesis.

In using Körper-Kappe analysis, one must remember that the cap of a growing root is always in flux. The older, distal ends of the columella files and the older, outer files of the periphery of the cap are being abraded away or sloughed off. Simultaneously, the basally-directed extremities of the inner peripheral files become outer cortex or rhizodermis, or a functional equivalent of these (R in Fig. 17.1).

In the proximal part of the columella, where orientation of divisions is Körper-like, divisions are mostly transverse, whereas there must be a large component of longitudinal divisions at the heads of the adjacent peripheral files (Kappe). Nevertheless, symplastic growth is maintained.

Cell-file delineation and Körper-Kappe analysis do not reveal the location of the functional initial cells. Presumably, they are at or near the border of the QC. A plausible location for the QC in the apices shown in Figs. 17.1 and 17.3 is indicated by the dotted outline in Fig. 17.1. QC borders change during root development, especially in the proximal part of the columella.

17.2 Differentiation of Primary Root Tissues

The ability of many higher plants to propagate vegetatively by generating roots on shoots and vice versa indicates that large fractions of the genetic information in developing cells become operative only under certain conditions. We have already discussed (Sect. 14.1) the premise that genetically identical cells in various regions of a shoot apex can follow different developmental programs in response to different positional information (Sects. 14.1.3.1; 14.1.3.2). This premise applies to root apices also.

"Determination" of the developmental pathway that a cell will follow may be established by a quantal cycle (Sect. 12.2.2.1) in an initial cell or "mother cell". (Quantal "determination" is not necessarily irreversible; Sect. 17.1.3.) The persistence, through many cell generations, of characteristics that initially arose as responses to environmental stimuli, including positional effects, may be regarded as a kind of somatic cell heredity (Brink 1962).

As a root grows, the region of rapid cytogenesis migrates forward with the tip, leaving behind a region of declining mitotic rates, cell enlargement, and incipient cytodifferentiation. Cell division slows and stops first in the future central-cylinder region. Indeed, in files of cells differentiating into metaxylem, mitosis fails to occur only a few cell diameters from the initials, even though divisions may continue in neighboring files. Next, mitotic rates decline in the cortex. Simultaneously, cell elongation growth becomes prominent. Because growth is symplastic, the cells in files in which mitosis stops first become the longest (Clowes 1976).

In some tissues of some taxa, the cell cycle stops rather abruptly in G_1 in cells that begin to differentiate, while in other tissues DNA synthesis continues after a last mitosis, resulting in endopolyploidy. For example, in *Zea mays*, although most cells probably remain 2C or become 4C after a final mitosis, cells that differentiate into vessel elements often attain a level of 32C, or higher, before their protoplasts autolyze (List 1963).

Within a root tip, many processes occur simultaneously, at rates that are variously related to root elongation. Some are closely coupled with growth in the spatial or volumetric sense, whereas others seem more time-dependent. An example of the former kind of coupling is provided by roots of *Libocedrus decurrens*, in which the elongation rate commonly is uniform if root diameter is uniform (Wilcox 1962). Further, *Libocedrus* root tissues differentiate and mature at a distance from the stelar pole that depends greatly on root diameter, elongation rate, and branching order. Therefore, it is prudent to assume that data obtained from rapidly growing, primary roots may not be applicable to roots of higher order, which typically are more slender and more slowly growing.

Much of the literature about primary differentiation in roots pertains to *Zea*, *Triticum, Oryza, Hordeum,* and other monocots having "closed" root apical meristems and fibrous root systems. Development in these taxa may not be typical of seed plants in general. We do not yet know what is representative for roots of woody species. Most of the woody species for which data are available have "open" apical meristems and dendritic root systems with several orders of branching.

17.2.1 Development of the Central Cylinder

Rigorously, a "central cylinder" consists of vascular tissue and a boundary layer of pericycle. The cylinder is ensheathed in cortical tissue. The term "central cylinder" is nearly synonymous with the old term "stele", which we use only with reference to the "stelar pole". The central cylinder is initiated when files of provascular cells differentiate within the "ground" meristem near the QC and are separated from the nascent cortical tissues by an incipient pericycle.

In longitudinal sections, one can see that primary vascular tissues differentiate continuously in acropetal progression. These tissues may not at first be radially or tangentially continuous, but, rather, tend to differentiate in patterned *groups* of cell files, commonly extending an established pattern within a matrix of ground-meristem files. These groups of files are the basis of "archy" numbers, or numbers of loci of protoxylem differentiation.

The earliest, readily detectable, evidence of vascular differentiation is increased vacuolation in the cell files that will become metaxylem vessels. Usually, some dozens of such files are arranged in a compact, central bundle. The cells of these files soon stop dividing and become endopolyploid and very large. Metaxylem typically differentiates closer to the initials in roots than in shoots (Esau 1965b). In *Libocedrus*, metaxylem differentiation is detectable about 100 μm basal to the root initials (Wilcox 1962).

The shape, in transection, of the early metaxylem region typically is a narrow rectangle, or a hypocycloid having three, four, or five cusps. The first protoxylem elements differentiate just abaxial to the two narrow sides of the rectangle (the diarch condition) or at the vertices of the hypocycloid (the tri-, tetra-, or pentarch conditions). Thus, in a hypocycloid-shaped metaxylem area, the number of cusps or of vertices predicts the archy level of the central cylinder. Protophloem differentiates in the cusps.

Root metaxylem differentiates early but matures slowly. In an actively growing *Libocedrus* root, most metaxylem elements are not functionally mature until they are 10 to 15 cm behind the tip. This requires several weeks of moderate growth. If growth is more rapid, the distance to mature metaxylem is greater because the rate of cell maturation is more steady (time-dependent) than is the rate of cell production and expansion. Some metaxylem elements mature very late, concurrently with differentiation of secondary-xylem derivatives (Wilcox 1962). This may be due to their large, endopolyploid nuclei, which must be autolyzed before the metaxylem elements can function in conduction.

Early protoxylem elements are much narrower than metaxylem elements, apparently because protoxylem precursor cells continue to divide longitudinally for a longer time than do metaxylem precursors. The protoxylem precursors also retain the diploid condition longer. They then expand rapidly, to about the same diameter as metaxylem elements, and mature quickly. As a result, in a particular locality, protoxylem generally matures and becomes functional before metaxylem does. Thus, primary xylem is usually exarch, maturing centripetally and forming "radial plates" of protoxylem.

Distance from the initials to the first maturing protoxylem in *Abies procera* is less than 0.5 mm when growth is slow and more than 7 mm when it is rapid (Wilcox 1954), suggesting again that time-dependent processes are involved in maturation. Distances in *Libocedrus decurrens* are similar (Wilcox 1962). This resembles the situation in shoot apices of *Picea abies* seedlings, in which differentiation is more closely related to chronological time than either to morphogenic age of the apex, as measured by internode number, or to linear distance below the base of the apical dome (Gregory and Romberger 1977).

Protoxylem and protophloem usually differentiate along different radii of the root body, yielding a pattern of alternating xylem and phloem sectors (Sect. 17.9). In stems, though, these tissues usually differentiate along the same radii, forming collateral, or sometimes bicollateral, vascular bundles. Mature sieve elements generally occur closer to the QC (well within the meristematic zone) than do mature xylem elements.

As the protoxylem becomes discernible (in transections), so does an annulus-like region of pericycle. It consists of densely cytoplasmic cells not showing vacuolation, in contrast to cells in both the nascent central cylinder and the cortex. The pericycle is less specialized than the vascular tissue it ensheathes. Being in a sense a residual "ground" tissue, it maintains a diploid condition and permanent meristematic potential. It may become multiseriate. The pericycle may give rise to secondary roots, phellogen, and, in part, vascular cambium (Sect. 17.9).

Early protophloem, which in many gymnosperms is preceded by a large-celled, thin-walled, precursory phloem, differentiates between the cusps of the metaxylem and the nascent pericycle. Early protophloem cells are conspicuous because they are larger and less cytoplasmic than pericycle cells, and, in transection, seem radially elongated. Later primary phloem elements, which are subject to different tissue stresses, are smaller and tend to appear tangentially elongate in transection.

A comprehensive explanation of the control of root vascular-tissue differentiation must allow for establishment of a pattern and for changes of pattern in time and under changing conditions. The possibility, that maturing cells of xylem or phloem induce younger cells in the same file to undertake similar developmental pathways, would explain the stable tissue patterns common in roots over long growth periods (Torrey and Wallace 1975). However, induction by maturing cells cannot explain the origin of additional loci of protoxylem differentiation, which are not extensions of existing files (Charlton 1980). The process by which archy number changes as root diameter increases under natural conditions has been studied in only a few species. In *Libocedrus decurrens* (Wilcox 1962), additional loci of protoxylem initiation appear in the larger cusps of the central metaxylem body, adaxial to a large protophloem body. The possibility that certain initials on the border of the QC are "determined" via quantal cycles to generate progeny "programmed" to follow a pathway to vascular development better explains the origin of new strands than does induction by maturing cells (Meins and Binns 1979).

However, the developmental situation is more complex than either of these propositions alone can accommodate. For example, if *Pisum* root tips, which are triarch, are excised just distal to a point where the vascular pattern has become established, and are then cultured on agar medium, most remain triarch. Some, though, become diarch or monarch, then revert to a triarch pattern — by differentiation of new provascular strands that are not extensions of preexisting ones (Torrey 1955).

Archy number is correlated with the cross-sectional area of the primary vascular cylinder where the youngest discernible vascular elements are differentiating (Torrey and Wallace 1975). This implies that vascular strands can arise spontaneously only if the vascular cylinder is large enough to accommodate them. In turn, the size of the vascular cylinder — and of the whole primary root — is related to QC size and to the number of formative divisions of the initials on its boundary (Barlow 1984). Thus, changes in complexity of the primary vascular system in a root are related to shifts in QC size (Feldman and Torrey 1975; Feldman 1984).

17.2.2 Development of the Rhizodermis

In gymnosperms generally and in angiosperms having "open" root-apical-meristem organization, there is no well-defined protoderm. As a result, the rhizodermis is not sharply delineated from the cortex, which, in turn, is poorly demarcated from the trailing peripheral tissues of the root cap.

In contrast, in those angiosperms having "closed" root apical meristems, the progeny files of a definable set of initial cells on the border of the QC at the stelar pole constitute a protoderm. Via proliferative and anticlinal formative divisions, followed by differentiation, a protoderm produces a rhizodermal layer, which is somewhat analogous to the epidermis of shoots. Via periclinal formative divisions, the protoderm may also produce one or several hypodermal or exodermal layers.

In gymnosperms, there is a de facto, functional rhizodermis on the surface of the cylindrical part of the root (i.e., behind the tip). Its cell files, if followed acropetally into the region where peripheral cap layers persist, are not distinguishable from neighboring files. Where a de facto rhizodermis is exposed, it is not a smooth or regular sheath of cells. In many genera, several layers in addition to the outermost one can form root hairs (see later). This raises the possibility that cells of a de facto rhizodermis have not been "determined" toward rhizodermal differentiation by quantal division, but that they function in response to positional factors. In some locations, root-hair-bearing hypodermal cells may actually be rhizodermal cells that are still covered by one or more layers of persistent root-cap cells, as reported, for example, in *Pseudotsuga menziesii* (Bogar and Smith 1965).

In many monocots, especially herbaceous forms, and in some dicots, the early protoderm can be recognized in the meristematic zone, even where it is still covered by root cap. In both groups, the rhizodermis often consists of **trichoblasts**, or root-hair-forming cells, and **atrichoblasts**. Treatment with exogenous auxin, however, demonstrates that atrichoblasts are *potential* trichoblasts.

In some plants, differentiation into trichoblasts and atrichoblasts occurs very early. The first visible sign of differentiation is commonly an obviously polar distribution of cytoplasm, before the last transverse division of the precursor cell. In *Phleum*, which has a protoderm, the final transverse division produces an apical cell that is shorter and more densely cytoplasmic than the basal one (Avers 1963). As development proceeds, the length difference increases because the shorter cell, which in *Phleum* is the trichoblast, elongates more slowly than does the longer cell, the atrichoblast (Avers 1957).

Within a trichoblast, the nucleus and masses of cytoplasm move to the site of root-hair initiation. Through localized wall expansion, a thin-walled, tubular evagination rapidly develops in the area that will be the root-hair tip. The nucleus and some cytoplasm, which forms a thin parietal layer in the cell and a mass around the nucleus, move into the growing hair, commonly to the tip.

Although there are exceptions (Cormack 1949; McDougall 1921), notably in *Gleditsia* and related species, root hairs typically function only a few hours or days.

The fate of the nucleus and the main body of the trichoblast as the root hair collapses and dies is not known in detail. Their fate may be of little consequence, however, because in most roots the entire rhizodermis is ephemeral.

17.2.3 Early Differentiation of the Cortex and Endodermis

A traditional view is that all cell files of the cortex can be traced acropetally to their origin in procortical "parent files". The rapid increase in root diameter in the meristematic region has been ascribed mostly to repeated periclinal and longitudinal anticlinal divisions among the initials heading the procortical cell files and/or among their early derivatives. In some roots, all the innermost procortical files, or all such files around the root circumference at similar distances from the initials, undergo repeated periclinal and longitudinal anticlinal "T" (formative) divisions (Sect. 17.1.4) in synchrony. As seen in transection, these divisions suggest "cambium-like" activity in the nascent cortex (Williams 1947), but, in the context of the QC concept, the founder-cell hypothesis, and Körper-Kappe analysis, a more comprehensive concept of cortical development can be proposed.

Founder cells, or initial cells derived from them, in a ring-like set around the distal boundary of the QC, undergo quantal divisions, in which they and their progeny are determined to develop into cortical cells. Files of cells derived from them, radiating outward and backward, undergo repeated anticlinal and periclinal formative T divisions, as well as many transverse, proliferative divisions. The extent of synchronization of T divisions at a particular distance from the stelar pole determines the regularity of radial files and circumferential continuity of layers of cortical cells, as seen in transections (Williams 1947; Heimsch 1960). The progeny files derived from a single cortically determined founder or functional initial cell, in effect, constitute a clonal developmental compartment occupying a sector of the cortex. By the same logic, the entire nascent cortex, derived from the determined cortical initials, can be regarded as a polyclonal compartment (Sect. 12.4.4). This concept depends on the existence of a QC that is a source of initials, but not on the existence of discrete histogens or other groups of dedicated initials, in which determination is considered irreversible. The concept is relevant to cortical development in both "open" and "closed" root apices.

The endodermis, which is generally uniseriate, and its immature precursor stage, the proendodermis, are conventionally regarded as the innermost cell layer of the cortex (Peterson 1989), and we so consider them here. However, in dormant or slowly growing (but not in rapidly growing) roots of *Libocedrus decurrens*, the proendodermal cell files can be traced acropetally to the border of the QC and give no evidence of being derived from inner files of the cortex (Wilcox 1962). This suggests that, although a cortical origin of proendodermis combined with acropetal induction from existing tissues cannot be generally excluded, it is possible that founder cells in a ring-like set just proximal to that producing the cortex are determined to produce the endodermis. The progeny of these cells would undergo many transverse and anticlinal formative divisions, but very limited, if any, periclinal divisions, and would also differ physiologically and cytologically from cortical cells. A further possibility is that both the endodermis and a so-called "phi layer", which may occur just abaxial to it (Wilcox

1962), are derived from the same set of initials. This may be why two endodermal layers have been reported in some gymnosperm roots, though only the inner one has authentic Casparian bands (Boureau 1939).

In slowly growing roots of *Libocedrus*, the proendodermis may be discernible only 0.1 mm behind the initial cells. The distance increases with increasing growth rate (Wilcox 1962), suggesting that differentiation is time dependent (Sect. 17.2). In transections of roots of many taxa, the proendodermis can be recognized within 0.2 mm of the QC as a rank of cells with tangential axes longer than radial axes. This shape contrasts strongly with that of the nascent pericycle cells, which are adaxial to them (Clarkson and Robards 1975).

Except in relation to lateral-root initiation (Sect. 17.3), there are no mitotic figures to indicate that the pericycle contributes cells to the endodermis, or vice versa. Thus, the two tissues, separated by common walls that form uncrossed boundaries between clonal files or packets of cells, seem to constitute different developmental compartments. The common walls become polylamellate and notably thicker than walls nearby (Williams 1947), which aids early recognition of the proendodermis.

17.2.4 Later Development in the Cortex and Endodermis

In roots having extensive secondary growth, the cortex is ephemeral. Typically, it dies and is lost to the soil after periderm develops from the pericycle. A short-lived cortex may remain parenchymatous until it dies. However, if secondary growth is delayed or absent, the outermost cortical layer may differentiate into an **exodermis**. Exodermis begins to differentiate basal to the region of elongation and of root hairs, after the death and disintegration of the rhizodermis in this region has exposed the outer cortical (hypodermal) cells to the external environment. These cells respond by reorganizing themselves, forming a layer that may become thicker via periclinal cell divisions and that lacks intercellular gas spaces. Subsequently, suberin and related substances are deposited (Guttenberg 1968).

We use the term "exodermis" in the sense of Peterson (1989), who applied it to hypodermal layers that have both Casparian bands and lamellae of suberin. She found that deposition of suberin in the hypodermis is preceded or accompanied by development of Casparian bands similar to those in endodermis (see later) in about 90% of the species surveyed in more than 50 angiosperm families; in those taxa in which the hypodermis was not suberized, there were no Casparian bands. Generally, Casparian bands in exodermis occupy the entire width of radial and transverse walls and thus are more massive than analogous, endodermal bands (Peterson 1989).

The continuous lamellae of suberin that typically are deposited onto the primary walls of exodermal cells resemble those deposited in the later developmental stages of endodermis. Cellulose lamellae then cover the suberin lamellae, forming a secondary wall. Lignin may be deposited in primary and secondary walls. Even after lignification and suberization, the exodermal protoplasts typically remain alive. The wall and the living protoplasts allow the exodermis to control both apoplasmic and symplasmic transport of solutes between the external, nonliving environment and the root interior

(Sect. 1.5.3). Though less ephemeral than the rhizodermis, the exodermis remains alive and functional only as long as does the underlying cortex.

In the proendodermis, deposition of Casparian bands in radial and transverse walls is slow to begin. The distance basal to the QC at which the bands appear is greatly influenced by rate of root elongation and by root diameter. The distance is typically 7—14 mm in *Libocedrus decurrens* (Wilcox 1962) and is probably less in the graminaceous crop plants (Clarkson and Robards 1975).

Casparian bands are deposited not *on*, but *in*, the common walls and middle lamellae between proendodermal cells. These bands fill and seal the microspaces and form a continuous matrix. The properties of the bands, and the tenacious adherence of the plasmalemmae to the inner surface of them, almost totally prevent apoplasmic transport across the endodermis, though they do not affect symplasmic transport.

In most taxa, after Casparian-band deposition, suberin lamellae are deposited on tangential, or on all, faces of the cell wall. In some taxa, there is also deposition of a cellulosic secondary (or tertiary) wall that may become lignified or suberized. These suberin and cellulose lamellae do not seal off the numerous plasmodesmata that connect the endodermal cells with neighboring cortical and pericyclic cells. Thus, symplasmic transport across the endodermis is little inhibited.

Deposition of Casparian bands in proendodermal cells, which typically occurs in the same axial region of the root tip as does maturation of early protoxylem elements, is usually well synchronized circumferentially. In contrast, the second, or suberization, stage begins abaxial to areas of protophloem, in the cusps of still immature metaxylem, then progresses circumferentially towards endodermal cells opposite the protoxylem "poles". Typically, suberization opposite the protoxylem lags so much that these cells have been interpreted, on the basis of transections, as "passage cells". However, because these cells already have Casparian bands, their walls are not actually passageways for apoplasmic transport (Clarkson and Robards 1975; Peterson 1989).

When vascular-cambial activity increases the diameter of the vascular cylinder, the endodermis is not always immediately ruptured and destroyed. It is at first stretched, and may then increase in girth by radial cell divisions (Clarkson and Robards 1975). In roots with extensive secondary thickening, the endodermis, and any abaxial cortical tissue, senesce after being isolated by peridermal tissues, which arise adaxially.

In some monocots, the inner cortical layers may differentiate into sclerenchyma, which, like exodermis, may become suberized. In other monocots, including many palms, the hypodermal or other outer cortical cells may divide periclinally, forming radial files of cells that become suberized and sclerified. The next inner layers of cortical cells may undergo the same process. In this way, a persistent "root bark" of **storied cork** of cortical origin is formed (Philipp 1923). This type of secondary tissue is derived neither from a vascular cambium nor from a true phellogen (Sect. 21.3.6).

The root cortex of some aquatic plants, both herbaceous and woody, develops an anastomosing system of schizogenous intercellular gas spaces and persists as aerenchyma. This tissue can greatly increase the oxygen supply to submerged roots. It occurs in some mangroves, including *Sonneratia* (Sect. 9.5.2).

17.3 Genesis and Development of Lateral-Root Apices

Roots usually branch by forming lateral (secondary, tertiary, etc.) root apices. A mature oak, as a result of repeated branching, may have as many as 500 million living root tips, mostly of higher-order laterals, which are very small (Lyford 1975), hardly visible to the eye. Yet, because of their vast numbers, they constitute an appreciable fraction of the most metabolically active, or potentially active, tissues in the plant.

A lateral root commonly originates in the pericycle of the parent root, in the region basal to the elongation zone. In general, the more slowly the parent root is growing, the closer to the tip the lateral-root primordia will be initiated (Wilcox 1968). The effect of growth rate on position of lateral initiation is similar to that on differentiation of vascular, cortical, and other tissues in the root (Sect. 17.2). It is possible that, due to an internal "clock" (Sect. 12.4.3), the time required for pericyclic and endodermal tissues to become mature enough to initiate a lateral root is relatively constant. Thus, if root growth is rapid, the pericyclic and endodermal tissues may not initiate a root until the root apical meristem has left them far behind; but, if growth is very slow, a lateral primordium may arise so close to the tip that a dichotomy seems to have occurred. True dichotomous branching in nonmycorrhizal roots of seed plants is rare, however.

We briefly discuss lateral-root development in *Malva sylvestris*, as reported by Byrne and Heimsch (1970a) and Byrne (1973), and then compare it with development in some other taxa. Though *Malva sylvestris* is an herb, the order Malvales includes many woody taxa, such as *Elaeocarpus*, *Hibiscus*, *Ceiba*, *Tilia*, and *Theobroma*. Furthermore, lateral-root development in *Malva* is similar in broad outline to that in the gymnosperm, *Pseudotsuga menziesii*, as sketched by Bogar and Smith (1965). The root apex of *Malva* is also similar in structure to that of *Fagus* (Byrne and Heimsch 1970b).

Primordia of lateral roots of *Malva sylvestris* are initiated in the pericycle, 2 to 3 mm behind the apex for second-order roots, and 1 to 1.5 mm behind for third-order roots. Metaxylem is not yet mature in these regions. The pericycle becomes locally biseriate due to periclinal divisions in three to five cells opposite a protoxylem pole. Because initiation involves several files of cells that apparently have no recent common ancestor, the lateral root probably is not initiated by a quantal event in a single pericyclic cell.

The biseriate pericyclic cells then divide periclinally, forming four initiating layers (outside to inside: IL-1, IL-2,...). Meanwhile, some contiguous endodermal cells divide anticlinally and enlarge radially. Next, cells of IL-1 and IL-2 divide anticlinally and enlarge, whereas those of IL-3 and IL-4 divide periclinally (transverse to the incipient central cylinder) and some periclinal divisions occur in the radially enlarged endodermal cells. Thus, the endodermis also becomes locally biseriate. It then is gradually modified into an **endodermal cover** (Tasche, poche, pocket, etc., in the older literature) of the emerging root.

IL-1 cells at the vertex of the primordium soon divide periclinally, initiating the central part of the root cap. The peripheral IL-1 cells divide only anticlinally, forming a uniseriate protoderm. IL-2 cells meanwhile divide periclinally, initiating a cortex.

Anticlinal divisions in derivatives of IL-3 and IL-4 produce a central cylinder. At this stage, the primordium is still embedded in the cortex of the parent root, beneath the biseriate endodermal cover. Emergence of the primordium through the parent cortex seems to depend primarily on mechanical forces (Byrne 1973). The endodermal cover, which is soon shed, may serve as a temporary root cap while a permanent cap is organized beneath it.

Young lateral roots of *Malva sylvestris* have a "closed" apical structure (Byrne 1973), though the primary root characteristically is of the "open" type (Byrne and Heimsch 1970a). In second- and third-order roots, QCs become detectable as the roots attain 0.5 cm in length. In lateral roots 1-6 cm long, the QC typically occupies about 40% of the width of the root. It includes 100-150 cells and tends to enlarge as elongation growth continues (Byrne 1973).

In some plants, the endodermis contributes substantially to the lateral root, while pericyclic derivatives contribute only the central cylinder (Popham 1955). In *Zea mays*, endodermal derivatives become a de facto rhizodermis of the lateral primordium. Some cells of this rhizodermis divide periclinally, producing a root-cap meristem (Bell and McCully 1970), which in this "closed" apex accounts for all of the Kappe. Similarly, in *Pisum*, endodermal derivatives of the parent root contribute the Kappe. Because *Pisum* has an "open" root apical structure, however, the Kappe may include the rhizodermis and part of the cortex as well as the cap.

Lateral roots are usually initiated in acropetal sequence, proximal to the meristematic zone of the parent root. Some data have been interpreted as indicating that an active apical meristem in a parent root inhibits the formation of lateral-root primordia nearby, but this could be a tissue-age effect related to growth rate (see earlier). Commonly, an existing primordium inhibits the genesis of other primordia in its vicinity. This is more prevalent axially than circumferentially. It is not known whether the axial spacing of primordia is more closely related to elapsed time or to numbers of cells along the axial pericyclic files. The axial spacing of lateral roots varies widely — from 1 or 2 to 20 or more per centimeter (Lyford 1975).

Siting of lateral-root primordia is also influenced by the vascular-tissue arrangement within the parent root. In diarch roots, initiation sites are just abaxial to the protoxylem bundles, or on both sides of the bundles abaxially. Thus, most laterals in these roots appear in two or four meridional ranks on the root surface.

As a lateral root ruptures the parent-root endodermis, local areas permitting apoplasmic transport from the cortex to the central cylinder of the parent root may appear. As the new root differentiates further, its own endodermis becomes linked to that in the parent root. Casparian bands appear in all its endodermal cells, thus restoring a barrier to radial apoplasmic transport (McCully 1975).

As tissues in a lateral apex are differentiating, bridging elements between the parental and lateral vascular systems are also differentiating. The number of vascular bridging elements presumably is related to the size of the lateral apex. The successive orders of lateral roots have decreasing diameters. The highest-order laterals are threadlike.

17.4 Structural Aspects of Indeterminate (Episodic) and Determinate Apical Growth

17.4.1 Long Roots Versus Fine Roots

In most perennial plants, two morphological types of roots are distinguishable: long roots, which grow indeterminately; and fine roots, or rootlets, which mostly grow determinately. Typically, the population of fine roots on a plant is highly dynamic. Many die while many others are being initiated (Persson 1983). In *Juglans regia*, for example, as many as 90% of the fine roots are shed each winter and replaced in spring. In gymnosperms, shedding of fine roots usually is accompanied by anatomical changes somewhat analogous to those occurring during periderm formation in petiole abscission zones. Sometimes, a continuous periderm differentiates on the surface of the vascular cylinder of the parent root, and the cortex is then shed along with the fine roots (Head 1973).

In perennials, the tip of a long root usually elongates episodically, as periods of growth alternate seasonally with periods of dormancy. Both the length and the diameter of the meristem decrease at onset of root dormancy. During growth, established patterns of cell files radiating outward and backward from the proximal border and periphery of the QC maintain the root structure in a near-steady state, though the archy number (Sect. 17.2.1) may change if diameter of the QC and central cylinder changes.

Dormancy in roots is poorly delineated both structurally and temporally, compared with that in shoots. Even the term "dormancy" has been controversial, because in a single tree some roots may continue to grow while others are "dormant" and while the entire shoot system is dormant (Romberger 1963).

17.4.2 Metacutization

In many species, a **metacutis** (Sect. 2.3) differentiates at onset of root dormancy. The **metacutization** process consists mostly of deposition of specialized secondary or tertiary wall lamellae in existing cells rather than generation of new cells. In effect, the exodermis and endodermis extend acropetally, and new, similar barriers are established in the distal and peripheral regions of the root cap. In addition, a bridging barrier may develop between the exodermis and endodermis across the cortex, just basal to the meristematic region. Plaut (1910, 1918) characterized four types of metacutization. However, the "type" is not necessarily a species characteristic. For example, three types of metacutization have been reported in *Polyalthia longifolia* (Sharma and Sharma 1987).

All types of metacutis isolate the apoplasm of the apical meristem, central cylinder, and inner root cap from the apoplasm of the external tissues (Sect. 2.3), which then commonly accumulate dark pigments and become necrotic. Thus, metacutized root tips are usually brown to black, as is exodermis or exposed endodermis, not whitish as are growing tips. In *Calycanthus* and *Buxus*, however, dormant root tips remain colorless, though histochemical tests reveal the presence of typical metacutized barriers (Plaut 1918).

Differentiation of a continuous metacutis, extending from the exodermis through intermediate layers of the root cap, as in *Libocedrus decurrens* (Wilcox 1962), must involve cells in many different lineage files, derived from many different initials. Thus, metacutization is almost certainly induced by local positional and environmental information that overrides earlier influences and determinations.

As a metacutized root breaks its dormancy, the apical region swells, the meta-cutized barriers are ruptured, and elongation growth resumes. Apical parts of the metacutis are usually sloughed off with root-cap cells, whereas basal parts of the bridging barrier that had joined the exodermis and endodermis may persist for a while. The beginning of new growth is marked by an abrupt change in tip diameter, and, usually, also in color.

17.5 Quantitative Aspects of Root Apical Growth

Extension growth patterns differ between root and shoot apices (Ivanov 1983). The entire growing zone may be > 10 cm long in a shoot, compared with < 1 cm in a root. In the root, the shortness of the growing zone minimizes the organ surface that moves relative to the soil.

The cell cycle lasts only a few hours in typical meristematic root cells outside the QC, compared with > 48 hours in typical meristematic shoot cells. This rapid cycling — and the rapid elongation of progeny cells — more than compensate for the shortness of the root growth zone and enable root systems to permeate large volumes of soil quickly. Thus, annual axial growth increments in roots may exceed those typical of shoots. Horizontally growing long roots of *Acer rubrum* and *Betula papyrifera* may grow in length by as much as 1.5 m seasonally (Wilson and Horsley 1970).

Comparing axial growth rates of roots and shoots, however, is much easier than comparing relative elemental rates of growth in volume, because of the structural differences between the two kinds of axes. A root lacks nodes and internodes, and its tip is much more sharply divided into regions of cell division and cell elongation than is a shoot tip. Relative elemental rates of growth in volume as high as 40% *per hour* have been recorded in *Zea mays* roots about 4 mm from their tips (Erickson and Goddard 1951). In contrast, similar measurements made in shoot tips of *Picea abies* seedlings showed growth rates of 15-20% *per day* (Romberger and Gregory 1977).

In a steadily growing root, the meristematic and elongation zones remain nearly constant in length, while the zone of mature cells and tissues continually increases. In some roots, the meristematic cells produce a predictable number of new cells during a given time interval. For example, the central-cylinder meristem of *Zea* includes about 125 000 cells, which leave behind about 170 000 new cells per day in the region of elongation. The expansion of progeny cells such as these is the major force pushing the tip forward. At the same time, in the *Zea* cap, about 6000 meristematic cells produce enough new cap cells to allow about 10 000 per day to be sloughed off without decreasing the cap size (Clowes 1971).

A quantitative description of the root meristem must take into account not only overall cell production, but also site-specific relative cell partitioning rates, P; specific relative rate of extension, E; and mean cell length, L. These three variables, as point functions, P, E, and L, are interrelated such that, if two are known, the third can be calculated (Green 1976). Thus, a descriptive model of root apical growth could be based on the point functions E and L. Erickson and Goddard (1951) presented a one-dimensional version of such a model. Bertaud et al. (1986) and Bertaud and Gander (1986) formulated a more elaborate model, encompassing two- and three-dimensional growth from perspectives both of site-specific variables (referring to positions in space) and cell-group-specific variables (referring to populations of cells and their progeny). Simulation models of cell growth and proliferation based on the latter approaches may exhibit stochastic behavior among the progeny "families" constituting cell files, but, if large numbers of files (determined along a particular developmental pathway) are considered, such behavior is smoothed out — as it would be in an actual root tip — and the models acquire interpretive value.

The rate of three-dimensional growth of a root can also be expressed as a growth tensor (Hejnowicz and Romberger 1984). The tensor can be specified so that its principal directions coincide with the pattern of periclinal and anticlinal cell walls in the apex and that the grid formed by particles aligned along the principal directions of growth trajectories maintain that alignment during growth (Sect. 12.3.2). Two growth tensors of special interest can be formulated for root apices. One expresses a minimum and the other a maximum of volumetric relative elemental growth rates in the topographic initial cells. With these tensors and appropriate computer-graphic techniques, cellular patterns in root apices can be simulated. The tensor formulated for maximal mitotic activity in the initials yields a pattern having a single apical cell, which generates merophytes. The tensor for minimal activity in the initials yields a pattern of lineage files diverging from a quiescent center (Hejnowicz 1989).

17.6 Initiation of Shoot Buds in Roots

Shoot meristems, or "**root-shoots**", arise spontaneously on the uninjured roots of many pteridophytes and angiosperms, but have not been reported in gymnosperms. Among woody angiosperms, root-shoots seem to be more common in taxa of the temperate zones than of tropical and cold zones, though perhaps only because more temperate-zone taxa have been studied. Root-shoots are relatively common on shallow roots of *Ailanthus, Artocarpus, Amelanchier, Aralia, Calycanthus, Castanea, Diospyros, Fagus, Gymnocladus, Paulownia, Pyrus, Malus, Populus, Prunus, Rhododendron, Rhus, Robinia, Salix, Viburnum, Xanthoxylum*, and other genera, but their occurrence is not general in woody taxa.

Root-shoots are also common among herbs, with a scattered occurrence in genera of about 40 families. Root-shoots are almost entirely confined to perennials (Holm 1925), including many troublesome agricultural weeds (Raju et al. 1966). Some of these weeds can survive repeated plowing by sending up root-shoots from severed roots below plow depth or from root fragments within the plowed soil.

The structural development of those root-shoots that have so far been investigated varies widely. In the numerous taxa in which root-shoot meristems occur in intact roots, the most common sites of initiation are (1) the pericycle or a phellogen derived from it, often near a lateral root, and (2) the cortex, often near a lateral root or lenticel. Detailed developmental studies are sometimes required to distinguish clearly between pericyclic and cortical origins. The common association of root-shoot meristems with lateral roots may be related to the disruption of the endodermal sheath (akin to mechanical wounding) when it is stretched or ruptured by the lateral-root apices pushing across the cortex (Priestley and Swingle 1929). This may induce some cells to undergo renewal of meristematic activity.

Root-shoot meristems of pericyclic or inner cortical origin arise in acropetal order. These are not generally considered "adventitious", in contrast to the widely occurring root meristems that arise in callus masses evoked by wounding. In addition, under some conditions, root-shoots arise de novo in groups of proliferating cells at the intersection of a parenchymatous ray and a phellogen — after extensive secondary thickening has already occurred (Siegler and Bowman 1939). Meristems having an outer cortical or a wound-callus origin often are ephemeral, because inadequate vascular connections are established with the parent system before the periderm isolates the cortex and its embedded primordial shoot meristems from the central cylinder.

When and how do cells deep within a mass of root tissues begin to recognize themselves as "shoot" rather than "root", and then behave as though they were superficial rather than deep-seated, assume a new polarity, establish a tunica, and initiate leaf primordia? The seemingly reciprocal question, regarding adventitious root apices on shoots (Sect. 17.7), is less complex, because root apices have a deep-seated origin, not only in the embryo and within the root, but also within shoot tissues. A cavity, or region of loosely arranged cells, may form abaxial to a mass of meristematic cells that is becoming organized into a shoot meristem in a root (Siegler and Bowman 1939; Peterson 1970). Perhaps this cavity is important as a gap in an otherwise continuous apoplasmic system, in that it provides a local superficiality for the developing shoot apex.

Information, on the processes that permit root-shoot development in some taxa and prohibit it in others, might be obtained by comparing structural/developmental aspects of lateral-root formation in taxa that form root-shoots and those that do not. The latter include the gymnosperms and some angiosperms. The question of control of initiation of root-shoots has many aspects of practical importance to forestry, agronomy, and horticulture.

17.7 Initiation of Roots in Shoots

The term "adventitious" is often loosely applied to all roots that arise on shoots (including roots on shoot-borne organs), or out of acropetal sequence on older roots, to distinguish these roots from ones formed from the embryonic radicle or in acropetal sequence on the primary root or its branches. However, in both gymnosperms and dicots, root primordia that are organized early in shoot development and then remain latent for long periods are less truly adventitious than are roots that are newly initiated after a drastic change in the local environment or physiology of the shoot. We follow Girouard (1967a) and Haissig (1974) in referring to these two kinds of roots on shoots as **preformed** and **induced**, respectively. This situation is partly analogous to the reciprocal one of shoots on roots in angiosperms (Sect. 17.6).

Preformed root primordia occur in shoots of many angiosperms and some gymnosperms, both at nodes and along internodes. They may occur singly or in groups, and in some taxa are arranged in regular axial files in predictable locations. In twigs of *Salix fragilis*, root primordia develop from initial cells near the nodes, shortly after the internodes stop elongating. By autumn, they are present near all but a few of the youngest nodes (Carlson 1938).

In stems of some *Populus* species, root primordia may remain latent between the secondary phloem and periderm for many years, until sequent periderms form adaxial to them (Sect. 21.3.3). They then die and are relegated to the rhytidome. Secondary preformed (latent) root primordia may then develop in living tissues beneath deep fissures in the rhytidome. Latent *P. nigra* root primordia may emerge if stems are shielded from light for about six days (Shapiro 1958). Latent preformed root primordia have also been reported in some, but not all, species of *Acer, Pyrus, Ulmus, Populus, Thuja, Cupressus*, and other genera (Priestley and Swingle 1929; Carlson 1938, 1950).

In the relatively few species in which their histological origin has been studied, preformed root primordia in shoots tend to be initiated on the margins of leaf and branch traces, and on the margins of medullary rays. The initials of the primordia near rays are recent derivatives of a young vascular cambium in regions of ray-parenchyma precursor cells. Pericyclic origins, while not unknown, are less prevalent (Priestley and Swingle 1929; Haissig 1974).

A root primordium may arise from a single initial or from a small group of initials. As the primordium grows, it crushes tissues obstructing it. A cavity, possibly formed hydrolytically, may develop around the advancing tip. Analogous cavities have been reported around shoot apices developing in roots (Sect. 17.6). Such cavities may provide isolation from the different root-shoot polarity of the host tissue.

The question of how cells within shoot tissues become "determined" to develop into roots is not yet answerable. Is there a quantal cell cycle and division, which produce progeny that then follow a new pathway? Why is a vascular cambium often involved? When is the root-shoot polarity of the cells reversed? How do hormonal factors differ between primordial and host tissues? Similar questions can also be asked about the initiation of shoot buds on roots (Sect. 17.6).

Induced, as distinguished from preformed, root primordia have mainly been studied in relation to attempted rooting of "cuttings" of stems as a means of propagation. Root primordia typically arise near the basal end of a stem cutting, after the cutting has been incubated in a suitable medium. These primordia generally arise in the same tissues as do preformed primordia in the same species.

Induction of roots in tree branches that have become embedded in forest litter or alluvium ("natural layering") has been less studied than has rooting of cuttings. Most studies of natural layering have been in conifers, in which root induction is a slow process. The histological site varies considerably. In naturally layered *Thuja occidentalis* stems, the root primordia are initiated in the vascular cambium, sometimes in both ray and fusiform initials. In this species and in several *Juniperus* species, the primordia arise from recent derivatives of cambial initials that produce xylem rays that are much larger than ordinary rays and that have a greater proportion of tracheary elements (Bannan 1941b). The root primordia are radially continuous with these rays (see also Sect. 17.9).

Bannan (1942) found that, in *Abies balsamea*, *Picea glauca*, and *P. mariana*, induced root primordia arise from basal regions of latent shoot buds, but found no indications of induced roots in *Tsuga* or *Pinus* species. It is well known, however, that cuttings of very young shoot or hypocotyl segments of some *Pinus* species can be induced to root. Interestingly, root initiation in the *Pinus radiata* hypocotyl begins with enlargement of a single inner cortical cell. Influences from this cell induce asymmetric divisions in nearby cells. The smaller progeny of these divisions form a meristemoid that differentiates into a root primordium (Smith and Thorpe 1975).

It is not clear whether there is a structural basis for the differing rootability of similarly aged shoot segments of the various woody taxa. Although the ease of rooting in some species is inversely correlated with the extent of sclerification of phloem parenchyma and perivascular fibers (Mahlstede and Watson 1952; Beakbane 1961), this association is not widespread (Girouard 1967a,b). Critical factors may be the relative continuity of fiber bands and the timing of their sclerification rather than their mere presence. In *Carya illinoensis* stem cuttings, for example, discontinuous bands of perivascular fibers do not seem to hinder development of induced roots (Brutsch et al. 1977).

Roots that are more truly "adventitious" (i.e., arising de novo) than preformed or induced roots on shoots may arise in callus formed in response to wounding. Typically, these roots arise after a cambium has been organized (Heaman and Owens 1972; Montain et al. 1983) and a condition similar to that at intersections of medullary rays and cambium has been simulated.

17.8 Symbiotic Relationships in Roots

17.8.1 Root Nodules and Dinitrogen Fixation

The term "root nodule" is applied to several types of complex structures formed by some taxa in response to invasion (infection) by endophytic microsymbionts. The terms bacteriorhizal, actinorhizal, cyanobacteriorhizal, and mycorrhizal are used to specify the kind of endophyte that is present. The first three terms refer to dinitrogen-fixing nodules. While recognizing no sharp structural or developmental demarcation between "dinitrogen-fixing" and "mycorrhizal" nodules, we concentrate on the former here and consider the latter in Section 17.8.2.

Long coevolution was required for host plants and endophytes to produce jointly, in a nodule, the structures, conditions, and biochemical and energetic systems necessary to reduce atmospheric dinitrogen and fix it into amino compounds. In legumes (formerly Leguminosae, now Fabaceae), the dinitrogen-fixing, endophytic symbionts are bacteria of the genus *Rhizobium* (Dart 1975). In other angiosperms having dinitrogen-fixing nodules, the endophytes are mostly actinomycetes of the genus *Frankia* (Torrey 1978; Bond 1983). In some gymnosperms, especially the cycads, the dinitrogen-fixing endophytes are cyanobacteria (blue-green algae) of the genera *Nostoc* and *Anabaena* (Bond 1983).

17.8.1.1 In Legumes. Symbiotic dinitrogen fixation involving various species and strains of *Rhizobium* has been widely studied in legumes of agronomic interest (Dart 1975; Postgate 1987). It has been much less studied in the hundreds of species of leguminous trees that are a major component of many tropical forests and occur widely in mesic temperate and semiarid regions (Domingo 1983; Postgate 1987).

In nature, neither the endosymbiont, *Rhizobium*, nor the leguminous host can, by itself, reduce dinitrogen. A key enzyme in the process, nitrogenase, is contributed by *Rhizobium*, and an essential protein, hemoglobin (commonly called leghemoglobin in legumes), can be synthesized only with information from the host genome. Further, the structure and energy-producing systems of the nodule are essential.

The earliest visible host response to *Rhizobium* in the rhizosphere is the curling and branching of root-hair tips. Infection typically occurs in the concave part of a curl or kink. Before *Rhizobium* breaches the cell membrane, the host, by a response that probably originated as a defense, may seemingly attempt to wall off and isolate the invader. The isolating wall, which structurally is an invagination, has the same composition as the root-hair wall. Gradually, this invagination, which is about one-tenth the diameter of the root hair, becomes an **infection thread**. The thread contains the rhizobial cells, which at this stage are outside the host cells.

The thread usually grows along the root hair towards the nucleus of the trichoblast. A "successful" thread grows into the main body of the trichoblast (Sect. 17.2.2), but many threads abort before they reach the base of the root hair. Thus, only a small fraction of initial infection events result in nodule formation.

When a "successful" infection thread approaches the inner tangential wall of a trichoblast, the wall becomes perforated as far as the middle lamella. The walls of the infection thread become cemented to the margins of the perforation, and the middle lamella swells as rhizobial cells migrate into it. This swelling then bulges into a subrhizodermal cell, and the infection thread progresses into that cell. By continuing this process, the thread can, within a day, advance far into the cortex. After reaching the inner cortex, the thread can ramify and produce terminal vesicles. These vesicles, which contain the rhizobial cells, enter the cytoplasm of host cells. By interacting with that cytoplasm, they become **bacteroids** (Dart 1975), as is further discussed later. Nodule initiation, which never occurs close to an active root tip, begins with renascent activity of these cortical cells.

The chromosomes in the nuclei of the trichoblast and other cells through which the infection thread passes undergo endoduplication, and the nuclei may also divide. This may be a response to locally enhanced auxin and cytokinin activity evoked by growth of the thread. Nodule initiation may be influenced by hormones produced by the thread (Libbenga and Bogers 1974), as well as by polyploidy and cell division.

In some legumes, rhizobial infection involves neither root hairs nor infection threads but intercellular, **zoogleal strands**. Intercellular infection tends to occur where cortical tissues have recently been disrupted by emergence of a lateral root (Dart 1975).

The renewed meristematic activity that initiates a nodule soon becomes localized on the abaxial side of the inner cortical cells, where a hemispherical nodule cap of uninfected cells having large, presumably polyploid, nuclei is formed. In effect, a nodule is a short, thick, lateral root, though the nodule is likely to have a cortical origin rather than the pericyclic origin typical of an ordinary lateral root. A nodule usually grows slowly or determinately in length, though it may branch and become lobed. (For details of nodule development, which vary somewhat with the species, see Libbenga and Harkes 1973; Dart 1975).

Nodules commonly become cylindrical or spheroidal, but other forms are also common. Regardless of shape, at their periphery is a "cortex" of large, uninfected, vacuolate, thick-walled cells, which cover the nodular meristematic regions. Beneath this layer is a nodule endodermis, which becomes continuous with the endodermis of the parent root. This layer may also function as a phellogen. In the core are the rhizobia-containing cells, the most highly differentiated cells of the nodule. Between the core and the endodermis is parenchyma with embedded vascular bundles that include both xylem and phloem. Each bundle also has its own endodermis. Surrounding this vascular/parenchymatous inner sheath are densely cytoplasmic pericycle cells. As the nodule grows, its vascular tissues become connected with the vascular cylinder of the parent root. A system of dichotomous branching of the nodular body is also established.

Cortical and other nodular tissues peripheral to the nodule endodermis are pervaded by a system of intercellular spaces, seemingly continuous with the gas space of the soil. The nodule endodermis lacks intercellular spaces and thus is a barrier to gas exchange. In *Glycine max*, for example, the endodermis strongly retards oxygen diffusion into the nodule interior. This barrier, and the involvement of hemoglobin, provide a near-anaerobic environment, thus preventing denaturation of nitrogenase and permitting dinitrogen fixation in the nodular core (Tjepkema and Yocum 1974). In some nonlegumes, in contrast, nitrogenase is protected from oxygen in other ways (Tjepkema 1983).

Even after rhizobia have crossed much of the cortex within infection threads or in intercellular zoogleal strands, they are, strictly speaking, still extracellular. As the vesicles formed by ramification of the terminal parts of the zoogleal strands or infection threads enter the host cells, the rhizobia escape from them. The host cells surround the rhizobia, singly or in groups, with membrane envelopes (Newcomb 1981). Enclosed in these membrane envelopes, the rhizobial cells divide and grow, spreading through the host cytoplasm. These enveloped units become functional bacteroids. They cannot move between cells, but the number of bacteroid-containing cells may increase by mitosis.

Numerous plasmodesmata connect the bacteroid-containing cells, which can fix nitrogen, with neighboring, noninvaded cells. These plasmodesmata presumably transport sugar into the bacteroid-containing cells and amino acids out of these cells. A bacteroid-containing cell metabolizes about 10 mg of sugar in fixing 1 mg of dinitrogen. Transfer areas in cells along the margins of nodular vascular strands may also aid transport (Pate et al. 1969).

Leguminous nodules may persist for several years, but commonly, in leguminous crop plants, they function only a few months. Degenerating nodules are greenish or brownish due to breakdown of hemoglobin.

17.8.1.2 In Nonlegumes.

Most nodule-inducing endophytes of nonleguminous angiosperms are actinomycetes of the genus *Frankia*. Dinitrogen fixation involving *Frankia* has been reported in 200 species scattered among 20 genera in eight dicot families. Nearly all of these plants are perennial shrubs or trees. In several species of *Parasponia* (first reported to be *Trema*), of the family Ulmaceae, *Rhizobium* rather than an actinomycete is the endophyte (Trinick 1979; Bond 1983).

Dinitrogen fixation by actinorhizal nodules is ecologically and economically significant. Not only are the trees, shrubs, and other perennials having these nodules widely distributed, but many of them, such as species of *Alnus, Casuarina, Elaeagnus*, and *Coriaria*, colonize nitrogen-deficient sites, including rocky or gravelly soils, sand dunes, and bogs; others, including *Ceanothus* and *Cercocarpus*, grow on infertile soils in dry forest, chaparral, and subalpine regions (Becking 1975; Torrey 1978). Like the species bearing them, actinorhizal nodules are perennial. They tend to become larger and more complex structurally than the nodules on the widely studied leguminous crop plants.

Formerly, all nonleguminous root nodules that are actinorhizal and contain *Frankia* endophytes were referred to as being of the "*Alnus*" type. However, many actinorhizal nodules are now known to differ in ontogeny and structure from those of *Alnus*. Some authors have used terms such as "*Alnus*" and "*Myrica/Casuarina*" types to distinguish among them (Torrey and Callaham 1978). Other type designations could be used, based, for example, on differences in mode of infection. However, developmental information about actinorrhizal nodules is still too limited to justify broad generalizations about their ontogeny, mature structure, and function.

Early stages of infection of *Alnus* roots by *Frankia* resemble those of infection of legume roots by *Rhizobium*. In response to the presence of the potential endophyte in the rhizosphere, root hairs are deformed (Berry and Torrey 1983). *Frankia* hyphae enter the hairs at deformed sites. The host cells encapsulate them in polysaccharide, forming structures analogous to infection threads in legumes. Within a "thread", *Frankia* filaments quickly invade the cortex, passing through cell walls until stopped by the endodermis, which has walls that differ in chemical contents from cortical walls.

The invaded cortical cells, and neighboring cells, expand and may divide, causing the root to swell locally. The hypertrophy and cell division in this "prenodular" cortex activate an existing lateral-root primordium in the underlying pericycle (Taubert 1956), or induce initiation of an adventitious-root primordium there (Angulo Carmona 1974). As this root primordium grows outward through the parent cortex, it is secondarily infected by the endophyte, thereby becoming a primary-nodule-lobe primordium. The apical meristem of the root primordium produces the lobe, with little further involvement of the parent-root cortex (differing in this respect from leguminous nodules). Apical extension growth of the lobe is soon permanently arrested. Primary-nodule lobes, which show no geotropic response, branch repeatedly and extensively, forming complex, spheroidal, "coralloid" masses (Becking 1975).

Nodule lobes on *Alnus* generally consist of a central vascular cylinder surrounded by a pericycle and endodermis, a parenchymatous cortex, and a rhizodermis. As *Frankia* tends to live only in mid-cortical cells, the fractional volume of infected tissue is small. In contrast, in legumes, the endophyte (Rhizobium) lives in cells near the nodule center. These cells are surrounded by parenchymatous and vascular tissues, which are, in turn, surrounded by an endodermis (Sect. 17.8.1.1).

Newly infected cells just proximal to the apical meristem of *Alnus* nodule lobes are small and contain only the hyphal form of *Frankia*. In the zone of maturation of nodular tissues, infected cells enlarge and the endophyte hyphae fragment into vesicles with complex internal structures. These vesicles occupy a large fraction of host-cell volume. The host cell may enclose them in a membrane envelope (Lalonde and Knowles 1975; Torrey 1978). Dinitrogen fixation probably occurs in these vesicle/-membrane complexes. As in leguminous nodules, hemoglobin is present in actinorhizal nodules, including those of *Alnus* (Tjepkema 1983), and probably helps provide oxygen for metabolism while protecting nitrogenase from oxygen denaturation.

In *Casuarina* (Torrey 1976), *Myrica* (Torrey and Callaham 1979), and *Gymnostoma* (Racette and Torrey 1989), actinorhizal infection commonly occurs through deformed root hairs, as in *Alnus* (Callaham et al. 1979). However, discrete infection threads are not always detectable, and later nodular development differs from that in *Alnus*. For example, in *Casuarina* (Torrey 1976) and *Myrica* (Torrey and Callaham 1978), among other taxa, meristematic growth at the apices of nodule lobes is only temporarily arrested. Eventually, these nodule-lobe meristems produce **nodule roots**, having cells that contain no endophytes. A nodule root typically is slender, has no root hairs, branches very sparsely, has a poorly developed root cap, grows determinately, and is moderately to strongly *apogeotropic* (negatively geotropic). Its cortex is a loosely arranged aerenchyma, and sometimes, as in *Myrica*, a large fraction of the volume consists of "air channels" (Torrey and Callaham 1978).

Though the actinorhizal nodules of *Casuarina*, *Myrica*, and various other genera are perennial, individual lobes and the nodule roots they bear may be short lived. Lateral to the older lobes, a succession of new lobe primordia arise, thus maintaining nodular structure. These new lobes, after a period of apical quiescence, also produce determinate nodule roots.

We know of no reports, among gymnosperms, of root nodulation involving *Rhizobium* or *Frankia*. However, most cycads develop lateral roots modified into multilobed to coralloid nodules that superficially resemble those of *Alnus*. They are not actinorhizal. Though cycad nodules are typically near the soil surface, they can be as deep as 30 cm (Wittmann et al. 1965) and can develop in the absence of potential dinitrogen-fixing or other endophytes (Lamont and Ryan 1977; Webb 1982). In *Zamia*, these nodules are apogeotropic at low light intensities and may function as pneumatophores. At higher light intensities, these structures can foster intercellular invasion of specific tissues by dinitrogen-fixing cyanobacteria (Webb 1982, 1983). Though invasion by cyanobacteria is not essential to early development of the nodules, those nodules that are not invaded tend to be small and short lived, whereas those with high populations of endophytes become several centimeters in diameter and are perennial.

In general, the lobes of cycad coralloid nodules arise in the pericycles of apogeotropic secondary roots close enough to the soil surface to be influenced by light. Though details of nodule development have been variously interpreted, it seems that at relatively high light intensities, such as would pertain at the soil surface, lobe elongation is inhibited and the original root cap of the lobe senesces. However, persistent outer tissues of Kappe origin (Sect. 17.1.4) become modified into a "secondary cortex" overlying a nascent rhizodermis (Milindasuta 1975). The rhizodermal cells then enlarge radially, and intercellular spaces filled with gas or mucoid material appear between them. These spaces are invaded by cyanobacteria (Wittmann et al. 1965; Nathanielsz and Staff 1975), typically *Anabaena* or *Nostoc*. The cyanobacteria multiply, forming a deep-green zone just beneath the "secondary cortex", as seen in cross sections of the lobe (Milindasuta 1975; Webb and Slone 1987). Dinitrogen fixation in cyanobacterial nodules and its distribution to other parts of the plant have been demonstrated in *Macrozamia* by Bergersen et al. (1965). They also noted the significance of nitrogen contributed by a *Macrozamia* understory to some Australian forest ecosystems.

17.8.2 Mycorrhizal Associations

Mycorrhizal short roots (**mycorrhizae**) are manifestations of a form of symbiosis between plants and fungi. These associations are much more widespread than are those involving dinitrogen-fixing nodules. Mycorrhizae seem to be almost ubiquitous, occurring both in natural habitats and on croplands (Harley and Smith 1983). Plant roots provide hospitable sites and accessible carbohydrates for the fungal symbionts, while the fungi increase the absorptive capacity of the roots. Mycorrhizal short roots are especially prevalent on trees. Though each weighs little, collectively they constitute a significant fraction of the total biomass in some forests (Fogel 1983). Because mycorrhizal short roots typically are ephemeral, their development, maintenance, and necrosis account for a large fraction of biomass turnover and nutrient cycling in certain forest stands (Fogel 1980).

The basic concepts of mycorrhizal symbiosis were introduced in Section 3.2.3.3, which emphasizes function in enhancing absorption. Here, we emphasize development.

A mycorrhizal association typically begins near the tip of a young, slowly growing, lateral root. The first host tissues to be involved are usually the superficial, senescing parts of the root cap; the rhizodermis, if present; and the hypodermis or exodermis, if present. Except in the so-called "superficial" ectomycorrhizae, in which the fungal symbiont is mostly confined to the rhizodermis (Malajczuk et al. 1987), subsequent stages of mycorrhizal establishment, and most symbiotic exchange, occur in the cortex. Thus, a mycorrhiza generally lives as long as does the cortex. However, the apical meristems of active mycorrhizae may continue producing cells that differentiate into new cortical cells, thus maintaining a juvenile zone at the distal end of the mycorrhiza while the proximal end is senescing. This acropetal migration along a growing root increases the longevity of the mycorrhiza. It is especially characteristic of endomycorrhizae (Sect. 17.8.2.2).

The development of the mycorrhizae on an individual plant is poorly synchronized. Environmental factors explain some of this variation. For example, mycorrhizal development may vary with differences in moisture availability and edaphic factors at different positions and depths, and with changing carbohydrate supply from the shoot. Further, various kinds of mycorrhizal associations may occur in the same species, and even in the same host plant. A succession of mycorrhizal fungal symbionts may accompany growth of a seedling into a tree (Mason et al. 1983), and there may be a succession of fungal associates or symbionts in what seems to be a single mycorrhiza. Possibly, also, the same fungus can assume the role of symbiont, benign parasite, pathogen, or saprophyte, depending on nutritional or hormonal factors as they vary with root vigor and soil conditions.

Ectomycorrhizae and endomycorrhizae use different means of exchange between symplasms of the symbionts. Both types occur on woody plants.

17.8.2.1 Ectomycorrhizae. Ectomycorrhizae, known for more than a century (Frank 1885), have been studied extensively. They are almost ubiquitous on trees in cool to cold boreal or montane forests and are also common on trees of temperate zones, where there are large seasonal fluctuations in temperature and available water (Meyer 1973). Taxonomically, ectomycorrhizae are much less widely distributed than are endomycorrhizae. Ectomycorrhizae are mostly restricted to trees in a few families, which include many common north-temperate genera. Probably fewer than 5% of woody genera, including most members of Pinaceae, Fagaceae, and Betulaceae, are exclusively ectomycorrhizal. Members of Cupressaceae, Salicaceae, Juglandaceae, Tiliaceae, Myrtaceae (including *Eucalyptus*), and Fabaceae (in part) may be either ecto- or endomycorrhizal; both types may even occur on the same individual. Most other families of vascular plants are endomycorrhizal (Harley and Smith 1983). A very few groups are nonmycorrhizal, notably those having proteoid roots (Sect. 3.2.3.3).

Fungal symbionts of ectomycorrhizae include genera of many families of Basidiomycetes and Ascomycetes, some Fungi Imperfecti, and many unnamed fungi of uncertain taxonomic affinity. On sites where a host species has been growing, residual inoculum of its fungal symbiont(s) is apt to be present in the rhizosphere. The fungal hyphae grow toward polysaccharides and possibly specific fungal attractants released by the root tips.

If a root is growing vigorously (as a long root might), its susceptible, subdistal zone may move forward so rapidly that the fungus cannot establish an effective pre-invasion envelope. If the root is growing slowly, however, the fungus within a few days forms a loose, open hyphal envelope around the tip. This envelope seems to be a prerequisite to invasion. It is different from the dense hyphal mantle formed later (see later).

After another few days or weeks, hyphae from the envelope grow between dead cap cells or between living rhizodermal or cortical cells, often at several points. Presence or absence of root hairs is not relevant. Though there may initially be some lysis of middle lamellae (Foster and Marks 1966; Warrington et al. 1981), hyphae seem to penetrate between host cells mostly by mechanical action, then expand osmotically. They, however, seem unable to penetrate between endodermal cells (Marks and Foster 1973; Warmbrodt and Eschrich 1985b; Kottke and Oberwinkler 1986). Commonly, perhaps generally, the growing hyphae do not separate the walls of adjacent cells at pit fields. Thus, most plasmodesmatal connections between cortical cells are maintained (Nylund 1980; Warmbrodt and Eschrich 1985a). Growth of fungal hyphae between cells may distort cell shapes somewhat, but does not detectably change cell-wall structure (Nylund and Unestam 1982; Duddridge and Read 1984a).

After penetrating the cortex intercellularly, fungal hyphae branch, often like fingers on a hand (Sect. 3.2.3.3), and begin forming a **Hartig net**. This net, intercellular with respect to host tissues, has itself been described as being cellular and pseudoparenchymatous; as consisting of either septate or aseptate anastomosing hyphae (Duddridge and Read 1984a,b); or as composed of masses of partially septate hyphae having labyrinthine walls (Nylund and Unestam 1982; Massicotte et al. 1986). One interpretation, based on in vitro studies of *Amanita muscari* mycorrhizal on *Picea abies*, is that Hartig-net hyphae are mostly aseptate, very profusely branched, and tightly adpressed into a complex mass, and that walls between branches are sometimes misinterpreted as septae. Thus, in some planes of sectioning, the hyphae may seem to be cellular, though they probably are not (Kottke and Oberwinkler 1986, 1987).

In some angiosperms, the Hartig net is primarily confined to spaces between rhizodermal cells (Massicotte et al. 1986). For example, in some *Eucalyptus* species, the hypodermis is a barrier to hyphal penetration of the cortex (Massicotte et al. 1987; Malajczuk et al. 1987).

Collectively, the hyphae of the Hartig net have an enormous wall area in close contact with living cells. As a result, the hyphae, though probably acellular, seem to function as composites of transfer cells, with translocation within the hyphae relatively unhindered by septae. It is possible that the presence of fungal hyphae between cortical cells, rather than evoking a pathological response, induces the host cells to generate signals or signal substances that elicit morphogenic changes (e.g., profuse branching) in the hyphae and, thus, development of the Hartig-net "transfer organ". As has been observed in *Alnus crispa/Alpova diplophloeus* ectomycorrhizae (Massicotte et al. 1986), transfer areas may also form in host cells adjacent to intercellular blankets of Hartig-net hyphae. This dual development marks the genesis of a new organ, the mycorrhiza, adapted to exchange between the symbionts.

The Hartig net does not form synchronously throughout the cortex. After the first-formed parts become functional, growth vigor of the fungus increases and the cortical intercellular spaces become more completely occupied by hyphae. In *Picea abies*, a compact layer (sheath) of hyphae then begins forming on the root surface (Nylund and Unestam 1982). This layer gradually becomes the **mantle**, a characteristic feature of most mature ectomycorrhizae. In some tree species, under some nutritional conditions, the mantle forms prior to the Hartig net (Harley and Smith 1983).

In transections of ectomycorrhizae, the mantle typically appears as several to many irregular layers of small, variously shaped, compactly arranged "cells", forming a sheath or rind of "pseudoparenchyma". As in the Hartig net, these "cells" probably are, in reality, mostly aseptate hyphae, many of which are not parallel to the root axis.

The mantle, which typically extends over the entire root tip (though the Hartig net is absent from the meristematic zone), isolates the host tissue from the soil. Thus, all transport between the soil and the host tissue within the mycorrhiza must pass through fungal-hyphae symplasm or interhyphal apoplasm. A relevant observation is that mycorrhizal roots tend to have a well-developed endodermis, either with Casparian bands as in *Larix decidua* or with a suberin layer as in *Picea abies*, this endodermis blocks apoplasmic transport between the cortex and central cylinder (Kottke and Oberwinkler 1990).

In some associations — for example, *Suillus variegatus* ectomycorrhizal on *Pinus sylvestris* — hyphae in the outer part of the mantle are more loosely arranged than are those in the inner part. Interhyphal spaces and spaces within the mycorrhizal root are occluded by a matrix material (Warmbrodt and Eschrich 1985a), possibly derived from the degradation of root-cap cells. Hartig-net and mantle hyphae, which collectively constitute the **matrical mycelium**, may account for 40% or more of the mycorrhizal mass (Harley and Smith 1983).

The matrical mycelial system, especially its mantle component, is continuous with an **extramatrical hyphal system**, which is essential to mycorrhizal function. Extramatrical hyphae include single hyphae as well as **rhizomorphs**, bundles of hyphae that look like slender rootlets (Zak 1971). Both forms may branch. Dozens to hundreds of extramatrical hyphae typically extend in various directions from the mantle, permeating large volumes of soil and giving the ectomycorrhiza an absorptive capability much greater than that of an otherwise comparable, nonmycorrhizal, short root. The fungus benefits from the extramatrical system's storage depots and capability of organizing sporophores rapidly. Extramatrical hyphae can interconnect neighboring mycorrhizae on the same root system, and possibly also mycorrhizae of adjacent trees.

A dense mantle that covers an ectomycorrhizal root tip may alter normal root development, but usually does not completely arrest apical-meristem activity. For example, in a typical *Fagus sylvatica* ectomycorrhiza, initials at the QC margins (Sect. 17.1.3) continue to produce cells after the tip is encased in a mantle. These new cells grow and push the whole mantle forward. The mantle in turn pulls the outermost parts of the Hartig net with it. This can result in obliquity of transverse walls of cortical cells (Clowes 1951).

In ectomycorrhizae of some *Pinus* species, the original root apical promeristem becomes inactive and is replaced by two new promeristematic centers on its diametrically opposed flanks, resulting in dichotomous branching (but see Faye et al. 1980). This may be repeated several times, forming a cluster of small mycorrhizae of coralloid or nearly tuberculate appearance (Marks and Foster 1973; Piche et al. 1982; Duddridge and Read 1984a). In a few other taxa, the mycorrhizal clusters produce a common mantle or rind and become distinctly tuberculate, with complex internal structures (Zak 1971; Dell et al. 1990).

In some taxa, especially among conifers, a root cap that is covered by a mantle may become very small or disappear (Wilcox 1968). In others, notably *Fagus sylvatica*, cap initials may continue producing cells, which seemingly are digested by hyphae of the inner mantle and do not accumulate (Clowes 1954b, 1981b).

Generally, the "infection" of a root system by a fungal symbiont is permanent, though individual mycorrhizae (both ecto- and endo-) have short lives. Ectomycorrhizae, like root tips, function for varying lengths of time, typically 6 to 24 months (Harley and Smith 1983). They commonly senesce as host cortical cells within them become isolated from the host symplasm by normal developmental processes. At this stage, the fungal symbionts may enter the dying host cells and become parasitic or saprophytic. Nutrients from declining host tissues may possibly be transported through the extramatrical hyphal system and used by nascent mycorrhizae elsewhere.

17.8.2.2 Endomycorrhizae. Unlike ectomycorrhizae, which are mostly restricted to arborescent members of a few families in temperate to cool regions, endomycorrhizae occur worldwide and in most plant groups. They occur in plants on sites ranging from arctic to tropical and aquatic to desert. Endomycorrhizae are especially prevalent in gymnosperms other than the Pinaceae, and in tropical angiospermous trees.

Development of endomycorrhizae produces no conspicuous changes in root morphology. Therefore, the prevalence of these mycorrhizae was slow to be recognized, though they were surveyed and cursorily studied long ago (Janse 1897; Rayner 1926-1927). Intensive study of endomycorrhizae began only recently compared with that of ectomycorrhizae (Sanders et al. 1975), and research data are still fragmentary (Harley and Smith 1983).

The fungal hyphae of endomycorrhizae may be septate or nonseptate. Septate symbionts are mostly confined to the Orchidaceae, Gentianaceae, and some members of Ericales. In the Ericales, mycorrhizae usually have a Hartig net and mantle, like the ectomycorrhizae, yet the fungal symbiont enters the host cells, as in endomycorrhizae. This entry may be confined to late stages of mycorrhizal development, however (Harley and Smith 1983). Most endomycorrhizae have nonseptate fungal symbionts, characterized by intracellular **arbuscules** and by intra- or intercellular **vesicles** that develop within host roots. Mycorrhizae of this least conspicuous of all types are known as **vesicular-arbuscular**, or **VA,** mycorrhizae. We focus on this type here.

In VA mycorrhizae, those fungal symbionts that have been identified so far seem mostly to be nonseptate phycomycetes of the genera *Glomus, Gigaspora,* and *Sclerocystis,* among others, of the Endogonaceae. The taxonomic status of this family, which includes the poorly defined genus, *Endogone,* is controversial. These fungi, especially *Endogone* species, have extremely broad host ranges as mycorrhizal symbionts. For example, *Endogone lactiflua* reportedly forms ectomycorrhizae in *Pseudotsuga* and *Pinus* (Fassi et al. 1969), while *Endogone fasiculata* forms VA endomycorrhizae in *Zea* and *Liriodendron* (Gerdemann 1965).

Anatomical or cytological changes induced by VA mycorrhizal fungi are recognizable only microscopically. Fungal invasion is confined to the rhizodermis, any exo- or hypodermal layers, and cortex, as in ectomycorrhizae. The endodermis and the meristematic region are unpenetrated. But, in contrast to ectomycorrhizae, many, though usually not all, hyphae in VA mycorrhizae may be intracellular, even in early developmental stages.

Infection of angiosperms begins with contact between extramatrical hyphae (derived from spores or existing mycorrhizae) and the root surface. Mode of penetration varies with root anatomy, typically involving both mechanical and enzymatic action. Hyphae may directly invade the hair or body of a rhizodermal cell, or traverse several outer cell layers intercellularly before invading cells. As it invades a cell, a hypha causes the host plasmalemma to invaginate, and, in a manner reminiscent of *Rhizobium* (Sect. 17.8.1.1), grows within a plasmalemmal tube. The fungal wall, however, is always separated from the host plasmalemma by a thin "interfacial matrix".

An infecting hypha may form a loop or "coil" in the first cell invaded, or may pass through several cells before coiling. The mid- to outer cortex develops into a mosaic of cells with and without hyphal coils, which are connected by intra- or intercellular linear hyphae (Bonfante-Fasolo 1984). In *Liriodendron tulipifera*, hyphal coils occupy most of the lumen volume of some cortical cells (Kinden and Brown 1975). The function of these coils of completely walled hyphae, containing a large fraction of the fungal cytoplasm, is uncertain.

Commonly, in angiosperms, once the infection is well established, branches of some intercellular hyphae located between axial cell files penetrate scattered cells in the inner cortex. As in the initial invasions, the host cell wall is perforated and the host plasmalemma invaginates around the invading intracellular hyphal branch. This branch, by rapid and repeated bifurcation, then forms an intracellular tree-like structure, the **arbuscule**. All branches of an arbuscule are surrounded by invaginated host plasmalemma (Kinden and Brown 1975). In some instances, arbuscules arise directly from linear intracellular hyphae, as in *Glomus tunicatum/Acer saccharum* VA mycorrhizae, in which intercellular hyphae are rare (Yawney and Schultz 1990). This direct mode of arbuscule development seems more common in gymnosperms than in angiosperms (Mejstřík and Kelly 1979; Fontana 1985). In both angiosperms and gymnosperms, arbuscules, more than vesicles (see later), are the defining features of VA mycorrhizae.

At the tips of the ultimate branches of an arbuscule, fungal cytoplasm and host cytoplasm, each within its own plasmalemma, attain their closest association. They are separated only by a very thin, interfacial zone. In contrast to the intracellular coils formed earlier, there is no continuous fungal wall to impede, even slightly, symplasm-to-symplasm exchange of polyphosphate granules and other nutrients across the interfacial zone (Cox et al. 1980) Arbuscule branch tips are the ultimate transfer areas in VA mycorrhizae.

A VA mycorrhizal association is dynamic and continually migrating forward with root growth. The life span of an arbuscule is short, probably less than a week. During this time, it functions efficiently in transfer for only a few days (Bonfante-Fasolo 1984). Senescent arbuscules tend to become partly septate, then to degenerate into "clumps", which may be digested as the host cell reverts to its preinvasion state and new arbuscules arise in more distal cortical cells.

Vesicles occur in some VA mycorrhizae, but are lacking in others, such as the *Glomus etunicatum/Acer saccharum* association (Yawney and Schultz 1990). Vesicles vary in size, shape, and location. They probably are depots of reserve lipids and/or glycogen that sustain extramatrical hyphae and fungal reproductive processes after the mycorrhizal cortex dies.

Though VA mycorrhizae, with arbuscules, occur in gymnosperms, including *Ginkgo* (Fontana 1985), *Sequoia* (Mejstřík and Kelly 1979), and *Taxus* (Prat 1926), these are less well known than their angiosperm counterparts. In gymnosperms, compared with angiosperms, coiled hyphae seem to be more common, intercellular hyphae more rare, and vesicles, if present, smaller. The significance of VA mycorrhizae to forest and orchard tree development, as well as to agronomic productivity throughout the world, is probably still underestimated.

17.9 Vascular Cambium and Secondary Tissues in Roots

Research on the differentiation and development of secondary tissues from derivatives of vascular cambium (Chaps. 18, 19, 20) has been strongly biased towards stems. In view of such shoot bias in much of this book, we here briefly focus on the relatively scant information available on secondary development in roots.

Vascular cambium originates differently in roots and shoots. In a meristematic shoot apex or caulis, strands of primary xylem and phloem typically differentiate along a common radius, within collateral sectors of the same vascular bundle ("fascicle", in earlier terminology). A strip of "fascicular" vascular cambium becomes organized from procambium between the xylem and phloem sectors of each bundle. In taxa having strong secondary growth, strips of "interfascicular" cambium subsequently differentiate between vascular bundles, resulting in a complete cambial sheath around the primary xylem.

In contrast, in most roots, collateral vascular bundles such as occur in shoots are atypical, and concepts of fascicular and interfascicular cambium are thus irrelevant. Instead, axial files of protoxylem and protophloem elements differentiate in separate, alternating strands, each along a different radius. These constitute the peripheral tissues of the nascent primary vascular cylinder; the central part of the cylinder is composed of nascent metaxylem.

As already noted (Sect. 17.2.1), precociously differentiating metaxylem of a root tip, as seen in transection, often occupies a narrow rectangular area, or a hypocycloid-shaped area with three, four, five, or more cusps. The nascent, exarchic protoxylem poles differentiate abaxial to the two narrow sides of the rectangle or at the vertices of the hypocycloid. These define diarch, triarch, etc., primary vascular patterns. Protophloem strands differentiate in the provascular tissue abaxial to the long sides of the rectangle or to the cusps of the hypocycloidal metaxylem area.

Vascular cambium differentiates acropetally within strands of procambium, to within several centimeters or decimeters of the root tip. The distance varies with species, root diameter, elongation rate, and, in heterorhizal systems, with root "type" (Esau 1943; Wilcox 1964; Fayle 1968, 1975). As seen in transections at a particular locale, ribbon-like strips of procambium first appear adaxial to the protophloem strands in the cusps (or, in diarch roots, just abaxial to the broad faces) of the metaxylem area and extend acropetally and laterally (tangentially) within tissues variously interpreted as ground meristem, residual provascular, or pericyclic. This "wave" of differentiation, progressing through contiguous tissues, might be termed "contagious" differentiation, in that similar cells are formed as a set or group (Sect. 1.6).

Because of the spatial and temporal variations in the differentiation of the sheath of procambium and later of cambium, this sheath is irregular in shape. As seen in transection, it is a sinuous ring. If it could be viewed from its surface, in three dimensions, it would appear as a corrugated and slotted tube with an irregular leading edge.

In many roots, as already mentioned, cambium is slow to differentiate abaxial to the protoxylem poles. For example, in *Pinus resinosa*, many two-year-old roots still lack a circumfluent vascular cambium (Wilcox 1964). Transections of such roots may reveal two annual increments of secondary growth on some radii and none on others.

The local tardiness in cambial development can result in roots with surface grooves (Wilson 1964b). These often bear axial rows of lateral roots, which mark the locations of strips of meristematic pericycle tissue abaxial to the protoxylem poles.

In *Pyrus communis* roots, which are typically tetrarch or pentarch, the pericycle becomes multiseriate abaxial to the protoxylem poles as lateral differentiation of vascular cambium approaches these poles. The pericycle derivatives near the poles then participate in organizing the cambium. They show a stronger tendency than nascent cambium elsewhere to become ray rather than fusiform initials. In *P. communis*, this results in broad, multiseriate rays opposite the poles, contrasting with the uniseriate rays elsewhere (Esau 1943). A similar pattern occurs in roots of many other dicots (Barghoorn 1940b; Philipson et al. 1971). In some dicots, sectors of secondary xylem between pairs of rays originating near protoxylem poles may include no vessels for several years, though vessels differentiate in other sectors (Wilson 1964b; Fayle 1968).

These observations suggest possible lingering developmental consequences of the heterogeneous origin of root cambium. Pericyclic cells tend to remain diploid and meristematic while radially neighboring cells differentiate and become polyploid (Sect. 17.2.1). Do such differences have any bearing on the seemingly variable behavior of root cambium in different sectors of a root, and different parts of a root system (see later)?

The occurrence of vascular cambium is not as general in roots as in shoots of the same species. The smaller the diameter of a root tip compared with that of its parent root, and, especially, the smaller the diameter of its primary xylem body, the more likely that root is to be shed before it develops a circumfluent sheath of vascular cambium (Horsley and Wilson 1971).

Further, a newly organized cambial sheath tends to be notably smaller in circumference in a root than in a shoot tip of the same species. This reflects both the slender vascular cylinders of young roots and the general lack of pith in roots. Slenderness of root tips can be interpreted as facilitating soil penetration.

Not only may many short slender roots be shed before they begin to develop a vascular cambium or before that cambium becomes circumfluent, but also, once it has become circumfluent, its activity may be less regular than that of typical shoot cambium. In some segments of some roots, however, radial growth increments are produced as regularly as in shoots. This regularity is especially characteristic of the transition region between the stem and large lateral roots, and also of tap roots and other vertical roots. Tap roots and other main roots in some species may have sequences of xylem increments that are regular enough to be useful in dendrochronology (Schulman 1945).

In contrast, radial growth of long horizontal and small subsidiary roots is generally less regular. In these, both in angiosperms (Wilson 1964b) and gymnosperms (Fayle 1968, 1975), xylem increments may be discontinuous or of variable thickness both longitudinally and circumferentially. As a result, serial transections along the lengths of these roots may show eccentricity in various directions (Bannan 1941a; Wilson 1975; Seth et al. 1989). These roots are often not useful in dendrochronology. In *Picea sitchensis*, for example, roots known to be 30 years old have been found to have as few as seven xylem growth increments, and there are long roots in which cambial activity was exclusively on the phloem face for more than a decade (Coutts and Lewis

1983). This is reminiscent of the unifacial production of phloem in leaf veins of various conifers — for up to 30 years in *Pinus longaeva*, for example (Ewers 1982a,b).

The gross structure of secondary root vascular tissues, especially in large tap roots and other vertical roots, is fundamentally similar to that of shoots, though there are differences in proportions of cell types present and in cell dimensions. Such differences are mostly interpretable as morphological adaptations to the divergent functions of roots and shoots.

Root xylem usually has more parenchyma and less sclerenchyma than does stem xylem. In angiosperms, the number of vessels per unit transectional area is usually smaller in the root than in the shoot. Within growth increments, the mean number of cells per radial file is smaller in the root. Differences between earlywood and late-wood, especially in distribution and diameter of vessels in angiosperms, are less pronounced, and thus boundaries between radial growth increments tend to be less well defined.

Xylem cells in roots are also generally larger, have thinner walls, and are less heavily lignified than their counterparts in shoots (Riedl 1937; Patel 1965). Thus, root xylem is typically lighter and mechanically weaker than stem xylem. Specific-gravity differences tend to be largest in ring-porous angiosperms and smallest in conifers (Riedl 1937; Fayle 1968).

Despite the anatomical differences between root and stem wood, a given area of cambium can produce xylem that is root- or stem-like, depending on the local micro-environmental conditions prevailing when the xylem is produced. Typically, the wood of exposed roots gradually becomes stem-like, and that of buried stems becomes root-like. The much higher concentrations of CO_2 in the soil atmosphere than in the aerial atmosphere (except where the "soil" is scree or coarse gravel) may be a controlling factor, though light and mechanical pressure may also be involved. Whatever the cause, developmental changes in the wood are pronounced enough and occur quickly enough to allow their use, along with dendrochronology, to establish dates of floods (Sigafoos 1964) or rates of erosion on slopes (LaMarche 1968).

Control of secondary growth in roots is poorly understood. Some auxin probably is produced by active root apical meristems (Åberg 1957) and is transported basipetally. However, during much of each growing season, far larger amounts of auxin, synthesized in shoot meristems, probably are transported through the cambial sheath toward the root tips. Exogenously applied auxin has been shown to migrate acropetally in cuttings of some woody roots (Fayle and Farrar 1965). It is possible that, in distal parts of root systems, an intermittent acropetal flux of shoot-derived auxin is super-posed on a smaller but more continuous basipetal flux of root-derived auxin. Perhaps root-derived auxin is adequate to maintain minimal cambial function, with slow production of phloem derivatives, and larger amounts of auxin arriving from the shoot are needed to induce bifacial cambial activity, with production of xylem as well as phloem. This superposition of activity evoked by auxin of shoot origin upon a low level maintained by auxin of root origin may, at least partly, explain some of the puzzling aspects of irregular and discontinuous secondary growth in roots.

18 The Vascular Cambium

18.1 Cambium: Terminology, Time, and Behavior

The term "cambium" is applied to two secondary, lateral meristems. These are the "vascular cambium", which produces secondary vascular tissues; and the "cork cambium", or phellogen, which produces phellem and phelloderm. For brevity, we use "cambium" for vascular cambium, and "phellogen" for cork cambium (Sect. 21.2.1).

The **vascular cambium** is a thin meristematic layer, located between the xylem and phloem. By periclinal division, cells of this meristem produce derivative cells that remain, to a greater or lesser extent, arranged in radial files — xylem adaxially and phloem abaxially.

Authors emphasizing developmental and functional aspects describe the cambium as a *uniseriate* layer of initial cells. Authors having a more structural and morphological orientation describe the cambium as a *multiseriate* layer of meristematic, undifferentiated cells (Wilson et al. 1966; Schmid 1976; Catesson 1980; Larson 1982).

According to the uniseriate concept, each radial file of cells includes *one* initial cell, which, by periclinal division, can produce two daughter cells, one of which differentiates (before or after further divisions), while the initial function devolves to the other. Although it is easy to imagine a uniseriate cambium, no known laboratory methods can render cambial initials in prepared section distinguishable, morphologically or cytologically, from their recent derivatives (except some ray initials; Sect. 18.2.1). Generally, from an anatomical and morphological perspective, only a multiseriate, **cambial zone** can be identified. However, in each radial file, a single cambial initial cell (and thus a uniseriate cambium) can be identified by comparing developmental attributes of the various immature cells in the file.

In Section 14.1.2, in the context of a discussion of shoot apical meristems, we explained the ontogenic, time-dependent meaning of "initial cells". Such initials may have no special anatomical-morphological characteristics, and can usually be reliably identified only through developmental analysis. Recall the imaginary model of the shoot apex in which each cell at time zero has a different color. As the cells divide, clones of cells with the same color appear; then, gradually, most colors disappear from the meristem. Eventually, the apex consists of a few sectors, in each of which all cells have the same color. The initial cells are those that gave their color to every cell in a sector. Periclinal chimeras (Sect. 14.1.5) confirm the validity of the imaginary "color" model. In a similar sense, a cambial initial is the one cell that, with time, gives its "color" to a radial file.

Table 18.1. Terminology of the vascular cambium and its progeny: meristematic, developing, and derivative tissues, arranged to indicate actual spatial relations. (Modified after Wilson et al. 1966)

Mature phloem		
Differentiating phloem	Maturing phloem	
	Radially enlarging phloem	
	Dividing phloem (phloem-mother cells)	Vascular cambial zone
Vascular cambium	Vascular cambial Initial (dividing)	
Differentiating xylem	Dividing xylem (xylem-mother cells)	
	Radially enlarging xylem	
	Maturing xylem	
Mature xylem		

With these provisos, we use the terminology proposed by Wilson et al. (1966; see also Iqbal and Ghouse 1990), as summarized in Table 18.1. In this scheme, the cambial zone includes the xylem-mother cells, the cambial initials, and the phloem-mother cells.

In the following sections, we use the term "cambium" often in the anatomical sense of "cambial zone", and sometimes in the ontogenic sense of "cambial initials". If it is not clear from the context, we specify which sense is intended.

18.2 Cambial Cells

18.2.1 Size and Shape

The cambial meristem, in the anatomical sense of the term, usually includes two morphological kinds of cells: **fusiform cells** and **ray cells**. Fusiform cells typically are axially elongated and tapered at each end. Ray cells are usually more nearly cuboidal, but may be moderately elongated in the radial direction. Ray and fusiform cells are the progeny of fusiform and ray initials, respectively. The relative numbers, sizes, and arrangements of the two kinds of initials vary greatly among taxonomic groups.

Figure 18.1 shows the basic geometry of fusiform cells, though most such cells would be relatively much longer, because they are shaped somewhat like flat shoelaces. In a ray-less cambium, a typical fusiform cell would tend to be an elongated tetrakaidecahedron (Fig. 18.1A; Sect. 1.3). The shaded cell shown in Fig. 18.1A, in which periclinal walls are parallel to those in neighboring files, has seven contact facets; seven others are implied but not shown. Most woody plants, however, have rays, and because each fusiform cell ordinarily contacts several ray cells, there are typically more than 14 sides per fusiform cell (Fig. 18.1C). Furthermore, if the periclinal walls of a fusiform cell are not parallel to the walls of fusiform cells in neighboring files, then the fusiform cell may have more than 14 contact facets with other such cells (Fig. 18.1B; Włoch 1981). In Fig. 18.1D, in which periclinal walls are nonparallel to walls in neighboring files, eight facets are visible; the implied total is 16. If the fusiform cells are long and sinuous, the number of contacts can be much greater.

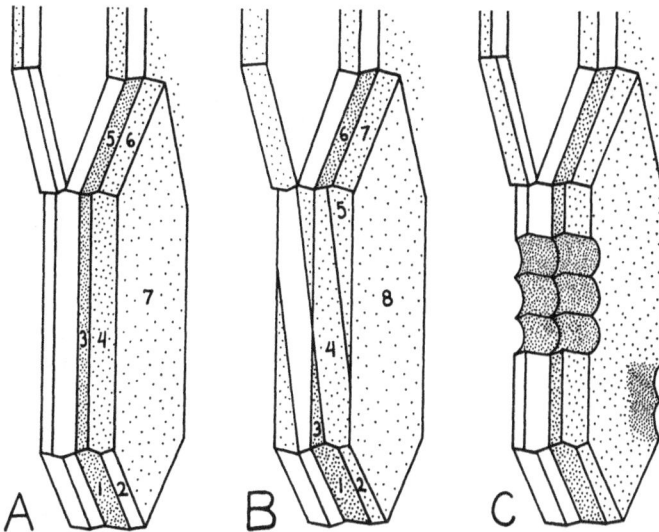

Fig. 18.1. Schematics of models representing the essential geometric features of typical cambial-fusiform cells, except that their axial dimensions are greatly shortened. **A** The cell to the *right* has seven visible contact faces. Of seven others implied by the geometry, six would be visible on the obverse side of the model. **B** The number of contact faces can be increased by nonparallelism between tangential neighbors in the cambium. The cell to the *right* has eight visible faces. Eight others are geometrically implied. **C** The total number of faces of fusiform cells can also be increased by ray contacts (*heavily shaded*)

Mean length of fusiform cells varies, among taxa and within an individual plant. It tends to increase with plant age. After many decades, some long-lived species may attain a "plateau" length, which is typically greater in gymnosperms than angiosperms (Bailey 1923). As senescence approaches, length may decline slowly from the plateau. Even after 2200 years, however, fusiform cells of some *Pinus longaeva* trees have not reached a plateau length (Baas et al. 1986). Mean fusiform-cell length is greater in those angiosperms having primitive vessel members than in those with more advanced ones, and storied cambia typically have shorter fusiform cells than do nonstoried (Carlquist 1975). These relations are summarized in Table 18.2.

Table 18.2. Mean length of adjacent fusiform initials in random samples of vascular cambium in old stems. Species are arranged in order of decreasing length of fusiform initials. (Data from Bailey 1923)

Species	Length, μm
Gymnosperms	
Agathis robusta	6800
Sequoia sempervirens	6600
Larix decidua	4000
Podocarpus nageia	3800
Picea abies	3300
Pinus strobus	3200
Cedrus libani	2900
Juniperus virginiana	2200
Dicotyledonous angiosperms	
Lacking vessels	
Trochodendron aralioides	4400
Drimys winteri	3300
Having vessels, cambia nonstoried	
Altingia excelsa	1900
Dillenia phillippinensis	1600
Gordonia lasianthus	1300
Liriodendron tulipifera	1100
Betula populifolia	940
Barringtonia racemosa	720
Mangifera monandra	570
Carya ovata	520
Acer rubrum	490
Prunus serotina	460
Having vessels, cambia storied	
Diospyros virginiana	410
Kleinhovia hospita	360
Tarrietia sylvatica	280
Grewia multifloria	250
Robinia pseudoacacia	170

In sections tangential to the stem, most ray cells in the cambial zone appear almost quadrilateral and much smaller than fusiform cells. In such sections, ray cells usually appear in lens-shaped groups of ray initials constituting ray initial units (for discussion of terminology see Iqbal and Ghouse 1990), which typically are from 3 to 50 cells high, and from one to ten cells wide. As seen in sections transverse to the stem, a typical ray cell in the cambial zone has a cross-sectional area notably greater than the area of a typical fusiform cell. Ray cells may be moderately elongated radially.

In a transection of a typical woody stem, a radial file of ray cells can be followed abaxially, from mature xylem, across the cambium, into the mature phloem. The cambial zone is recognizable by the thin walls and small radial dimensions of the fusiform and ray cells. Commonly, in a radial file of ray cells in the cambial zone, the arrangement of periclinal walls is such that a radially short cell lies between two longer cells. Comparing cell-wall thicknesses (Sect. 12.3.1) and alignments usually reveals that the short cell and one of its longer neighbors are daughter cells of a recent division. The "mother" cell of that division was a cambial-ray initial. The smaller of the daughter cells usually retains the initial position and function, whereas the larger daughter will become a xylem- or phloem-ray cell (or mother cell), depending on whether it is adaxial or abaxial to the ray-initial cell (Wodzicki and Brown 1973).

A ray-initial cell may elongate radially before its next, unequal division. As a result, at any given time, some ray initials are radially long, and some short. An initial may elongate more on the phloem side or on the xylem side, or perhaps equally on both sides. Thus some may extend adaxially, and some abaxially, from the mean cambial-initial plane (Wodzicki and Brown 1973; Fig. 18.2).

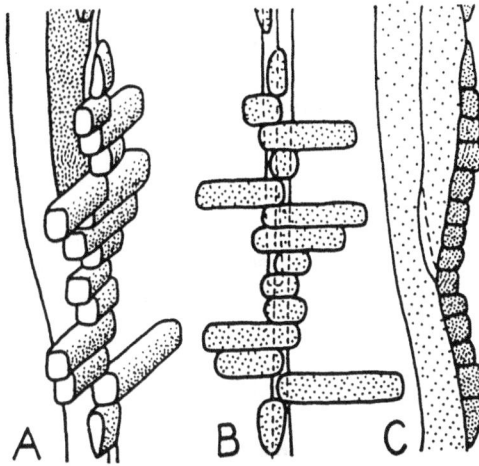

Fig. 18.2 A—C. Drawings of a model of a set of typical ray-initial cells in a Pinaceae stem. Progeny cells are not shown. **A** Three-dimensional perspective view. **B** Radial view (with respect to stem). The *vertical lines* represent the axial course of adjacent fusiform-initial cells. **C** Tangential view (with respect to stem). (Redrawn after Wodzicki and Brown 1973)

18.2.2 Cell Walls and Cytological Structure

The several tangential (T) and radial (R) walls in a radial file of cambial cells range widely in thickness. Even in the same cell, the thicknesses of R and T walls tend to differ. T walls are especially variable. This is usually explained by assuming that, after a periclinal division, each daughter protoplast deposits new primary-wall lamellae not only on the newly formed partition, but also on walls derived from the mother cell. Thus, the daughter protoplast is considered to be "emboxed". In principle, a primary

T wall contains a layer from each periclinal division in which the protoplast participated since the wall was first laid down. The primary walls, though, do not become extremely thick, because the cells are displaced into regions of maturing xylem or phloem, where secondary-wall deposition prevails.

Within a radial file, a characteristic pattern of T wall thicknesses may allow groups of related cells to be identified as pairs, quartets, etc. This may permit identification of cell lineages and of the probable initial cell of the file (Mahmood 1990). However, the development of cambial cell walls may be more complex than the emboxing concept implies (Catesson and Roland 1981). If a cambial-initial cell produces derivatives repetitively on one side, a thick T partition accretes on its opposite side. As growth continues, T and R walls of the cell may become different in structure and composition. T walls undergo almost no expansion, while R walls expand greatly, which counters the deposition of new primary-wall lamellae. In a dormant cambium, T walls are relatively thin, whereas R walls become thickened. In *Platanus*, R walls thicken as mitotic activity declines in late summer. The thickening disappears just before spring cambial activity resumes (Catesson 1980).

Active cambial cells are richly cytoplasmic, and are not wholly undifferentiated. They are obviously different in shape from embryonic ground-meristem and apical-meristem cells, and also differ cytologically from these other meristematic cells. They are more highly vacuolated, have larger mitochondria, and often have more highly differentiated plastids (Catesson 1990). In a cambial fusiform cell, the nucleus is quite elongated, whereas in a ray cell it usually is more nearly spherical.

Ultrastructural changes accompany some seasonal changes in hormonal balance and in general meristematic activity (Catesson 1980). Active fusiform cells commonly have one or two large vacuoles traversed by many slender cytoplasmic strands, and small vacuoles in the peripheral cytoplasm. In autumn, possibly in relation to sugar accumulation, there is a gradual transformation from large to small vacuoles.

18.2.3 Fusiform-Initial-Cell Arrangement: Storied and Nonstoried Cambium

Two major morphological types of cambium occur. These are usually called **storied** and **nonstoried**. The terms "stratified" and "nonstratified" have also been used. In storied cambium, the fusiform cells are arranged in tiers, or stories. That is, the ends of large tangential groups of cells are aligned at the same level of the axis. The ends of cells in axially adjacent stories generally overlap only slightly, making a zig-zag pattern if viewed tangentially. The ends of nonstoried cambial-fusiform cells typically overlap much more extensively, and in a seemingly random manner (Fig. 18.3).

The cambia of some species are "double storied", in that rays as well as fusiform cells are storied (Sect. 18.5.2; Fig. 18.3B). Ray stories always coincide with fusiform cell stories.

A storied arrangement, recognized as phylogenetically advanced, can have evolved only after fusiform initials had become much shorter than the initials of primitive forms. However, not all taxa that have short fusiform initials have storied structure. Storied cambia occur in about 50 families of dicots, but commonly not in all genera of a family. Storied cambia do not occur in gymnosperms.

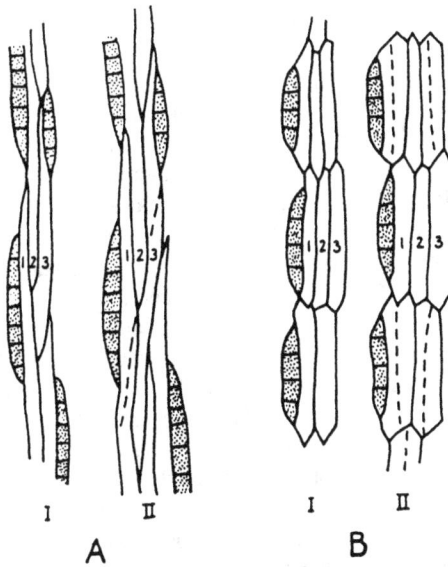

Fig. 18.3. Cell arrangements in small regions of **A** nonstoried and **B** storied cambium, as seen in tangential sections. Drawings labelled **II** represent the same locality of cambium as those labelled **I**, but weeks or months later, after many periclinal divisions (not portrayable in such drawings) and a few anticlinal divisions (*dashed lines*) have occurred among the fusiform initials. Periclinal divisions have little effect on this aspect of cambial development. Anticlinal divisions, however, allow lateral expansion and cell rearrangements. Ray cells are *stippled*

The procambial meristems that give rise to storied cambia are not usually storied. Rather, during cambial development, sets of cells derived from single cells in the early cambium remain in contact and in parallel alignment because of transverse growth and longitudinal, anticlinal divisions. The number of cells in a tangentially extended, tier-like set increases as the cambial girth increases. However, because of displacement of cell ends via intrusive growth and the complex dynamics of anticlinal divisions, cells of different lineages may ultimately be present in the same tier.

18.2.4 Ray-Initial-Cell Arrangement

The local groups of cambial-initial cells that produce rays may be only one cell wide (uniseriate) or may be two or more cells wide (bi- to multiseriate). Rays in most woody dicots are multiseriate. Gymnosperms generally have uniseriate rays, though they sometimes have bi- or triseriate rays beneath lenticels. In addition, gymnosperms may have specialized rays containing resin ducts. In these rays, the width of a group of initials is usually four cells plus the duct lumen. The two innermost initials produce epithelial cells lining the duct. Although duct-containing rays are fusiform in shape, and therefore are termed **fusiform rays**, these rays, like other rays, are derived from ray, not fusiform, initials.

Ray-initial-cell groups also vary in height, or axial dimension, as expressed by cell number. Generally, rays having a height of 12 or fewer cells are considered "low". Multiseriate rays typically are higher than uniseriate rays.

The pattern of ray-initial-cell groups within the cambium is quite dynamic (Sect. 18.4.7), as indicated by changes in ray size. The height and width of ray-initial-cell groups, and of the rays they produce, typically increase for a decade or two in young stems and then stabilize. Environmental changes may also have an effect. For example, release of long-suppressed conifers from competition can evoke large increases in ray height. These increases are consequences of changing population dynamics among ray- and fusiform-initial cells (Gregory and Romberger 1975).

Large aggregations of small rays, termed **aggregate rays**, occur in some genera of dicots, notably *Carpinus* and *Alnus*. The individual ray-initial subgroups within aggregate rays are separated from each other by slender strands of fusiform initials. The derivatives of these fusiform initials are so different from fusiform derivatives outside the ray that the group of aggregate-ray initials, with included fusiform initials, must be regarded as a single population of initial cells.

Despite the continually changing pattern of ray and fusiform initials in the cambium (Sect. 18.4.1), the ratio of the tangential areas occupied by the two cell types is nearly constant within a tree, suggesting that rays, which are major translocation routes between xylem and phloem, are essential to secondary growth of woody stems. In temperate-zone trees, a significant fraction — typically 15 to 25% — of the cambial area seems to be occupied by ray initials. The range may be wider in tropical species, from nil as in *Alseuosmia*, to about 75% in *Dillenia indica* (Iqbal and Ghouse 1990).

18.3 Production of Cambial Derivatives

18.3.1 Periclinal Divisions in Fusiform Cells

Periclinal (tangential) divisions predominate in the cambium. Though it has been considered a "law" of cellular behavior that the area of a new partition wall be minimal (see Thompson 1942), partitions in periclinally dividing fusiform initials tend toward maximal rather than minimal area. This is a radical departure from the behavior of most meristematic cells. Periclinal divisions of fusiform cells are specialized divisions in specialized meristematic cells. Yet, divisions of these cells conform to one of the basic principles of organ growth. This principle is that the orientation of the partition walls be orthogonal to a principal direction of growth (Sect. 12.3.2). In this case, the partition is orthogonal to the radial direction, in which the growth rate is maximal.

Cytokinesis in cambial-fusiform cells is slow because the phragmoplast (Sect. 12.2.1) must migrate a long distance to reach the cell tips. In other highly vacuolate meristematic cells, the site of the incipient partition wall is usually indicated by a plate of cytoplasmic strands, the *phragmosome*, within which the phragmoplast functions (Sinnott and Bloch 1941). In periclinally dividing fusiform cells, there is no phragmosome, although cytoplasm that accumulates in front of the advancing phragmoplast could be deemed a local phragmosome (Evert and Deshpande 1970).

The phragmoplast appears towards the end of mitosis, and separates the telophasic nuclei. It directs the organization of the first part of the cell plate while it expands laterally. The margins of the phragmoplast soon contact the radial walls, usually midway between the tangential walls. By the time this radial contact occurs, a cell plate has formed from aggregated vesicles in the midpart of the phragmoplast. The phragmoplast then has two advancing fronts, which begin their long migrations to opposite ends of the cell. The two sister nuclei usually remain near their original positions while the phragmoplast fronts migrate symmetrically away. In long fusiform cells, the cytokinesis phase of periclinal division may take most of a day, or longer. For example, in *Pinus strobus*, mitosis may take five hours, and phragmoplast-front migration 19 hours (Wilson 1964a). Intervals between successive periclinal divisions may be as short as ten days.

Most migrating phragmoplast fronts eventually reach the cell ends, but some deviate from their narrow course. A deviating phragmoplast may attach itself to one of the periclinal walls of the mother cell. One of the daughter cells thus retains the length of the mother cell, while the other is shorter. If the initial function devolves to the shorter cell, part of the original cambial initial is, in effect, eliminated, and part of the cambial-initial "map" temporarily vanishes. Another variation during phragmoplast migration is a partial rotation, resulting in a partition shaped like a propeller blade (Włoch 1981). Such an originally "periclinal" phragmoplast may become attached to the inner or outer periclinal wall (rather than the radial walls) of the mother cell. This distortion can be important in the rapid changing of grain inclination, because its effect is similar to that of inclined anticlinal divisions (Sect. 18.5.6).

18.3.2 Thickness of the Cambial Zone and Rate of Cell Production

Although active cambial cells undergo repeated periclinal divisions and radial growth, the width of the cambial zone does not increase indefinitely. Conversely, although differentiation of cambial derivatives into xylem and phloem continually removes cells from the cambial zone, the zone does not vanish. If the rates of radial growth and periclinal division are just balanced by the rate of cell loss through differentiation, cambial-zone thickness is constant. However, the balance is often imprecise, and cambial-zone thickness tends to vary during the active season. The rate of production of cambial derivatives depends on the number of cells in the cambial zone and on the duration of the cell cycle. Trees vary greatly in the relative importance of these two factors.

If, in a radial file, N cells are produced during a time, t, then the rate of cell production per unit time is N/t. This rate equals the sum of rates of periclinal divisions of all cambial-zone cells in the file. The distribution of mitoses (determined microscopically) is roughly uniform across the cambial zone (Catesson 1964; Wilson 1964a). Thus, we can assume, without great error, that the rate of division, r, per cell is also similar across the cambial zone (Wilson 1964a). This variable r is also the fractional number of cell divisions that the normative cambial cell undergoes per unit time, and the reciprocal of r equals the time between two successive divisions of the normative cell — that is, $r = 1/t$ and $1/r = t$. Further, if at some moment the percentage of the cambial-zone cells in the process of dividing, either mitotically or cytokinetically, is determined, then the duration of cell division can be estimated. If the percentage of cells undergoing mitosis is determined, the mean duration of mitosis can also be calculated.

Examples of data derived from these relations are given in Table 18.3. Note that tree No. 3 in the table has a cambial zone only about half as thick as the zones in trees 1 and 2; yet, the duration of the cell cycle in this tree is similar to that in the others. Thus, the rate of cambial cell division in these trees does not depend on cambial thickness. A thicker cambium would be expected to produce more derivative cells than would a thinner one. Variations in thickness of the cambium between different parts of the tree, and between different trees in a stand, are important in explaining variations in xylem production.

Table 18.3. Activity of vascular cambium in three *Pinus strobus* trees during May and June. (Data from Wilson 1964a)

Attribute	Tree number		
	1	2	3
Number of samples	16	15	35
Typical number of cambial cells per radial file	13-16	12-14	6-8
Typical number of derivative cells produced per file per day	1.3-1.5	1.3-1.5	0.7-0.9
Percent of cells in mitosis and cytokinesis	10.6	11.7	9.1
Percent of dividing cells that are in mitosis	22.0	18.2	24.5
Duration of cell cycle in days	10.4	9.3	10.4
Duration of cell division in hours	26.4	26.4	21.6
Duration of mitosis in hours	5.8	4.8	5.2

Production of xylem derivatives is typically greater than of phloem (Sect. 20.4.2). For example, in *Carya*, the increment of xylem per growing season usually is three to five times the phloem increment (Artschwager 1950). In general, phloem production predominates early in the season and xylem later (Bannan 1955; Evert 1963a; Alfieri and Evert 1968; Tucker and Evert 1969).

The boundaries of an active cambial zone may be difficult to delineate. In conifers, the cambium reportedly has a relatively steady mitotic index for a long period, and, though mitotic activity usually decreases notably at zonal borders, it is not entirely confined to a narrow zone of cambial cells (Bannan 1962; Wilson 1966). Similarly, in *Acer pseudoplatanus*, the distribution of mitoses in spring is such that one cannot readily distinguish cambial-zone cells from derivatives (Catesson 1974). As expected, the mitotic rate is highest in the cambial-zone proper, where the cells are radially narrow and have thin walls. However, a significant number of mitoses occur beyond this obvious "cambial zone". Where, then, are the "true" borders of the cambial zone? This is a matter of definition, and cannot be resolved here.

18.4 Population Dynamics Among Cambial Initial Cells

18.4.1 A Long-Term Record of Cambial-Initial Dynamics: Wood as an Archive

Secondary xylem is an archive of information about the developmental dynamics of the cambium that produced it. This information is encoded in the dimensions, numbers, and arrangements of the wood cells. Because wood is durable, this information may be preserved for centuries or even millennia. Some similar information is also encoded in the phloem, but because cell arrangements in phloem become distorted after only a few years, and because old phloem is generally degraded and lost, the useful record in that tissue usually is limited to a thin layer. We briefly describe here the basis of decoding the information in the xylem and phloem, so that information can be used to reconstruct the behavioral dynamics of the cambial initials that produced the tissues in earlier years. More details are given in Hejnowicz and Romberger (1973, 1979).

A basic element of the information encoded in xylem and phloem is the arrangement of cell walls as seen in tangential section. Additional information is provided by radial sections, which expose cell lineages. This is because, as fusiform initials rapidly divide periclinally, they compete for space in the cambial layer; some gain and some lose. The gainers extend their tips or edges between others by intrusive growth, while the losers become shorter or narrower. These gradual changes are recorded in the altered shapes and sizes of the successive cell generations in a radial file. The rate of these changes in initial cells, relative to the rate of periclinal division, also is recorded.

In nonstoried cambium, there is, superposed on the record of gradual changes deriving from space competition, a record of infrequent, disruptive events of short duration but permanent consequence. These are anticlinal, oblique divisions in fusiform initials. By such a division, a long fusiform initial becomes two shorter initials, which, almost as soon as their anticlinal partition wall is completed, begin to divide periclinally. Each daughter initial cell thus becomes the head of a new radial file. These two new files replace the older file headed by the mother cell. The slow intrusive growth of the daughter cell toward (or even beyond) the length of the mother cell is recorded in the graded lengths of the cells produced by successive periclinal divisions. The records of these events can be read and dated from sections in appropriate planes.

Interpreting the long-term record of cambial dynamics preserved in wood is easier in conifers than in dicots. In conifers, differentiation of cambial derivative cells into xylem mainly involves limited radial growth, secondary-wall deposition, and autolysis of protoplasts. Intrusive growth of the derivative cells (as distinct from that of cambial initials) is so limited that cellular alignment and arrangement in mature tissues quite accurately reflects that prevailing in the cambium when those derivatives were formed. Cell patterns in serial tangential sections of secondary xylem are analogous to images on successive motion-picture frames. Thus, photomicrographs of successive tangential sections can be used to make simulated motion pictures of some developmental changes in cambium.

In vigorously growing dicot trees, in contrast, much of the record is quickly masked or severely distorted by extensive intrusive growth of fiber tips and extreme lateral intrusive growth of vessel elements. In some taxa, though, the last layer of xylem deposited in a growing season, or the first layer of the next season, does not contain large vessels and is relatively free of distortion. It is a good replica of the pattern of cambial initials that formed that layer.

This "terminal" or "marginal" xylem consists of septate fibers or parenchyma strands that are of nearly the same length as the fusiform initials from which they arose. Typically, each strand is a discrete set of four to eight short cells probably formed by septation of a late-season cambial derivative. Each strand preserves the outline of the fusiform initial cell from which it arose. Thus, in trees in which there is a distinct annual dormant period, a record of the arrangement of cambial initials is left at least once a year. This record includes evidence of ray splitting and uniting events and of elimination or genesis of rays. Study of the marginal layers of a series of annual xylem increments enables one to reconstruct the long-term developmental dynamics of the cambium of these trees. Short-term dynamics can be reconstructed from studies of the cambium itself. In the following sections we attempt to present an integrated view of developmental dynamics of the cambium independent of the way it was studied (see Iqbal and Ghouse 1990).

18.4.2 Anticlinal Divisions in Fusiform Initials

Unlike periclinal divisions, anticlinal divisions in cambial fusiform initials are multiplicative. They increase the number of initials and, consequently, the number of files of progeny cells. Because girth of the cambial sheath increases with stem growth, some anticlinal divisions are essential if the sheath is to remain continuous.

The mechanics of anticlinal division differ between storied and nonstoried cambia. In storied cambia, there is moderate lateral growth of some cells, which then divide **longitudinally**, producing daughter cells of nearly the same length as the mother cells. These then grow in width, and the cycle continues. In nonstoried cambia, in which the fusiform initials are much longer than in storied cambia (Table 18.2), the anticlinal divisions are **pseudotransverse** (oblique). New partition walls formed in these divisions are generally much shorter than the mother cell — often only 15 to 30% as long, compared with partitions that are 80 to 95% of the mother-cell length in storied cambia. Nevertheless, in an absolute sense, the partitions in nonstoried initials may be longer than the partitions in storied initials. In nonstoried cambium, daughter cells grow intrusively in length, eventually attaining the length of the mother cell. They thrust their tips between neighboring cells above and below, seemingly pushing them apart and gradually increasing the cambial area.

Some truly transverse anticlinal divisions occur in fusiform-initial cells, especially in conifers. They tend to be traumatic responses to wounding, and the subsequent intrusive growth at the site of cytokinesis does not cause the transverse partitions to rotate; it only adds characteristic extensions on the right and left as seen in tangential view. In nonstoried cambium of some dicots, a small percentage of anticlinal divisions may be transverse, even in the absence of trauma. Oblique orientations, however, strongly prevail.

The partition formed by anticlinal, pseudotransverse division of a fusiform initial may be inclined to the right or to the left, as viewed in tangential section. These directions are conventionally symbolized as S (left) and Z (right). If these letters were imprinted on the outside of the cambium, their middle strokes would be inclined in the direction referred to. On the basis of many analyses and mappings, we know that S and Z divisions occur in about equal numbers over large areas of cambium, and that they are not randomly distributed. The Z or S chiral type predominates in local areas, or domains (Sect. 18.5.1). We can relate the dynamics of these domains to wave mechanics and to attributes of populations of coupled oscillators (Sect. 18.5.5).

An anticlinal division of an initial cell can be distinguished from an anticlinal division of a derivative cell. This is because division in an initial is usually recorded, in both the xylem and phloem, as a permanent doubling of radial files of derivative cells, whereas an anticlinal division in a xylem- or phloem-mother cell produces only a temporary doubling of the file, and this doubling occurs on only one side of the initial — i.e., in either the xylem or phloem.

A typical conifer fusiform initial divides anticlinally once during the formation of a xylem increment of about 5 mm. Assuming a mean of 25 tracheids/mm in a radial file, we can thus expect one anticlinal division per 125 xylemward periclinal divisions. However, in our experience, only 10 to 20% of periclinal divisions occur in the fusiform initials themselves; the rest occur in their derivatives. We estimate that a fusiform initial will divide periclinally an average of 12 to 25 times on the xylem side for each time it divides anticlinally. The anticlinal division rate is highest in young cambium, especially during formation of the first few annual increments (Brański 1970). In old cambium, the rate is much lower generally, though it seems higher if the data are reported in terms of xylem increments rather than years. In temperate-zone trees, anticlinal divisions tend to occur late in the growing season.

High rates of anticlinal division in cambial-zone cells occur in some anomalous types of wood. For example, Bannan (1957) found that in fluted stems of *Thuja*, the rate was higher (per millimeter of wood formed) in concave than in adjoining, convex sectors of the circumference. Further, areas of cambium producing tumorous wood in *Picea* typically have much higher rates of anticlinal division (both per radial unit of xylem formed and per unit time) than do adjacent areas of normal cambium. The abnormally active areas in this wood apparently begin as single "transformed" initial cells, which, by aggressively competing for space in the cambium, gradually form lens-shaped groups of transformed initials. This cambium is a good model for general study of meristems having "tumorous" initial cells, because the developmental history of the tumor is recorded in the wood, and many dated, long-term records are available.

Although fusiform initials generally divide anticlinally much less frequently than periclinally, nonstoried cambium still has many more anticlinal divisions (followed by intrusive growth) than necessary to compensate for growth in girth. This evokes severe competition for space among initial cells. Only the "fittest" survive. Thus, the control system regulating the rate of anticlinal division also influences the selection pressure on initials.

18.4.3 Intrusive Growth of Fusiform Initial Cells

The active advance of a cell edge or tip into a microspace as it opens along the middle lamella between neighboring cells is **intrusive growth**. It cannot be accommodated by the old concept of sliding growth (Priestley 1930; Sinnott and Bloch 1939), which postulated shearing of plasmodesmatal connections. Rather, the cell seemingly releases enzymes that promote swelling and weakening of the middle lamella, along with dissolution of plasmodesmatal connections in advance of its intruding edge. Though intrusive growth is most extensive and obvious along the thin edges at cell tips, it probably can occur along any fusiform-cell edge.

A precondition for intrusive growth seems to be tension across the middle lamella adjacent to the intruding edge. If there is local tension, no wedging force need be invoked. The intruding edge can advance along the neighboring walls without sliding (just as an amoeba flowing over a substrate does not slide).

In nonstoried cambia, the average local rate of intrusive growth of fusiform initial cells, as determined from serial tangential sections of xylem, can be expressed as millimeters of cell-length increase per millimeter of xylem increment produced. A typical rate in young conifers is 0.3 mm (total for both tips)/mm of xylem produced. Intrusive-growth rates vary widely among individual fusiform initials. Not all tips grow during any one season. Intrusive-growth rates are highest at tips of vigorous initials contacting initials that are being eliminated. Tips of the latter may retract.

Rates of intrusive growth at the two tips of a cell commonly differ. In *Pinus sylvestris*, the rate is generally higher at the apically directed tips. In samples taken from a series of 16 annual increments, cumulative growth was found to be greater in the apically directed tips of 14 increments, and slightly greater in the basally directed tips of two increments (Brański 1970).

In a nonstoried cambium, what happens if the tips of two cells, growing intrusively in opposite directions, meet? Momentarily, there may be a slowing and a slight flattening, as they establish contact. Then, as two earthworms might pass in a narrow tunnel, a direction of overlap is established, with one tip pushing forward on the left, the other on the right. Significantly, the direction of overlap is not random. It generally conforms to the chiral type (S or Z) of the local chiral domain (Sect. 18.5.1).

Likewise, an intrusively growing fusiform cell tip that meets the upper or lower edge (margin) of a ray, usually passes on the right or the left, in conformity with the prevailing domain type. If an intruding tip meets the flank of a large ray, however, it may curve in the direction dictated by the chiral domain type, grow intrusively between ray initial cells, and eventually split the ray.

Fusiform initials can grow intrusively not only at the geometric cell tips, but also, especially in storied cambium, along radially aligned edges that are not the apparent "leading" edges of cell tips or ends. Thus, some obtuse, "non-leading" edges may become acute, "newly leading" edges, and the cell ends become temporarily forked (Fig. 18.4). The non-leading edge that grows intrusively is "selected" from among several edges, such that the direction of growth conforms to the chiral domain prevailing in that locality of the cambium. An appreciation of the intrusive-growth potential of aggressive cell tips and edges is basic to understanding chiral domains and their dynamics in the cambium (Sects. 18.5.1; 18.5.2).

The three sets of cell outlines in Fig. 18.4 represent successive stages in the increasing Z inclination of cells in a storied cambium. In this example, the first stage is intrusive growth along the right-side, obtuse, nonleading, radially oriented edges (normal to the plane of the paper and marked with arrows in Fig. 18.4A). This growth

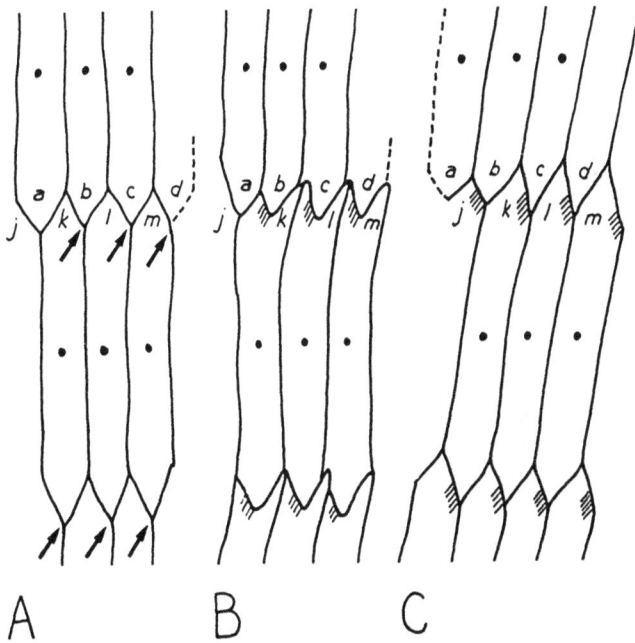

Fig. 18.4 A—C. Diagrams of an idealized example of changing orientation of fusiform-initial cells in storied cambium. A, B, and C represent the same sets of cells (*a, b, c, d* and *j, k, l, m*) at successive temporal stages. Cell inclination is increasing in the Z direction. In A, *arrows* mark nonleading, radially oriented edges. Very active intrusive growth along these edges results in formation of forked tips, as in B. The former nonleading right tips (or edges) then become leading as their left counterparts (*shaded*) retreat, producing the arrangements shown in C. The new, nonleading, radially oriented edges on the *right* (*shaded*) may then repeat the process, through stages similar to those in A and B, as inclination of the cells increases. Thus, during the tenure of the Z domain, the cells slowly rotate in the Z direction about their centers (*black dots*). In an S domain, rotation would be in the opposite direction. Over many years, initial cells may oscillate between S and Z inclinations

causes the initials in each tier to become forked at their upper ends (Fig. 18.4B). Then, the "old" branches of the forked ends are withdrawn. This withdrawal, which may be passive on the part of the disappearing tip, is probably actively driven by the intrusive growth of the adjacent edge. As withdrawal progresses, the cells gradually reacquire simple, pointed tips, though their inclination has increased (Fig. 18.4C). Only the ends of the cells have moved, however; the cell centers have not been displaced. In actual storied cambia (see Hejnowicz and Zagórska-Marek 1974), the synchrony between related events is seldom as close as the figure portrays.

As an initial cell is being eliminated from the cambium, intrusive growth is usually detectable along longitudinal edges of more vigorous, neighboring cells. This growth may, in part, be stimulated by a decrease in the contribution of the failing cell and its immediate derivatives to radial growth of the file. While neighboring files continue to grow, the declining file may come under radial tension. This tension puts tension across the tangential middle lamellae, which in turn favors intrusive growth of adjacent cell edges.

18.4.4 Elimination of Fusiform Initials

Anticlinal divisions in nonstoried fusiform initials produce many more daughter initials than necessary to compensate for growth in girth (Sect. 18.4.2). There is severe population pressure as daughter cells intrude their tips between their neighbors and grow to about the length that their mother cells had at the time they divided. For example, as mentioned in Section 18.4.3, a typical rate of intrusive growth of a cambial fusiform initial in young conifers is about 0.3 mm (total for both tips)/mm of xylem increment; but, because a rate of 0.03 mm/mm of xylem would accommodate cambial girth increase in a stem 15 cm in diameter, the typical rate exceeds the sufficient rate by about tenfold. Because of the resulting, severe competition for space, only the most vigorous and "aggressive" cells can maintain themselves as initials.

The elimination of a fusiform initial is gradual. As an initial lags in competing for cambial space, its length decreases and its vigor declines. According to Bannan (1951, 1953), a declining cell cannot compete with more vigorous neighbors for water. It thus loses turgor and cannot grow normally. However, in our experience, initial-cell elimination is ascribable not only to failings of the "victim" cell, but also to aggression by neighboring cells. Vigorous cambial cells grow intrusively at their tips and also along their axial edges, encroaching on the space of less vigorous cells and hastening their decline.

Periclinal divisions of declining fusiform initials commonly are unequal. Partition walls do not reach the cell ends, possibly in part because encroachment of neighboring cells changes stress distribution and water availability. The initial function commonly devolves to the shorter daughter cell, which never regains the length and cambial area commanded by the mother. Successive unequal divisions can eventually result in elimination.

Most declining fusiform initials are displaced towards the phloem or xylem, and differentiate. Some undergo further divisions as xylem- or phloem-mother cells. The radial files of fusiform derivatives that these initials once headed terminate. Some former fusiform initials survive as ray initials, commanding much less space than they did previously. The transformation of a fusiform into a ray initial is recorded in the xylem and phloem as a merger of a radial file of fusiform cells with a ray.

Some fusiform-initial cells are eliminated during the first season they function. Even when the relative rate of cambial girth increase is maximal, there is an excess of daughter initials and of intrusive growth. The rate of elimination (expressed as number of failing initials, per initial existing at the beginning of the season, during the formation of 1 mm of xylem) is markedly higher in young cambium than in older.

Geometry dictates that, in thick stems, the rate of girth increase per year will be small relative to the existing girth. Accordingly, the rate of elimination of initials in large stems should be close to the rate of anticlinal division, because the net gain in number of initials is very small. This has proved to be a realistic expectation. A study of the fates of 18 fusiform initials in a randomly chosen cambial area in a thick trunk of *Pinus sylvestris* showed that, after two years, 18 initials were still present; though there had been 26 anticlinal divisions, there were also 26 eliminations (Brański 1970).

Other factors being equal, a fusiform initial having no, or only one, ray contact has a lower probability of survival than has an initial with several contacts (Sect. 18.4.6). However, having many ray contacts does not insure survival. For example, in broadleaved trees having large rays, those strands of fusiform cells separating two rays, and having an inclination opposite that of the prevailing chiral domain (Sect. 18.5.2), tend to be eliminated despite other ray contacts.

18.4.5 Changes in Length of Fusiform Cells During Development

Initial-cell length may change during cambial development, due to orderly, predictable, ontogenic processes, or to environmental factors. Change in length of the initials can be detected by analyzing the length and arrangement of cells in the xylem. Cell lengths can readily be measured in macerates of wood samples, and cell arrangements can be studied and analyzed from serial microtome sections in appropriate planes.

In conifers, mean length of cambial fusiform initial cells is closely related to, but commonly slightly less than, that of tracheids. Bailey (1920), after extensive studies of the cambium and xylem of conifers, reported mean fusiform initial length to be only about 1.1 % less than mean tracheid length. Our own studies of conifers revealed that, with few exceptions, cell ends are relatively closely aligned along radial files from the cambium into the xylem. The differences between mean lengths of tracheids and cambial cells are so small as to be ascribable to differential swelling or shrinking during fixation or maceration. Further, measurements in stem samples of various dicots have shown that the tip-to-tip length of vessel members is very nearly the same as that of cambial fusiform cells.

These relations make it valid to use tracheid length in conifers, and vessel-member length in dicots, as approximations of the length of the cambial initials at the time when the measured elements were initiated. Accordingly, year-by-year records preserved in the xylem can be used to study dynamics of length change in cambial initial cells.

Mean length of cambial fusiform cells changes with age of the cambium in the axial segment being examined. Typically, the length gradually increases with radial distance from the pith until it reaches a plateau. The rate of increase is greatest early in cambial development, then diminishes. In dicots, a plateau length is sometimes attained in 5 to 10 years, more usually in 20 to 80 years. In conifers, length increases slowly for 50 or more years (Bailey 1920), and in some trees, for centuries (Sect. 18.2.1).

In nonstoried cambium it can be shown mathematically (Hejnowicz 1967b) that, as a radial increment of xylem is deposited, the mean length of cambial initials changes at a rate that depends on: the intrusive growth of the normative fusiform initial during the formation of the increment; the number of oblique anticlinal divisions per fusiform initial during production of the increment; the fraction of the original existing initials that are eliminated during the interval; the fraction of the mean length of cambial initials representing the mean length of an oblique anticlinal partition; and the ratio of the mean length of the eliminated initials to the mean length of cambial initials. (Cell elimination influences the mean length only if it acts selectively.)

After analyzing the arrangement and dimensions of cells in a xylem increment, one can estimate all of these variables for the cambium that produced the xylem. Mathematical analysis of these variables indicates that fusiform-cell length is directly proportional to the intrusive growth rate and inversely proportional to the rate of anticlinal division (Hejnowicz 1967b). That is, intrusive growth tends to increase the mean length of a population of cambial cells. At the same time, but in a saltatory manner, anticlinal divisions reduce mean cell length. The prevailing mean length is mostly determined by the relative rates of intrusive growth and anticlinal division. Internal or external factors affecting mean cell length presumably operate by modulating one or both of these rates.

In spite of the general relation between fusiform-cell length and radial distance from the pith, mean cell length in a specific population may fluctuate somewhat irregularly. In addition, if examined in detail, the mean length of fusiform cambial cells, in temperate-zone trees, is seen to follow an annual "sawtooth" pattern. This is because anticlinal divisions, which reduce mean cambial-cell length, mostly occur late in the growing season, and intrusive growth cannot then quickly counter the decrease.

After a disturbance, such as a change in environmental conditions, several seasons may elapse before a new equilibrium cell length is established. There may be strict relations between long-term actions of environmental variables and the rates of intrusive growth and anticlinal division.

However, because the mean length that would be in equilibrium with short-term deviations in rates of intrusive growth and anticlinal division cannot be attained quickly, there may be no simple relations between short-term environmental changes (i.e., lasting for months) and the mean length of fusiform cells formed during that same time. If an environmental factor that modulates the rates of intrusive growth and anticlinal division shows a change that persists for only one or two growing seasons, cell length will not be equilibrated to that change during its tenure. In conifers, it is not unusual to find that rates of intrusive growth and anticlinal division vary in annual increments of different widths (Bannan 1965), so that the length of fusiform cells may seem to be far from the equilibrium length, H_{eq}, if that length is calculated from data taken in a very wide or very narrow increment. Because the mean fusiform-initial length approaches H_{eq} only asymptotically, the end of the equilibration period is indefinite and the length of the period cannot be determined accurately.

18.4.6 Transformation of Fusiform Initials into Ray Initials, and Vice Versa

Fusiform-initial cells can be transformed into ray-initial cells, and vice versa. Transformation can be gradual or sudden. During normal cambial development, some fusiform cambial cells are gradually transformed into ray initials by shortening. This occurs mostly by repeated oblique or transverse anticlinal divisions, after which one of the daughter cells fails to undergo active intrusive growth and compete for space in the cambium.

Sudden transformation of a fusiform initial into a ray initial can occur in several ways. A ray initial may originate as a small cell that is partitioned off laterally from a fusiform initial. Alternatively, the septation of a fusiform initial may result in cells that become ray initials. The direct partitioning off of a ray initial from the tip of a long fusiform initial has traditionally been considered a common method of ray-initial genesis. However, this requires a very unequal transverse cell division, which seems to occur only rarely.

In *Liriodendron*, Cheadle and Esau (1964) found that most ray initials arose by gradual transformation. In *Pyrus communis*, though, Evert (1961) found that about 40% of new ray initials arose from segments partitioned off at the end of fusiform initials, and an additional 13% from segments partitioned off from the side; the others arose more gradually.

As mentioned in Section 18.4.4, fusiform initials with few or no ray contacts tend to decrease in length and eventually to be transformed into ray initials. This may be part of a regulating system whereby rays are initiated in cambial areas where their incidence is low, thus keeping the distribution of rays in the cambium uniform.

Transformation of ray initials into fusiform initials is less common than the reverse process. Sometimes, during ray splitting, a ray initial or a small group of these initials is physically isolated (within the cambial-initial layer) from the main body of the ray. These isolated initials may then begin axial intrusive growth, while their polarity changes from radial to axial. They thus gradually attain the shape and function of fusiform initials.

The frequency of transformation of ray initials into fusiform initials can be greatly increased by making oblique or transverse cuts through the cambium. Ray initials adjacent to fusiform initials killed by the wounding grow intrusively, under conditions of decreased space competition. Their tips grow across the lines of cutting, and, as they lengthen, they gradually assume the form and function of fusiform initials. Similar processes are significant in establishing tissue unions between graft components.

18.4.7 Ray Development

Four kinds of developmental change ascribable to dynamic aspects of initial-cell behavior occur in rays within the cambium: (1) increase in ray height and width, (2) splitting of rays by intruding fusiform cells, (3) fusion of rays to form larger rays, and (4) decline and elimination of rays. All these changes can occur simultaneously, even within a small area of cambium.

An increase in ray height usually results from intrusive growth of those ray initials along the upper and lower margins of the ray-initial group, followed by transverse cell division. These divisions are multiplicative, each adding a cell to ray height. Marginal ray cells, because they tend to grow intrusively, are usually taller than their nonmarginal counterparts, as seen in sections transverse to the ray. Rays may also increase in height by incorporating cells at their upper and lower margins. These cells, which originate from neighboring fusiform initials, are usually axially elongated and may maintain that shape through many periclinal divisions. This is another source of heterogeneity among ray cells.

As a result of anticlinal divisions transverse to the stem axis among ray initials, the height of uniseriate rays in conifers may gradually increase over the years. In addition, if a suppressed coniferous tree is released from competition, ray initials may multiply rapidly during the resulting general increase in cambial activity. As a result, mean ray height may increase markedly, though ray number per unit of cambial area does not increase (Gregory and Romberger 1975). Increases in ray width generally arise from tangential (with respect to the cambium) growth of ray initials, and only rarely from lateral additions of cells derived from failing fusiform initials.

A ray may split into smaller rays by several means: (1) by transformation of a ray initial located in the middle of a uniseriate ray into a fusiform initial, (2) by intrusive growth of the tip of a neighboring fusiform initial through the body of a ray, or sometimes (3) by a combination of these two. As discussed in more detail in Sections 18.5.1 and 18.5.2, ray-splitting events generally are of two chiral types, S (left) and Z (right).

Ray fusion in the cambium is essentially the reverse of ray splitting. Typically, fusion occurs between two nearby sets of ray initials, one of which is just above and slightly to the side of the other, with edges overlapping. The two groups of ray initials are separated by one or several strands of fusiform-initial cells. Fusion occurs if these intervening cells are eliminated. These fusions may be either S or Z, depending on the inclination of the fusiform initials that were eliminated relative to a line joining the centers of the two rays.

Two other kinds of ray fusions are nonchiral. One kind occurs between rays axially aligned such that the upper margin of one and the lower margin of the other impinge on the same middle lamella between radial walls of neighboring fusiform cells. As the marginal cells of the rays grow intrusively along this middle lamella, they come into contact and fuse — with no directional component.

A second, rare kind of nonchiral fusion occurs between rays that are side by side and are separated by only a single radial file of (nonchiral) fusiform cells. The fusiform initial heading that file is more likely to be eliminated from the cambium than is a fusiform initial that contacts a ray on one side only. If it is eliminated, the two rays that it had separated will fuse.

Ray initials may be eliminated in ways similar to those in which fusiform initials are eliminated. For example, ray initials originating from declining fusiform initials often are soon eliminated themselves, as the lineage continues to decline. Early phases of ray establishment and development are "high-risk" ventures for the ray initials. For example, of 83 new rays in *Liriodendron*, 25 soon were completely lost, while many others at first declined, but survived (Cheadle and Esau 1964). Even after a ray has reached maturity, individual ray initials or small groups of them may be split off and eliminated, as a result of aggressive intrusion by tips of neighboring fusiform cells. Piecemeal elimination of ray initials is intensified during changes of the locally prevailing inclination of fusiform initials (Sects. 18.5.2; 18.5.3)

18.5 Dynamics Within the Cambial-Initial Layer

18.5.1 Chiral Events in the Cambium

As mentioned in Sections 18.4.2 and 18.4.7, several kinds of cambial morphogenic events occur in left or right orientations. These are **chiral events**. Events inclined to the left are S events; those inclined to the right are Z events. In our experience, chirality in vascular cambium commonly is nonrandom.

The major types of chiral events in the cambium are: oblique anticlinal divisions of fusiform initials, overlapping of intrusively growing cell tips, splitting of rays, uniting of rays, and twisting of periclinal partition walls along the length of fusiform cells (Włoch 1981). These chiral events, except for the twisting of periclinal partitions, cycle between S and Z orientations, in the same cambial locality. The twisting of periclinal partitions is more nearly constant in orientation. Typically, S and Z events, though occurring simultaneously within a sheath of cambial cells, are spatially segregated into local areas called **S domains** and **Z domains**.

In Z domains, fusiform cells become more inclined to the right with time. In S domains, they become more inclined to the left. Once the grain has reached the maximum inclination in, say, the S direction, and is changing back toward the axial direction, chirality of the oblique divisions is already of the Z type, and inclination *change* is in the Z direction. However, actual local inclination is still S, though decreasingly so.

Cyclical alternation in domain type at a cambial locus is a manifestation of migration (usually acropetal) of patterns of domains. Fusiform initials remain in place, while the domains flow through them. Patterns vary widely in scale and in shape of the constituent domains. On a time scale of days and a spatial scale of centimeters or meters, domain patterns seem static. On a time scale of months or years and a spatial scale of fractions of millimeters, domain patterns seem dynamic. In our experience, it is not unusual for domain borders to migrate axially over several fusiform cell lengths and laterally by dozens to perhaps 200 cell widths per growing season.

18.5.2 Chiral Domain Dynamics and Fusiform Cell Inclination

In nonstoried cambium, the cumulative effects of chiral events change the inclination of fusiform cells relative to the stem axis. These events are of two kinds: oblique, anticlinal cell divisions, and meeting and overlapping of oppositely directed, intrusively growing cell tips. Changing fusiform-initial inclination is accompanied by elimination of strands of fusiform-initial cells that separate vicinal rays and are oriented in a direction opposite that of the domain type, with resulting uniting of rays. In addition, there is splitting of rays by aggressive intrusive ingrowth (S or Z) of tips of fusiform-initial cells in the direction of the prevailing domain type.

If pseudotransverse, anticlinal divisions are frequent, domains can be delineated either on the basis of orientation of ray splitting and ray fusion, or of the orientation of newly formed anticlinal partitions. If such divisions are infrequent, then ray splitting and uniting are too infrequent to be useful in delineating domain borders. However, pseudotransverse divisions may still be frequent enough to allow domains to be delineated on the basis of orientations of newly formed partitions. Using photomicrographs of serial tangential sections of xylem to detect and map either the orientation of pseudotransverse divisions, or ray splitting and uniting events, is a laborious but effective method of charting domains and following their migration over time.

In storied cambium, anticlinal divisions of fusiform initials are nearly longitudinal, in contrast to the oblique divisions in nonstoried cambium. Thus, daughter fusiform-initial cells in storied cambium are only slightly shorter than was the mother cell. Also, because intrusive growth is much more limited than in nonstoried cambium, there is little aggressive advance of tips of new cells between their neighbors. Rather, cell inclination changes by migration of the upper tips of cells of one story with respect to the lower tips of those of the story above. Because the upper and lower tips of cells of the same story migrate in opposite directions, the cells rotate slightly about their centers (Fig. 18.4). Tip migration is mediated by *localized* intrusive growth at an appropriate edge near the tip. This usually causes the tip to fork, after which the branch that was the "old" tip withdraws.

In some storied cambia, rays are entirely within stories, as storied rays (Sect. 18.2.3). That is, they are not split at story boundaries. Rays that are arranged in stories do not interfere with migration of cambial-fusiform-cell tips and thus permit the inclination of the fusiform cells to change. This changing inclination is the anatomical basis of wavy and interlocked grain, which occurs in some tropical trees (Hejnowicz and Zagórska-Marek 1974).

The specific way in which the inclination of fusiform initial cells is maintained or changed depends on the frequency of chiral events and the spatial and temporal characteristics of the domains. Four kinds of patterns occur:

1. Frequency of chiral events is so low throughout the cambium that the domain pattern does not affect the inclination of the cambial cells, and straight-grained wood is produced.
2. Frequency of chiral events is high. Domains are well delineated and are of similar width. If the domains are axially short, they may migrate only a millimeter or so per year, forming wood with wavy grain. If domains are axially long, they move rapidly, forming wood with interlocked grain (Sect. 18.5.3).
3. Frequency of chiral events is higher in one kind of domain than in the other, though the relative frequency in the two kinds of domain may change with time and eventually reverse. The inclination of cambial cells changes in the direction of the type having the higher frequency. If the higher frequency is still relatively low, the effect may be a slow increase in inclination in one direction for some decades, before a reversal, if any. This leads to spiral grain. Over several decades, S spiral grain may gradually become straight, then Z spiral, or vice versa.
4. Noncycling chiral events predominate, and grain inclination develops in the direction (S or Z) of these events (Sect. 18.5.6).

18.5.3 Domain Dynamics in Cambia Producing Wood with Wavy or Interlocked Grain

The dynamics of cambial domains are complex. Two components may be distinguished: (1) unidirectional, slow movement of the whole domain pattern, which we call

domain-pattern migration; and (2) **pulsation of domains,** in which domain borders advance and retreat relative to the domain center.

Migration (usually acropetal) of a pattern of S and Z domains is common in cambia with a high frequency of chiral events, in both gymnosperms and arborescent dicots. The oscillation of inclination of fusiform initials in a particular locality is a function of this migration of domain patterns and is the anatomical basis of **wavy, interlocked,** and some forms of **spiral grain.**

We could understand the spatial arrangement of domains and their changes over time more clearly if we could make ourselves very small and, first, rapidly move acropetally in the cambium, and then remain at a single cambial locus for a number of years. As we moved acropetally, we would find ourselves passing through domains of alternating types — S, then Z, then S, etc. That is, we would see reversals of domain type along the cambium at a fixed time. The domains would be arranged as alternating bands, each tangentially oriented in the plane of the cambial sheath. Then, as we sat on a radius to observe a single cambial area year after year as it pushed us outward, we would see a cyclic change of domains (S, then Z, then S, etc.), as they passed through our cambial locus with time.

We can best interpret these characteristics of migrating domain patterns in terms of wave phenomena. Even if we do not understand the biological basis of the waves, it is reasonable to propose that waves of different behavioral attributes are propagated through the cambium and are manifested by nonrandom distribution of S and Z types of morphogenic events in the fusiform initials. The alternation of domain types axially along the cambium at a fixed time represents the spatial aspect of the waves. The change of domain type at a specific cambial locus represents the temporal aspect.

"Waves" visible in wood are not true waves in a physical sense. They are actually static. They are archival records of biophysical and physiological waves that were a part of the control systems of the cambium when the wood was formed.

On a smoothed tangential surface of "wavy-grained" wood, the fibers or tracheids are seen to be arranged in axially alternating, tangential bands of S and Z inclination — as already described. This surface has well-defined areas in which the grain is inclined to the right or to the left of the mean axial direction, giving the grain its "wavy" appearance. In contrast, on a radially split surface of this wood, exposing the record of many years of xylem accretion, one sees a different kind of "wave": three-dimensional undulations. The undulations slope upward from older to more recent increments (see later), indicating the upward migration of domains over time. Thus, "wavy"-grain patterns on a tangential wood face show the spatial distribution of domains at an earlier time, whereas undulation patterns on a radially split face have a temporal component in the radial direction and a spatial component in the axial direction. Radial-face undulation patterns are records of systematic domain migrations during the years in which the wood was formed (Fig. 18.5). The "wave" patterns on tangential and radial faces of small blocks of wavy-grained wood are represented in Fig. 18.6.

As seen in a radially split sample of "wavy"-grained wood, the small angle formed by the fronts of the undulations with the radii and thus with the xylem rays indicates the rate of upward migration of the domain pattern. That is, the domain pattern that underlies the "wave" pattern migrates acropetally within the cambium (Fig. 18.6), and

Fig. 18.5. Twelve-year series of maps (based on tangential sections) of distribution and frequency of S and Z ray splitting and uniting events occurring in the same cambial locality in a wavy-grained sample of *Fraxinus* wood. Nonrandom distribution of marked events in a map indicates chiral domains. *Map 1* records the events occurring during the formation of the 2nd annual ring in the sample (based upon comparison of rays in the 2nd to those in the 1st terminal-xylem-parenchyma layer). Successive maps record events of successive years. *Left* and *right semicircles* denote ray splitting and ray uniting events, respectively. *Solid semicircles* indicate S events, *open semicircles* Z events. Because of lagging effects of prior domains on current grain inclination, S and Z grain-inclination zones (as marked by *near-horizontal lines* and *by letters* along the *right edges* of the maps) do not precisely coincide with the mapped chiral domains. Groups of rays (*small vertical dashes*) used as position indices are indicated on the maps in their respective positions. Note that the pattern migrated upward by more than one mean domain height (about 3 mm) in 12 years. (Hejnowicz and Romberger 1973)

the time during which this migration occurs is represented by annual increments of xylem counted along a radius of the stem, represented by a vascular ray. The faster the domain pattern migrates, the steeper is the inclination of the undulation fronts seen on a radial surface. The mean angle that these fronts make with a radius in relation to the increment of xylem formed per year is a record of domain migration rates.

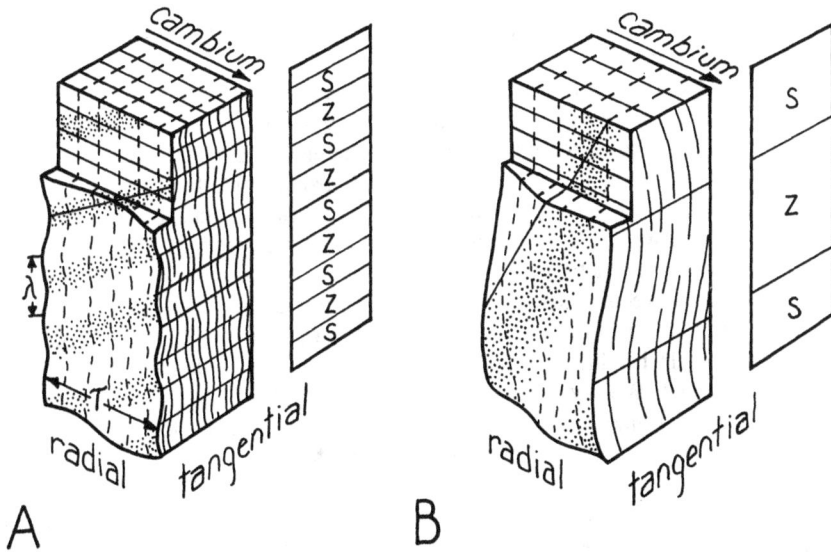

Fig. 18.6 A,B. Perspective drawings of blocks of wavy-grained wood. As the wood accreted, the cambium moved in the direction of the *arrows*. The wavelength is λ, and τ indicates the period in years. Panels to the *right* of each perspective drawing represent domain maps of the tangential faces of the wood blocks. **A** The gently sloping undulations of the radially split face are typical of wavy-grained *Fraxinus* wood and indicate slow upward migration of domains. **B** The more steeply sloping undulations on the radial face of *Platanus* wood indicate more rapid upward migration of domains. (Redrawn after Hejnowicz 1980)

The various attributes of wavy, interlocked, and some kinds of spiral grain patterns can be considered as archival data stored in the wood. They are static records left by dynamic, chiral domains in the cambium. In wavy grain, the angle of grain inclination changes cyclically, and, because the domains are axially short, the variable inclination is obvious on tangential surfaces, and more so in the gently sloping undulations on radial split surfaces. In interlocked grain, the domains are axially longer, and their manifestations as undulations on radial surfaces have steeper upward slopes. Thus, interlocked grain can be interpreted as wavy grain with very steep wave fronts (Hejnowicz and Romberger 1979). On smoothed tangential surfaces, interlocked grain appears as "locally spiral" grain, but, in contrast to true spiral grain, the direction of the spiral changes cyclically in successive groups of annual increments (see figures in Krawczyszyn and Romberger 1979). Wavy- and interlocked-grain patterns differ only quantitatively.

In most tree stems investigated, the frequency of S and Z morphogenic events in the cambium is quite balanced and too low to evoke obviously wavy or interlocked grain. However, in a wide range of species, small percentages of individuals have a high frequency of chiral events and show pronounced wavy or interlocked grain. Such individuals occur, for example, in *Fraxinus*, *Acer*, *Prunus*, and *Quercus*. In some other genera, most or all individuals typically have a high enough frequency of chiral morphogenic events to evoke interlocked grain, as in *Platanus*, *Liquidambar*, and *Nyssa*, among temperate-zone genera, and in *Entandrophragma*, *Swietenia*, and various other tropical genera.

By studying cambial ray splitting and uniting events or the orientations of cambial anticlinal divisions, as recorded in the wood, we and our associates have learned that, in general: (1) the axial height of domains varies among species over a 1000-fold range (from millimeters to meters); (2) the greater the domain height, the faster the domain pattern migrates axially; and (3) domains of widely differing height and migration rates may coexist in a cambium, and, by their superposition, may produce grain patterns of great complexity and beauty.

The relation between domain height and domain-migration velocity is such that the tenure of one kind of domain in the locale of a particular cambial initial does not depend on domain height. That is, the period of tenure of a domain seems to be relatively invariant. We say "seems to be", rather than "is", relatively invariant because both calendar time (years) and amount of xylem produced (radial increment), as measures of morphogenic time, may be only approximate measures of the actual "biological time" that prevails in the cambium.

To justify calling the period of tenure of a domain "relatively invariant", we must compare the radial distance between grain inclination maxima with the axial distance between undulation crests. In trees of different species or in different individuals of the same species, the number of annual increments (counted along a radius) between grain-undulation crests on a radially split surface usually ranges between 10 and 20. The number of annual increments between undulation crests is an estimate of the time needed for one full cycle of domains (one S and one Z) to pass through a particular cambial locality. In these same samples, the linear distance, measured along a radius, between undulation crests usually ranges between 2 and 5 cm. Thus, when estimated either as years or as thickness of xylem produced, the developmental time between undulation crests varies within a range of 2- to 2.5-fold. It is as though the inclination reversals underlying both wavy and interlocked grain were the record of some long-term endogenous cycle typically having periods of 10 to 20 years (Hejnowicz and Romberger 1979; Krawczyszyn and Romberger 1980).

If we measure the axial distance between undulation crests (that is, along a single annual increment of xylem) in a variety of wood samples, we find values from about 2 mm to 2 m or more. The range may thus be greater than 1000-fold. Compared with this 1000-fold variation, the 2- to 2.5-fold range of variation in radial distance between crests, which is a time-dependent distance, is "relatively invariant". That is, in general, in wood having wavy or interlocked grain, the radial distance between wave crests tends to vary within a narrow range, while the axial distance varies widely.

The possibility that several chiral waves are superposed in the same region of cambium is the only plausible explanation for some phenomena. Consider, for example, the complex wavy-grain patterns of some samples of *Guibourtia ehie* (Fig. 18.7). In the cambium that produced the wood shown in Fig. 18.7A, domains at level *d* were shorter and more slowly moving than those at level *e*. Though waves differing in length and velocity were relatively well separated spatially in this cambium, some superposition is evident. Below level *e*, note that, aside from steep undulations, there are others that are less distinct and less steep. In another sample of wood of the same species (Fig. 18.7B,C), the superposition of different undulations in some regions is evident. We infer that, in the region of *e*, a single wave was moving upward (as the wood was accreting on the right). However, the checkered pattern near *d* can be in-

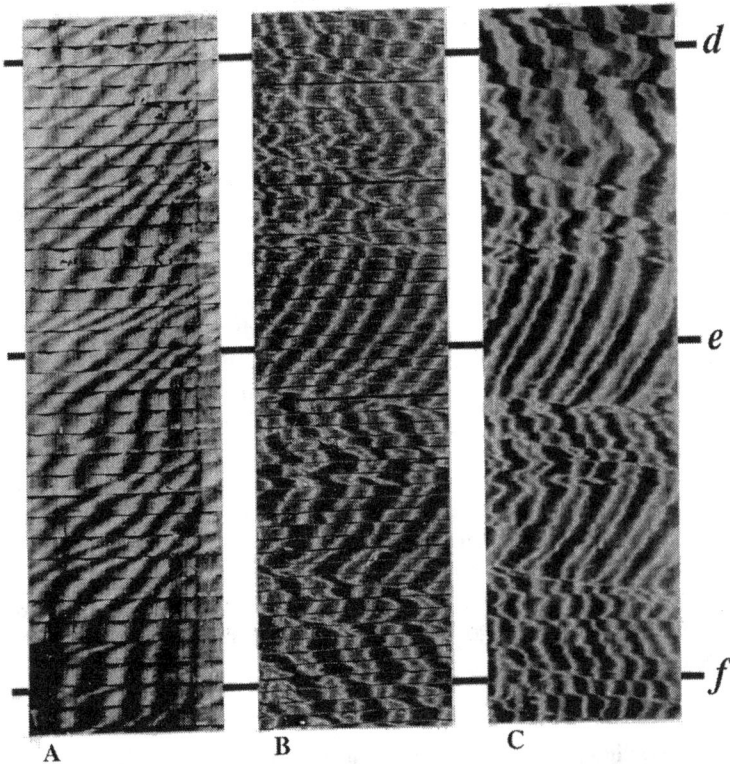

Fig. 18.7. Complex wavy grain figures on radial faces of wood of *Guibourtia ehie*. **A** and **B**. Undulations on radial split faces were revealed by cutting sawn boards into transverse sticks and splitting each stick radi-longitudinally. The split sticks were then arranged in their original orientation and sequence. **C**. The same board as in B, but with its surface photographed before cutting into sticks. The grain figure is visible even in the planed surface because differences in grain inclination evoke differences in light reflection. The checkered and moiré areas visible in **C** assume a different appearance after splitting (**B**). The cambial side is on the right side in each photograph. Depicted portions of the boards are about 1 m long and 26 cm wide. For further interpretation see text. Adapted from Hejnowicz and Romberger (1979)

terpreted as being derived from the superposition of two chiral waves in the cambium. These waves possibly had lengths and velocities similar to those of the single waves at e and f, but were moving in opposite directions. The resultant would be somewhat like a standing wave (Hejnowicz and Romberger 1979; Sect. 12.4.3). The reader is invited to seek other plausible explanations for these patterns.

The concept of cambial waves that are superposable and have near-constant periods, but have different lengths and velocities, offers a plausible explanation for a range of chiral phenomena in cambium. Even the pulsation of heights of domains can be explained in this way.

Certainly, one of the most intriguing properties of chiral waves in cambium is their long periods — typically 10 to 20 years! How can a quantity or value in an organism oscillate so slowly and precisely that a small but systematic difference in phase between successive points is maintained over the years? We have no answer.

18.5.4 Chiral Events in Cambia Producing Wood with Straight or Spiral Grain

In most tree species investigated, stems with straight- or spiral-grained wood are much more common than those with wavy- or interlocked-grained wood. This indicates that, in these species, cambia having low frequencies of chiral events are more common than ones having high frequencies.

However, even in stems in which chiral events are so infrequent that domains can hardly be mapped, these events are still nonrandomly distributed within the cambia. This nonrandomness of low-frequency chiral events has been observed in nonstoried cambia, and especially in storied cambia, in which groups of cell ends may undergo chiral "creeping" (Sect. 18.4.3) along story borders (Zagórska-Marek 1984). These groups of cell ends may not exhibit the domain migrations described in the preceding section. Rather, their chirality changes in a manner that may be described as a flickering between S and Z (Włoch and Bilczewska 1987). Nevertheless, in some taxa or individuals, migrating domains can and do occur in cambia producing straight-grained wood. We suggest that, in cambia producing spiral- and straight-grained wood, there may be a continuum of modes of change of domain type (not necessarily in the same individual), with flickering change at one extreme and regular migration of alternating types at the other.

18.5.5 An Hypothesis About Wave Dynamics and Chiral Events

To help understand the various patterns of chiral domains and their differing rates of migration, we can consider phenomena resulting from coupling of oscillators (Winfree 1980). Three such phenomena are of special interest:

1. Phase coordination upon coupling can lead to synchronization or to a phase gradient. If synchronized, all oscillators are in the same phase and the system behaves as a single oscillator. A phase gradient implies a wave with a front normal to the gradient direction.

2. Coupling may result in an overall oscillation period longer than that of any component. In combination with a phase gradient, this phenomenon can yield a moving wave with a long period.

3. Coupling may result in a phase shift in some of the oscillators; that is, an oscillator is either advanced or delayed by a perturbing factor concomitant with coupling. The magnitude and sign of the phase shift may depend on the phase prevailing when the factor first acts. Such shifts make possible an abrupt change of phase and thus an abrupt change of duration of the cycle in which it occurred. Phases of neighboring oscillators may become coordinated in this way.

With reference to phenomena such as these, we propose the following hypothesis.

A population of cambial initial cells acts as a system of coupled oscillators. The unit oscillator is a small group of cells or cell ends, not a single cell or cell end. The oscillation pertains to an unknown behavioral attribute or process within the cell group. This process is manifested in two chiral types of events, but need not be limited to only two states.

The tightness of coupling of the unit oscillators affects the phase and period prevailing in the system as a whole. The tightness of coupling in turn depends on the amplitude of the oscillations — that is, on frequency of chiral events, with high frequency and amplitude favoring tight coupling. Tight coupling evokes either synchrony, which leads to interlocked grain, or a constant gradient of phase (that is, a linear change of phase with distance), which produces wavy grain. The steepness of the phase gradient determines the wavelength, but does not affect the period; therefore, it also determines the wave velocity. The less steep the gradient, the longer and faster the wave. Tight coupling increases the duration of the oscillation cycle, due to "drag" and inefficiency as the fast oscillators are retarded by the slower ones. The migrating morphogenic waves of chiral phases in the cambium then have relatively long periods.

If coupling is loose, neighboring oscillators may be in different phases, and may display the same chirality some of the time, though their chirality does not change synchronously. Consequently, a cambial domain characterized by a particular chirality may change its shape and size with time. Rapid cyclical changes in chirality and in area of S or Z domains may be the basis of the flickering chirality in some cambia producing straight-grained wood. We cannot exclude the possibility of factors that can shift the phases of neighboring groups of oscillators to the same value, thus uniting them, or shifting part of a large group of oscillators to a different phase, thus splitting them into two groups.

According to our hypothesis, groups of chiral oscillators in normal cambium producing straight-grained wood are loosely coupled. Such cambium is also characterized by a low frequency of chiral events. In contrast, cambium producing wavy-grained wood has a high frequency of chiral events and its oscillators are tightly coupled.

18.5.6 Noncyclic Chiral Events

Not all types of chiral events show patterns of cyclic change between S and Z. A predominance of chiral events of one type occurs during the formation of spiral grain.

Harris (1989) considered that spiral grain may be evoked by a chiral cytoplasmic stress which determines the direction of skewing of the mitotic spindle during periclinal divisions. Our studies, which included extensive analysis of serial transections of old conifer xylem, revealed that complicated distortions of radial rows of tracheids occur near tracheid extremities. We concluded that such distortions result from pronounced twists of certain periclinal partitions. These twists in *Picea* and *Larix* are of the S type regardless of the chirality of the anticlinal partitions themselves (which change cyclically). Such twisted periclinal walls were described by Włoch (1981): an originally periclinal partition may become increasingly anticlinal in orientation as it twists near the cell ends, like the blades of an aircraft propeller. It becomes attached to one of the periclinal walls of the mother cell. As the twisting increases towards the cell ends, the attachment site becomes closer to the center of the periclinal wall. Within a stem, chiral twisting of this type may be either S or Z (Sect. 18.3.1), but there is no evidence of cycling.

Because xylem production involves numerous periclinal divisions, the cumulative effect of even slight S or Z twists in periclinal partitions may cause the grain to become inclined in that direction. S twisting of periclinal partitions is common in coniferous wood and could be a basis of the frequent development of S spiral grain in young conifers, as in *Larix* and *Picea*. The twisting of periclinal partition walls may produce effects similar to those of oblique, anticlinal partitions, directing subsequent intrusive growth and thus contributing to inclination of cell axes which is characteristic of spiral grain (Kubler 1991).

18.6 The Origin of Vascular Cambium, with Its Ray Pattern

The vascular cambium usually, but not always, has a dual origin within the primary caulis: from provascular strands, and from the "ground"-meristem tissue between those strands. These two modes of origin are termed **intrafascicular** (within fascicles) and **interfascicular** (between fascicles). In most stems with spiral phyllotaxy, provascular strands are commonly so compactly arranged that, very early in stem development, they form a primary vascular cylinder, dissected only by leaf gaps.

As we use the terms, "provascular tissue" is the precursor of all vascular tissue (Sect. 14.3.1), and "procambium" is that part of the provascular tissue that is the precursor of the vascular cambium (which may also produce some metaxylem). The communities of meristematic cells to which the terms "provascular tissue", "procambium", and "cambium" refer are not sharply delineable from one another; nevertheless, these terms are all valid and useful.

In an intact plant, procambium grades continuously into cambium, both temporally and spatially, although use of the most convenient sampling techniques is apt to reveal only parts of the vascular meristem that are widely separated in space and in developmental time. Larson (1976) denoted the transitional stages between procambium and cambium as **metacambium**.

Both procambium and its successor, metacambium, differentiate acropetally within provascular bundles. In transections of *Populus*, metacambium can first be detected by a series of radial, anticlinal divisions in laterally extended sets of tangentially aligned cells. Further divisions in this layer are mostly periclinal, producing metaxylem and metaphloem (Sect. 14.3.2.2). The cells between the metaxylem and metaphloem eventually begin to function as cambial initials.

How do cambial initial cells come to consist of two morphological types — axially long, slender, fusiform cells, and axially short, blocky, ray cells? Procambium at first consists of short cells, from which longer cells may arise in two ways (Soh 1990). By one way, different cell lengths result from locally different rates of transverse and/or pseudotransverse cell divisions during growth. The shorter cells become ray initials and the longer become fusiform initials. By the other way, all procambial cells first become quite elongated; then, some of them, by nonrandom transverse and/or pseudo-transverse divisions, are secondarily transformed into sets of axially short, ray initials. The two ways differ in whether the nonrandom, transverse cell divisions occur early or late in cambial development.

In the "early" way, studied in *Ginkgo, Robinia, Syringa,* and *Canavalia* (Soh 1972, 1974), early transverse divisions produce uniseriate strands of short cells between strands of longer cells. During rapid internodal elongation, long cells become still longer, while short cells seemingly are pulled apart and assume lens shapes. Eventually, the short cells are present mostly as singles or small groups. These cells then begin to function as ray initials.

The "late" way has been observed in *Aucuba* (Soh 1974), and probably also occurs in *Acer* (Catesson 1974). In *Acer*, all future cambial cells of an internode are about 200 μm long when the internode has attained half its final length. Then, some of the cells divide transversely, producing sets of short cells that become ray initials. We do not know how certain cells are determined to divide transversely, in either "early" and "late" developmental patterns.

18.7 Control of Cambial Activity; Spatial and Temporal Variation

Meristematic activity of the cambium is regulated by a complex of systems involving carbohydrates and several types of hormonal and other growth regulators. Among the growth regulators, auxin (IAA) is probably primal, followed by gibberellins, other hormonal substances, and various inhibitors (some of which may also be hormonal). Auxin and some inhibitors may be directly involved in regulating cambial activity, whereas gibberellins are probably more involved in differentiation of cambial derivatives. Water also influences cambial activity —directly because all growth is based on inflation of cells by water, and indirectly because water deficits inhibit photosynthesis and thus synthesis of structural carbohydrates (Creber and Chaloner 1990; Fahn and Werker 1990).

Seasonal activation of the cambium begins at the bases of swelling buds, or bud-borne extending shoots, and is propagated basipetally in all kinds of tree stems. In ring-porous species, propagation is very rapid, reaching the base of the trunk before the buds open. In diffuse-porous species, propagation is much slower, and the cambium at the base of the trunk may still be dormant when bud-opening in the crown is already far advanced. This difference between ring- and diffuse-porous trees is ascribable to differing degrees of dependence of the cambium on hormone supplies from the buds or shoot apices (Sect. 19.5.1). In ring-porous trees, a reserve of auxin or its precursors is probably present throughout the cambium, even during dormancy (Wareing 1951). In contrast, cambial reactivation in diffuse-porous trees depends on arrival of auxin translocated from reactivated buds or from more apically located cambium that has already been reactivated (Digby and Wareing 1966). The duration of cambial activity varies within a tree. In young branches, activity is likely to be brief and intense. Activity lasts longer in older branches, and longest in the main stem (see also Lachaud 1989).

18.8 Cambium and Regeneration Phenomena

Cambium is susceptible to various kinds of wounding, and, almost without exception, wounded cambial-fusiform cells die. Neighboring cells then divide transversely into strands of many short cells, resulting in a localized, callus-like cambium. Aggressive intrusive growth of these cells (Sect. 18.4.3), and the elimination of those cells of low vigor or of certain sizes or positions, make possible cambial rebuilding in a pattern much like the original.

Warren Wilson and Warren Wilson (1961) concluded, partly on the basis of the work of Janse (1921), that: (1) regenerating cambium always develops internally within a callus mass, and begins from an existing cambium; (2) cambium has a pronounced adaxial-abaxial polarity, usually manifested as a xylem-phloem polarity; (3) the "free" edge of a regenerating cambium tends to fuse with another cambial edge of the same polarity. It should be emphasized that cambium always forms beneath an open wound surface, never at the very surface — even if there is no periderm development that might possibly be competitive.

Warren Wilson and Warren Wilson (1961) also proposed a set of hypotheses to explain the regularities of cambial regeneration: (1) between the wounded (open) surfaces of an organ and the interior, there is a gradient of an unidentified factor; (2) cambium may develop only where this gradient has a certain range of values; (3) cambial polarity (xylem-phloem, or adaxial-abaxial) is determined by the direction of the gradient; (4) differentiation of phloem on one side and xylem on the other side of the cambium fixes the gradient; and (5) an established gradient in a cambial field evokes a similar gradient in neighboring tissues. The conceptual framework of these hypotheses is still useful, though little progress has been made in understanding the gradient that they invoke, or in understanding the basis of xylem-phloem (adaxial-abaxial) polarity.

Cambial regeneration has been studied in "tongues" of "bark" and differentiating xylem that were cut away from the stem from below and remained attached at their upper ends (Brown and Sax 1962). If a tongue of tissue, enclosed in a plastic bag, is left hanging free, the exposed undifferentiated xylem cells proliferate and form a thick callus pad on the "back" of the tongue. Within this callus, a cambium differentiates, spreading laterally from the cambium already present within the tongue, until it establishes a ring of cambium within the tongue. However, if the tongue is held against the stem with a bandage, young xylem cells proliferate until any spaces between the older wood and the repositioned tongue are filled. The cells then come under radial pressure, due to growth against the constraint of the bandage. In response to this pressure, they form secondary walls, which become lignified. The original cambium meanwhile continues its normal activity. From such data, one can conclude that pressure is significant in directing cambial development and regeneration.

19 Secondary Xylem

19.1 Introduction

As secondary xylem (wood) is a dominant tissue in the structure and function of arborescent plants, we discuss its development in some detail. This chapter focuses on stem wood, which is of particular technological and economic value and which has been more thoroughly studied than root wood. Root wood is discussed in Section 17.9.

In gymnosperms and large angiosperms, secondary xylem is more persistent than secondary phloem. As a tree grows, new increments of secondary xylem are added to most segments of most axes. This xylem enables the axes to transport the additional water required and also to support the increasing weight of the growing crown. Transport occurs only in the outer increments of xylem, whereas all the xylem contributes to the support function, though not equally so.

A large fraction of xylem consists of tracheary elements. A common feature of these elements is that, in their functional state, they are the remains of dead cells. The lumina of these cells constitute the super apoplasm (Sect. 1.5.2), which is the major route of water transport in higher plants. The arrangement of the tracheary elements and other cells in the secondary xylem is of great interest developmentally, because it is a record of the dynamics of the cambium during the time when the wood was produced (Sect. 18.4.1). This record is much more stable than the record encoded in the secondary phloem (Sect. 20.5.1), because, in contrast to that of phloem cells, the arrangement of xylem cells changes little after they differentiate from cambial derivatives. The cessation of conduction by the tracheary elements, and the eventual transformation of sapwood into heartwood, do not disturb the cellular arrangements in these tissues. Thus, such cambial attributes as the chirality of anticlinal divisions and of intrusive growth of daughter cells (Sects. 18.4.2; 18.4.3) are recorded permanently in the wood of many species. These cell-arrangement patterns, by determining grain characteristics of the wood, affect its technical properties and economic value.

As already discussed (Sects. 18.2.1; 18.2.4), the spatial arrangement and dimensions of fusiform- and ray-initial cells greatly influence the three-dimensional structure of wood. The details of this pattern, however, are, to varying degrees, modified and distorted by morphogenic processes that occur during differentiation and early maturation of the xylem derivatives. These processes include additional periclinal (and sometimes anticlinal) cell divisions; intrusive growth of tips of fibers, fiber-tracheids, and, to a small extent, other kinds of tracheids; and, in most dicots, extensive transverse intrusive growth of vessel members. Because xylem of most gymnosperms lacks vessels and fibers per se, the cell-arrangement pattern that the xylem derives from the cambial-initial-cell pattern is much less distorted during maturation than it is in dicots. Accordingly, xylem cells in gymnosperm woods, as seen in transection, are arranged in fairly regular radial files. Each file is a lineage of cells derived from the same cambial initial.

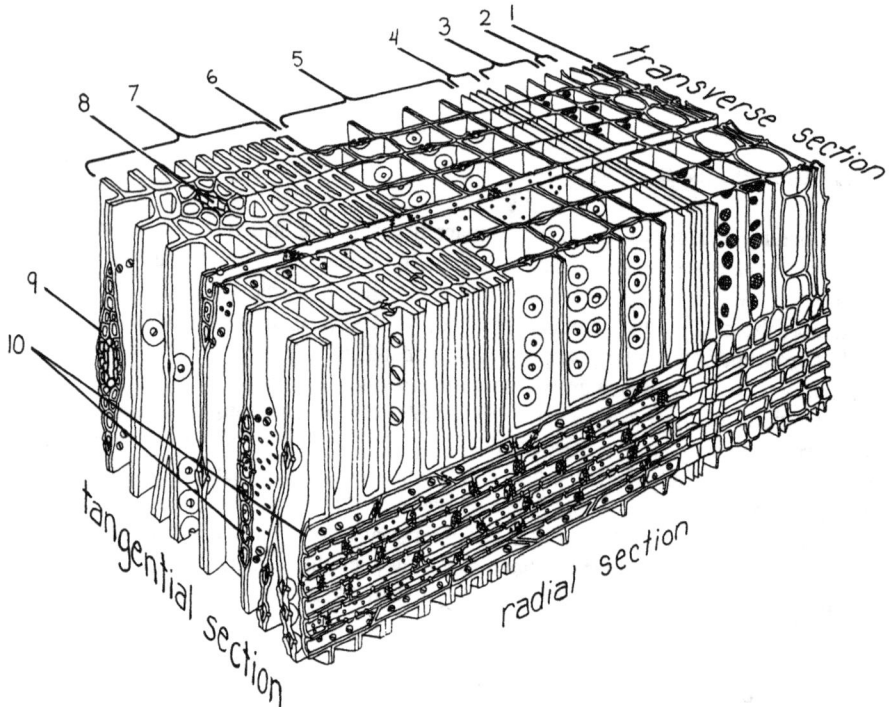

Fig. 19.1. Perspective rendering of cellular detail of block of secondary tissue of stem of *Larix*, as a representative gymnosperm genus. The sample was taken in early summer. Radial, tangential, and transverse sectional views show various aspects of: (*1*) collapsed sieve cells; (*2*) phloem parenchyma; (*3*) functional sieve cells; (*4*) cambial zone; (*5*) current-year xylem (earlywood); (*6*) annual-ring boundary; (*7*) xylem of preceding year (latewood), with (*8*) axial resin duct; (*9*) fusiform ray and its radial resin duct; and (*10*) uniseriate rays. (Drawn from samples and after Mägdefrau 1951)

In contrast to gymnosperm woods, many dicot woods have large numbers of vessels. Cell pattern disruptions during later stages of differentiation and maturation of these vessels evoke so much distortion that the cambial-initial-cell pattern is hardly discernible in tangential sections of mature wood (Sect. 18.4.1). In some dicots, however, a layer of marginal xylem-parenchyma cells undergoes little distortion during development and thus reflects the cambial-initial-cell pattern prevailing at the end (or beginning) of a growing season (Sect. 18.4.1). Similar layers of marginal parenchyma are formed in the secondary phloem of some species (Sect. 20.5.1). Examples of cell arrangements in secondary tissues of stems of typical gymnosperms and woody dicots are shown in Fig. 19.1 and 19.2.

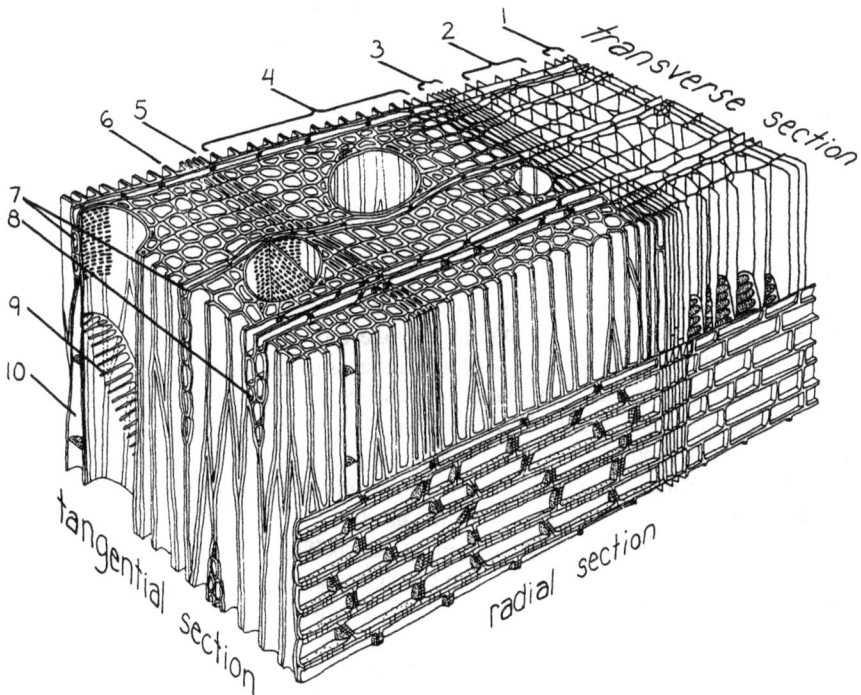

Fig. 19.2. Perspective rendering of cellular detail of block of secondary tissue of stem of *Betula*, as a representative genus of arborescent dicotyledonous angiosperms. The sample was taken in early summer. Radial, tangential, and transverse sectional views show various aspects of: (*1*) collapsed phloem tissues; (*2*) functional phloem with sieve tubes; (*3*) cambial zone; (*4*) current-year xylem (earlywood); (*5*) annual-ring boundary; (*6*) xylem of preceding year (latewood); (*7*) uniseriate rays; (*8*) biseriate ray; (*9*) large xylem vessel with scalariform perforation plate between vessel members; and (*10*) axial xylem parenchyma. (Drawn from samples and after Mägdefrau 1951)

In this chapter, we first examine the differentiation and structure of the various cell types and ducts in wood of gymnosperms and dicots; then, we examine cyclical changes and other variations in wood structure; finally, we focus on the possible controls of cellular differentiation in the xylem.

19.2 Axially Oriented Xylem Elements

In stem transections, the tracheary elements produced by a young cambium that is just becoming active may be difficult to distinguish from primary tracheary elements.

However, in axial sections of many long-shoots, tracheary elements of secondary xylem are identifiable because they are much shorter than primary, metaxylem elements. These two kinds of elements differ in length because of their different origins. Cells that differentiate into metaxylem tracheary elements in long-shoots stop dividing rather early, while the internodes in which they are located are still elongating. Thus, metaxylem elements elongate as they grow symplastically with neighboring cells. Meanwhile, the procambial cells — those provascular cells that will give rise to the vascular cambium — also elongate, but they undergo so many transverse or pseudotransverse anticlinal divisions that their mean length remains short. Because the first secondary tracheary elements (tracheids or vessel members) that differentiate from cambial derivatives do not undergo significant intrusive growth, they are the same length as the fusiform cells in young cambium, and thus are also short.

The differences in length between tracheary elements of primary and early secondary xylem are determined not only by the timing of internodal elongation in relation to the duration and frequency of transverse or pseudotransverse division in the metaxylem and procambial cells, but also by the extent of this internodal elongation. For example, in short-shoots, in which there is almost no internodal elongation, metaxylem and secondary-xylem elements differ little in length.

In gymnosperms, the axially oriented derivatives of the cambium consist of tracheids and, commonly, parenchyma strands. In addition, there may be epithelial cells of axial resin ducts. In gymnosperms, tracheids perform both support and transport functions. During the evolution of angiosperms, the two functions became separated, as tracheids became specialized along two lines. One line led to vessel elements, arranged to form continuous, tube-like vessels that are adapted to efficient water transport. The other line led to fibers, adapted to mechanical support (Carlquist 1975). Cells intermediate between these elements contribute further to the diversity of axial elements in dicot woods. The axial xylem elements in these woods consist of vessel elements, tracheids, fiber-tracheids, fibers, and parenchyma.

Tracheary elements include vessel members, tracheids, and fiber-tracheids. The lumina of all of these function in water conduction after the cells have died. The walls are variably thickened, commonly in elaborate patterns.

Xylem fibers, which occur only in angiosperms, are not tracheary elements in a rigorous sense. Those that are nonliving can store some water super-apoplasmically in their lumina, and all xylem fibers can store water apoplasmically in the microspaces within their walls, but xylem fibers cannot contribute significantly to water conduction. Confusingly, the term "fiber" is often applied to tracheids of gymnosperm wood, especially in literature about wood pulp and paper. Those so-called "wood fibers" really are tracheary elements.

The parenchyma-cell component of the axially oriented system of xylem tissues in most gymnosperms and dicots functions primarily in storage, and secondarily in support. Unlike tracheary elements, axial parenchyma cells commonly remain alive for years. Together with the ray parenchyma, they constitute the living component of sapwood.

19.2.1 Tracheary Elements

19.2.1.1 Cytodifferentiation. Differentiation of tracheary elements occurs in three phases: (1) attainment of characteristic physical dimensions and arrangements — which in angiosperm vessel elements may involve enormous transverse growth and alignment of the elements into continuous, tube-like channels; (2) deposition of secondary wall in a pattern of lignified reinforcements, which protect the elements against collapse; and (3) protoplast autolysis and digestion of the primary-wall areas not covered by secondary-wall deposits. We discuss the first phase — cell enlargement and formation of tubular channels — in Section 19.2.1.3, which focuses on vessel elements and vessels; we discuss the second and third phases here.

After a tracheary element has completed volumetric growth, secondary-wall deposition begins. Secondary wall (S) generally consists of three sequentially deposited layers (S_1, S_2, and S_3), though S_3 is usually lacking in tracheids of compression wood (Barnett 1981). There may be some infra-layering within each of the three S layers. The S_1 and S_3 layers typically are only 0.1 to 0.3 μm thick and hardly resolvable by light microscopy, whereas the S_2 layer may be as thick as 5 μm. The microfibrils of all the S layers are arranged helically. The helices are steepest in the S_2 layer. Arrangement of the microfibrils may be influenced by physical strains that arise during differentiation of the tracheary element (Boyd 1980).

Microtubules seem to be involved in determining the pattern of secondary-wall thickening. In a cell just beginning to differentiate into a tracheary element, the microtubules are uniformly distributed throughout the peripheral cytoplasm. Soon, they become grouped in patterned bands of arrays adjacent to the plasmalemma. Secondary thickening of the wall occurs just outside the plasmalemma, in the same pattern as the microtubule aggregations inside the plasmalemma (Hepler 1981; Falconer and Seagull 1985). If colchicine, which disrupts microtubular structure, is supplied to the cells early in differentiation, microtubule aggregation is inhibited, and the secondary-wall thickenings are deposited chaotically over the primary-wall surface (Hepler 1981).

During secondary-wall deposition in differentiating tracheary elements, the nucleus commonly enlarges, possibly due to endoduplication of its DNA complement. The nucleus tends to remain intact well into the autolytic phase, during which the protoplast with all its organelles gradually loses its integrity and is digested. Autolyzed protoplastic materials are salvaged and partially reused in the synthesis of secondary-wall materials.

The process of autocatalysis and reuse of materials is not well understood. Wodzicki and Humphreys (1973) proposed that, early in the autolytic phase, a network of cytoplasmic strands traverses the vacuolar sap, thereby exposing some of the cytoplasm to digestion by that sap. The cytoplasmic strands then resorb the digestion products and transport them to the peripheral layers of cytoplasm, where they are used to synthesize secondary-wall materials. During the last phases of secondary-wall deposition, the cytoplasmic strands disintegrate, opening all remaining cytoplasm to autolysis, as active hydrolases gain access to the whole interior of the cell. These hydrolases also attack some primary-wall constituents, but not in areas protected by secondary thickening and lignification.

Hydrolysis of unprotected primary wall is a patterned, controlled process (Benayoun 1983; Singh 1987), which leaves a meshwork of cellulosic microfibrils suspended between the protected parts of the wall (O'Brien 1974). The closing membranes of bordered pits also develop in this way.

19.2.1.2 Tracheids in Coniferous Wood.

Structurally, tracheids of conifers and other gymnosperms resemble cambial fusiform initials much more closely than do vessel elements of angiosperms. The only appreciable transverse growth of tracheids is radial. Tracheids accumulate in the order in which they were partitioned off by periclinal divisions of the fusiform initials. Thus, as is apparent in cross sections of coniferous woods, tracheids are arranged in beautifully regular radial files. Typically, the lumina of the tracheids are smaller in latewood than in earlywood. Tracheids also have thicker secondary walls in the latewood than in the earlywood (Sect. 19.5.3), and latewood is more dense than earlywood.

Because they undergo little intrusive growth, tracheids of coniferous woods are of about the same length as the fusiform initials from which they arose. This length usually ranges between 0.5 mm and 8 mm, depending on the species and on the part of the tree. The longest tracheids occur in the outermost increments of old stems of *Sequoia*, *Araucaria*, and related genera.

In wide growth increments, the tips of tracheids may curve in the radial axial plane. Because gymnosperm woods have nonstoried structure (Sect. 18.2.3), tips of tracheids may be adjacent to the median parts, rather than to the tips, of tracheids in tangentially adjacent files, and, if curved, these tips are strikingly nonparallel to the midparts of those neighboring cells. The tips always curve away from the cambium. This curvature arises because, during periclinal divisions in long fusiform initials, there is radial growth as the phragmoplasts are migrating toward the distant cell tips (Sect. 18.3.1). The curvature is most pronounced in wide annual increments, in which the many successive divisions amplify the initially small deviations from the planar condition. At the end of each growing season, the geometry of the fusiform initials is restored to a near-planar condition by periclinal divisions in which the phragmoplasts no longer follow a median path between existing cell walls, but cut off the deviating tips (Włoch 1981). This may cause some decrease in the mean length of fusiform-initial cells, aside from that caused by the oblique, anticlinal divisions that occur late in the season (Sect. 18.4.2).

Tracheids that overlap each other have bordered-pit pairs in their common walls. These pits join the tracheids into continuous translocation routes. The bordered pits of tracheids typically have a torus-and-margo arrangement. The general structure of pits was described in Sections 1.4.2 and 7.4.1.2; here, we mention only the details of development of the torus-and-margo structure.

The **margo**, which is the marginal part of the pit membrane of a bordered-pit pair, becomes porous, while the central part, the **torus**, remains solid. The torus is primary-wall material, even if it is thickened. It is composed mainly of pectic compounds, with small amounts of cellulose and hemicellulose (Bauch et al. 1968). During autolysis of the protoplast, a precisely controlled system removes pectic compounds from the margo but not from the torus. As a result, the margo becomes a mesh-like annulus. The control system probably involves localized protection, by methylation, of the pectins in the torus. In the margo, there is selective hydrolysis of that part of the pit membrane that was unlignified and consisted largely of nonmethylated pectins. In effect, cell-wall matrix materials are digested away, leaving only the cellulose microfibrils.

Trabeculae are rather rare, radially oriented bars within a cell. They are most common in tracheids. Groups of these bars cross the lumina of groups of successive tracheids in a radial file. In sectioned material, if one is able to follow the trabeculae back to the cambium, one can see a precursor area of trabeculae in an initial cell. If a cell containing a trabeculae-precursor area divides periclinally, a part of this area is inherited by each daughter cell, thus propagating the feature radially. Trabeculae are probably pathological responses to fungal hyphae that have invaded the cambium. Cell-wall materials deposited around these areas may be a defense mobilized by the cell. This is reminiscent of some processes occurring during invasion of root tissues by fungal symbionts (Sect. 17.8.2). Trabeculae have been found not only in tracheids, but also in the phloem parenchyma and interfascicular parenchyma of *Vitis vinifera* plants infected with leafroll virus (Hoefert and Gifford 1967).

19.2.1.3 Tracheary Elements in Wood of Dicots.

Vessel members are the major water-conducting elements in most dicot woods. though tracheids may also be present. Unlike tracheids, vessel members have perforated end walls (Sect. 7.4.1.2). These so-called **perforation plates**, or perforation partitions (Meylan and Butterfield 1981), form when cell-wall material is partially hydrolyzed during cytodifferentiation (O'Brien 1970). In this process, primary-wall areas not protected by lignin or methylated pectins are digested. The protoplast is involved both in controlling the pattern of secondary-wall deposition and subsequently in the attack on the exposed primary wall (Sect. 19.2.1.1).

In primary-wall areas not protected by lignin or methylated pectin, the noncellulosic polysaccharides are degraded, leaving a meshwork of fine cellulose fibrils sometimes called **hydrolyzed wall**. Water can pass through this meshwork. The wall areas that later become perforated are so weakened by the partial hydrolysis that they could easily be ruptured by streaming water; however, cellulase activity may also be involved (Butterfield and Meylan 1982). In some species, the perforation plates are fully open, or "simple", whereas in others, a pattern of bars or bands of lignified secondary-wall material remains — evidence of patterned protection of some areas from hydrolysis. The number of bars present, or the lack of them, is a species characteristic (Meylan and Butterfield 1981).

The perforated areas of vessel-member end walls are always much larger than pits, though these two kinds of perforations may be formed at the same time and in much the same way. In *Populus*, hydrolysis of the intervascular pit membranes (making them porous) begins before deposition of the secondary wall is completed (Benayoun 1983). The structure of pit membranes is apparently more diverse than earlier recognized (Sauter 1986). Thus, the control of wall deposition and of selective hydrolysis during cell maturation is probably complex.

End walls of vessel members in the secondary xylem always begin as anticlinal partition walls. They are at first sharply oblique, because the shape of an immature vessel element resembles that of the cambial-fusiform cell from which it was derived. As the vessel element grows transversely, the inclination of its end walls and of the developing perforation plates decreases. In taxa having short cambial-initial cells, and mature vessels that are so wide as to be barrel-like, the perforation plates are likely to be simple and nearly transverse. Such plates are usually present in species having storied cambia. In contrast, vessel members that are long and narrow when mature typically have scalariform perforation plates, oriented obliquely. In these elements, the total perforated area of the plate exceeds the cross-sectional area of the element, thus compensating for the resistance to flow caused by the bars in the plate.

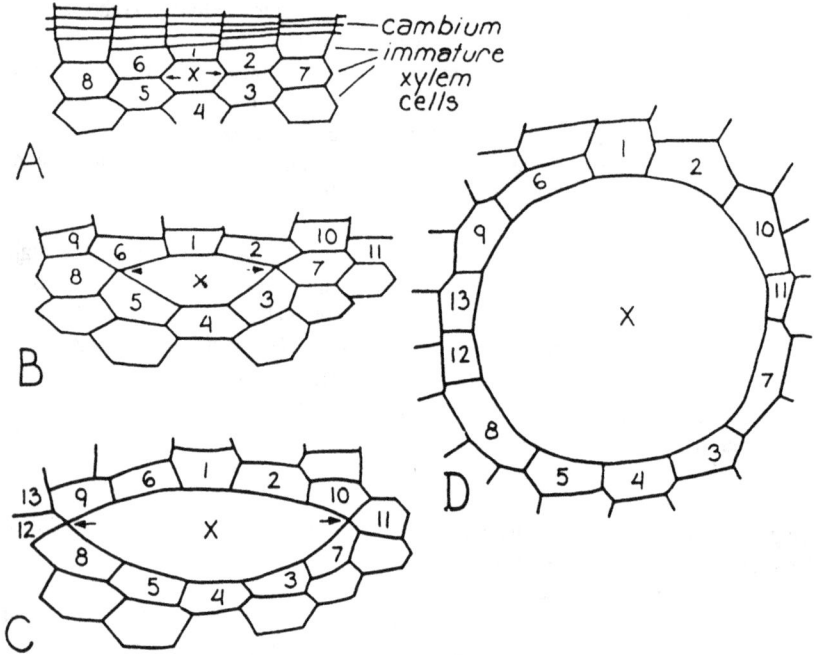

Fig. 19.3 A—D. Cell outline drawings of successive stages (A,B,C, and D) of development of a xylem derivative, X, into a very wide vessel element. *Arrows* mark localities of most extensive lateral intrusive growth. Note that new cell contacts are established, but that cells do not actually migrate past one another. (Redrawn after Hejnowicz 1973)

Vessel members generally have very nearly the same overall length as have the cambial-fusiform initials from which they were derived, but their transverse dimensions, both radial and tangential, typically increase greatly during differentiation. In the earlywood of *Fraxinus*, *Robinia*, and some other ring-porous genera (Sect. 19.5.1), the transverse dimensions become very large, as the axial edges of the vessel members grow intrusively between cells in neighboring radial files. The matrix of surrounding cells must accommodate this extensive growth. Formerly, it was assumed that a differentiating vessel member, after somehow developing a greater internal pressure than neighboring cells, physically pushed away these other cells, thus putting them under compression. However, our own observations are that in spring, when these large vessels are developing, the cambium and differentiating xylem seem to be under radial and tangential tension instead of compression. As hydrolases from differentiating vessel members enzymatically dissolve the middle lamellae between neighboring cells, the tension helps to separate those cells along their axial walls. This process allows not only extensive intrusive growth of the axial edges of the vessel-member wall, but also readjustment of cell contacts around the periphery of the expanding element (Fig. 19.3).

Radial tension in the cambium and differentiating xylem may result from collapse of phloem sieve elements and associated cells that functioned during the previous growing season (Sect. 20.5.3). However, tension generated in this manner may be inadequate to allow vessel elements to grow to their final radial size, because, as the cambium produces new cells, radial tension is dissipated and compression develops. We suggest that this new compression load is borne by "piers" of numerous, radially growing and differentiating xylem cells located tangentially between the vessel elements, which are analogous to arches between piers. If the collective radial growth rate of these piers of many cells is even slightly greater than the radial growth rate of a young vessel member, then the vessel member, instead of pushing against its radial neighbors, is stretched radially by their growth. The entire process of vessel-element expansion needs further study. Methods permitting such study are available (see: Burggraaf 1973; Zakrzewski 1983).

Vessels in secondary xylem differentiate basipetally, as continuations of metaxylem vessels of young shoots or of "secondary" vessels of the midvein of a leaf. So rapid is this basipetal propagation, that differentiation occurs almost simultaneously throughout the vessel's length. Because the signal that evokes basipetal differentiation originates in leaves or buds, the most abaxial (i.e., youngest) vessels in the secondary xylem of the stem lead to the primary-xylem vessels of the most recently formed leaves. In a somewhat analogous manner, the primary-xylem vessels of young, thin roots are in end-to-end contact with the most recently formed secondary-xylem vessels in older roots. Thus, vessels of the secondary xylem connect the vessels of primary xylem of young roots with vessels of primary xylem of young shoots, and thereby form a continuous water-translocation channel from root to leaf. The signal from leaves or buds that normally evokes basipetal differentiation of vessels can be replaced by exogenously applied auxin in some experimental systems (Zakrzewski 1983).

Though vessel elements compose the major water-translocation system in woody dicots, tracheids also contribute. These plants have three kinds of tracheids: vessel-tracheids, vasicentric tracheids, and fiber-tracheids.

Vessel-tracheids are fusiform in shape and commonly are arranged in axial strands. Members of a strand are interconnected by bordered pits (with porous membranes), arranged scalariformly on the anticlinally oblique, "end" walls. Members of a strand may be connected with elements of laterally adjacent strands by small circular pits, also having porous membranes. Walls of vessel-tracheids are often reinforced by helical secondary thickenings, as in *Carpinus*, *Morus*, *Robinia*, *Aesculus*, and *Tilia*. In transections, groups of vessel-tracheid strands resemble clusters of small vessels. To ascertain that they are not vessels, one must determine that the "end" walls of the elements of the strands do not have the perforations characteristic of vessel elements. Because the tips of vessel-tracheids do not grow intrusively, these elements have about the same mean length as the cambial-fusiform initials from which they were derived.

Vessel-tracheids are commonly located between vessels (radially and tangentially), and provide a pathway for inter-vessel flow of water. Vessel-tracheids are less susceptible to embolism (Sect. 7.4.3.4) than are vessels, because they are generally narrower and have only pit connections with neighboring cells.

Vasicentric tracheids are located around the periphery of very wide vessel members. As the developing vessel members undergo their extreme transverse growth, these tracheids, growing symplastically, may assume grotesquely irregular shapes. On the sides opposite their vessel contacts, vasicentric tracheids are commonly in contact with parenchyma cells, through numerous small pits. These specialized tracheids divide transversely and/or obliquely as they differentiate, and thus tend to be shorter than other tracheids. A sheath of vasicentric tracheids can function as armor for the vessel wall, damping shocks or vibrations that could break adhesive bonds between the sap solution and that wall.

Fiber-tracheids are structurally and functionally intermediate between tracheids and fibers (Sect. 19.2.2). However, because there is also a series of transitional forms between tracheids and fiber-tracheids, and between fiber-tracheids and fibers, the nomenclature of these elements is not rigorous.

Fiber-tracheids are generally longer than ordinary tracheids, and also longer than vessel members, because their tips undergo appreciable intrusive growth during differentiation. Their walls, which have bordered pits, are obviously thickened and can function mechanically. Their lumina, though, are still large enough to serve as water-translocation channels. In some genera, including *Populus*, fiber-tracheids are the major cellular component of the wood.

19.2.2 Fibers and Axial Xylem Parenchyma

In addition to tracheary elements, there are two other axial cell types in the wood of dicots — fibers and axial parenchyma.

There is overlap between the two major kinds of xylem fibers — living fibers and libriform fibers. Both kinds, like fiber-tracheids, develop from cambial-cell derivatives that undergo extensive intrusive growth as they differentiate. Having lost the ability to transport water through their lumina, they are more specialized than are fiber-tracheids.

Living fibers, though they have some of the characteristics of tracheary elements, cannot transport water super apoplasmically, because of their living protoplasts. Living fibers are more adapted to mechanical than to transport functions. They may be septate, appearing as discrete strands of short cells. Because these fibers commonly contain starch grains, their protoplasts presumably have a physiological function and are not merely vestigial. In *Tamarix aphylla*, these fibers live as long as does the xylem parenchyma — about 20 years (Fahn and Arnon 1963).

Libriform fibers, which are very long and thick-walled, with simple pits, are the extreme form of xylary fiber. They resemble fibers of phloem (bast, or "liber," from Latin: the inner bark, which explains the term "libriform"). Because of their narrow lumina and small pits, libriform fibers could not function efficiently in super-apoplasmic transport even if their protoplasts died earlier than they ordinarily do.

In some woods, libriform fibers are so numerous that they constitute the cellular matrix in which strands of tracheary elements and parenchyma are embedded. In woods in which they are less numerous, they may form strands between layers or strands of tracheary elements and parenchyma. Libriform fibers are characteristic mainly of species having short cambial fusiform initials. Because of extensive intrusive growth, they become longer — even several-fold longer — than the initial cells. In some stems, they are the longest cells of all, but in most stems, the phloem (bast) fibers are as long or longer (Sect. 20.3.4).

Unlike most other cellular constituents of the secondary xylem of dicots, parenchyma occurs in two orientational types: **axial parenchyma**, arranged parallel to the axis of the stem; and **ray parenchyma**, arranged radially, forming the bulk of the xylem rays. The secondary phloem also contains both axial and ray parenchyma.

Axial xylem-parenchyma cells, like other axial elements, are derived from cambial-fusiform initials. Commonly, they divide transversely during differentiation, thus forming septate, **xylem-parenchyma strands** consisting of about eight cells. Because of differences in thickness of transverse and axial walls, and a "packet" arrangement of cells, the profile of the parent fusiform cell is evident in a parenchyma strand. Axial xylem-parenchyma cells that do not divide transversely constitute **fusiform parenchyma**. Both kinds of axial parenchyma usually have secondary walls, which may be as thick as those of fibers. Pit pairs may be simple, half bordered, or fully bordered (Frost 1929). Mature axial xylem-parenchyma cells retain living protoplasts, often for many years. As they are mechanically strong, fusiform parenchyma cells have been called **substitute fibers**. These occur in Fabaceae, among other groups.

In some taxa, including *Populus tremuloides*, an isotropic layer of wall material of uncommon texture and composition is deposited on the primary walls of some axial and ray parenchyma cells where they abut on tracheary elements (Sect. 19.3.3; Chafe 1974a). This layer is rich in hemicellulose; has a loose, vesicular structure; and in later stages is covered by secondary wall. This unusual layer had been thought to protect the parenchyma cell against hydrolytic damage during the final, autolytic phase of tracheary-element maturation (O'Brien 1970). However, it has properties similar to those of the wall ingrowths in transfer cells of primary xylem, suggesting that a cell having such a layer has a role in translocation (Chafe 1974a).

The amount of axial xylem parenchyma varies from none in some species to a large fraction of the wood in others. In many species, a thin layer of parenchyma is formed at each annual-increment boundary. This may be either **terminal parenchyma**, formed at the end of a growing season, or **initial parenchyma**, formed at the beginning of a growing season; or it may include both kinds. In some species — *Populus tremula*, for example — parenchyma cells are formed at the end of a growing season but do not mature until the beginning of the next growing season. In such woods, the term **marginal parenchyma** is more appropriate than "terminal" or "initial" parenchyma. Marginal parenchyma, whether it consists of septate strands or of fusiform cells, replicates the sizes and arrangement pattern of the cambial-fusiform initials, and thus, in trees having a single growing season annually, constitutes a year-by-year record of cambial dynamics (Sect. 18.4.1).

If axial parenchyma is definitely vessel-associated, it is **paratracheal**; if not, it is **apotracheal**. Each of these kinds of parenchyma can be divided into several subtypes, which have diagnostic value in wood identification. In some species, apotracheal axial xylem parenchyma is arranged in tangential bands, or layers, alternating with bands of fibers. This arrangement poses interesting questions of control of xylem-cell differentiation (Sect. 19.5.3). Among north-temperate trees, thin bands of axial xylem parenchyma alternate with wider bands of fibers in *Fagus, Tilia, Cornus, Diospyros, Carpinus, Ostrya,* and *Carya.* In genera having diffuse-porous wood, such as *Fagus, Cornus, Carpinus,* and *Ostrya,* the bands of parenchyma are quite evenly spaced, regardless of radial growth rate. In those species of *Carya* that have ring-porous wood, however, the spacing of the parenchyma bands is strongly influenced by growth rate (Hill 1982, 1983; Sect. 19.5.3).

Braun and Wolkinger (1970) proposed a functional classification scheme for axial wood parenchyma, recognizing three kinds on the basis of position, pit contact, and probable function (review Fig. 7.4). These are:

1. **Paratracheidal:** this kind of parenchyma partially surrounds neighboring tracheids. Well developed pit pairs connect the parenchyma with the tracheids. The parenchyma cells function in storage and help to regulate the carbohydrate and ionic composition of the solution flowing through the tracheids. However, by separating the tracheids from one another, this parenchyma may interfere with the axial transport of water through the tracheids. It is the least common kind of parenchyma.

2. **Paratracheal:** though similar in structure and function to paratracheidal, this parenchyma is in contact with vessels rather than tracheids. It is very common and is often involved in formation of tyloses (Sect. 19.5.2.3). This parenchyma can act as a buffer against hydrostatic pressure oscillations in vessels (Bel and Schoot 1988).

3. **Interfibral:** this parenchyma commonly consists of large aggregates of cells in contact with ray parenchyma and with paratracheidal or paratracheal parenchyma. It functions in storage and has little pit contact with tracheary elements.

Because marginal parenchyma (either terminal or initial) is in contact with tracheary elements of earlywood, it is a special manifestation of paratracheidal or paratracheal parenchyma.

19.3 Radially Oriented Xylem Elements: Rays

19.3.1 General Features of Xylem Rays

The radially oriented parenchyma system of wood consists of xylem rays. A few of these rays seem to originate in the pith, but, if the axis is large, "pith" rays compose only a very small fraction of the rays. Most xylem rays are produced by ray initials in the cambium and have no close relation to the pith. In a stem cross section that includes both xylem and phloem, one can see that a xylem ray is continuous, along a radius, with a phloem ray, and that the xylem and phloem segments of the ray arise from the same cambial locus — indeed, from the same set of cambial-initial cells. This continuity makes rays the main route of radial symplasmic transport between xylem and phloem (Bel 1990).

Ray length may be expressed in linear units or in units of annual xylem increments across which the ray extends. If a ray is as old as the cambium that produced it, it may be almost as long as the radius of the axis in which it is located. Most rays, however, are younger and shorter. Accurate determination and expression of ray length and age are complicated by morphogenic events in the cambium, such as ray splitting and uniting, and ray elimination and formation. The xylem is the archival record, the physical evidence, of those events (Sect. 18.4.1).

The width and height of a ray, in terms of cell numbers, are usually determined in the cambium (Sect. 18.2.1). Rays may be only a single cell wide, as are the ordinary rays of gymnosperms, or two to many cells wide, as in most dicots. Hence, one can speak of **uni-, bi-,** and **multiseriate** rays. In addition to uniseriate rays, gymnosperms may have specialized, multiseriate rays containing resin ducts. These are several to many cells wide and are termed fusiform rays, due to their shape as seen in tangential sections of the axis.

A ray may vary in measured width along its length, even though its width in number of cells is constant. Its measured width is commonly less in the xylem than in the cambium, because its cells become more narrow as they mature. One might attempt to relate this narrowing to tangential (circumferential) compressive stress that arises in growing woody tissues as tension develops along the grain (Kübler 1959a,b, Kubler 1987). However, in the early spring, there is tangential tension rather than compression in the cambium and differentiating xylem (Sect. 19.2.1.3). Our experience suggests that the tangential narrowing of xylem rays is probably due to active shrinkage as the ray cells mature. This narrowing may help make space available for the intrusive growth of other kinds of differentiating xylem elements.

In some taxa, the rays are wider in the terminal layer of xylem than in the midpart of the ring, and nearly as wide as in the cambium. The lack of intrusive growth in terminal xylem may be a factor in this broadening of rays.

A xylem ray may change in height along its length. For example, height usually increases gradually with increasing ray length. In addition, splitting or uniting of sets of ray initial cells in the cambium may cause ray height to change abruptly, by several to many cells (Sect. 18.4.7). Ray height may also fluctuate somewhat regularly with the seasons: in temperate-zone trees, it is apt to be maximal at an annual-increment boundary (as is ray width), to decline during earlywood production, to be minimal during the main part of the growing season, and then to increase as the growing season ends. This fluctuation, which seldom exceeds two cells, can be ascribed to intermittent activity of ray initials along the upper and lower ray margins (Barghoorn 1940a; Gregory and Romberger 1975).

The protoplasts of ray and axial parenchyma cells form a continuous, symplasmic system of living cells embedded in a matrix of dead cells, the tracheary elements in the sapwood. Rays themselves can also form an anastomosing system, if ray splitting and fusion events in the cambium are frequent enough (Sect. 18.4.7). Such anastomosing systems are especially well developed in wood having interlocked grain, as in *Platanus* and *Nyssa*.

19.3.2 Rays in Coniferous Wood

As already mentioned, rays in coniferous wood are uniseriate, with the exception of those containing **resin ducts**. Rays having resin ducts are fusiform, as seen in transection, and therefore are called **fusiform rays**. The parenchymatous, epithelial cells lining the resin ducts are cytohistologically distinguishable from the surrounding ray cells (Sect. 19.4).

The walls of ray-parenchyma cells of conifers may be primary only (as in Araucariaceae, Cupressaceae, Taxaceae, and Taxodiaceae, among others) or they may also have secondary layers (as in most members of Pinaceae). The walls may be unlignified in the sapwood, then become lignified as the sapwood is transformed into heartwood. Alternatively, the walls of ray cells may become lignified even before those of tracheids do.

In conifers, the walls of ray cells generally have simple pits, even if a secondary wall is present — except in some cross-fields (i.e., where a ray-parenchyma cell abuts an axial tracheid). Chafe (1974b) noted that walls of ray- and axial-xylem-parenchyma cells in *Cryptomeria* are "typically crossed polylamellate". This polylamellate structure was interpreted by later workers as helicoidal (Sect. 1.4.5).

In coniferous wood, rays consisting entirely of parenchyma are known as **homocellular**, whereas rays that also include some tracheids are **heterocellular**. Ray tracheids, which resemble other tracheids except that they are radially rather than axially oriented, occur in some species of Cupressaceae and Taxodiaceae and in most genera of Pinaceae. They are usually arranged in single or double rows along the ray margins; the mid-part of the ray consists of parenchyma cells. A ray that is less than four cells high may, when it is first formed, consist entirely of either tracheids or parenchyma cells, and only later become heterocellular. It is not clear when xylem-ray derivatives become determined to differentiate as tracheids.

19.3.3 Rays in Wood of Dicots

Width and height of xylem rays vary much more among dicots than among gymnosperms. Some dicots have only uniseriate rays, as in *Aesculus*, *Populus*, and *Salix*, and some have mostly multiseriate rays, as in *Platanus* and *Liriodendron*; the majority have rays of a range of widths. The number of multiseriate rays commonly increases as the cambium increases in girth, generally as a result of intrusive growth of fusiform initials into rays (ray splitting; Sect. 18.4.7), followed by increase in size of the ray fragments. Uniseriate rays may also be split by intrusively growing fusiform initials, though increase in number of these rays is more likely to occur through transformation of fusiform initials into ray initials (Sect. 18.4.6).

Some dicot genera (*Carpinus*, *Alnus*, *Corylus*) have **aggregate rays** (Sect. 18.2.4). Narrow strands of axial elements — usually tracheids or parenchyma — separate the small, individual rays that compose an aggregate ray. The large proportion of parenchyma in aggregate rays makes them look, macroscopically, like large, ordinary rays.

Quercus stem wood has both uniseriate and multiseriate rays. The multiseriate rays are many cells wide and many cells high, and are readily visible without magnification. Their height commonly exceeds 2 cm. Multiseriate rays arise from aggregate rays that persist for only a short radial distance (a few years) before the many uniseriate rays within them fuse. The resulting multiseriate rays may persist for many years. In contrast, *Quercus* root wood has no large, multiseriate rays, because ray development stops at the aggregate-ray stage.

Rays of most dicot woods consist only of parenchyma cells. However, in some taxa, including *Annona*, *Combretum*, *Guatteria*, and *Rollinia*, among others, rays also contain *radial vessels* (Vliet 1976; Botosso and Gomes 1982), just as rays of some coniferous woods contain *radial tracheids*.

Parenchyma cells in xylem rays of dicots are of two morphological types: radially elongated, or **procumbent**; and axially elongated, or **upright**. Some ray cells that appear to be axially elongated in tangential sections are, in radial sections, seen to be large in both the radial and the axial dimension, and thus nearly square. In contrast, those that are radially shorter appear as truly axially elongated. Ray initials that are newly formed from transformation of fusiform initials tend to be "taller" than typical ray initials, and may be expected to produce derivatives of the upright type.

Xylem rays consisting entirely of either procumbent or upright cells are **homocellular**, whereas rays composed of both kinds of cells are **heterocellular**. (Note that these terms have different meanings for rays of dicots than for rays of gymnosperms; Sect. 19.3.2). The upright cells in a heterocellular ray are usually located at the upper and lower ray margins. After ray fusion, though, this simple arrangement tends to become confused. In species having rays of a variety of widths and heights, the high rays may be heterocellular, whereas the lower ones may consist entirely of upright cells. The majority of arborescent genera of the temperate zones have homocellular rays; those having heterocellular rays include *Salix*, *Cornus*, and *Carpinus*. Heterocellular rays are common in tropical trees.

Functionally, there are two kinds of ray cells. **Contact ray cells** have pit contacts with neighboring tracheary elements, and can release solutes directly into these elements. **Isolated ray cells** do not have pit contacts with tracheary elements (Braun 1967, 1970, 1982). The pits of contact cells are relatively large and simple, and, in cross-fields, are paired with bordered pits in the tracheary-element walls. In addition, in some taxa, the primary walls of contact cells are covered by a so-called "protective" layer (Sect. 19.2.2), which may not become lignified until the wood is transformed into heartwood. This layer may function in translocation rather than being merely protective (Chafe 1974a).

Isolated ray cells can symplasmically exchange solutes with other living cells, but cannot release them directly into tracheary elements. In *Populus*, the number of plasmodesmata per unit area of tangential wall is higher in files of isolated cells than in files of contact cells, possibly indicating the specialization of isolated cells for symplasmic, radial translocation within the ray (Sauter and Kloth 1986).

Upright cells are always of the contact type, but procumbent cells may be of either the contact or the isolated type. A ray that consists entirely of contact cells is a **contact ray**. A ray that has both contact and isolated cells is a **contact-isolated ray**. Small rays, including uniseriate rays, commonly consist only of contact cells. Rays that begin as small contact rays may enlarge as isolated cells form along the ray edges or in the ray interior. Typically, once a cambial-ray initial begins to produce isolated rather than contact cells, it continues to do so until some locally catastrophic change occurs.

Isolated cells in uniseriate rays generally have well-developed air channels along their edges, but such channels are lacking along the common edges of pairs of contact cells. Air channels are also present in multiseriate rays of some species (Back 1969).

19.4 Resin Ducts and Gum Ducts

The wood of many gymnosperms contains resin ducts. Except in their earliest developmental stages, these ducts generally contain resin (Sect. 8.5.2). Analogous ducts in some dicots contain resins, gums, or mucilages. These may be called resin ducts if, as in *Mangifera indica*, they contain resins (Joel and Fahn 1980a,b). However, in much of the literature, all ducts in dicots have been, and still are, collectively referred to as "gum ducts", regardless of their contents.

Resin ducts are a normal anatomical feature in xylem of *Pinus*, *Picea*, *Larix*, and *Pseudotsuga*, whereas they probably form only as a response to injury in other conifers. Even in those taxa that form resin ducts as a normal feature of the wood (Sect. 8.5.2), the number of ducts can be modified by mechanical stress and injury. Thus, many taxa form both traumatic and nontraumatic resin ducts, though the two types may not be readily distinguishable. In many trees, the intensity of the stimulus for duct initiation influences the number formed. Apparently, most conifers have the genetic information needed to develop resin ducts and synthesize resins, but that information may not be used unless specific stimuli are received.

Traumatic resin ducts in conifers differentiate within masses of callus and regenerating cambium evoked by wounding (Sato and Ishida 1983). In a particular region of regenerating xylem, the orientation of the ducts may be either axial or radial, but generally not both. In contrast, resin ducts that are a normal feature of the wood are oriented both axially and radially in the same region.

Resin ducts have no structure other than that imposed by the epithelial cells that line them. Thus, an account of the origin and development of these cells explains the development of the ducts. Not surprisingly, this development is different for the radial and axial types.

The epithelial cells of axial resin ducts differentiate from recent derivatives of fusiform initials of the cambium. The first anatomical evidence that an axial duct is forming is transverse division of putative xylem-mother cells, followed by division of the daughter cells in several axially oriented planes. The progeny differentiate into epithelial and associated parenchyma cells of the duct. The duct cavity begins to open schizogenously (with a probable lysigenous component) between axial files of differentiating epithelial cells while the cambial initials are still only a few cell diameters away. Groups of three or four cells separate along their common middle lamellae, which seem to be under tension. During early stages, the resin-duct lumen is filled with a water solution. Thus, the lumen cannot be formed by resin pressure pushing apart the epithelial cells. Axial resin ducts vary greatly in length, both within and among species (Sect. 8.5.2).

A radial resin duct develops within a uniseriate ray that enlarges and is modified into a specialized multiseriate ray. This multiseriate ray, as seen in tangential sections of xylem, is fusiform in shape and is therefore called a fusiform ray (Sect. 19.3.2), though it is not derived from fusiform initials but from a special set of ray initials — or possibly from phloem-ray mother cells. Duct initiation may begin when a few cells near the center of a group of cambial-ray initials divide in several planes, in addition to the usual periclinal plane. Some of the daughters of these divisions retain the cambial-ray-initial function, while the others differentiate into duct epithelium and its surrounding parenchyma, within the body of the ray. Intercellular spaces do not necessarily open between the cambial ray initials themselves. Rather, a lumen is formed on both the xylem and the phloem sides of the cambium, probably by a combination of lysigenous and schizogenous processes, as in axial ducts. Thus, though the epithelium is continuous from the xylem across the cambium into the phloem, the lumen may be closed across the cambium (Werker and Fahn 1969).

Every radial duct, at its apparent inner end in the xylem, is connected with an axial duct. Differentiation of an axial duct is apparently evoked by differentiation of an inner terminus of a new radial duct within a xylem ray, just adaxial to the cambium. The lumen of the new radial duct somehow becomes continuous with that of the new axial duct, though ordinary ray cells would seem to intervene between them.

Differentiation of an axial duct, from xylem mother cells, is only a brief interruption in the mode of differentiation of lineages of tracheids in several contiguous radial files. The accretion of radial files of tracheids abaxial to the axial duct indicates that the fusiform initials have not been permanently transformed. In contrast, the stimulus leading to initiation of a radial duct seems to evoke a long-term change in a set of ray-initial cells. For years afterward, their derivatives differentiate into epithelial and related cells of the radial duct.

It is possible that the radial-duct-evoking stimulus evokes its first anatomically detectable response not in the cambial-ray initials themselves but among the phloem-mother derivatives of a set of ray initials. This response would be propagated adaxially through the ray initials to their recent xylem-side derivatives, and then to axially contiguous, fusiform xylem-mother cells — which differentiate into the epithelial lining of an axial resin duct (Werker and Fahn 1969).

As a new axial duct is propagated from its site of initiation, it generally establishes contacts with existing radial ducts. By an unknown process, a small open passage develops between the lumina of the two kinds of ducts. In *Pinus halepensis*, an axial duct contacts about 10 rays/cm. Because about one ray in 20 contains a resin duct, an axial duct should contact a radial duct about every 2 cm. In addition, a radial duct may be in contact with axial ducts on both sides. Thus, a network of ducts can form in the sapwood.

A radial duct, within a fusiform ray, is propagated in both directions from the cambium. Thus, as time passes, it lengthens on both the xylem and phloem sides of the cambium. The lumen becomes continuous across the cambium if the cambial-ray initials separate as do their xylem and phloem epithelial derivatives. Near a bark injury, the lumen is more likely to be continuous, allowing resins to drain from the xylem into the injured area.

Epithelial cells of resin ducts vary in wall thickness, lignification characteristics, and longevity. These cells generally are surrounded by a sheath of other parenchyma cells. In *Pinus*, the sheath cells have thin walls, as do the epithelial cells. In some species, the walls of sheath cells reportedly become lignified, and their protoplasts then may die before those of the overlying epithelial cells. However, sheath-cell longevity can vary even within a species. In *Pinus halepensis*, sheath-cell walls are not lignified, but become encrusted with a thin layer of suberin (Fahn and Zamski 1970), which may hasten the demise of overlying epithelial cells.

As already mentioned, the xylem of many arborescent dicots contains ducts analogous to the resin ducts of gymnosperms. These are generically referred to as **gum ducts**, even though they may contain substances other than gums. Like resin ducts of gymnosperms, gum ducts may be either axial or radial. Some taxa characteristically have only radial gum ducts, some have only axial ducts, and a few have both. The radial ducts are located within rays. Unlike gymnosperm fusiform rays, dicot rays may contain more than one duct. Because dicot rays generally are wide and multiseriate, the presence of ducts does not alter their shape, as it does in gymnosperm rays.

Gum ducts of dicots have not been as thoroughly studied as have the resin ducts of gymnosperms. Gum ducts apparently are a normal feature in some species and a traumatic response in others, as are the resin ducts of gymnosperms. Due to continuing lysigenous processes, some features that begin as gum "ducts" rapidly enlarge into extensive, irregular cavities, especially just adaxial to traumatized areas of cambium.

There is evidence that increased ethylene production is a factor in inducing the formation of traumatic gum ducts and cavities and resin ducts. These form after wounding, after certain chemical treatments, or after invasion by fungi (Hillis 1975; Gedalovich and Fahn 1985; Yamamoto and Kozlowski 1987). Ethylene may interfere with the differentiation of ordinary xylem elements from axial cambial derivatives. The ducts or cavities develop within the traumatic axial parenchyma that accumulates instead of ordinary xylem (Babu et al. 1987).

Axial cavity initiation is lysigenous in *Ailanthus* and schizogenous in *Citrus*. In *Ailanthus*, in which the lysis of epithelial cells is extensive, much of the wall substance is converted into polysaccharide gums. Eventually, most of the axial parenchyma in the affected area disintegrates, forming an extensive, tangential, anastomosing system of cavities. In some unknown way, ray cells are protected from lysis. Thus, the rays, which may be surrounded by gum cavities on all sides, remain as intact bridges to the cambium and phloem (Babu et al. 1987). The survival of rays is probably a factor in the resumption of normal cambial activity across the affected area. This activity results in the sequestering of the gum- or resin-filled ducts or cavities beneath younger layers of xylem. Further research on this point seems justified.

The so-called "kino veins" of many *Eucalyptus* species are also formed by lysigenous breakdown of bands of parenchyma newly formed by the cambium (Skene 1965). Whether the veins become part of the xylem or of the phloem depends on the radial location of the subsequently resumed, normal cambial activity (Tippett 1986). In some species, resurgent cambial activity is localized in surviving parenchyma on the pith side of the kino veins; the veins then become part of the phloem and ultimately are shed with the outer layers of the bark. In other species, the reactivated cambium is localized in parenchyma on the bark side of the kino veins; the veins then become permanent features in the wood.

19.5 Development of Secondary Xylem as a Tissue

19.5.1 Cyclic Aspects of Xylem Formation

The secondary xylem of some dicots shows cyclic patterns in the structure and arrangement of its component cell types. For example, trees of temperate zones and other areas having pronounced seasonal variation, such as annual wet and dry periods, form variably distinct annual increments of wood known as **annual rings**. A major factor in the formation of these rings is cyclic variation in cell sizes and wall thicknesses (Fahn and Werker 1990).

The term "annual ring", though firmly established in the literature, is a misnomer, because the growth increments are sheaths rather than "rings". Nevertheless, because much of the information about wood structure has come from the study of cross sections, in which the increments appear as "rings", we use the term here in the conventional sense.

The inner, or adaxial, part of an annual ring of a temperate-zone tree consists of early xylem, or **earlywood** (springwood). The outer part consists of late xylem, or **latewood** (summerwood). In many species, the transition from earlywood to latewood is anatomically gradual. However, the transition from the latewood of one ring to the earlywood of the next is abrupt, making the annual-ring border obvious. An understanding of these transitions enables one to distinguish the inner from the outer side of even a small block of wood of a tree that grew in a seasonal climate.

Within the annual rings of dicotyledonous trees, vessel distribution is visible on cross sections, often without the aid of magnification. In some species, the vessel elements produced early in a growing season are distinctly larger in diameter than are those produced later. If the diameters differ greatly and the transition is sharp, a ring of "pores" can be seen along the inner (adaxial) margin of the annual ring. This **ring-porous** condition is macroscopically visible in the wood of such genera as *Quercus* (deciduous species only), *Fraxinus*, *Ulmus*, and *Vitis*. In numerous other genera, including *Betula*, *Acer*, *Populus*, and *Fagus*, vessel-element diameter is small and relatively uniform across an annual ring. In woods having this **diffuse-porous** arrangement, annual-ring boundaries are more difficult to discern.

Ring porosity might seem to be related to rapid leaf unfolding and expansion and to accompanying high levels of auxin production early in the growing season. However, there is no strong correlation between the development of ring-porous wood and the length of the phase of leaf unfolding and expansion. For example, both *Quercus* and *Robinia* have ring-porous wood, but in *Quercus* the leaves develop rapidly, in the spring, whereas in *Robinia* leaf development continues throughout the growing season. Formation of the very large vessels characteristic of ring-porous wood may be a consequence of particular mechanical stresses within the cambial expansion zone.

In humid tropical regions, trees of many species do not produce discernible xylem growth rings. Others produce rings correlated with periods of bud flushing, which may occur several times yearly, as in *Monodora tenuifolia* (Amobi 1972). Still others, including *Swietenia heterophylla* do produce annual rings.

In addition to distinctive, annually recurring growth increments, there is another kind of repeating pattern in the wood of some temperate-zone and tropical dicots. This pattern, which is on a finer scale than that of annual growth rings, consists of alternating tangential bands of differing cell types, such as parenchyma cells and fibers (Sect. 19.2.2). Repeating tangential bands of various cell types also occur in the secondary phloem (Sect. 20.4.1).

The general features of annual-ring formation in gymnosperms are similar to those of dicots, but the relations between characteristics such as cell dimensions, rate of wall thickening, and duration of radial expansion and wall thickening are less complex. In dicots, intrusive growth in both the radial and axial directions occurs near the inner (adaxial) edge of the cambial zone. In gymnosperms, in contrast, there is little axial intrusive growth, and radial intrusive growth is much less pronounced than in the differentiating vessel members of dicots. The differentiation of gymnosperm xylem is therefore easier to analyze than is that of dicots.

19.5.2 Long-Term Changes and Other-Than-Annual Variations in the Secondary Xylem

Several long-term changes and other noncyclical structural variations occur in the secondary xylem. These changes may be consequences of changes in cambial cell arrangement (Sect. 18.5), or may result from changes occurring during differentiation of the xylem. For example, wood structure changes with the age of the cambium producing the wood. In addition, the structural variant known as reaction wood may form if the cambium is oriented other than parallel to the gravity vector (Sect. 10.3.2). Further, xylem tissues undergo long-term changes as they cease to function in conduction and are converted into heartwood.

19.5.2.1 Juvenile and Adult Wood.
The wood formed in small, young, foliage-bearing twigs differs somewhat from wood formed simultaneously at a greater distance from the foliage. This implies that, within an axial segment of a trunk or branch, the inner wood differs from the outer wood, because the inner wood was formed by a young cambium, and the outer wood by an older cambium. This change in xylem structure is different in origin from the eventual transformation of older xylem into heartwood. Although the inner core of wood produced by a young cambium in or near a leafy shoot is, anatomically, "juvenile wood", this wood is chronologically the oldest, having been produced earliest in calendar time and by the youngest cambium. The "adult wood" is produced later, by an older cambium, but is itself younger. Thus, during a growing season, the cambium of the main stem produces adult wood, while that of the small branches in the crown produces juvenile wood.

In general, juvenile wood consists of the inner five to ten or more annual rings of any segment of a woody axis. In the cambium producing this wood, the fusiform cells are short but increasing in length, and the ray-initial-cell groups are small and few but increasing in size and number. Accordingly, the dimensions of cells in juvenile wood change progressively from ring to ring. In addition, the lignin and cellulose content of juvenile wood is lower than that of adult wood, while the content of noncellulosic polysaccharides is higher. This explains the considerable axial shrinkage of juvenile wood during drying, especially in conifers, as well as its low specific gravity and tensile strength (Rendle 1960).

19.5.2.2 Reaction Wood. The formation of reaction wood is a response to the position of the cambium relative to gravity. Because it changes its dimensions as it matures, this specialized wood can bring about bending and orientation movements in the axes within which it occurs. Reaction wood develops in lateral branches that are not oriented vertically, and in leaning stems. The reaction wood of gymnosperms, known as **compression wood**, typically forms on the lower sides of branches and leaning stems, whereas the reaction wood of dicots, which is **tension wood**, forms on the upper sides of such axes. However, these relations may be reversed in a plagiotropic stem that is bent upward from its normal position. The physiological anatomy of both types of reaction wood is described in Section 10.3.2.

In compression wood the major structural abnormalities concern tracheids (Westing 1965). They are thick-walled, even in earlywood, are nearly circular, and being quite round in cross section, are accompanied by large intercellular spaces. The spaces contain an amorphous, water-imbibing substance. The microfibrils in secondary walls are arranged in helices with flatter pitch than in normal tracheids. These walls have "fissures", paralleling the microfibrils, that are filled with "laricin" or similar substances. Relative to those of normal tracheids, the walls are more heavily lignified; pits are smaller, slit-like, and infrequent. Xylem parenchyma in compression wood appears to be normal. Typically there is an increased frequency of periclinal divisions in the cambium in the locale where the compression wood is being formed. This results in a thick increment of this wood.

The sensitivity of cambium to the gravity stimulus is remarkable. For example, *Sequoia sempervirens* seedlings grown for several months on a horizontal clinostat did not develop compression wood when the rotation rate was 120 revolutions/h, but did develop such wood around the entire stem when the rate was only 0.25 revolution/h (White 1908).

Tension wood is characterized by a remarkably low content of lignin and has fibers with a "gelatinous" cellulosic inner wall layer. Tension wood (somewhat like compression wood) is produced by the cambium at an elevated rate; thus annual rings including a sector of it are eccentric, being thicker in that sector. It is noteworthy, though, that in leaning stems of arborescent dicots, activity of the cambium on the upper side is greater than on the lower even if the wood produced there is anatomically normal. Tension wood is undesirable commercially because it is difficult to machine and can cause severe warping as sawed lumber including it dries.

19.5.2.3 Heartwood. Cessation of xylem vessel function in transport is commonly ascribable to gas embolisms. However xylem parenchyma typically continues to live and function in storage and local transport (particularly in rays) long after the vessels in the same increments have ceased conducting. Wood having living parenchyma is **sapwood**. During the extended period of xylem parenchyma activity in the sapwood, reserve materials, mainly starch, are metabolized or converted into "heartwood substances" which, after deposition in senescent or dying tissues, enhance resistance to microbial, fungal, and insect attacks (Hillis 1987). As these xylem parenchyma cells die, the sapwood is converted into **heartwood**. The latter is often distinguishable by a darker color. Heartwood typically comprises the inner annual rings of stems and branches and sometimes of larger roots. It is spatially continuous within a tree. Sapwood, with its living parenchyma and metabolizable reserves, is also spatially continuous.

The temporal and spatial aspects of sapwood-to-heartwood conversion vary considerably among species and individual trees within a species, depending upon environmental conditions. The number of annual rings of sapwood present after a tree has begun heartwood formation is often more consistent than the radial thickness of that wood, but the number is also influenced by tree age and by environmental conditions. For example, heartwood begins forming in *Pinus sylvestris* at an age of 25 to 70 years, but one 230-year-old tree growing in Norway reportedly had sapwood 35 mm wide containing 100 rings, while a 48-year-old tree growing in Germany had 60 mm of sapwood with only 35 rings. *P. radiata* typically has about 14 rings of sapwood, whereas *P. banksiana* usually has 30 to 35. In *Quercus robur* and *Acer pseudoplatanus* averages are about 14 and 40, respectively.

Transformation of sapwood into heartwood is relatively rapid, but there may be a long phase of preparatory changes. The zone of wood in which these changes prevail is the **transition zone**. This zone may be visible as a weakly-pigmented band having a lower moisture content than the peripheral sapwood. The living, but senescent, cells in the transition zone, before they die, tend to undergo a phase of increased metabolic activity during which "heartwood substances", mainly flavonoids, are produced (Magel et al. 1991). These diffuse into the walls and lumina of tracheary elements as well as accumulating in situ. In angiosperms increasing numbers of vessels undergo embolism in the transition zone, and neighboring parenchyma cells produce tyloses (balloon-like ingrowths through pits into vessel lumina). Parenchyma cell protoplasts may migrate into the tyloses, which then expand further until vessel lumina are completely occluded (Klein 1923; Zürcher et al. 1985).

Formation of tyloses is a normal part of heartwood development in angiosperms, though it may be affected by environmental conditions. However, in some trees, including *Robinia pseudoacacia*, tyloses form "precociously" as a normal developmental process in early-wood vessels of the outer sapwood. Heartwood tyloses are very abundant in *Castanea, Juglans, Robinia,* and *Quercus.* They occur to a lesser extent in many other angiospermous genera. Heartwood tyloses seem not to occur in normal gymnosperm wood (though some tracheid lumens are as wide as those of small vessels in angiosperms), but have been found in wound-response tracheids (Raatz 1892; Peters 1974).

As sapwood is transformed into heartwood it ceases to function in transport, but continues to function mechanically for indefinite periods. Formation of heartwood can be considered as an evolutionary adaptation that enhances long-term mechanical functions of secondary xylem. Nevertheless, some few timber-producing species exist that are not known to form heartwood (Büsgen and Münch 1931).

19.5.3 Developmental Correlations in Cellular Differentiation in the Xylem

Secondary xylem is a complex tissue. Differentiation of its various component cells, and the arrangement of these cells into beautifully regular patterns, necessarily require complex controls. Although anatomical and physiological studies have led to a better understanding of certain aspects of the control of xylem differentiation, that understanding is still quite incomplete. We will discuss some of the factors involved, while recognizing that their roles and their positions in the hierarchy of control are only partly understood.

Within the annual growth increments of some temperate-zone trees, certain cellular attributes seem to occur concomitantly — that is, they change in a seemingly parallel manner from the inner part of an increment to the outer part. There is apparent linkage of structural aspects of single cell types, and also of different cell types. Are these characteristics, which show a developmental correlation, closely linked physiologically during development, or are they jointly regulated only at a high level of the hierarchy of control?

For example, in some species, the radial diameter of vessel members is generally correlated with the radial length of xylem-ray cells, but this linkage is not tight (Süss et al. 1973). Furthermore, although wide vessels are generally longer than narrow ones, the vessel diameters within a species follow a normal distribution, whereas the distribution of vessel lengths is asymmetric, making the calculation of a mathematical correlation between them inadvisable (Zimmermann and Jeje 1981).

There is evidence that in some species control of the radial growth rate of the xylem as a tissue is exerted from a relatively high level in the hierarchy of control of development of the axis. For example, within annual rings of the ring-porous species, *Carya tomentosa*, three cellular characteristics change in a near-parallel manner from the inner, earlywood zone to the outer ring boundary. Vessel diameter decreases, as does radial distance between successively formed, tangential bands of axial-parenchyma cells, while vessel frequency increases (disregarding the zone of large, earlywood vessels). These cellular changes across the rings are not tightly coupled: there is a definite linear decrease in vessel diameter, compared with somewhat less strictly linear trends in the other two variables. However, even small changes in radial growth rate influence each of these characteristics in predictable ways. Thus, radial growth rate may be controlled from a higher hierarchical level than are the physiological systems that more immediately modulate trans-ring trends in these cellular dimensions and spacings (Hill 1983).

The alternating tangential bands of fibers and axial parenchyma cells in the ring-porous species, *Carya glabra* and *C. tomentosa* (Sect. 19.2.2), may form in response to signals that emanate from endogenous wave phenomena in the differentiating xylem.

Radial growth rate affects the temporal frequency at which xylem-mother cells respond to the signals that induce them to form parenchyma bands. A rapid growth rate shortens the time interval between formation of successive parenchyma bands, while a slow growth rate lengthens it (Hill 1982, 1983). This effect of growth rate is not evident in north-temperate-zone, diffuse-porous species having similar alternating arrangements of fibers and axial parenchyma.

There is additional evidence that radial growth rate (in terms of xylem cells produced per radial file per season) is an important factor in xylem-cell differentiation. For example, there is a correlation between annual-ring width and radial diameter of vessels (Desch 1932). Further, along the length of a dicot ray, cells of the same type (that is, procumbent or upright) may have different dimensional proportions in wide and narrow annual rings. Thus, the extent of radial growth of these ray cells depends on radial, symplastic growth in the cambial zone and in the differentiating and maturing xylem.

There is generally an inverse relationship between radial diameter and cell-wall thickness in conifer tracheids, as the pattern of their differentiation changes from that in earlywood to that in latewood. Radial growth rate influences these cellular characteristics, as it does some of those already mentioned. The relationship between tracheid diameter and wall thickness, however, has been uncoupled experimentally (Wodzicki and Witkowska 1961). Wodzicki (1971) further demonstrated that these two variables are not determined by the same factors.

Radial growth and cell-wall thickening of tracheids actually occur as two distinct stages, in two different regions of the cambial zone. The **zone of radial growth** of derivatives of cambial-fusiform initials is the more abaxial of the two zones, and is widest during the formation of earlywood. After a tracheid has attained its final radial dimension in that zone, it undergoes wall thickening, in the **zone of maturation**. Its protoplast deposits lamellae of secondary material on the inner surfaces of the walls, thereby not only increasing their thickness but also reducing lumen volume. Lignification also proceeds inwardly, from the middle lamella toward the inner wall surface. The wall-thickening phase ends with autolysis of the protoplast, which also occurs in the zone of maturation. In *Pinus sylvestris*, the maturation zone is much wider during the formation of latewood than of earlywood (Wodzicki 1971).

The radial dimension of the normative mature tracheid depends on both the rate of radial growth, V_r, and its duration, T_r. In *Pinus sylvestris* (Wodzicki 1971), *P. radiata* (Skene 1969), and *Tsuga canadensis* (Skene 1972), which have been studied in some detail, V_r varies about four-fold during the active growing season (from 1 to 4 μm/day), depending on the temperature and other, unknown factors. In comparison, T_r, under unknown control, varies by only about two-fold, being about half as long during formation of latewood as of earlywood. One interpretation of the available data is that the characteristically greater radial dimension in tracheids of earlywood than of latewood is established mainly by a longer T_r. This view seems justified if temperature effects on V_r are small to moderate. However, by similar reasoning, if temperature trends are such that V_r is notably decreased late in the season, V_r could then appear to be more influential than T_r. This interpretation seems to agree with Wodzicki's (1971) analysis of extensive data collected during a two-season study of *Pinus sylvestris* in central Poland. He concluded that variation in radial dimensions of tracheids was "probably dependent" on seasonal changes in V_r, in spite of data showing that T_r was 20 to 40 days in earlywood, but only about 15 days in latewood. Comparable durations in *Tsuga canadensis* in Massachusetts, as reported by Skene (1972), were 18 and 9 days.

The wall thickness, like the radial dimension, of the normative tracheid depends on the rate, V_w, and the duration, T_w, of a growth process. Large increases in T_w as the growing season progresses have been reported: 14 days to 55 days in *Pinus sylvestris* (Wodzicki 1971), and 10 days to 50 days in *Tsuga canadensis* (Skene 1972). It thus seems likely that T_w is mainly responsible for the large differences in tracheid wall thicknesses between earlywood and latewood in these species.

The length of the "maturation phase" of tracheid differentiation, during which secondary-wall material is deposited, is closely regulated by a system that controls the onset of autolysis of the protoplast. Within a cell, the period of autolysis is short — about four days in *Tsuga canadensis* (Skene 1972).

A major difficulty in interpreting data such as these is that little is known about the hierarchical relations of the various factors and control systems that may, for example, regulate V_r and T_r, or V_w and T_w. Of the various component systems and variables, which are controlling and which are controlled? At what level does auxin exert its influence, and what in turn controls its concentration or efficacy? How do the various phytohormones and physical factors interact? The dimensions of the problem are just beginning to be understood (Roberts 1988).

Control of cellular differentiation in the secondary xylem may also be linked to developmental events at the shoot tip. There is some evidence that differentiating secondary-xylem elements grow radially only if the shoot apical meristem(s) of the axis in which the elements are located are initiating foliar primordia and the latter are growing. These effects are probably largely mediated by auxin, though nonhormonal factors also are involved (Savidge and Wareing 1981). In general, if apical-meristem activity is inhibited during the growing season, the maturing tracheids are radially narrow, as in normal latewood, especially in the upper part of the stem, but their walls remain similar to those in earlywood. Apical-meristem inhibition may lead to so-called **false rings**, which are usually most pronounced in the upper stem and seldom extend to the base in large trees. In contrast, the development of normal latewood begins in the lower stem — far from the apical meristems.

20 Secondary Phloem

20.1 Secondary Phloem: Origin and Organization

Secondary phloem has developmental features that parallel those of secondary xylem, though the two tissues differ strikingly in ontogeny and in structure at maturity. Both tissue systems differentiate from derivatives of the vascular cambium. Radial files of immature phloem cells are continuous, across the cambium, with radial files of immature xylem cells. In a radial file, the immature phloem derivatives reflect the size, shape, and orientation of the cambial initial that gave rise to them as faithfully as do the immature xylem derivatives. This is true of both fusiform and ray derivatives.

Differences between xylem and phloem, however, are what enable these tissues to perform different physiological and mechanical functions; the xylem elements functioning in apoplasmic or super-apoplasmic space, and the phloem elements largely in symplasmic space (Sect. 1.5). There is a structural basis for their functioning in different spatial systems: conducting xylem elements are *dead*, have highly sclerified walls, and lack protoplasts, whereas conducting phloem elements are nonsclerified *living* cells, though commonly without nuclei and certain other organelles.

Time and space relations in the phloem are inverse to those in the xylem. In the xylem segment of a radial file, the most recently differentiated elements are at the outer (abaxial) end, next to the cambium. In contrast, in the phloem segment of the file, the youngest elements are at the inner (adaxial) end, also next to the cambium.

Xylem increments accumulate in place, increasing the diameter of the central woody cylinder and pushing the vascular cambium, phloem, and periderm outward. Many xylem elements due to their sclerification are mechanically durable. In contrast, many phloem elements are nonsclerified and partly hydraulically supported. Therefore, the phloem elements are easily deformed or crushed after a short functional period, and the kind of structural detail that is preserved almost indefinitely in the xylem (Sect. 18.4.1) is soon distorted or obscured in the phloem.

Because this book is focused on plants having extensive secondary growth — especially on arborescent forms with circumfluent vascular cambia — we consider "external" secondary phloem (i.e., phloem constituting a sheath abaxial to a cambium and to a central cylinder of xylem) as typical. This arrangement, though, is not prevalent in arborescent monocots and is not general in lianas, in which there may also be inter- and intraxylary phloem (Singh 1943; Esau 1969). Furthermore, because much more information is available on the secondary phloem of stems than of roots, this chapter is focused on stems. We recognize, of course, that secondary phloem of roots is vitally important, and that it differs from secondary phloem of stems in structural details and developmental dynamics (Sect. 17.9).

20.2 Terminology of Phloem, Bast, "Bark", Periderm, etc.

Secondary phloem of arborescent plants can hardly be discussed without also considering periderm (Chap. 21). The complex interrelations between these two tissue systems in older axes has led to terminology that is overlapping and nonrigorous. In particular, nontechnical terms such as "bark" and "bast" have been used loosely.

Some of the terminological confusion pertaining to "bark" has arisen because periderms may form in primary and in secondary stem tissues — in epidermis, subepidermis, cortex, and secondary phloem. Further, woody stems typically have living periderm and living secondary phloem, as well as dead periderm and dead secondary phloem.

The botanical terms "periderm", "secondary phloem", and "rhytidome" have been used more consistently than nontechnical terms such as "bark". "Secondary phloem", of course, is vascular tissue containing large numbers of sieve elements that differentiated directly from abaxial derivatives of the vascular cambium. The sieve elements generally function in conduction for only one or two years. In conducting, or "functional", phloem, the sieve-area pores (Sect. 7.2.2) are generally open, whereas in aged, "nonfunctional" phloem, the sieve-area pores are filled with callose and the companion or Strasburger cells are quiescent (Sect. 20.5.3). In most plant axes, nonfunctional phloem elements greatly outnumber functional ones. In zones of nonfunctional phloem, some parenchyma cells that are not specialized as companion cells or Strasburger cells remain alive and capable of storage until they are isolated from more adaxial living tissues by the formation of periderm (Sect. 21.1).

"Periderm" refers collectively to phellogens (cork cambia) and their derivative tissues. The major derivative tissue commonly is phellem, or cork, produced on the abaxial side of the phellogen. Adaxially, a parenchymatous tissue, phelloderm, is commonly formed. The phellem cells become suberized and die soon after they mature, thus isolating any living cells abaxial to them from water and nutrient supply from within the plant axis. As a result, tissues abaxial to a well differentiated periderm generally are dead.

Periderms commonly are initiated sequentially. The first phellogen generally arises in the epidermal and/or subepidermal layer. Later, a series of sequent periderms may arise within the cortex or secondary phloem. Generally, each sequent periderm arises more deeply within the cortex or phloem than did the preceding one. As a result, the older (outer) increments of phloem are sequentially isolated from the younger phloem.

As the living tissues (cortex, secondary phloem, phelloderm) abaxial to the latest periderm die, they become part of what is technically termed "rhytidome" (Esau

1965a). The term "bark" has been applied to the rhytidome. In recent decades, however, many botanists have instead applied the term "bark" to all tissues, either primary or secondary, abaxial to the vascular-cambial zone (Esau 1977), and sometimes including part of this cambial zone. We use the term in this sense.

The rhytidome has, nontechnically, been called the "outer bark". By default, then, the term "inner bark" includes part of the vascular cambium, the conducting or functional phloem and, external to that, the nonconducting phloem that still contains living parenchyma cells (Sect. 21.1; Zimmermann and Brown 1971; Esau 1977).

The term "bast", like "bark", has been used inconsistently in the literature. Before research on plant anatomy began, "bast" referred to bark fibers used for binding. Later, the concept of bast diverged along two anatomical lines. Along one line, it came to include all extraxylary fibers. Along the other line, it became limited to phloem but was broadened to include nonfibrous elements as well as fibrous. In the latter usage, phloem fibers became known as "hard bast", and nonfibrous elements as "soft bast" (Esau 1969).

We use the term "bast fiber" for thick-walled cells that originate directly from abaxial cambial-fusiform derivatives. Bast fibers differentiate close to the vascular cambium and hence within the region of functional phloem. In contrast, the term "fiber-sclereid" designates cells that arise from phloem-parenchyma cells by renascent growth and sclerification. Both bast fibers and fiber-sclereids tend to be very elongated and fusiform in shape. Collectively, they are nonrigorously termed "phloem fibers" (Sect. 20.3.4).

20.3 Cellular Constituents of the Phloem

Secondary phloem is more complex and dynamic than secondary xylem, in that phloem development does not end with the differentiation of functional conducting tissues. Instead, there may be complex changes, even after the tissue no longer functions in translocation. Besides sieve elements, the secondary phloem includes companion cells in dicots — and the analogous Strasburger cells in gymnosperms (Sect. 7.2.1) — as well as parenchyma cells, fibers, and sclereids. We first discuss each cell type that differentiates directly from derivatives of the vascular cambium (Sect. 20.3.1 to 20.3.4); then, we consider long-term developmental changes in phloem tissues (Sect. 20.5.1 to 20.5.4).

Before discussing the several kinds of phloem cells, we point out that details of early differentiation vary among taxa. For example, in some species, a phloem-mother cell may divide anticlinally and give rise to more than one cytotype. Of the 91 dicot species studied by Esau and Cheadle (1955), 35 showed evidence of anticlinal division of the phloem-mother cells producing axial elements, whereas the other 56 species showed no such evidence. In *Pyrus communis*, most axial phloem-mother cells were found to divide periclinally. Their progeny either differentiated directly into sieve elements and their ontogenetically related companion cells, or, by septation, differentiated into strands of clonal parenchyma cells. Of those few phloem-mother cells that divided anticlinally, some produced daughter cells that then divided further and differentiated into small packets of ontogenically related axial phloem of various cytotypes (Evert 1960).

20.3.1 Sieve Elements

20.3.1.1 General Features. Sieve-element structure is the major difference between the phloem of gymnosperms and angiosperms. In angiosperms, sieve elements (sieve-tube members) are arranged in axial series (analogous to the arrangement of vessel elements in xylem), constituting relatively straight, continuous transport routes known as **sieve tubes**. The cellular elements of a sieve tube are **sieve-tube members**. Adjoining end walls of these members develop **sieve plates**, which consist of one or more sieve areas having pores that are generally larger than pores in sieve areas on lateral walls. In gymnosperms, in contrast, sieve elements are not arranged in discrete tubes, and the pores of all sieve areas are similar in size. Sieve elements of gymnosperms are called **sieve cells**. The arrangement of sieve cells in the secondary phloem is distantly analogous to the overlapping tracheid routes characteristic of gymnosperm xylem.

A phloem derivative of a cambial-fusiform initial undergoes several morphological changes as it develops into a functional sieve element. These are: increase in diameter (but not length) relative to the initial cell; in angiosperms change in orientation and length of end walls; and development of sieve areas, which, in angiosperms, are more highly developed on the end walls than on the lateral walls.

Sieve-element length in secondary phloem varies from less than 100 μm in arborescent dicots having storied cambia, to more than 4000 μm in gymnosperms having long cambial-fusiform initials. Sieve-element length can sometimes be determined from macerates, but measurement from tangential sections is more reliable. However, because there is no known developmental process by which sieve elements can become appreciably longer than the cambial-fusiform initials and phloem-mother cells from which they are derived, sieve-element length can, in many species, be approximated from the length of cambial-zone cells or of vessel members, unless the cells differentiating into sieve elements undergo extensive secondary partitioning (Esau 1979).

Radial dimensions of sieve elements in the secondary phloem generally range from 10 to 50 μm and tangential dimensions from 15 to 70 μm (Holdheide 1951; Chang 1954b; Zahur 1959). Sieve-element diameter is commonly no greater in arborescent dicots than in conifers, which are phylogenetically more primitive. This relation differs markedly from that in the secondary xylem, in which vessels, the major conducting channels in dicots, generally are much wider than tracheids, the conducting elements of gymnosperms.

In angiosperms, axially contiguous sieve-element-precursor cells just outside the cambium undergo moderate, coordinated, intrusive edge growth (without appreciable change of length), and thereby lose the fusiform shape of cambial-initial and phloem-mother cells. They rapidly develop discrete end walls, commonly with near-transverse sieve plates. This process results in sieve-tube formation. What controls the multicellular alignment and growth integration processes in sieve-tube development? How long are sieve tubes? Do they differentiate along an entire axis simultaneously? These are open questions.

In gymnosperms, unlike in angiosperms, the elongate, fusiform shape of the cambial initial generally persists in the sieve element. Thus, anticlinal lateral walls are difficult to distinguish from the highly oblique, anticlinal "end" walls. Nonetheless, walls near the cell tips tend to have more sieve areas than do other walls.

The length and inclination of the end walls of sieve-tube members affect the characteristics of the sieve plates. If an end wall is near-transverse, the sieve plate tends to be "simple"; that is, it has a single sieve area, as in *Ulmus* and *Acer*. If the end wall is longer and more oblique, the sieve plate is generally "compound", that is, it has more than one sieve area. This relation is not strict, however. Some transverse or only slightly oblique end walls also have more than one sieve area. A distant analogy may be drawn between the pattern of end walls in sieve-tube members and that in vessel elements, in that transverse end walls tend to have a single opening (in vessels) or a single sieve area (in sieve-tube members), whereas oblique end walls tend to have several or many openings or sieve areas (Sect. 19.2.1.3; Chang 1954a,b).

The diameter of sieve-area pores of gymnosperms, which is generally less than that of sieve-plate pores of angiosperms, may be effectively reduced further due to partial occlusion by tubular ER (Evert 1984). Logically, then, flow velocities through gymnosperm sieve pores should be lower than velocities through angiosperm sieve-plate pores. This expectation was confirmed by measurements of rates of exudation from excised aphid stylets that had penetrated active sieve elements of angiospermous (Dixon 1975) and gymnospermous (Kollmann and Dörr 1966) trees. Further, in experimental systems in which translocation of fixation products of $^{11}CO_2$ was measured, rates in *Picea mariana* and *Pinus banksiana* were found to be only 40% or less of rates in *Fraxinus americana* and *Ulmus americana* (Thompson et al. 1979).

Groups of interdigitating gymnosperm sieve cells may, indeed, translocate metabolites less rapidly than groups of sieve tubes of equivalent cross-sectional area — but there are compensating differences. For example, in dicots, the annual phloem increment is only about one-tenth as great as the xylem increment, whereas, in conifers, the phloem increment is about one-third of the xylem increment (Holdheide and Huber 1952; Braun and Outer 1965). Because the xylem increments in the two groups of plants are roughly comparable, conifers thus generally produce more phloem than do dicots. In addition, the small pore size and lack of axial alignment of sieve elements into tubes in gymnosperms may be somewhat countered by the many sieve areas on lateral walls and by the relatively greater length of sieve cells than of sieve-tube members.

An evolutionary trend toward shorter sieve-tube elements is also ascribable to an evolutionary shortening of fusiform initials (Bailey 1920). As fusiform initials become shorter (and storied arrangement evolves), the end walls tend to decrease in area, and thus to become more discrete and more nearly transverse (Zahur 1959). The evolutionary trend toward short sieve elements is relevant to understanding the mechanism of mass flow through sieve tubes (Sect. 7.2.5).

20.3.1.2 Ultrastructure and Cytodifferentiation. We will discuss sieve-element differentiation first in angiosperms, then in gymnosperms. Our discussion is generally applicable to both primary and secondary phloem, because sieve-element differentiation is not known to differ notably between these tissues (Evert 1984). The progeny of phloem-mother cells in angiosperms undergo drastic cytological change as they differentiate into sieve-tube members. An initial **synthetic phase** overlaps, to some extent, a subsequent **phase of selective autophagy**, or **controlled autolysis** (Srivastava 1974). The *autophagic* phase, also known as "maturation", produces a cell that can function in translocation. This phase is very different from the much later, *autolytic* stage, which accompanies death of the element and loss of its conductive function.

The protoplast of a presumptive young sieve element just entering the synthetic phase of differentiation is similar to the protoplasts of other cells ending a meristematic lineage. During the synthetic phase, there are increases in the amount of rough ER, in dictyosome activity, and in wall synthesis, as well as morphological changes in the ER and plastids. Species having P-protein synthesize it late in the synthetic phase.

During the autophagic stage, many cellular components, including some products of the synthetic phase, disappear or change. There is destruction of the tonoplast, microtubules, dictyosomes, ribosomes, and usually the nucleus. A mature sieve-tube element still has a plasmalemma surrounding its parietal cytoplasm, which contains mitochondria, plastids, some ER, and in some cases remnants of the nuclear envelope. The plasmalemma maintains its integrity and differential permeability throughout the life of the sieve element.

The tonoplast, which usually degenerates during maturation (Esau 1969), persists in *Ulmus americana* until late stages of maturation (Evert and Deshpande 1969). When a tonoplast degenerates, delineation between vacuole and cytoplasm is lost.

Sieve elements, lacking functional nuclei, generally retain close symplasmic contact with companion cells (in angiosperms) or Strasburger cells (in gymnosperms). Companion and Strasburger cells remain active in transcription and translation throughout their lives — and they live as long as do their sieve-element associates.

In an immature sieve-tube element, the ER is cisternal, rough, and dispersed. As the nucleus begins to degenerate, the rough ER loses its microsomes and becomes smooth, and cisternae become stacked. The stacks of ER cisternae then migrate to the parietal cytoplasm and seem to become wall associated. As the cell nears maturity, the ER constitutes a network adjacent to the plasmalemma. This network is directly continuous with the ER of companion cells, via cytoplasmic strands that traverse sieve-area pores and plasmodesmata in the common wall between the cells (Evert 1984).

In immature sieve elements, the plastids are grossly similar to the mitochondria. As the elements mature, the mitochondria change but little, whereas the plastids enlarge and acquire inclusions. If the inclusions are of starch, the plastids are of the "S-type"; if of protein, they are of the "P-type". The various types and forms of sieve-element plastids provide ultrastructural criteria useful in classifying angiosperms (Behnke and Barthlott 1983). Both P and S plastids occur in dicots; only P plastids have been found in monocots (Evert 1984).

Sieve-tube-element walls are commonly considered primary. However, many are thickened and some are even lignified (Esau 1969; Kuo and O'Brien 1974), and thus by some criteria they could be considered secondary. The walls, which commonly vary in thickness, both temporally and spatially as the element develops, generally consist mainly of cellulose and pectin.

In many species, the wall of a developing sieve-tube member has discernible layers — a thin, outer one and a thicker, inner one. The latter, in recognition of its pearly luster when fresh, is known as the **nacreous layer** (Esau 1979). This layer commonly is distributed unevenly, and may be so thick that it almost occludes the lumen. The nacreous layer is highly hydrated and has less cellulose and pectin than has the outer wall (Botha and Evert 1981). In some species, it is ephemeral, forming while a sieve element is maturing, then becoming thinner and disappearing. This course is common in primary phloem, less so in secondary (Esau and Cheadle 1958).

The singular optical properties of the inner, nacreous layer may be due to peculiarities of its three-dimensional structure. In *Annona* and *Myristica*, the luster of this layer may be related to parallel arrangement of microfibrils (Behnke 1971). This arrangement is not inconsistent with helicoidal wall structure, in which the microfibrils within each lamella are parallel (Sect. 1.4.5). Sieve-tube members of *Populus x euamericana* and *Acer pseudoplatanus*, in fact, have polylamellate microfibrillar walls (Catesson 1982), that in ultrathin sections appear helicoidal.

During the early development of sieve elements in dicots, microtubules are few and randomly oriented. As the walls begin to thicken, microtubules may increase markedly in number (Thorsch and Esau 1982). The microtubules are generally transverse to the long axis of the cell and parallel to the microfibrils being deposited. In those species in which the thickening is ephemeral, the microtubules gradually disappear as wall thickening slows, stops, and is reversed (Evert 1984).

During the late synthetic and early autophagic phases, local areas of end walls and of many radially oriented lateral walls become defined as incipient pore fields (or incipient sieve areas), through which the symplasmic continuity of contiguous sieve elements is enhanced. Generally, the sieve areas on the lateral walls of adjacent sieve-tube members are apposed. These apposed areas constitute a sieve-area pair, analogous to a pit pair. Each sieve-area pore develops at the site of a plasmodesma. A sieve area may also be apposed to a field of plasmodesmata in an adjacent companion cell or ordinary phloem-parenchyma cell rather than to another sieve area (Sect. 20.3.2).

In dicots, early in the development of a pore between two sieve elements, ER cisternae on each side of the common wall are positioned close to the plasmalemma, around a plasmodesma. Soon, a pair of collars, or platelets, of callose (a ß-1,3 glucan) is deposited around the plasmodesma — one collar on each side of the common wall. The callose platelets at first may appear either as thickenings or as depressions, depending on the extent of nacreous material, if any, deposited around them. The platelets gradually thicken, while the primary-wall material that separates them becomes thinner. Eventually, all the original wall material at the pore site disappears. Then, the callose is hydrolyzed also, leaving an open plasmodesmatal channel that enlarges into a sieve-area pore (Deshpande 1974, 1975).

The development of sieve-area pores in secondary phloem of gymnosperms is not well known. However, in *Pinus radiata*, localized callose deposition appears to be involved, at least in early stages (Barnett 1974), and, in *P. pinea*, the later stages have been interpreted as a gradual enlargement of the narrow pores of primary pit fields (Wooding 1966).

It is uncertain whether sieve-area pores in the undisturbed, natural condition are open or occluded, because study of the pores necessarily injures the sieve elements and disrupts the translocation process. For example, callose, which may be a minor constituent of functioning, intact sieve elements, is commonly observed to plug the pores of putatively functional sieve elements prepared for microscopic study. Callose is probably rapidly synthesized and deposited in response to the wounding caused by tissue sampling (Eschrich 1975; Evert 1984). Callose is also deposited in intact, undisturbed sieve elements as they approach senescence or dormancy (Sect. 20.5.3).

Mature sieve-tube members of many angiosperms contain some form of P-protein. This substance was discovered during the mid-nineteenth century, when sieve elements were first studied in detail. It was known as "slime". After its polymorphous and proteinaceous nature became known, during the century following its discovery, it was renamed "P-protein" (Esau and Cronshaw 1967). The term "P-protein body" (formerly "slime body") now denotes an aggregation of P-protein. The detailed biochemical nature of P-protein and its metabolism are still unclear.

By middle to late stages of sieve-element development in dicots, P-protein is a characteristic component of the protoplast (Cronshaw 1974a). It occurs even in the relatively primitive, Ranalean taxa, *Liriodendron* and *Magnolia* (Friis and Dute 1983), though it is uncommon in monocots. After EM study, it was reported as lacking in many, but not all, palms, and is lacking in the common crop genera *Zea*, *Hordeum*, *Oryza*, and *Triticum* (Evert 1977, 1982). P-protein is not known in gymnosperms (but see Parthasarathy 1975).

P-protein may be amorphous, filamentous, tubular, or crystalline, depending on the stage of development of the sieve-tube member and on the species. P-protein filaments, as seen in electron micrographs, traverse pores of mature sieve areas in many dicots and some monocots. However, P-protein does not seem to have a direct role in assimilate transport (Evert 1982). Rather, along with wound callose, it may serve to plug sieve pores almost instantly when sieve-tube members are injured (Eschrich 1975).

P-protein bodies, some of which may be crystalline, generally enlarge as differentiation of the sieve-tube member progresses. As the nucleus degenerates and ER becomes parietally located, the P-protein bodies become dispersed in the parietal cytoplasm. In a functional sieve-tube member, P-protein may form a loose parietal meshwork of filaments, or may be dispersed throughout the cytoplasm. Later in development of a sieve-tube member, P-protein is likely to be parietally distributed (Evert 1982, 1984).

The intermediate stages of sieve-element differentiation in secondary phloem are less well documented in gymnosperms than in angiosperms. Detailed information is available for only a few taxa (Srivastava 1963, 1969; Wooding 1966; Timell 1980). This limited information indicates that intracellular organelles may be more persistent in elements of gymnosperms than of angiosperms. The nucleus and other organelles may still be present at functional maturity, and may persist until definitive callose has been deposited (Evert and Alfieri 1965). "Slime bodies" have been reported in some gymnosperm sieve cells, but these have not been confirmed as analogous to the P-protein of angiosperms (Parthasarathy 1975).

In most gymnosperms, as in angiosperms, sieve-element walls have been regarded as primary, even if thickened (Esau 1969). However, in the Pinaceae, there is thickening that seems to qualify as secondary (Abbe and Crafts 1939; Srivastava 1969). In *Pinus strobus* and *Abies balsamea*, the secondary wall is polylamellate (Srivastava

1969; Chafe and Doohan 1972; Timell 1980) and possibly helicoidal in structure (Sect. 1.4.5). Such a wall may be partly analogous to the rather ephemeral nacreous walls of angiosperm sieve elements (Srivastava 1969).

When does a sieve element become conductive? The answer is not entirely clear. However, in general, laterally overlapping sieve cells, or sieve elements in an end-to-end axial series (sieve tubes), differentiate and become functional very rapidly, early in a growing season. Conduction may begin even before the elements are mature. In *Salix fragilis*, studies using dyes, ^{14}C, and aphid-stylet techniques, revealed that recently differentiated sieve elements located adaxial to the most recently formed band of bast fibers, very close to the cambium, are already active in translocation (Lawton 1977). These findings are consistent with those of Kollmann (1965, 1967), who, using ^{14}C and aphid-stylet techniques in *Metasequoia glyptostroboides*, found demonstrably functional sieve elements very close to the cambium. The elements still had well-organized protoplasts and even intact nuclei.

20.3.2 Companion Cells and Strasburger Cells

Sieve elements of secondary phloem of both angiosperms and gymnosperms are mostly enucleate and characteristically are closely associated with ancillary, nucleate, parenchyma cells. There is no wide consensus on the nomenclature of these ancillary cells.

Wilhelm (1880) recognized "Geleitzellen", translated as "companion cells", as specialized parenchyma cells intimately associated with sieve elements of dicots. Such cells were soon found in the phloem of monocots also, and **companion cells** came to be recognized as integral components of the secondary phloem of angiosperms. It was also established that a companion cell is derived from the same fusiform initial as its associated sieve element, usually via a longitudinal division of a phloem-mother cell. A little later, Strasburger (1891) described "Eiweisszellen", translated as "albuminous cells", in axial and ray tissues of secondary phloem of gymnosperms, and recognized their morphological, and possible functional, similarity to companion cells of angiosperms. He believed them to have a further, special role in storing proteins and supplying them to their associated sieve elements. Although it is now known that these cells are not especially rich in protein, the English-language literature has widely used the term "albuminous cell" for the gymnosperm counterpart of the angiosperm companion cell. There is little doubt that the angiosperm and gymnosperm types of ancillary cells are functionally equivalent (Sauter 1980). Therefore, we follow numerous precedents in calling the gymnosperm counterparts of the angiosperm companion cells **Strasburger cells** rather than albuminous cells.

Companion cells and Strasburger cells have: (a) rich cytoplasm, generally with prominent nuclei and many mitochondria; (b) a functional tenure that is correlated with that of the associated sieve element; (c) a lack of starch when the associated sieve element is fully functional, though starch may be present when the sieve element is immature, or during dormancy or senescence; and (d) connections with sieve elements by a system of conspicuous cytoplasmic strands, each group of strands traversing branched plasmodesmata in the wall of the ancillary cell and passing through a median cavity to a sieve pore in the sieve-element wall. In addition, the partition wall between a companion cell and sieve-tube member, as seen in transection, is positioned within the constraints of the walls of a common mother cell. This is a consequence of the common origin of the companion cell and sieve-tube member. We will treat companion cells first, then discuss Strasburger cells in comparison and contrast.

Companion cells commonly are not easily distinguishable from other phloem-parenchyma cells, partly because companion cells and these other cells may intergrade both functionally and structurally. Accordingly, it is difficult to determine that a taxon lacks companion cells. It is sometimes better to say "companion cells not identifiable" than "companion cells lacking" (Esau 1979).

The walls of a functional companion cell are mostly unlignified and have no obvious secondary thickening. In some taxa, the walls become sclerified just before or during senescence, as in *Tilia americana* (Evert 1963b) and in the monocot, *Smilax rotundifolia* (Ervin and Evert 1967). Thus, companion cells may intergrade with sclerenchyma as well as with parenchyma cells.

The number of companion cells associated with a sieve element varies. In secondary phloem, most sieve elements have at least one companion cell. Sieve elements of *Tilia americana* generally have two such cells, and may have up to seven (Evert 1963b). In *Pyrus communis*, 69 of 70 sieve elements examined had identifiable companion cells (Evert 1960), and, in *P. malus*, 242 of 245 sieve elements had them (Evert 1963a). Protophloem sieve elements do not necessarily have companion cells (Esau 1969). Thus, these sieve elements can apparently function for short periods without enhanced access to nuclear and cytoplasmic products from other cells.

Although Bailey and Swamy (1949) found no companion cells at all in the woody vine *Austrobaileya*, further studies revealed such cells to be present in significant numbers. In this genus, as in gymnosperms, sieve elements and their associated companion cells are not necessarily closely related developmentally, and functional relationships can be confirmed only by detailed study (Srivastava 1970).

As a daughter cell of a phloem-mother cell (or sieve-element precursor cell) differentiates into a companion cell, its protoplast becomes more dense than neighboring protoplasts, because of increased numbers of organelles. Companion cells remain nucleate during their functional lives and generally have enhanced complements of mitochondria, rough ER, and plastids that are poor in starch (Evert 1984).

The numerous cytoplasmic connections between a sieve element and its companion cell are generally structured as follows: There is a single canal from a sieve-area pore to the middle-lamella region, where there may be a median "nodule" or "cavity". On the companion-cell side, there are seemingly multiple plasmodesmatal channels (Behnke 1983). Though these channels have been referred to as "branched", the process of branching is obscure. The presence of "branched" plasmodesmata is only weakly diagnostic of companion cells, however, because there are similar plasmodesmata between sieve elements and phloem-parenchyma cells that are not fully functioning companion cells (Esau 1979).

A companion cell probably helps maintain vital functions in its associated sieve element, and generally dies when that element becomes nonfunctional and dies (Sect. 20.5.3). In addition, many companion cells abut on phloem rays. Each of these companion cells is probably involved in symplasmic translocation between the adjacent ray and the sieve element with which the companion cell is associated.

Strasburger and companion cells differ in origin. Whereas a companion cell commonly is derived from the same phloem fusiform derivative as its associated sieve element, a Strasburger cell generally is not. According to Outer (1967), Strasburger cells arise from ray parenchyma in primitive taxa and from axial parenchyma in more advanced groups. In the Pinaceae, Strasburger cells commonly are derived from marginal ray initials and declining fusiform initials. However, it is sometimes difficult to distinguish between these two kinds of initial cells (Sect. 18.4.6).

Variability in origin of Strasburger cells helps to explain the wide range of sizes and shapes of these cells, and also partly explains their variable location. They may be axial or within rays, or both (Outer 1967). But, because no cambial initials are determined as Strasburger-cell initials, radial files of derivatives seldom consist entirely of these cells. Strasburger cells, like companion cells, are rich in organelles. Strasburger cells of *Pinus strobus* and *P. nigra* have organelles similar to those of other phloem-parenchyma cells, but their mitochondria are more abundant (Srivastava 1963).

The cytoplasm of a Strasburger cell, like that of a companion cell, is continuous with the cytoplasm of its associated sieve element via strands passing through pores in their common wall. The structure of the intercellular connections is similar to that in angiosperms (Behnke 1983). The sieve-area pores are larger in diameter than are the plasmodesmatal channels — and are also longer, because the sieve-cell walls generally are thicker than the ancillary-cell walls. As in angiosperms, a sieve-area pore extends from the sieve-element lumen to the middle-lamella region, where there may be a median "nodule" or "cavity". Several plasmodesmatal channels may lead from this nodule to the ancillary-cell lumen (Wooding 1966).

One might expect gymnosperm sieve elements to have more associated ancillary cells than do angiosperm sieve elements, for two reasons. First, sieve elements of gymnosperms tend to be longer than those of angiosperms because they arise from longer cambial-fusiform initials. (Because they are so long, they commonly have contact with many rays.) Second, the cambial initials from which Strasburger cells are derived are commonly marginal ray initials or declining fusiform initials, both of which tend to be shorter than the initials that give rise to companion cells. Yet, although individual Strasburger cells commonly are associated with several sieve elements, and although a sieve element along its length may be associated with several Strasburger cells, some gymnospermous secondary sieve elements seem to have no associated Strasburger cell. In *Metasequoia*, for example, of 46 sieve cells examined, 11 had no discernible contacts with Strasburger cells; 35 had symplasmic contacts with from 1 to 14, with the average being 4.4 such cells (Schumacher 1967; Kollmann 1967). More data on this point are needed.

Even if there are numerous contacts between a gymnospermous sieve element and Strasburger cells, the total contact area is small. In fact, it is not clear that the contact is adequate to allow Strasburger cells to perform ancillary-cell functions to the same extent that companion cells do in angiosperms (Schumacher 1967). Thus gymnospermous sieve elements may depend less on their ancillary cells for maintaining vital functions than do angiospermous sieve elements (Sauter 1980). This lesser dependence may be related to the greater persistence of organelles in the sieve elements of gymnosperms than of angiosperms (Sect. 20.3.1.2).

20.3.3 Parenchyma Cells: Axial and Ray

Aside from companion and Strasburger cells, which are specialized parenchyma cells, the secondary phloem of both angiosperms and gymnosperms includes variable numbers of other parenchyma cells. These have a range of structural and functional relations to sieve elements. In fact, undifferentiated phloem parenchyma and Strasburger or companion cells intergrade. Physical position is highly significant in determining whether a cell can assume specific functions. For example, generally only an axial parenchyma cell that abuts on both a sieve element and a phloem ray can become a companion cell, though these contacts alone do not determine a parenchyma cell to become a companion cell. Similar relations hold for Strasburger cells.

Phloem-parenchyma cells may originate from either fusiform or ray initials. Like their counterparts in the xylem (Sects. 19.2.2; 19.3.1), those phloem-parenchyma cells that are derived directly or indirectly from fusiform initials are *axial* parenchyma, and those arising from ray initials are *ray*, or *radial*, parenchyma. At some distance abaxial to the cambium, axial or ray phloem-parenchyma cells may undergo renascent growth and cell divisions, obscuring their ordered arrangements and origins (Sect. 20.5.4).

A cambial-fusiform derivative may differentiate into a single axial parenchyma cell, or may first divide, with the partition walls being either transverse or inclined. Ordered transverse division (septation) produces an axial phloem-parenchyma strand. Cell number per strand is variable (Esau 1979). Older strands may become crystalliferous (Hu et al. 1985), and their cell walls commonly become secondarily thickened and lignified, though many strands do not undergo these changes.

Axial phloem-parenchyma cells cannot always be distinguished from sieve elements in a single, thin cross section. Serial sections in various planes indicate that, in some taxa, axial parenchyma cells other than companion or Strasburger cells are irregularly intermixed with sieve elements, though they are rarely scattered singly (Holdheide 1951). In other taxa, axial parenchyma forms regular tangential bands one to several cells thick (Sect. 20.4.1). Axial parenchyma generally is more plentiful in late-season than in early-season phloem. In some genera, it forms the terminal layer of the annual phloem increments. Similarly, a layer of axial parenchyma terminates the annual xylem increments in some genera (Sect. 19.2.2; Esau 1979).

Much of the phloem-parenchyma consists of ray cells. These cells do not constitute a continuous matrix, as the axially oriented phloem elements commonly do, but rather form discrete, compact groups or sets of files of cells. Each set makes up a phloem ray, which is continuous, through the cambium, with a xylem ray. Phloem rays are dispersed in the matrix of axial phloem tissues, as xylem rays are dispersed in axial xylem tissues. The differences in grouping and geometry between ray and axial phloem (or xylem) cells are a direct consequence of the segregation of cambial initial cells into fusiform and ray initials.

As in xylem rays, radial elongation of phloem-ray cells is generally more characteristic of cells within the body of a ray than of marginal cells. In most arborescent species, the parenchyma cells in the body of a phloem ray are at least slightly elongated radially (that is, "procumbent"). Marginal cells, though of similar radial dimensions, commonly are axially longer than their procumbent neighbors and thus in radial sections do not appear radially elongated.

Most phloem-parenchyma cells are conspicuously vacuolated. Their organelles generally include starch-storing plastids and, in young shoots, functional chloroplasts. Both xylem- and phloem-ray cells have a role in starch storage and remobilization and may show pronounced polarity in localization of certain enzymes in the most recently formed annual increments (Sauter and Braun 1968).

In the later stages of their development, some axial and ray phloem-parenchyma cells become specialized in form or content. These idioblasts include tannin cells, oil cells, and mucilage cells. Because of the contents of these cells, the inner bark in many species has long been of medicinal interest. Phloem-parenchyma cells that ultimately develop into sclereids or fiber-sclereids are also idioblasts (Sect. 20.3.4).

20.3.4 Sclerenchyma Cells: Fibers, Fiber-Sclereids, Sclereids

Three kinds of sclerenchyma cells are common in the secondary phloem of large higher plants. Two of these, bast fibers and fiber-sclereids, are very long and generally fusiform in shape. In a nonrigorous sense, they are collectively designated as **phloem fibers**. The third kind of phloem sclerenchyma includes various thick-walled cells that are commonly irregular in shape and only moderately elongated. These are sclereids. We consider bast fibers and fiber-sclereids first, then sclereids.

Phloem fibers constitute the major fraction of the axial sclerenchyma in the secondary phloem of many gymnosperms and dicots. Generally nonrandomly distributed, they may form tangential bands that are commonly several cells deep, alternating with bands of thin-walled cells, as in *Castanea dentata* and *Fraxinus americana*. This gives a layered appearance to phloem transections (Sect. 20.4.1), especially if the rays are narrow. Phloem fibers may also form short tangential rows, as in *Liriodendron tulipifera* and *Magnolia korbus*. In *Campsis radicans* and *Litsea calicaris* phloem fibers are scattered singly. They are lacking in *Austrobaileya scandens*, *Paeonia suffruticosa* (Esau 1979), and *Bursera copallifera* (Gómez-Vasquez and Engleman 1984), and *Pinus strobus* phloem seems to have no sclerenchyma at all (Chang 1954a).

The two kinds of phloem fibers differ developmentally. Though both have thick walls with few pits, only **bast fibers** originate directly from cambial-fusiform derivatives. Bast fibers usually arise close to the cambium, within the region of functional phloem. In secondary phloem formed by an elongating stem, bast fibers may grow symplastically during the final stages of internodal elongation, and then grow intrusively in addition. In older axes of some taxa, intrusive growth of bast fibers is so extensive that mature fibers may be more than five times as long as the fusiform-initial cells (Ghouse and Sabir 1974).

The second kind of phloem fiber, the **fiber-sclereid**, is generally not present in functional phloem, but rather differentiates from a parenchyma cell by renascent growth and sclerification, usually during the second or a later growing season. This difference between bast fibers and fiber-sclereids in time of origin explains why, in some taxa, phloem fibers are scarce or lacking in the most recently formed increment, but are plentiful in older increments. By growing intrusively, a fiber-sclereid may approach the length and approximate the shape of a bast fiber, but it remains a specialized sclereid (Holdheide 1951; Srivastava 1964; Esau 1979).

Bast fibers generally have diameters of 10 to 30 μm, and, as seen in transections, have three to six sides. They are extremely variable in length, but generally are moderately longer than the xylem fibers in the same species. A length of somewhat more than 1 mm may be typical (Roth 1981). Thus, the length-to-diameter ratio exceeds 20 to 1 (Chang 1954b), and is generally 30 to 1 or greater. Fiber-sclereids have a similar length-to-diameter ratio. Sclerenchyma cells having ratios < 20 to 1 are considered sclereids.

Bast fibers are not scattered singly, but are arranged in bundles two to five cells wide and deep. Within these bundles, there are no intercellular spaces and no sieve elements. Fiber-sclereids, in contrast, may be arranged singly or in bundles that may include collapsed sieve elements. Bast fibers are present in Taxaceae, Taxodiaceae, Cupressaceae, and Araucariaceae — but not in Pinaceae (Srivastava 1963). In many gymnosperms and some dicots, tangential bands, or layers, of bast fibers alternate regularly with nonsclerified cells (Sect. 20.4.1).

If we designate as bast fibers only those sclerenchyma cells having length-to-diameter ratios > 20 to 1 and arising directly from recent cambial derivatives, then we can say that bast fibers are present in a minority of woody, temperate-zone species, including *Salix, Populus, Tilia, Quercus*, and *Robinia*. If we designate as fiber-sclereids those sclerenchyma cells having similar ratios and arising via renascent growth and development, then we can say that fiber-sclereids probably are present in a majority of woody, temperate-zone taxa (Holdheide 1951), including some that have bast fibers also (Chang 1954a,b).

Bast fibers are commonly accompanied by strands of crystalliferous cells that arise from fusiform cells by septation (Holdheide 1951; Roth 1981). Such strands have the elongated, fusiform shape of bast fibers, but their walls are not always thickened. Each mature cell of the strand may contain a single crystal, commonly of calcium oxalate.

Like fiber-sclereids, **sclereids** arise from renascent growth and development of axial or ray parenchyma cells in the region of the nonfunctional phloem. There are no widely accepted criteria for sharply distinguishing between sclereids and fiber-sclereids (Esau 1969). We use the term "sclereid" for a sclerenchyma cell that fails to attain the length typical of bast fibers of the species.

As already mentioned, the length of sclereids is generally < 20 times the diameter, whereas the length of fiber-sclereids is > 20 times the diameter. Sclereids and fiber-sclereids have similarly thickened, lignified walls, but sclereid walls tend to be more pitted than walls of either bast fibers or fiber-sclereids. The pit canals of mature sclereids commonly appear "branched". However, the "branches" actually arise from the fusion of canals as secondary-wall deposition reduces lumen dimensions and inner-wall surface area. Thus, these canals are more accurately interpreted as "coalescing" (Esau 1969).

As phloem sclereids differentiate, they generally enlarge only negligibly to moderately. Because of irregularly distributed intrusive growth, many develop with projections and branches that give them a contorted appearance. Others, generally scattered individuals that enlarge but little, become brachysclereids, or "stone cells". The protoplasts usually remain alive until a periderm isolates them from younger tissues. Stone cells tend to be numerous in dilatation zones (Sect. 20.5.4), though their distribution there is highly variable among taxa (Roth 1981).

20.4   Cellular Patterns in Functional Phloem and Their Possible Developmental Controls

20.4.1 Evidence of Endogenous Cycles in Patterns of Cytodifferentiation

In the functional secondary phloem of many gymnosperms and some dicots, the major kinds of axially oriented cells are arranged in radial series of regularly alternating layers (appearing as bands in transection). These series appear to be analogous to the regularly alternating layers seen in xylem transections in some species (Sect. 19.5.1).

In many gymnosperms, tangential bands of bast fibers alternate regularly with tangential bands of sieve cells or parenchyma cells, or both. Similarly, in *Tilia* and some other dicots, tangential bands of bast fibers (interrupted laterally by rays) alternate regularly with bands of other phloem elements. Tangential bands (or short rows) of phloem fibers alternate, in radial series, with thin-walled cells in at least some species of *Castanea*, *Fraxinus*, *Liriodendron*, and *Magnolia* (Sect. 20.3.4).

Such radial series do not extend uninterruptedly all along an axis, and they do not provide a basis for assuming that, at any given time, conditions were the same along that axis. Cyclical processes commonly are in different phases in different parts of a plant, as is demonstrated by migration of cambial domains along a stem (Sect. 18.5.2).

The abrupt changes that must occur in the differentiation of tangential rows of phloem-mother cells (as well as of xylem-mother cells; Sect. 19.5.1) to produce banded patterns suggests the operation of an endogenous biological rhythm.

Control systems operating via endogenous rhythms may be most highly developed in highly evolved taxa. This hypothesis is based partly on a putative evolutionary trend toward increasing organization, regularity and repetition in the arrangement of functional phloem tissues among gymnosperms. These tissues can be grouped into three types (Outer 1967):

1. *Pseudotsuga-Tsuga* type (Pinaceae): the axial system consists mostly of sieve cells, with a few scattered axial parenchyma cells and sclereids.
2. *Ginkgo* type (includes Ginkgoaceae, Cycadaceae, Araucariaceae, and parts of Podocarpaceae and Taxaceae): the axial system includes nearly equal proportions of sieve cells and parenchyma cells, arranged in regularly alternating tangential bands one to three cells wide. There are a few bast fibers or sclereids.
3. *Chamaecyparis* type (includes Cupressaceae, Taxodiaceae, and parts of Podocarpaceae and Taxaceae): the axial system is composed of a regularly repeating series of tangential bands of monolayers: bast fibers, sieve cells, parenchyma cells, sieve cells, bast fibers, etc.

In this series of types, the level of organization of the axial phloem system increases. It begins with about 90% sieve cells and a scattering of parenchyma cells and sclereids in the *Pseudotsuga-Tsuga* type, and culminates in an orderly, repeating sequence of monolayers of sieve cells, bast fibers, and parenchyma cells in the *Chamaecyparis* type (Outer 1967).

20.4.2 Effects of Radial Growth Rate; Relative Amounts of Xylem and Phloem

In most temperate-zone species, the most recently formed phloem increment is much thinner than its xylem counterpart. In northern-hemisphere deciduous trees, which have been relatively well studied, the annual phloem increment is generally fewer than a dozen cells thick, though there are numerous exceptions. This increment attains its greatest metrical thickness just before leaf fall; after that, most of the sieve elements and some other components shrivel and collapse (Huber 1939).

The relative thinness of the phloem increment may be due to a shorter duration of phloem production than of xylem, to a slower rate of phloem production, or to both. In temperate-zone trees, the relatively invariant annual increment of phloem contrasts with the widely variable annual xylem increment (Sect. 19.5.1). Cellular characteristics also tend to vary less from year-to-year in the phloem than in the xylem.

Nonetheless, some cellular characteristics change across annual phloem increments, and it is possible that these are influenced by growth rate. For example, in some trees in which the annual phloem increment is produced in two or more episodes, a "spring band" can be delineated within the increment. In *Robinia pseudoacacia*, a spring band of sieve elements, demarcated by bands of bast fibers, is the major phloem band, and its cells have the widest lumina (Huber 1939). Further, in many taxa, the phloem-mother cells that become ray cells grow symplastically with the ambient matrix of radially expanding axial phloem elements, and thus tend to elongate radially, though some then divide transversely. Does radial growth rate influence the lumen size of the sieve elements, or the radial length of the ray cells? Does it affect the numbers of cells laid down in the regularly alternating, tangential bands of different cytotypes that are characteristic of the phloem of some species (Sect. 20.4.1)? Further research on intra-increment variations in cellular characteristics is justified.

In tropical and subtropical trees having no distinct annual growth increments of xylem and phloem, it is difficult to determine the relative amounts of these tissues produced per season. However, by using $^{14}CO_2$ to label xylem and phloem constituents produced by a tree during a known, short period and then felling the tree, one can assay the relative amounts of the two kinds of vascular tissue produced. This method revealed that, in two-year-old *Eucalyptus camaldulensis* trees growing under controlled-environment conditions, the phloem-to-xylem ratio was about 1:4. The ratio was little influenced by common environmental variables (Waisel et al. 1966).

There are exceptions to the general pattern of much greater production of xylem than of phloem. For example, a study of the evergreen tropical tree *Mimusops elengi* revealed that only slightly less phloem than xylem was produced per season (Ghouse and Hashmi 1983). Further, in *Juniperus californica* growing on desert sites, about equal numbers of xylem and phloem derivatives were produced per radial file per year, though xylem was produced at a more rapid rate, during a shorter time (Alfieri and Kemp 1983).

In many species, it is valid to compare the thicknesses of the xylem and phloem only in the most recently formed increments, because the width of an annual phloem increment increases during the second and some subsequent growing seasons (Sect. 20.5.4), whereas the width of the xylem increment is stable. In *Pseudotsuga menziesii*, for example, a phloem increment is roughly 25% thicker in its second season than in its first (Braun and Outer 1965).

20.5 Phloem as a Dynamic Tissue

20.5.1 Overview of Long-Term Ontogenic Changes

Secondary phloem is a highly dynamic tissue system. Developmental changes occur over weeks, months, or years in its various components, and are drastic and extensive compared with those in xylem. There commonly is renascent growth and differentiation of parenchyma cells into fiber-sclereids and sclereids. This change may be accompanied by active shrinkage or passive collapse, under compression, of various nonsclerified cells, such as sieve elements. Compression and collapse of sieve elements may distort nearby phloem rays, particularly narrow ones. Indeed, narrow rays commonly are thrown into irregular zig-zag or sinuous folds, whereas broad rays have enough hydraulically based stiffness to remain straight, as long as they are alive (Holdheide 1951). In addition, ray- and axial-parenchyma cells in the nonfunctional phloem commonly enlarge and divide anticlinally, enabling the phloem to adjust to the increase in circumference accompanying secondary growth.

First-year, second-year, and succeeding-year phloem increments may differ greatly from each other in both relative and absolute volumes occupied by sieve elements, parenchyma, and sclerenchyma (Sect. 20.5.2). Additional, drastic, local changes occur in older phloem increments, as parenchyma cells differentiate into a phellogen. As the outer progeny of the phellogen become suberized, die, and inhibit translocation, they isolate the abaxially located phloem and living periderm components, which then die and become part of the rhytidome (Sect. 21.3.4).

Due to widespread cell collapse and other major changes, including formation of periderms, secondary phloem tends quickly to lose its usefulness as a record of cambial dynamics. In contrast, secondary xylem in many species is an archive of long-term information about these dynamics (Sect. 18.4.1). In some species, however, certain phloem tissues do not become badly distorted for several years. Włoch and Bilczewska (1987), in fact, were able to use the tangential patterns of cellular outlines in the terminal phloem parenchyma of *Tilia cordata* to study cambial domain dynamics. In addition, studies of phloem elements along radial files, and of proportions of cell and tissue types in sequences of annual increments, can provide some insight into some aspects of phloem development and function (Srivastava 1963; Esau 1969).

In most arborescent taxa, additional long-term changes in the secondary phloem arise from an increase in length of the fusiform initial cells of most arborescent taxa over many decades (Sect. 18.4.5). This length may or may not eventually reach a "plateau" (Sect. 18.2.1). Accordingly, the fibers and other elements in adaxial (younger) phloem increments may be significantly longer than comparable elements in abaxial (older) increments. This is true even though bast-fiber length depends on the extent of intrusive growth as well as on the length of the cambial-fusiform derivatives from which the fibers arose (Ghouse and Yunus 1976; Roth 1981).

20.5.2 Changing Proportions of Cytotypes During Ontogeny

Because secondary phloem develops over a number of years, data about relative areas or volumes occupied by specific kinds of cells in the various increments are pertinent only at the time of sampling. In addition, identification of cells is often difficult. For example, identification of a sieve element in a transection can be confirmed only by the presence of a sieve area or sieve plate. Confirmation is difficult to achieve in taxa having nonstoried cambia. However, in taxa having storied cambia, and thus storied sieve plates also, transections through a boundary region between stories may allow direct recognition of all the sieve plates in an area (Lawton and Canny 1970).

Such methods allow calculation of the percentage of transectional area of active phloem occupied by sieve tubes at different seasons as well as during different years. For example, in *Tectona grandis* growing in Nigeria, 32% of the active phloem area was occupied by sieve tubes during the February dry season, 57% in the rainy season in April, and about 18% in December. Trees of several other genera growing in the same region (*Albizia, Bombax, Sterculia*) also showed similar large seasonal changes in sieve-tube area, which was high during the rainy season and lower during the dry season. This variation indicates changing rates of differentiation of new sieve elements and rates of senescence and collapse of older elements. In *Antiaris africana*, in contrast, relative sieve-tube area stayed near 20% all year (Lawton 1972).

The percentage of the transectional area of the active phloem occupied by axial parenchyma varies with time and position, because of late developmental changes in the parenchyma cells. Rigorously speaking, all early derivatives of fusiform phloem-mother cells are parenchymatous. Some rapidly differentiate into sieve elements or bast fibers, while others remain parenchymatous for all or most of the first growing season. At the end of the first season, or during the second season, with or without renascent expansion growth, some of the remaining parenchyma cells develop thickened secondary walls. Thus, the relative proportion of sclerenchyma increases late in the first season and in subsequent seasons, at the expense of the parenchyma. Nonetheless, living axial- and ray-parenchyma tissue is still plentiful after several or many seasons. Additional parenchyma may eventually be formed, as dilatation tissue (Sect. 20.5.4).

Within a phloem increment, the relative volume occupied by bast fibers, which differentiate directly from phloem-mother cells, is likely to decline notably over the years, though the fibers do not collapse and are not digested. Rather, their relative volume decreases, due first to renascent growth of phloem parenchyma, including dilatation growth, and, subsequently, to meristematic activity of phellogens of first and sequent periderms. The phellogens, which are intercalated within the original phloem increments, "dilute" the phloem tissues by producing significant volumes of phellem and phelloderm (Sects. 21.3.1 to 21.3.3). If large quantities of phellogen derivatives and dilatation tissues are intussuscepted, then the change in bast-fiber volume can be substantial. For example, in *Sequoia*, bast fibers were found to occupy about 18% of the phloem volume in recent (inner) increments, but only about 6% in the "middle bark" or "outer bark" (Isenberg 1943).

20.5.3 Decline, Senescence, and Collapse of Sieve Elements and Associated Cells

One of many long-term changes in the secondary phloem is the eventual senescence, and often collapse, of functional sieve elements and associated cells. In discussing these aspects of sieve elements, it is important to note that there is a need for experimental verification of function in living elements (Ewers 1982a,b). There seems to be a lag between the time the elements stop functioning in conduction and their actual death. Yet, much of the pertinent literature does not distinguish between functional tenure and longevity of the elements. In general, we feel that the term "functional tenure" is more useful than "longevity" for sieve elements and their companion (or Strasburger) cells. In contrast, the term "longevity" is the more appropriate one for that fraction of the phloem-parenchyma cells that remain alive and capable of renascent growth and development long after sieve elements and their ancillary cells have ceased to function in conduction. The eventual death of those phloem-parenchyma cells that do not collapse with the sieve elements is discussed in the context of sclerification (Sect. 20.3.4) and periderm formation (Chap. 21).

Before their final decline, sieve elements may have periods of diminished function associated with dormancy. The temporary or permanent waning of functional competence of a sieve element is usually associated with deposition of callose on the sieve areas, and sometimes elsewhere in the cell. This callose is called either **dormancy callose** or **definitive callose**, depending on the extent to which the cytoplasm has degenerated before deposition begins. If degeneration is far advanced, the callose is definitive, in the sense that the sieve element will never again become functional. If the cytoplasm has degenerated so slightly that function may be resumed when the callose is digested, the callose is dormancy callose. Definitive callose is also eventually digested, when a senescent element becomes necrotic. Definitive and dormancy callose are presumed to be biochemically the same (Evert 1984).

In north-temperate-zone gymnosperms, deposition of definitive callose on the sieve areas of sieve cells that overwintered from the previous season usually begins in May or June. Subsequently, callose is deposited on sieve areas of sieve cells produced early in the current season. By mid-December, all but the last-formed elements have deposits of definitive callose. The callose that is deposited in the last-formed elements which may resume function in the spring, is dormancy callose (Sect. 20.6).

In many taxa, the cessation of function of sieve elements is accompanied by decline of companion cells (Strasburger cells in gymnosperms) *and* of closely associated, ordinary parenchyma cells. In some other taxa — *Mimosa pudica*, for example — there is decline of companion cells but not of other parenchyma cells that are developmentally related to the sieve elements (Esau 1973).

It is not known what causes sieve elements and some associated cells to collapse (or undergo "obliteration", as the process is sometimes called). The cells do not simply lose turgor, because sieve elements that are punctured or cut do not necessarily collapse. The walls may possibly become soft and plastic with age, thus becoming susceptible to collapse under the pressure generated by expansion of cells in nearby tissues (Holdheide 1951).

As already suggested, cessation of transport function in sieve elements does not necessarily lead to prompt cell death, nor is death necessarily followed immediately by collapse. In *Populus*, for example, sieve elements are so embedded in, and supported by, bast fibers that they may not yet have collapsed even several years after they have ceased to function. Similarly, in some gymnosperms, nonfunctional sieve elements, and sometimes phloem-parenchyma cells also, do not collapse (Outer 1967). Thus, duration of transport function cannot always be determined by light-microscopy alone.

20.5.4 Dilatation Growth and Dilatation Tissues

In vigorously growing axes, parenchyma cells may undergo renascent growth and division and contribute to increase in girth of the phloem. There may be **diffuse dilatation growth** of all phloem parenchyma, or strongly localized tangential growth in certain sectors, which produces **dilatation tissues.** Dilatation tissues, as seen in transections of actively growing sectors, form wedges that are wide abaxially and narrow near the cambium. Dilatation is due to radial anticlinal divisions of cells of either the phloem rays or the axial phloem parenchyma. It is seldom due to divisions in both axial and ray parenchyma in the same species (Borger 1973). In *Citrus limon* (Schneider 1955), dilatation may occur before a first periderm is organized, but in many other species, including *Populus tremuloides* (Rees and Shiue 1957-1958), dilatation is evident only later. Dilatation generally occurs only in the older phloem that no longer functions in long-distance symplasmic transport but that still includes many live, metabolically active parenchyma cells (Borger 1973). The significant increases in phloem volume that dilatation produces are usually partly offset by collapse of some sieve elements and axial parenchyma cells.

Dilatation generally proceeds only after collapsing sieve elements have caused notable distortion and displacement of tissues. This suggests that mechanical strains may have a role in provoking mature phloem-parenchyma cells into renascent growth and renewed meristematic activity. Dilatation begins when cells under tangential tension accumulate strain and then divide anticlinally, inserting radially oriented partition walls. These divisions may become so prevalent that radially oriented, cambium-like meristematic zones form, especially in wide rays, as in *Celtis*.

Holdheide (1951) localized the genesis of dilatation in 49 species of European trees. He reported it to be based in axial phloem parenchyma in 28 species (9 broadleaved trees, 12 shrubs, and 7 conifers) and in phloem rays in 21 species (14 broadleaved trees and 7 shrubs). In 28 other species, he found no evidence of dilatation. In *Tilia*, *Juglans*, *Populus*, and *Celtis*, dilatation is generally confined to the phloem rays. In *Castanea*, in contrast, only the axial phloem parenchyma undergoes dilatation, whereas the rays, which are uniseriate, are merely displaced and distorted (Holdheide 1951).

The extent of dilatation may vary within a genus and even among the rays of a single tree. In *Eugenia*, for example, some species have abundant dilatation tissue and others have little (Wyk 1985). In *Tilia*, dilatation may increase the width of some rays 10- or even 20-fold, though nearby rays, as seen in transection, seem unaffected (Holdheide and Huber 1952). Tangential sections may provide more information than single transections in this regard, however, perhaps revealing little change near the margins of a ray, but considerable dilatation within the ray body.

In species in which sclereids are the only phloem sclerenchyma, rays tend to undergo conspicuous dilatation. In species such as *Salix* and *Liquidambar*, which have bast fibers but no sclereids, there is little dilatation (Chang 1954b). The significance of these relations is not known.

20.6 Phenology of Phloem Development and Duration of Transport Function

The phenology of initiation and cessation of phloem production, and the functional tenure of sieve elements, are highly variable among taxa. Functional tenure, which is difficult to determine, generally is confined to one season or parts of two seasons, but in some taxa, it may last for several or many seasons.

The time lag between the end of functional tenure of sieve elements and their death poses a terminological problem (Sect. 20.5.3). Here we focus on "functional tenure". Although we cannot review the entire subject, the following examples suggest some patterns that seem to be emerging among plant groups.

The common temperate-zone angiospermous and gymnospermous trees differ in the functional tenure of their sieve elements. In angiosperms, sieve tubes commonly function for only one growing season, or, especially in taxa having ring-porous xylem, for the last part of one season and the first part of the next (Huber 1939, Esau 1969). After they have ceased functioning, they may either collapse or remain open. Common, deciduous, arborescent species having sieve elements that function only a short time include *Pyrus communis* (as discussed later), *P. malus*, *Robinia pseudoacacia*, and *Populus tremuloides* (see Davis and Evert 1970 for specific references). *Tilia* and *Liriodendron* are exceptions, in that their sieve elements generally function for a number of years. In gymnosperms, sieve elements tend to function for several years, especially if a new annual phloem increment does not have enough sieve cells to replace all of the eight to ten functional elements per normative radial file (extending across several annual increments) — which may be the minimal complement necessary for survival (Holdheide 1951).

Pyrus communis, growing in central California, has no overwintering sieve elements, but does have one or two rows of overwintered phloem-mother cells near the thin, cambial-initial zone. In early March, these cells expand radially, and rapidly differentiate into sieve elements and companion cells. Concomitantly, the cambial initials begin to produce new phloem-mother cells, which begin differentiating into phloem elements. By mid-June, differentiation of phloem elements has slowed. Meanwhile, in early May, xylem has begun to differentiate. The new, mature phloem functions during the spring and summer of its first season. Definitive callose is deposited in September. By early January, callose has disappeared from most sieve elements, leaving them devoid of contents and presumably nonfunctional (Evert 1960).

In contrast in the woody vine, *Vitis riparia*, growing in Wisconsin, sieve elements generally function more than one growing season. Many elements overwinter in a dormant state. Cambial cell division does not begin until late June, and phloem and xylem differentiation not until early July. By late July, functional competence of the overwintered, reactivated sieve elements begins to wane. Differentiation of the new phloem and xylem is completed by mid-August. During October, callose occludes the sieve pores of the remaining active elements, some of which resume activity in spring as the callose is digested (Davis and Evert 1970). A similar phenological pattern occurs in many north-temperate arborescent species, including *Pseudotsuga menziesii, Tilia americana, Quercus alba, Pinus strobus,* and *Ulmus americana.* In *Tilia,* sieve elements may function as long as ten years (Holdheide 1951).

In *Pinus banksiana, P. resinosa,* and *P. strobus* growing in Wisconsin, sieve cells that form late in the growing season overwinter and then function until new sieve cells differentiate in the spring. Overwintered phloem-mother cells on the outer margin of the cambial zone begin to differentiate into sieve elements late in March. The cambium begins to produce new phloem in early April. Spring phloem production ends as differentiation of new xylem begins in early May. Additional phloem, and much more xylem, are produced during the summer. Most of the sieve elements produced in summer differentiate and mature within a few weeks. The last ones, maturing in late September, overwinter in the functional or potentially functional state. By mid-December, all other sieve elements have deposits of definitive callose. Functional deterioration begins in late May and June in elements that overwintered, then progresses to those elements that matured during the spring and summer (Alfieri and Evert 1968).

Phloem phenology in gymnosperms growing on warm, xeric sites is different from that in the previous examples. In *Juniperus californica* growing in the southern California desert, all the sieve elements formed during a growing season remain in a dormant state during the following winter. Dormancy callose is digested in late winter. By mid-March (late in the rainy season), the cambium begins producing phloem-mother cells, which rapidly differentiate into functional sieve elements and other phloem components. The overwintered sieve elements (i.e., all the previous increment's complement) and their Strasburger cells die as the cambium is producing the new phloem derivatives. In June, as water stress and temperatures increase and cambial activity declines, dormancy callose begins to accumulate on the sieve areas of the first-formed elements of the new increment. By early July, the cambium and phloem are dormant (Alfieri and Kemp 1983).

Phloem phenology in tropical and semitropical trees varies with seasonality of temperature and moisture availability and also may differ between evergreen and deciduous taxa growing in the same region. In the tropical evergreen *Mimusops elengi* (Sapotaceae) growing in Aligarh, India, the cambium is active from late May (beginning of the rainy season) through mid-November (beginning of the cool, dry season). Xylem differentiates from early July through October. Phloem does not begin to differentiate until early August (though new leaves begin emerging in early May), and continues to differentiate through late November. The last-formed phloem derivatives remain immature during several months of dry-season dormancy, then become functional in early April, before the cambium is reactivated (Ghouse and Hashmi 1983).

The semitropical tree *Grewia tiliaefolia* (Tiliaceae) growing in Gujarat, India, has a two-month leafless, dormant period, just before the spring equinox. As in *Mimusops,* the last-formed phloem elements of a growing season remain immature and radially narrow during dormancy, then mature to the functional state in early April. Bud break and cambial reactivation occur in mid-April. New phloem then begins to differentiate and mature, and continues to do so, until early January. The sieve areas of potentially functional sieve elements are generally occluded by callose during the subsequent dormant period. The elements become functional as the callose is digested early in the rainy season. Many *Grewia* sieve elements function for more than one season (Deshpande and Rajendrababu 1985).

The tropical, deciduous trees *Albizia lebbeck*, *Dalbergia sissoo*, and *Terminalia crenulata* were found to have two episodes of cambial activity per year and two well-defined episodes of differentiation of new phloem. However, *Tectona grandis*, likewise deciduous, has only one such episode per year. Similarly, the evergreen species *Calophyllum inophyllum*, *Mangifera indica*, and *Morinda tinctoria*, which commonly grow in association with the deciduous species already mentioned, have only one episode of cambial activity and phloem differentiation per year (Venugopal and Krishnamurthy 1987). These data hint at a possible difference between deciduous and evergreen trees, but, obviously, too few of the vast number of tropical tree species have yet been studied to justify generalizations about the phenology of their phloem development.

In temperate-zone dicots having diffuse-porous xylem, phloem differentiation tends to precede xylem differentiation by several weeks. In contrast, in ring-porous species, xylem and phloem tend to begin differentiating almost simultaneously. *Tilia* is an exception; though diffuse porous, it initiates xylem and phloem at about the same time (Davis and Evert 1970).

In stems of some arborescent monocots, which lack circumfluent vascular cambia, sieve elements are very long-lived. They probably live more than 100 years in *Kingia australis* (Lamont 1980) and likewise in *Sabal* spp. (Parthasarathy 1974). In the studies of these taxa, sieve elements were considered to be living if their walls had the thickenings characteristic of sieve elements and if their protoplasts were intact. However, proof that these sieve elements are functional is difficult to provide.

21 Periderm

21.1 Periderm, Surrounding Tissue, and "Bark"

The term **periderm** applies collectively to the protective tissues **phellem** and **phelloderm**, and the meristem that lies between them and gives rise to them. This meristem, the **phellogen**, like the vascular cambium, is a uniseriate layer of initial cells that, by periclinal divisions, partition off derivatives from their adaxial and abaxial faces. Whereas a vascular cambium produces xylem adaxially and phloem abaxially, a phellogen produces phelloderm adaxially and phellem abaxially. Unlike a vascular cambium, which generally persists indefinitely in dynamic stability, a phellogen producing periderm on a major axis commonly is transitory. In time, it is supplanted by more adaxially located, sequent phellogens, which produce sequent periderms.

Periderms, along with remnants of cortex and/or phloem isolated between them, are major "bark" constituents (Sect. 20.2). The phellogen of a first periderm generally arises in subepidermal or cortical tissues. It usually develops as a local, tangential sheet of parenchyma cells exhibiting renascent growth and cambium-like meristematic activity.

In common usage, "bark" refers to all those living and dead peridermal and other tissues abaxial to (and sometimes including a part of) a circumfluent vascular cambium (Sect. 20.2). This tissue zone may be divided into **inner bark**, which includes the mostly living phloem and part of the vascular-cambial zone; and **outer bark**, which consists of the mostly dead periderms, the mostly dead phloem between periderms, and the mostly dead remnants of any persisting cortical and epidermal tissues. These terms lack rigor, however, unless a boundary between "inner" and "outer" is delineated. In the nomenclature we use here, the innermost periderm is that boundary. This periderm delineates that part of the phloem that is an uninterrupted, mostly living tissue mass (the inner bark) from the dead rhytidome (outer bark). The innermost periderm itself is not part of the inner or the outer bark.

In tangential aspect, periderms are commonly nonhomogeneous, in that they are differentiated into lenticular and nonlenticular regions. Because nonlenticular periderms cover much more area, we discuss them first, and treat lenticels and lenticular periderm later (Sects. 21.5; 21.6).

Phellem, a major constituent of periderm, generally is suberized and inhibits translocation of water and solutes to tissues abaxial to it. As a result, these tissues senesce and die. Thus, the most recently differentiated phellem layer is a boundary between living and senescent or dead tissues. Tissues adaxial to the boundary are still parts of the greater symplasm and apoplasm, whereas those abaxial to it lose that continuity and die. In time, an "innermost" periderm (with its phellem) usually, but not invariably, is displaced by another, more deeply located periderm. Displaced periderms, senescent or dead, are part of the **rhytidome** (Sect. 21.3.4).

Rhytidome is a composite of tissues of different kinds and origins. Although most rhytidome tissues are dead, necrotic, or senescent, tissues recently relegated to the rhytidome by differentiation of a sequent periderm may be briefly capable of further development, including dilatation growth, suberization, and sclerification.

In woody axes, the periderm that generally arises in the subepidermis or cortex during the first or second growing season replaces the epidermis as a protective barrier. In *Fagus, Carpinus, Ilex, Prunus*, and some other genera, a first periderm may persist for decades (Sect. 21.3.2). Its phellogen initials accommodate growth in girth by anticlinal divisions, while continuing to produce derivatives by periclinal divisions. In most woody genera, however, sequent periderms eventually develop, adaxial to the first periderm.

The earliest sequent periderms generally arise in the cortex and may be relatively superficial. In most taxa, later sequent periderms arise in progressively deeper tissues until they are being initiated in older, nonconducting phloem. In transection, these sequent periderms may appear as closely paralleling each other, but they more commonly appear as overlapping arcs, inwardly convex. The lunes of dying or dead tissue isolated between these periderms become part of the rhytidome (Sect. 21.3.4). In a few taxa, including *Abies balsamea, Alnus rubra, Betula papyrifera*, and *Populus tremuloides*, living cortical tissues may persist adaxial to superficial periderms for decades (Chang 1954b). In roots, a first periderm generally arises in the pericycle, after which the cortex dies and disintegrates.

21.2 Constituents of Periderm

21.2.1 Phellogen

Initiation of a first or sequent periderm begins with renascent divisions of parenchyma cells, yielding phellogen initials. Periclinal divisions of these initials produce adaxial derivatives that differentiate into phelloderm and abaxial derivatives that differentiate into phellem.

A phellogen is a single layer of initial cells. These initials are radially narrow, polygonal rather than fusiform in tangential aspect, and slightly elongated axially, regardless of the predominant sizes and shapes of cells in the tissues in which the phellogen arose. Because phellogen initials are only about 2% as long as nearby cambial-fusiform initials (Patel 1975), the spatial pattern of phellogen-initial cells, even in a phellogen that originates from the secondary phloem, shows little relation to that of the local vascular-cambial initials. The precursors of phellogen initials commonly are daughter cells of ray and septated, axial phloem-parenchyma cells, or of parenchyma cells that arose through dilatation growth. Ray organization is lost in the phellogen, except for occasional ray "ghost" patterns of anticlinal-wall arrangements.

Phellogen initials have thin, primary walls and dense cytoplasm. Their nuclei are prominent, especially when the cells are meristematically active. These cells commonly have chloroplasts and may have tannin-filled vacuoles. In transections of vigorous periderms of many taxa, phellogen initials are seen to head radial files of phelloderm derivatives accumulated adaxially and phellem derivatives abaxially. Phellogen cells divide anticlinally in some taxa, including *Citrus limon* (Schneider 1955), though little is known about the frequency of their division. However, in individuals or taxa having extensive dilatation tissue, phellogen initials in sectors that are dilating may undergo so many anticlinal divisions that radial files are obscured and seemingly replaced by tangential files (Roth 1981). The extent of intrusive growth among phellogen initials is not known.

21.2.2 Phelloderm

Phelloderm, which consists of adaxial derivatives of a phellogen, has been little studied as a tissue. It is well developed in many tropical angiospermous trees and moderately so in many gymnosperms, but is little more than vestigial in the common temperate-zone deciduous trees (Huber 1961).

In shape and content, phelloderm cells are similar to cortical or phloem parenchyma cells (Liphschitz and Waisel 1980). Typically, they can be distinguished from these adjacent cells only by their arrangement in radial files, each file originating from a phellogen initial. Phelloderm cells, especially in first periderms, usually are rich in chloroplasts and generally have intercellular spaces allowing gas exchange. Thus they may be photosynthetically active. This specialization is particularly common in conifers such as *Picea glauca, Tsuga canadensis,* and *Abies balsamea* (Godkin et al. 1983) and in associated deciduous species such as *Populus tremuloides* (Pearson and Lawrence 1958). The phelloderm may also be chlorophyllous in relatively old stems, especially near lenticels; in *Ginkgo,* this tissue may be as much as 40 cells thick (Stahl 1873).

Many tropical taxa have very little rhytidome. The protective functions that rhytidome performs in temperate-zone species are performed instead by a multi-layered phelloderm. In some *Ficus* species, phelloderm may constitute more than one-third of bark volume and, in some *Brosimum* species, as much as two-thirds (Roth 1981).

The protective capability of phelloderm is due to sclereids, which differentiate in many of the older (still living) cell layers. Sclerification is commonly preceded or accompanied by irregular intrusive growth, which causes the sclereids to be grossly enlarged and grotesquely shaped. Phelloderm also may develop into a "palisade" of radially elongated, sclerified cells. Chattaway (1955a) found phelloderm of this kind in 71 of 272 *Eucalyptus* species she examined.

In relatively old individuals of some species, notably the common pines of the southern United States, several layers of phelloderm cells very near the phellogen of a sequent periderm have thick, lignified walls, with numerous, simple pits. In contrast, the two to eight layers of phelloderm formed earlier by the phellogen are thin walled and radially elongated (Howard 1971; Fig. 21.1).

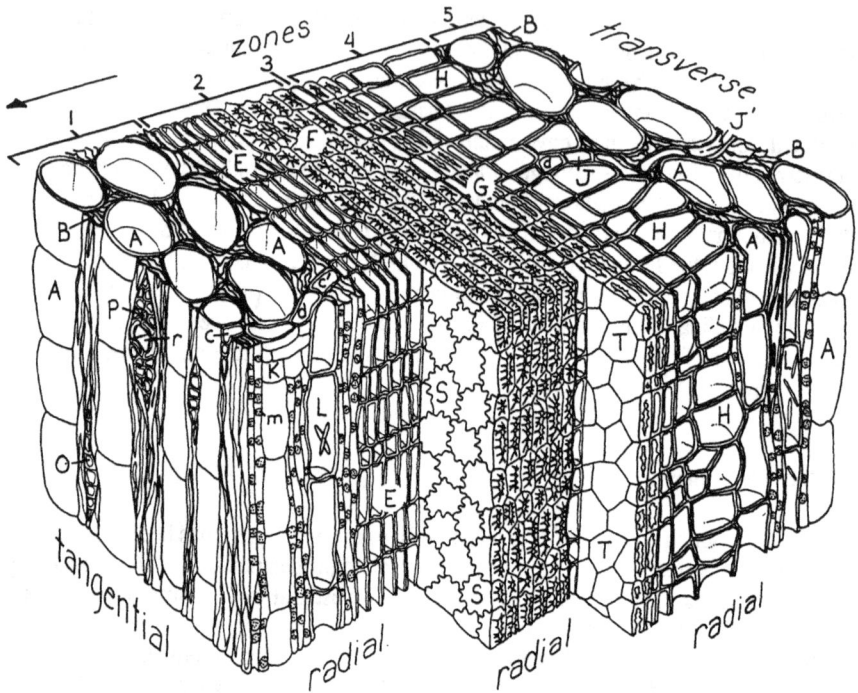

Fig. 21.1. Perspective drawing of a block of *Pinus* bark tissue centered on a sequent periderm embedded in nonfunctional phloem. The *arrow* points outward. **Transverse view:** zones *1* (older) and *5* (younger) are nonfunctional phloem originally contiguous but subsequently separated by interposition of the sequent periderm consisting of tissues in zones *2*, *3*, and *4*. Grossly expanded phloem-parenchyma cells (*A*) are surrounded by a matrix of collapsed ("obliterated") sieve cells (*B*); *C-C'* and *J-J'* mark a uniseriate phloem ray consisting of parenchyma cells (*d*). Zone *2* includes about six layers of thin-walled phellem cells (*E*) and five to six layers of interdigitated phellem sclereids having ramified pit canals (*F*). Zone *3* is a single layer of phellogen initial cells. Zone *4* is the phelloderm, including about three layers of thick-walled, unexpanded cells (*G*) and several layers of radially expanded, thin-walled cells (*H*). **Radial view:** Strasburger cells (*k*), and axial phloem parenchyma with styloid crystals (*L*) and sieve areas (*m*). **Tangential view:** uniseriate rays (*O*), a fusiform ray (*P*) with a radial resin duct (*r*), "cog-wheel-like" phellem sclereids with interdigitating "cogs" (*S*), and irregularly polygonal phelloderm cells (*T*). (Redrawn after Howard 1971)

The phelloderms of upland *Quercus* species of the southern United States (Fig. 21.2) are relatively inconspicuous. They are each six to eight cells thick, and distinguishable from phloem parenchyma only by their arrangement in radial files (Howard 1977).

As phelloderm cells differentiate, they generally separate along their edges, producing intercellular spaces. In contrast, in the phellem, there are spaces only in lenticular regions (Sect. 21.5.1; Sifton 1945).

In many taxa, recent phelloderm derivatives of a phellogen are greatly outnumbered by recent phellem derivatives. In *Nothofagus fusca*, for example, radial files of phellem

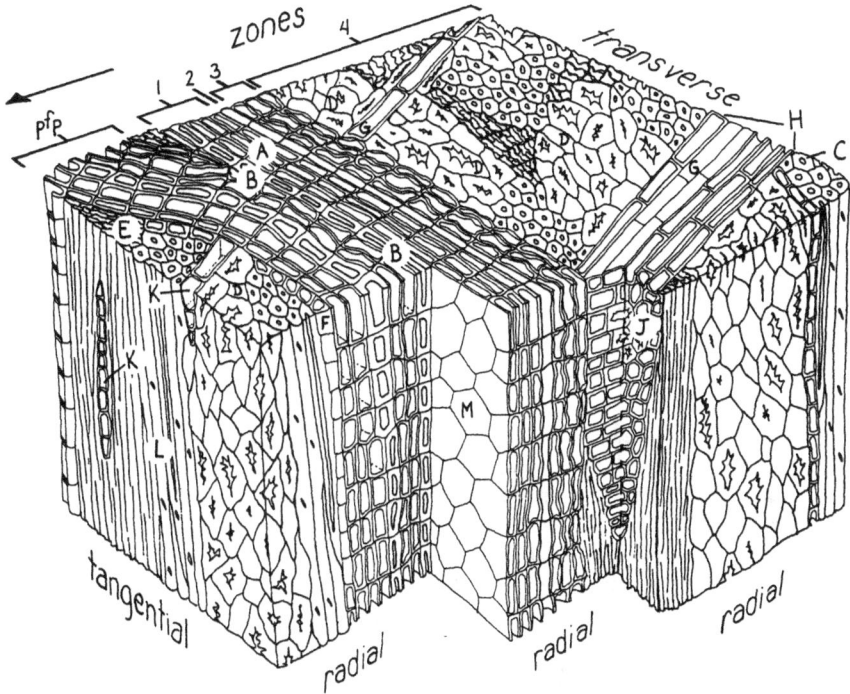

Fig. 21.2. Perspective drawing of a block of *Quercus* bark tissue including a recently formed sequent periderm and its confluence with a previously formed sequent periderm, both embedded in nonfunctional phloem. The *arrow* points outward. Peridermal tissues are interposed between younger nonfunctional phloem (zone *4*) and older nonfunctional phloem, shown on the *leftmost edge* of the leftmost radial face and the *rightmost edge* of the leftmost tangential face. *Transverse view*: zones *1*, *2*, and *3* represent the phellem, phellogen, and phelloderm, respectively, of the most recently formed sequent periderm. External to this periderm is the previously formed periderm (*pfp*), which consists of phellem and phelloderm only (as the phellogen initials have differentiated). Zone *1* consists of thin-walled (*A*) and thick-walled (*B*) phellem cells. Zone *2* is a single layer of phellogen-initial cells. Zone *3* consists of about three layers of phelloderm cells. Zone *4* is nonfunctional phloem tissue including bundles of bast fibers (*C*), groups of sclereids (*D*), collapsed ("obliterated") thin-walled phloem elements (*E*), crystalliferous phloem parenchyma adjacent to some phloem fiber groups (*F*), and ray parenchyma (*G*). *Tangential view*: features include a multiseriate ray (*H*) with some ray sclereids (*J*); two uniseriate rays (*K*); fibers with slit-like pits (*L*); and irregularly polygonal, tangential faces of phellem cells (*M*). *Radial view*: radial aspects of tissues and features already noted. (Redrawn, in part, after Howard 1977)

generally consist of six to eight times as many cells as the corresponding phelloderm files (Patel and Shand 1985). Yet, in some taxa, the phelloderm cells maintain a meristematic potential and may eventually differentiate a new phellogen just adaxial to the phellogen from which they originated. As a result, the accumulated phelloderm progeny cells may eventually far outnumber the persisting phellem progeny. Sequent periderms are commonly initiated in this way in *Citrus* (Schneider 1955).

21.2.3 Phellem

Phellem arises from abaxial derivatives of a phellogen. Immature phellem cells are parenchymatous and of the same tangential shape as phellogen initials and phelloderm cells. Early during maturation, the radial dimension of the cells increases. Mature phellem cells commonly have suberin layers. Intercellular spaces are lacking, except in lenticular regions. Generally, as cell-wall deposition ends, the lumina become repositories of tannins and/or resins. Then, within a few weeks, the protoplasts are autolyzed and dehydrated, and the lumina become irreversibly filled with gas (Liphschitz and Waisel 1980) of unknown composition. As a result of their gas-filled lumina, phellem cells form a barrier to exchange of water and gas and to heat transfer. Commercial cork, which is phellem tissue from *Quercus suber*, has particularly good sealing and insulating properties.

Though a simplification, it is conventional to consider a phellem cell wall as having primary, secondary, and tertiary layers. The primary layer, or outermost zone, consists of randomly oriented cellulose microfibrils. The secondary, or suberized, layer consists of alternating lamellae of phenolic-acid polymers and complex waxes that are not readily solvent-extractable. The tertiary layer, interior to the suberin zone, consists of waxes that are of different molecular orientation and are solvent-extractable.

Abaxial derivatives of a phellogen, though initially similar, may differentiate into several kinds of phellem cells: thick-walled, thin-walled, sclereid, crystalliferous, and phlobaphene. In many conifers, including *Tsuga canadensis*, the phellem includes both thick- and thin-walled cells. The thin-walled cells are generally arranged in continuous layers that may function as abscission zones during the eventual shedding of bark flakes (Sect. 21.7.2). In both thick- and thin-walled cells, thin suberin lamellae generally cover the pit fields in the primary layer, but may not seal all plasmodesmata (Godkin et al. 1983).

Thin suberin layers are not totally impermeable to water. However, the addition of a thin waxy layer can significantly reduce water loss through the walls. This waxy layer may also plug any remaining plasmodesmatal canals. In the phellem of *Tsuga canadensis* and *Abies balsamea*, plasmodesmata are probably functional when suberization is completed and become occluded only as waxy layers are deposited over them (Godkin et al. 1983).

Phellem cells seem to expand almost to their final radial dimension before suberin lamellae begin to be deposited. The nucleus remains large and active during suberization, then rapidly declines and degenerates as tertiary wall is deposited. By killing and fixing equivalent samples of an active periderm at short intervals, it should be possible to find living cells having some suberin lamellae but not having completed their suberization. Wattendorff (1974), however, found very few such cells in periderm of *Acacia senegal*. The indication is that suberization in any one cell is rapid and that phellem protoplasts die and disintegrate very soon after suberization is completed. Indeed, deposition of both secondary and tertiary layers is rapid in many taxa. In some, a newly formed phellem cell has already finished depositing suberin lamellae before the phellogen initial that produced that cell partitions off the next phellem derivative (Mader 1954).

Wall thickening in thick-walled phellem cells commonly is heavy on the abaxial walls, asymmetrical on the radial walls, and slight on the adaxial walls. As seen in transection, these walls appear "dome-shaped" or "U-shaped" (Grozdits 1982). Taxa may vary even within a genus, however, in the intracellular distribution of wall thickening. For example, in *Entandrophragma utile*, thickening is most prominent on the adaxial tangential walls, whereas, in *E. cylindricum*, it is most prominent on the abaxial walls (Parameswaran and Liese 1968). These unusual walls pose an interesting question of regulation of wall deposition.

In *Picea glauca*, *Tsuga canadensis*, and *Abies balsamea*, a phellogen produces thick-walled phellem cells at the beginning of each growth episode and thin-walled phellem later (Grozdits et al. 1982; Godkin et al. 1983). The phellem of most of the common pines of the southern United States also includes both thin- and thick-walled cells. The walls may be so thick that the cells become **phellem sclereids**, which are heavily lignified, have many ramified pit canals, and generally are arranged in layers five to six cells thick. As seen in tangential section, phellem sclereids in *Pinus* resemble irregularly rounded, interlocked cogwheels (Fig. 21.1). The finely controlled, patterned intrusive growth that presumably is involved in their development has not been studied. Though plentiful in periderm, phellem sclereids rarely occupy more than 10% of *Pinus* rhytidome volume. They strengthen the bark, which lacks bast fibers (Holdheide 1951; Howard 1971).

The wall thickenings of phellem sclereids of *Picea abies* are composed of lamellae of cellulose microfibrils that are alternately parallel and transverse to the cell's major axis. These lamellae, which may appear as light and dark bands in electronmicrographs (Parameswaran, et al. 1976), are possibly helicoidal in structure (Sect. 1.4.5).

Although most of the abaxial derivatives of a phellogen become heavily suberized, layers of phellem cells in some taxa may remain unsuberized or only lightly suberized. These **phelloid** layers alternate with heavily suberized phellem layers. In *Betula alba* and some related species, the phelloid cells contain a white pigment, betulin, and also serve as laminar abscission zones. Once abscission of outer periderms begins, the bark appears whitish, due to the betulin.

Living, mature phellem cells are characteristic of *Acacia seyal*. They seem to function both as a substitute epidermis on twigs and as chlorenchyma, helping the trees survive when they are leafless, during the long dry season (Hagerup 1930).

Crystalliferous and phlobaphene cells in the phellem are repositories of various types of crystals and of tannins, respectively. Little is known about development of these cells. In *Picea glauca*, a layer of crystalliferous phellem one to several cells thick commonly differentiates just before thick-walled phellem forms (Grozdits et al. 1982).

21.3 Development of Normal Periderm

21.3.1 Initiation of a First Periderm

Almost all woody-plant axes eventually develop a periderm that functionally replaces the epidermis (or rhizodermis in roots). A first periderm generally remains active for only a few months or years, and then is relegated to the rhytidome by differentiation of a more adaxially located, sequent periderm.

The site of origin of a first periderm varies from the epidermis and subepidermis to tissues deep within the cortex or secondary phloem. The most common site in stems is the outermost cortex (Douliot 1889; Robinson and Grigor 1963; Chiang 1978), whereas in hypocotyls and roots it is the pericycle (Waisel and Liphschitz 1975). However, in *Citrus sinensis* and some other taxa in which roots have little secondary growth, the first root periderm may arise in the outer cortex, as in stems, and later be replaced by a periderm arising in the pericycle (Hayward and Long 1942).

In seedlings, phellogen generally differentiates first in the hypocotyl (Borger and Kozlowski 1972c). In some taxa, including *Pseudotsuga menziesii*, it differentiates first in the root pericycle, only six to eight weeks after germination; differentiation then proceeds up into the hypocotyl and shoot (Smith 1958).

In twigs of *Robinia pseudoacacia*, the first phellogen appears in collenchyma in the second or third subepidermal layer (Arzee et al. 1968). In twigs of *Abies lasiocarpa*, it differentiates in the outer cortex and produces several cell layers of phellem and phelloderm within a month; the epidermis, nonetheless, may not be lost until the third or fourth growing season (Mogensen 1968).

In most taxa, the first nonlenticular periderm arises in the same tissue throughout the epicotyl, but in *Melaleuca leucadendra* it develops as a wavy band, varying in course from epidermal to deep within the cortex (Chiang 1978). Its variable depth of origin poses problems in understanding the controls of periderm initiation.

Twigs of some taxa may not initiate a first phellogen for several years. In *Negundo* (= *Acer negundo*), *Ilex*, and *Sophora*, for example, a first periderm is not initiated until a second or later xylem growth increment is formed (Borger 1973). In *Pyrus* and *Malus*, a first periderm may differentiate only in an axis that is five or more years old (Liphschitz and Waisel 1980). Environmental variables may influence the time elapsing before initiation (Sect. 21.3.5).

A first phellogen may be irregularly distributed (circumferentially) within a branch. In some species, such as *Euonymus alatus*, lateral strips of periderm, which may become aliform (Sect. 21.8.1.), arise in the first season, but periderm does not become circumfluent until the third or fourth season (Bowen 1963). In seven- to eight-year-old lateral branches of *Acer negundo*, the first periderm may still be incomplete on the lower side.

In many taxa, putative mother cells of an incipient first phellogen are distinguishable because they are rich in tannins or leucoanthocyanins. Through synchronized periclinal divisions, the mother cells produce phellogen initials. The partition walls formed in these divisions eventually constitute a near-continuous, tangential band. The first division of a new phellogen initial yields daughter cells with prominent nuclei. Commonly, the abaxial daughter becomes a phellogen initial and the adaxial one a phelloderm cell (Liphschitz and Waisel 1980). In *Fraxinus pennsylvanica* and *Pinus resinosa*, abaxial derivatives of the phellogen initials develop into phellem directly (Borger and Kozlowski 1972c), whereas in *Ailanthus altissima* and *Robinia pseudoacacia*, the abaxial derivatives function as phellem-mother cells.

The first periclinal divisions of new phellogen initials may be equal or unequal. If the phellogen is relatively superficial, first divisions tend to be unequal. Generally, the smaller daughter cell remains an initial and the larger differentiates into phellem or phelloderm. Superficial first phellogens commonly are at different developmental stages in different sectors around an axis. In contrast, if the first phellogen is deep seated, its first daughter cells tend to be of nearly equal size, and overall periderm development usually is synchronized circumferentially.

21.3.2 Persistent First Periderms

In some species, the first periderms, both in stems and roots, may persist for decades before being isolated from water and nutrients by sequent periderms. The first periderm in *Fagus* and some other taxa, including many tropical trees, may even persist for the life of the tree. In these taxa, sequent periderms generally do not differentiate, no obvious rhytidome accumulates, and, on a macro scale, the bark surface remains smooth. On a micro scale, layers of superficial phellem cells weather away in small flakes, or as powder. In some tropical trees the first periderms are persistent, but not in the usual sense (Roth 1981, see later).

In a few species, the persistence of the first periderm varies with the individual (Chang 1954b). For example, on lower stems of *Abies lasiocarpa*, this periderm may remain active for as few as 20 or as many as 100 years (Mogensen 1968). The age at which the first sequent periderm appears in this and other species is strongly influenced by environmental factors, notably light intensity (Mogensen and David 1968).

The phellogens of some "persistent" periderms are not persistent in the same way a vascular cambium is persistent. That is, the same phellogen initials may not remain continuously active over the decades. Instead, after an active growth episode, the initials may differentiate into phellem and phelloderm. Then, after a quiescent period, renascent meristematic activity in the last-formed phelloderm cells may generate new phellogen initials, which replicate the tangential arrangement of the earlier initials and continue the same radial files. In a sense, the first periderm persists, but its phellogen is episodically replaced. It is usually difficult to distinguish episodic replacement from true persistence.

21.3.3 Sequent Periderms

In most arborescent species, a first periderm is supplanted within a few years by a sequent periderm, which arises adaxial to it. In *Pinus sylvestris*, the first periderm is generally supplanted when a stem segment is 8 to 10 years old, but in *Abies, Carpinus, Betula*, and some *Quercus* species, it is supplanted only when the segment is 20 or more years old.

Of 280 tropical South American tree species studied by Roth (1981), mature specimens of about one-third had only a first periderm. Individuals of almost as many species had one or two periderms, and slightly more than a third of the total had three or more periderms. Individuals of only 3% of the species had ten or more periderms.

Temperate-zone trees tend to produce more sequent periderms than do tropical trees, but, on average, fewer than one per year. For example, in *Pinus radiata*, 25-

year-old trees had about 21 periderms near the base of the stem, and 31- and 45-year-old trees had 26 and 39 periderms, respectively (Sands 1975). In some trees, the time that elapsed between initiation of one sequent periderm and the next in the same locality is recorded by the number of annual phloem increments in the lunes of tissue between the periderms. These lunes generally consist of a single annual phloem increment in *Salix* and *Populus*, three to four in *Platanus*, four to six in *Quercus*, and five to 30 in *Acer pseudoplatanus* (Holdheide 1951).

The location of sequent periderms varies greatly. For example, in exposed twigs of *Abies lasiocarpa* sequent periderms may be formed within the cortex at intervals as short as a year, and six to eight may be formed before the phloem becomes involved (Mogensen 1968). In *contrast, in Citrus limon*, in which each sequent phellogen may arise from the youngest phelloderm layer derived from the preceding phellogen, the rhytidome may be composed almost entirely of tissues formed by a series of sequent periderms (Sect. 21.3.3; Schneider 1955). In this species, sequent periderms arise beneath small fissures in the phellem of the first periderm. The fissures are formed because of extensive local dilatation growth.

21.3.4 Periderms and the Origin of Rhytidome

Rhytidome consists of the tissues abaxial to the innermost periderm. Thus, in young plants having only a first periderm, the rhytidome consists of any primary tissues persisting abaxial to that periderm. After the first sequent periderm forms, the rhytidome consists of: the persistent, dead, primary tissues; the first periderm; and tissues between the first and the first sequent periderms.

Later in development, the number and spatial arrangement of sequent periderms have considerable effect on the properties of rhytidome, including its generally species-specific, gross appearance (Sect. 21.7.3). In many tropical genera, and some temperate-zone ones such as *Fagus*, it is not rigorously correct to speak of rhytidome, because there is little or no tissue abaxial to the single periderm. In many other taxa, there are only a few sequent periderms. For example, in *Nothofagus*, the first periderm remains active for many years; even in old stems, the rhytidome usually consists of this periderm alone, or of two overlapping periderms with thin lunes of included, dead phloem tissue (Patel and Shand 1985).

Though periderm is the delimiter of rhytidome, the lunes of cortical or phloem tissue relegated to rhytidome are not entirely passive while they are being transformed to periderm. Rather, each layer of living tissue that will be isolated by a new periderm is first "prepared" anatomically and physiologically. Generally, before the phellogen that will isolate these tissues appears, nearby parenchyma cells undergo renewed expansion growth (Sect. 20.5.4), which results in the crushing or "obliterating" of sieve elements that have not yet collapsed (Fig. 21.1). Accordingly, the transition from old phloem to rhytidome has been called "obliteration" (Howard 1971).

In some genera, including *Pinus* and *Larix*, phloem-parenchyma cells expand by an amount far greater than the volume made available by sieve-element collapse (Shimakura 1936). In *Pinus*, during this pre-necrotic water inflation, the cross-sectional area of many phloem-parenchyma cells increases 50- to 100-fold. Later in rhytidome development, these cells become gas-filled and supplement the insulating properties of the phellem. In some genera, including *Betula*, the phloem- or cortical-parenchyma cells may undergo renascent divisions rather than gross enlargement. Further, in the Cupressaceae and Taxaceae, and in *Liriodendron*, *Larix*, and *Pseudotsuga*, some of the parenchyma cells undergo renascent growth and cell division, whereas others become sclereids.

21.3.5 The Influence of Light and Phytohormones on Phellogen Initiation and Activity

Phellogens, like vascular cambia, are influenced by environmental variables. For example, light strongly influences both the age at which an axis first initiates a phellogen, and the activity of that phellogen. Seedlings of *Pinus resinosa*, *Fraxinus pennsylvanica*, and *Robinia pseudoacacia* kept in very dim light fail to develop periderm, but, with incremental increases in light intensity, the time elapsing before initiation of a first phellogen decreases and the production of phellem increases (Borger and Kozlowski 1972d). *Pinus resinosa*, though, requires at least ten times more light energy than *Fraxinus pennsylvanica* for induction of a first phellogen (Borger and Kozlowski 1972d; Borger 1973). Light intensity also influences the genesis and activity of sequent periderms. In general, a tree exposed to high light intensity has more periderm tissue than has another of the same age and species growing under a forest canopy (Zeeuw 1941; Mogensen and David 1968).

In *Fraxinus pennsylvanica*, *Robinia pseudoacacia*, and *Ailanthus altissima* seedlings, optimal light and temperature conditions are not necessarily the same for phellogen activity as for cambial activity. Thus, physiological controls over the two meristems seem to differ (Borger and Kozlowski 1972a,b,e; Sect. 21.6).

21.3.6 Tissue Functioning as Periderm in Monocots

Monocot stems that are markedly thickened, as in palms, do not form true phellogens, but develop a phellem-like protective tissue. This tissue is initiated by periclinal division of subepidermal parenchyma cells. Daughters of these divisions also divide periclinally, producing radial lineages of four to eight cells. These cells become suberized. Then, deeper lying parenchyma cells similarly divide and become suberized. As seen in transection, these phellem-like tissues are arranged in irregular tangential bands that are only approximately parallel to the surface of the axis (Philipp 1923). Nonsuberized cells become embedded in this suberized tissue. Thus, a tissue analogous to, but less regularly organized than, the rhytidome of dicots and gymnosperms is formed. This kind of tissue is formed in *Dracaena*, *Cordyline*, and *Yucca* (Esau 1965a, 1977; Fahn 1974).

Because monocots lack "true" phellogens, they must also lack "true" lenticels, according to some botanists. However, the surfaces of aerial roots of palms commonly develop long fissures that may function as lenticels (Devaux 1900; Tomlinson 1961).

21.4 Wound-Induced Periderms: Developmental Relations

Periderms may be induced by wounding that results from mechanical injury, as from hailstones, grazing animals, human activities, and insect attacks. The more subtle damage inflicted by pathogens may also lead to formation of periderms, as may the abscission of plant parts, and fissuring of tissues due to growth stresses. Once initiated, wound-induced periderms develop much as first and sequent periderms do.

In general, there are two main groups of periderms, based on differences in pigmentation and fluorescence properties (Mullick 1971). In gymnosperms, at least, sequent periderms delineating rhytidomes that include massive dead or dying tissues seem to have histochemical and fluorescence characteristics in common with wound periderms. Both are also reddish-purple, in contrast to some other sequent periderms and the first periderm, all of which are brown.

On this basis, Mullick and Jensen (1973) proposed that gymnosperms have two functional kinds of periderm, **exophylactic** and **necrophylactic**. Exophylactic periderms protect living tissues from the vagaries of the external environment. In contrast, necrophylactic periderms protect living tissues from adverse effects of proximity to dead or dying tissues. Both wound periderm and reddish-purple sequent periderms in gymnosperms may, on a histochemical basis, be considered necrophylactic periderms (Biggs et al. 1984). This terminology is also applicable to some angiosperms.

A possible functional basis for the histochemical similarity between reddish-purple sequent periderms and wound periderms is that peridermal and nearby tissues tear and fissure due to growth strains. Although these fissures are a normal part of growth, they may damage living cells and thus induce necrophylactic sequent periderms.

Undoubtedly, the ability of a plant rapidly to develop necrophylactic, or wound, periderms has great survival value. At the time of injury, the protoplast of an afflicted cell is part of the greater symplasm, and the microcapillary spaces within the walls are part of the greater apoplasm. As the cellular membranes lose their integrity, exposing the protoplast contents to the external environment, the integrity of the greater symplasm is breached. As the walls are torn, broken, and exposed to the environment, the integrity of the greater apoplasm also is lost. Thus, the living cells near a wound (or growth-stress fissure) are subjected to the trauma of uncontrolled loss of water and metabolites and to uncontrolled ingress of foreign substances.

Programmed cell death, or autolysis, is metabolically controlled and does not subject nearby cells to uncontrolled loss of water and metabolites or to uncontrolled ingress of foreign substances, as wounding does. Thus, autolysis does not threaten the integrity of the symplasm and apoplasm. For example, during normal ontogeny, local groups of sieve elements that collapse and die as the phloem ceases to function commonly remain embedded in a matrix of living cells and thus are not exposed to the external environment. They constitute a lacuna in the symplasm, but their walls may still be continuous with the greater apoplasm.

Processes culminating in wound-periderm development begin very soon after the tissue has been wounded or ruptured. Cells traumatized by wounding of their neighbors, but still physically intact, may deposit callose over their pits and plasmodesmata within minutes, and over the remainder of their walls within hours (Lipetz 1970; Kauss

1987). In gymnosperms, an early response is differentiation of a thin, impervious layer of tissue (which may involve some callose deposition) from the undamaged living cells closest to the wound. This layer was at first thought to be nonsuberized (Mullick 1975), but more recent evidence indicates that its imperviousness is ascribable to very thin suberin lamellae (Biggs 1985). In the uninjured tissues beneath this impervious layer, wound periderm arises (Mullick 1975). The impervious layer, or an equivalent suberized boundary layer arising a few cell layers adaxial to the wounded zone, as in *Prunus persica* (Biggs 1986; Biggs and Stobbs 1986), is another feature distinguishing necrophylactic from exophylactic periderm, in addition to histochemical differences.

21.5 Lenticels

Lenticels are highly differentiated lens-shaped areas of periderm. They are characteristic of most stems and roots that have a distinct periderm. Because of their many intercellular spaces, lenticels have a loose structure. On young axes, lenticels appear as somewhat raised, corky spots, seemingly centered on small epidermal fissures. These fissures, along with the intercellular spaces in the lenticular periderm, facilitate gas exchange between the external atmosphere and the interior of the axis.

21.5.1 Structure and Development

Each lenticel originates from a local phellogen that becomes continuous with the nonlenticular phellogen. Lenticular phellogens have more intercellular spaces and produce derivatives at a higher rate than do nonlenticular phellogens. Because of locally rapid production, lenticels are lens shaped (hence the term "lenticel", from Latin, *lens*: lentil). In most taxa that have been investigated, cells are produced mainly on the abaxial, or phellem, side of the lenticular phellogen. In some taxa, such as *Apeiba echinata*, phelloderm constitutes most of the lenticular tissue (Roth 1981).

The lenticular phellem in some taxa is composed only of **complementary (filling) tissue**, which consists of thin-walled, spheroidal cells separated by large intercellular spaces. The cells may or may not be suberized. In other taxa, layers of complementary tissue alternate with **closing layers**, which consist of more compactly arranged cells (Fig. 21.3). Each closing layer is one or more cells thick. The cells of the closing layers are generally suberized, but small intercellular spaces that penetrate the layers permit gas exchange.

Complementary cells may be of the **tissue** type, having extensive cell-to-cell contact, or of the **powder** type, with little contact. Complementary cells of the tissue type are generally suberized; those of the powder type generally are not. In taxa without closing layers, the complementary cells tend to be compactly arranged and suberized, and thus of the tissue type. In taxa having closing layers, complementary cells tend to be loosely arranged and nonsuberized, and thus of the powder type.

On the basis of development and structure, Wutz (1955) described four kinds of lenticular phellem in plants of the north-temperate zone:

1. Gymnosperm type — phellem cells are suberized and differ from nonlenticular phellem only in having intercellular spaces;
2. *Salix* type — phellem cells are suberized and grade from thin-walled spring cells to thick-walled autumn cells (as in *Liriodendron, Magnolia, Malus, Populus, Pyrus,* and *Salix*);
3. *Sambucus* type — most of the lenticular phellem is nonsuberized, but near the end of a growing season, a layer of suberized cells, a "closing layer", is produced (as in *Fraxinus, Quercus, Sambucus,* and *Tilia*); and
4. *Prunus* type — The lenticular phellogen each year produces several series of loose, non-suberized, complementary cells, in regular alternation with compact, suberized, closing layers (Fig. 21.3). Each closing layer may be several cells deep.

A dense closing layer that is formed at the end of each growing season in many temperate- zone taxa protects underlying tissues from desiccation during winter. In the spring, it is ruptured by formation of new lenticular tissue. In this way, lenticels change their permeability to gases with the season. Further, in taxa that form several closing layers during each growing season, the layers are successively ruptured by continued phellogen activity (Fig. 21.3). Each successive layer of complementary tissue then undergoes desiccation and necrosis.

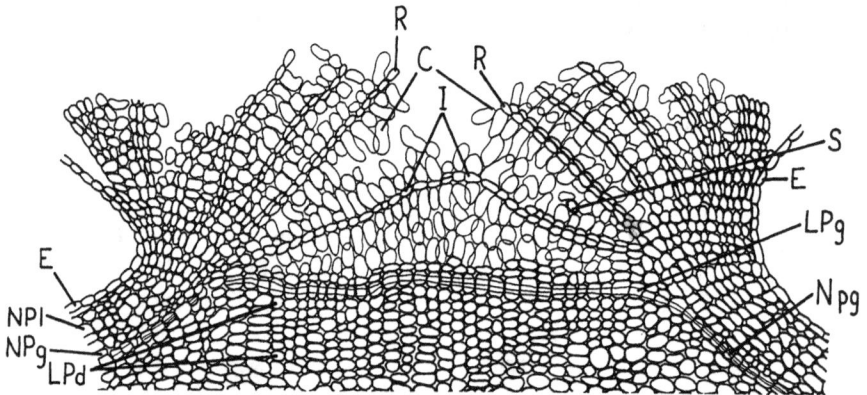

Fig. 21.3. Transection of a lenticel on a 1-year-old stem of *Coriaria myrtifolia* collected at the time of flowering. *C* Complementary (filling) cells, of which one is suberized (*S*); *R* the innermost of several ruptured closing layers; *I* the outermost of two intact closing layers; *LPg* lenticular phellogen, located adaxial to the innermost closing layer and giving rise mostly to phelloderm; *NPg* nonlenticular phellogen; *E* epidermis; *NPl* nonlenticular phellem; *LPd* file of cells in the lenticular phelloderm. (Redrawn after Devaux 1900)

Generally, lenticels of sequent periderms are similar in structure to those of first periderms, but are less well developed (Wutz 1955). In plants with fissured rhytidome, lenticels of sequent periderms form at the bases of fissures and are not easily seen.

In many dicots, the first lenticels are formed before nonlenticular periderm arises. Phellogen formation proceeds outward from the edges of the lenticular regions into nonlenticular regions. In species having long-lived epidermis, such as *Sophora japon-*

ica, *Acer negundo*, and *A. pensylvanicum*, lenticels appear in the first year of growth, long before nonlenticular periderm is formed (Bary 1877). In gymnosperms, in contrast, lenticels differentiate in regions where nonlenticular phellogens have already arisen.

In many taxa having superficial first periderms, lenticular phellogens are commonly initiated beneath cauline stomata that are considerably larger than ordinary, foliar stomata. The nature of the stimulus for formation of these lenticels is unknown, but it may be related to the perpetually open state of the pores of the large, "lenticular" stomata.

In plants in which the first periderm is deep seated, lenticels arise independently of the stomata, at the same time as the nonlenticular periderm. In stems of *Berberis*, *Ribes*, and many conifers, there are no stomata, and lenticels arise in already existing periderm by differentiation of lenticular phellogens within the nonlenticular phellogen.

The first-formed, or "primary", lenticels, along with the rest of the first periderm, may be shed from the rhytidome relatively early in development. Alternatively, they may persist for many years (Srivastava 1964), accommodating secondary growth of the axis by enlarging laterally via anticlinal cell divisions within the lenticular phellogen. As stem circumference increases in *Betula*, *Prunus*, and some other taxa, the shape of the lenticels changes from circular or axially elongated to transversely elongated, and the lenticels extend for varying distances around the circumference. In other taxa, including *Fraxinus* spp., *Ailanthus* spp., and *Quercus suber*, lateral enlargement is accompanied by differentiation of the peripheral cells of the lenticular phellogen into nonlenticular phellogen. Similar differentiation in the middle rather than at the periphery of a lenticel, as in *Pyrus malus*, *Rhamnus frangula*, and *Tsuga canadensis*, results in fragmenting of the lenticel.

In *Robinia*, formation of the first few sequent periderms begins at the margins of the first-formed (primary) lenticels. These early sequent periderms cut deeply into the secondary phloem lateral to a lenticel but do not form directly beneath the lenticel. Thus, the lenticel is not isolated by periderms and remains functional for several years. Eventually, however, a sequent periderm does form deeply beneath the lenticel, isolating it from the adaxial, living tissues. Years later, the primary lenticel is shed with the rest of the first periderm (Wutz 1955).

21.5.2 Distribution

Most woody plants have lenticels. Those that seem to lack them, including many vines, shed their outer layers of rhytidome annually and thus can maintain new tissues in close proximity to the external atmosphere.

The factors controlling lenticel distribution are complex and poorly understood. Distribution varies with the taxon, the growth rate, the developmental stage of the individual, environmental variables such as moisture, and the orientation of the plant organ in space.

In some conifers, lenticels develop on both sides of leaf bundle scars, and, in many taxa, on both sides of lateral roots (Devaux 1900, Wetmore 1926). In stems that have aliform outgrowths (wings) of cork, as in *Ulmus*, *Liquidambar*, and certain species of *Euonymus* and *Acer*, there are lenticels in the depressions between the wings.

Orientation of twigs may also influence the distribution of lenticels in some taxa. On erect twigs of *Gleditsia triacanthos*, lenticels are scattered uniformly over the surface, whereas they tend to be aggregated on the lower sides of horizontal twigs. A one-year-old horizontal twig 20 cm long had an average of 210 lenticels on the lower side and 72 on the upper. The asymmetric distribution was much less pronounced in older twigs (Haberlandt 1914).

21.6 Dynamics of Meristematic Activity and Differentiation in Lenticular and Nonlenticular Periderm

As might be expected on the basis of differing responses to environmental variables (Sect. 21.3.5), activity of nonlenticular phellogen commonly does not coincide with activity of vascular cambium, though both of these meristematic tissues may persist for many years. Nor is the meristematic activity of lenticular phellogen usually closely correlated with that of the vascular cambium or of the nonlenticular phellogen in the same region of the plant axis. Of the lateral meristems, the vascular cambium usually is the most active, followed by the lenticular phellogen and then the nonlenticular phellogen.

The activity of phellogens is both slower and less regular than that of vascular cambia. Generally, in seasonal climates, the vascular cambium each year has a single active period, lasting several months, whereas the nonlenticular phellogen is likely to have several active episodes per year (Arzee et al. 1970).

The vascular cambium in the trunk of a vigorous forest tree generally produces several to several dozen phloem derivatives, and fifty to several hundred xylem derivatives, per radial file per year. In contrast, nonlenticular phellogen usually generates a total of fewer than ten phellem and phelloderm cells per radial file per year (Liphschitz and Waisel 1980). For example, in 10- to 20-year-old *Abies alba* trees growing in Poland, a typical nonlenticular phellogen initial may undergo one division by June 1, then may not divide again until late summer or autumn, when it may produce two to four phellem cells and one to two phelloderm cells before becoming inactive for the season (Golinowski 1971).

Not only do the activities of the lateral meristems seem to be under separate controls, but the differentiation of derivatives of these meristems seems to be under different controls as well. Lenticels may be composed of alternating bands of complementary tissue and closing layers, whereas, in the same tree, there may be no alternating cellular patterns in the nonlenticular phellem or in the vascular-cambial derivatives. For example, in *Betula*, which has lenticels in which two kinds of cells are alternately produced each year, there is only a single annual increment of sieve elements (Wutz 1955), and no bands are formed in the nonlenticular periderm (Huber 1961) or in the xylem. The reverse pattern occurs in *Tilia* and *Populus*, which each year produce bands in the nonlenticular periderm (Huber 1961) and bands of sieve elements in the phloem, but no bands in the lenticels (Wutz 1955).

The alternating series of cell types that appear in the lenticular and/or nonlenticular phellem of some species suggest that endogenous, oscillating rhythms are involved in controlling the differentiation of these tissues. Similar rhythms may be involved in producing the regularly alternating patterns of cells seen in the xylem and phloem of some species (Sect. 19.5.3; 20.4.1).

21.7 Dynamics of Periderm and Bark

21.7.1 Tangential Strain

The phloem-periderm-rhytidome tissue system is dynamic. This dynamism, which is usually not apparent from the slowly changing surface appearance of the bark, is driven mostly by xylem production.

Accumulating xylem evokes tangential strain in all tissues abaxial to it. Vascular-cambial initials accommodate the strain by anticlinal divisions and tangential growth. Living bark components accommodate strain by general symplastic growth, and, in some taxa, by localized development of dilatation tissues in the phloem. The rhytidome, consisting primarily of dead tissues, accommodates tangential strain by fissuring.

Of course, anticlinal, multiplicative divisions in the vascular cambium result in increases in the number of phloem-mother cells and recent phloem derivatives. These phloem tissues, while themselves subjected to strain generated by xylem increments, contribute additional tangential strain to more abaxially located tissues. Phellogens, as they produce phelloderm and phellem derivatives, add still further tangential strain.

Attempts have been made to quantify relations between the xylem/phloem incremental growth ratio and the production of dilatation tissues. However, accurate determination of xylem/phloem growth ratios may be difficult, especially in species not having distinct annual xylem growth increments (Whitmore 1962a,b, 1963). If annual xylem increments are distinct, as in the Fagaceae, relative xylem/phloem growth rates and the mean phloem increment can be estimated, as follows: first, the amount of dilatation tissue is measured. Because the amount of this tissue is proportional to the xylem/phloem growth ratio, and because the mean xylem increment can be calculated directly from measurements of xylem increments, the mean phloem increment can then be estimated. By this method, Whitmore (1963) found that the phloem growth rate in *Fagus sylvatica* is very low, even lower than the rates in those Dipterocarpaceae that have similarly smooth bark.

Generally, as phloem incremental growth rates increase, the rhytidome becomes more fissured and furrowed and the number of sequent periderms increases. Thus, one would expect that those rare individual trees of *Fagus sylvatica* trees that have deeply fissured rhytidome (such as those growing in a forest near Rogów, Poland) would have much higher phloem incremental growth rates than would their smooth-barked neighbors. In theory, the degree of fissuring of rhytidome could be quantified to reveal the amount of tangential strain that had developed before the fissuring occurred. More research on these relations is needed.

21.7.2 Bark Thickness: Accumulation Versus Shedding

The radial thickness of xylem produced during the life of a tree is easily determined because xylem accumulates and is preserved. Phloem and periderm derivatives, in contrast, generally do not accumulate indefinitely. In most taxa, there is episodic or continuous flaking or weathering away at the rhytidome surface. This loss may be so slow that the bark accretes with time, but commonly the loss is nearly equivalent to the amount of new phloem and periderm tissues produced. The relation between bark thickness and diameter of the axis is highly variable, both among species and with the ontogenic stage and vigor of the individual.

Bark thickness generally ranges from 1 mm, as in *Clematis*, and 10 to 15 mm, as in *Juglans* and *Fagus*, to some tens of centimeters, as in some millennial conifers. Bark thickness can be expressed, but not measured, by the number of annual phloem increments incorporated therein. Generally, there are one or two in *Clematis* and *Vitis*, five to seven in *Robinia*, as many as 100 in *Carpinus*, and more than 200 in very old, smooth-barked *Fagus* specimens. Old phloem increments sequestered between old periderms, of course, are extremely compressed, distorted, and hard to count (Hold-heide 1951). Bark tends to be thickest in vigorous individuals, in all age groups. The extent to which bark patterns on different parts of a tree, and on different individuals of a species, can reveal the condition and relative vigor of a forest stand is under-appreciated.

Bark-thickness data covering many years for the same individual, or for individuals of the same species on similar sites, are lacking. In general, however, bark thickness seems to increase rapidly in young axes and then to approach a balance between accretion and loss. Among 77 European species, bark generally composed a smaller percentage of stem radius in older stems than in younger: 20% in five-year-old *Clematis*, 10% in eight-year-old *Fagus*, and 1% in 470-year-old *Quercus* stems (Hold-heide 1951). In *Sequoia sempervirens*, the rhytidome is so resistant to exfoliation and weathering that its thickness increases almost linearly with time (Borger 1973), and on old trees may exceed 30 cm (Isenberg 1943).

Bark tissues change in composition as secondary tissues accrete and primary tissues are shed. Composition changes even in species that have a long-lived first periderm. For example, the bark of a one- or two-year-old *Populus tremuloides* twig consisted of 10% phellem; 45% cortex, pericycle, and endodermis; and 45% secondary phloem. In a 30-year-old axis of the same species on the same site, the phellem composed only 2%; the phelloderm, cortex, pericycle, and endodermis combined only about 25%; and the secondary phloem about 73% (Rees and Shiue 1957-1958).

In most but not all temperate-zone trees, the rhytidome tends to exfoliate and be shed as macro-sized flakes, chunks, or strips. This tendency results in surface patterns and textures that on mature trees are quite characteristic of the species (Sect. 21.7.3). A majority of tropical tree species also have scaly, fissured, or sculptured rhytidomes, which exfoliate and are shed, as in temperate-zone species. However, about one-third of 267 species of the Venezuelan-Guyanan forest studied by Roth (1981) form such narrow, shallow arcs of sequent periderm that the exfoliating scales are commonly only 0.5 mm wide. Even old trees of these species generally accumulate little rhytidome, and appear smooth barked. The protective function in these trees may be assumed by a thick phelloderm.

21.7.3 Developmental Bases of Bark Types

The bark of a tree generally passes through developmental stages. The appearance of the bark differs at each stage. The surface pattern, color, and texture that are character-istic of the one or two stages that persist the longest are useful in identifying species and in assaying growth vigor. Commonly, these attributes in combination are so distinctive that an experienced person can identify a species on sight, with little awareness of the criteria he or she uses.

Bark appearance is determined by some or all of the following factors, depending on the developmental stage: the way in which tangential strain is accommodated in the epidermis and first periderm; the depth and frequency of sequent-periderm formation; the volume and arrangement of phloem fibers and sclereids; the disposition within a periderm of mechanically weak phellem or other tissues that can serve as fracture planes during exfoliation; the number, type, and arrangement of lenticels; and pigmen-tation of the various tissue layers, along with resistance of the pigments to bleaching and leaching. We will mention some of the developmental relations underlying the major kinds of bark.

Smooth bark develops if the first periderm is superficial and persistent, or if sequent periderms are shallow and narrow. These superficial periderms give rise to very small exfoliating scales. Smooth bark is characteristic of *Populus tremuloides* and *Fagus sylvatica*, which have first periderms that persist for many decades, commonly for the life of the tree (Kaufert 1937; Borger 1973). Although *Citrus limon* forms numerous sequent periderms, each differentiates in the phelloderm of its predecessor, thus maintaining the superficial position of the periderm (Schneider 1955). This develop-mental pattern is common in tropical species (Roth 1981). Superficial periderms persist without fissuring because the phellogens and some of their living derivatives, and the cortical cells beneath, undergo frequent radial anticlinal divisions. For example, a four-year-old twig of *Populus tremuloides* has about 18-fold more cortical cells around its periphery than has a one-year-old twig (Rees and Shiue 1957-1958). As a result of this expansion, the primary lenticels extend into tangential streaks. Such lenticels are also very prominent on *Betula* and *Prunus* (Sect. 21.5.1).

Rough, furrowed rhytidomes develop after some decades in species in which the sequent periderms are deeply located and in which the phloem has a large component of sclerenchyma, especially of bast fibers. These fibers may form a three-dimensional network that is continuous through the sequent periderms. In *Sequoia sempervirens*, for example, the interlocked fibers inhibit the exfoliation of "flakes" delineated by the peridermal arcs, and a thick, fibrous, furrowed rhytidome accumulates (Chang 1954a). The furrows develop because the rhytidome, being dead, can accommodate the tangential expansion of the stem only by fissuring. The pronounced ridges on *Quercus prinus* bark form in the same manner. Rhytidome of the various species of *Salix*, *Quercus*, *Ulmus*, and *Celtis* is furrowed because of the admixture and arrangement of phellem and bast fibers and other sclerified elements (Borger 1973).

A net-like or criss-crossed pattern of short ridges with rhomboid hollows between them is common in rhytidome of *Fraxinus*, *Tilia*, *Juglans*, and *Liriodendron*, and in some species of *Carya* and *Quercus*. This pattern is due mostly to the cohesive strength of phloem-fiber bundles and their adhesion to neighboring tissues. As the circumference of the axis increases, the strain in the rhytidome causes the fibers to separate along lines of least resistance, forming the net-like pattern. The overall, major orientation of the fissures in these patterns reveals the orientation of the phloem fibers, which, in turn, indicates the orientation of the fusiform initials that gave rise to the fibers. Thus, fissured patterns having a spiral component suggest spiral-grained xylem.

If phloem fibers are sparse or lacking, as in *Pinus*, *Picea*, some species of *Acer* and *Carya*, and many other taxa, then weathering may cause the rhytidome surface on old trees to break up into flakes or scales. However, because phellem has some cohesive strength, these flakes and scales may persist for many years (Holdheide 1951).

In "ring bark", which is present in Cupressaceae and in *Vitis*, *Clematis*, and some other genera, the first periderm is circumfluent and deeply seated in the secondary phloem. Sequent periderms are concentric to the first periderm. Thus, increments to the rhytidome, which are usually added annually, are concentric sheaths, as seen three dimensionally (though they appear as "rings" in transection). In some species, the tissues tend to exfoliate as narrow strings or ribbons (Chattaway 1955b; Borger 1973).

21.7.4 Annual Rings in Phloem and Rhytidome

Annual incremental growth rings may be discernible in the phloem adaxial to the innermost periderm, and also in some lunes of phloem tissue embedded in the rhytidome by periderms. Rings may also be discernible within the periderms, but these rings can rarely be counted with much confidence.

Of the 77 temperate-zone woody species studied by Holdheide (1951), 55 (all of which were trees) had bark with distinct annual phloem rings, 15 (all shrubs) had no recognizable phloem rings, and 7 could not be categorized. In the phloem, as in the xylem, the basis for distinguishing annual-ring boundaries is the contrast between the generally wide elements formed in the spring and the narrower ones formed later in the growing season (Sect. 20.4.2). In dicots, in particular, late phloem consists predominantly of rather narrow parenchyma cells, which contrast strongly with the wide-lumened sieve elements formed the following spring. In gymnosperms, in which late-season phloem-parenchyma layers are not always present, a more reliable ring-boundary indicator is an abrupt change in lumen width of the sieve cells.

In genera having a persistent first periderm, including *Fagus*, *Carpinus*, and *Betula*, the phloem increments are relatively the same each year, regardless of growing conditions. Generally, they consist of a narrow band of sieve elements, followed by a band of parenchyma one or two cells thick.

In a few species, differences among the phellem cells that develop within a season are so sharp that annual phellem increments can be distinguished. In *Abies lasiocarpa*, for example, after the first year, there is, within each annual phellem increment, a single layer of cells in which the abaxial tangential walls are markedly thickened (Sect. 21.2.3). The increments are not always discrete and continuous around the stem, but their age can nevertheless be estimated (Mogensen 1968).

21.8 Anomalous Periderm

21.8.1 Aliform Periderm

The twigs of various species of *Euonymus*, *Ulmus*, *Acer*, *Liquidambar*, *Aristolochia*, and a few other genera commonly develop ridges of superficial phellem tissue that are so pronounced as to justify the terms "wing" or "winged cork". The general term **aliform periderm** refers to the gross morphology of these structures, of which there are two known developmental types. In one type, best known in *Euonymus*, localized strips of suberized subepidermal and cortical cells, and subsequently strips of periderm, develop along the twig before the first periderm becomes circumfluent (Bowen 1963). In the other type, common in some species of *Ulmus* and *Acer*, the aliform ridges arise only after a circumfluent first periderm has formed and has developed fissures. Details of aliform-periderm development in other genera are little known.

Development of aliform ridges on first-year *Euonymus alatus* twigs is especially interesting, because notable suberization precedes phellogen differentiation. "Wing" development begins when a groove forms in the caulis above the axil of a lateral bud, soon after internodal extension ends. The groove is delineated by locally reduced cell division rates and by local precocious maturation of cortical cells. Subepidermal and cortical cells along the bottom of the groove become suberized. Periclinal divisions in cortical cells beneath the suberized cells then lead to organization of a local phellogen, which begins producing phellem in regular radial files. The phellem cells elongate radially and become heavily suberized. Their elongation fills in the groove and bursts the epidermis, exposing the phellem to the air and forming a ridge. After several years, the phellogens beneath the wings become inactive. Ordinary first periderms then develop between aliform ridges (wings) and gradually become circumfluent.

The aliform ridges on twigs of *Acer campestre* and *Ulmus suberosa*, which arise only after circumfluent first periderm has formed, usually develop on the lower internodes of vigorous shoots. Due to tangential growth strains, this periderm develops fissures over sectors of vigorous expansion growth, usually over the strongest leaf traces. The fissures separate strips of original periderm. The phellogens of these strips, by hyperactivity, then produce large volumes of radially elongated phellem cells, which become elevated into aliform ridges, or "wings" (Smithson 1952, 1954).

21.8.2 Interxylary Periderm

Interxylary periderm is a feature of at least ten families of dicots. Generally, it is related to the fission of perennating axes that occurs when ephemeral parts or shoots, commonly flowering shoots, become seasonally necrotic. This fission is significant in the ontogeny of some desert shrubs (Ginzburg 1963). A developmentally different kind of interxylary periderm arises in some woody Asteraceae, notably in various species of *Artemisia*. In these, there are distinctive suberized layers between successive annual xylem growth increments (Moss and Gorham 1953).

In *Artemisia tridentata*, the common sagebrush of the American West, the stem surface is protected by periderm derived from an ordinary phellogen. In addition, at the beginning of each growing season, the vascular cambium produces a single layer of xylemward derivatives that do not differentiate into xylem elements, but instead slowly become phellogen initials, while ordinary xylem elements accumulate between them and the vascular cambium. During the drought of mid-summer, while the vascular cambium may still be active, this interxylary phellogen produces several layers of derivatives. These derivatives and the phellogen initials then become heavily suberized and die as the ray cells of the preceding annual increment become suberized. Suberization in the interxylary periderm closely parallels that in the superficial periderm. By the end of summer, the most recently formed annual xylem increment is sealed off from the older wood, all of which is dead (Diettert 1938). Functionally, this suberization may isolate the apoplasm that is the immediate environment of the perennating symplasm of the plant from the apoplasm of the much greater mass of dead tissue of the old wood. This isolation probably helps conserve water in the arid environment in which *Artemisia tridentata* grows (Moss and Gorham 1953; Ginzburg 1963).

References

Abbe LB, Crafts AS (1939) Phloem of white pine and other coniferous species. Bot Gaz 100:695-722

Åberg B (1957) Auxin relation in roots. Annu Rev Plant Physiol 8:153-180

Albersheim P (1975) The walls of growing plant cells. Sci Am 232(4):80-95

Alfieri FJ, Evert RF (1968) Seasonal development of the secondary phloem in *Pinus*. Am J Bot 55:518-528

Alfieri FJ, Kemp RI (1983) The seasonal cycle of phloem development in *Juniperus californica*. Am J Bot 70:891-896

Allaway WG, Carpenter JL, Ashford AE (1985) Amplification of intersymbiont surface by root epidermal transfer cells in the *Prisonia* mycorrhiza. Protoplasma 128:227-231

Allen GS (1946) Embryogeny and development of the apical meristems of *Pseudotsuga*. I. Fertilization and early embryogeny. Am J Bot 33:666-677

Allen GS (1947) Embryogeny and the development of the apical meristems of *Pseudotsuga*. II. Late embryogeny. Am J Bot 34:73-80

Aloni R (1976) Polarity of induction and pattern of primary phloem fiber differentiation in *Coleus*. Am J Bot 63:877-889

Alosi MC, Calvin CL (1985) The ultrastructure of dwarf mistletoe (*Arceuthobium* spp) sinker cells in the region of the host secondary vasculature. Can J Bot 63:889-898

Ambronn H (1881) Ueber die Entwickelungsgeschichte und die mechanischen Eigenschaften des Collenchyms. Ein Beitrag zur Kenntniss des mechanischen Gewebesystems. Jahrb Wiss Bot 12:473-541

Amobi CC (1972) Multiple bud growth and multiple wood formation in the seedlings of *Monodora tenuifolia* Benth. Ann Bot 36:199-205

Aneli NA (1975) Atlas epidermy lista. Metsniereba, Tbilisi (in Russian)

Angulo Carmona AF (1974) La formation des nodules fixateurs d'azote chez *Alnus glutinosa* (L.) Vill. Acta Bot Neerl 23:257-303

Appleby RF, Davies WJ (1983) The structure and orientation of guard cells in plants showing stomatal responses to changing vapour pressure difference. Ann Bot 52:459-468

Archer RR (1986) Growth stresses and strains in trees. Springer, Berlin Heidelberg New York

Armacost RP (1944) The structure and function of the border parenchyma and vein-ribs of certain dicotyledon leaves. Proc Iowa Acad Sci 51:157-169

Armstrong JE, Heimsch C (1976) Ontogenetic reorganization of the root meristem in the Compositae. Am J Bot 63:212-219

Armstrong W (1968) Oxygen diffusion from the roots of woody species. Physiol Plant 21:539-543

Armstrong W (1982) Waterlogged soils. In: Etherington JR (ed) Environment and plant ecology, 2nd edn. Wiley, New York, pp 290-330

Arnott HJ (1962) The seed, germination, and seedling of *Yucca*. Univ Calif Public in Bot 35(1):1-164

Artschwager E (1950) The time factor in the differentiation of secondary xylem and phloem in pecan. Am J Bot 37:15-24

Arzee T, Liphschitz N, Waisel Y (1968) The origin and development of the phellogen in *Robinia pseudacacia* L. New Phytol 67:87-93

Arzee T, Waisel Y, Liphschitz N (1970) Periderm development and phellogen activity in the shoots of *Acacia raddiana* Savi. New Phytol 69:395-398

Ashworth RP (1963) Investigations into midvein anatomy and ontogeny of certain species of the genus *Ilex* L. J Elisha Mitchell Sci Soc 79:126-138

Avers CJ (1957) An analysis of differences in growth rate of trichoblasts and hairless cells in the root epidermis of *Phleum pratense*. Am J Bot 44:686-690

Avers CJ (1963) Fine structure studies of *Phleum* root meristem cells. II. Mitotic asymmetry and cellular differentiation. Am J Bot 50:140-148

Aylor DE, Parlange J-Y, Krikorian AD (1973) Stomatal mechanics. Am J Bot 60:163-171

Baas P, Schmid R, Heuven BJ van (1986) Wood anatomy of *Pinus longaeva* (bristlecone pine) and the sustained length-on-age increase of its tracheids. IAWA Bull 7:221-228

Babcock EB, Clausen RE (1927) Genetics in relation to agriculture, 2nd edn. McGraw-Hill, New York

Babu AM, Nair GM, Shah JJ (1987) Traumatic gum-resin cavities in the stem of *Ailanthus excelsa* Roxb. IAWA Bull 8:167-174

Back EL (1969) Intercellular spaces along the ray parenchyma — The gas canal system of living wood? Wood Sci 2:31-34

Bailey IW (1920) The cambium and its derivative tissues. II. Size variations of cambial initials in gymnosperms and angiosperms. Am J Bot 7:355-367

Bailey IW (1923) The cambium and its derivative tissues. IV. The increase in girth of the cambium. Am J Bot 10:499-509

Bailey IW, Swamy BGL (1949) The morphology and relationships of *Austrobaileya*. J Arnold Arbor Harv Univ 30:211-226 + 7 pl

Baird WV, Riopel JL (1984) Experimental studies of haustorium initiation and early development in *Agalinis purpurea* (L.) Raf. (Scrophulariaceae). Am J Bot 71:803-814

Balkema GH (1971) Chimerism and diplontic selection. Balkema, Rotterdam

Ball E (1952a) Experimental division of the shoot apex of *Lupinus albus* L. Growth 16:151-174

Ball E (1952b) Morphogenesis of shoots after isolation of the shoot apex of *Lupinus albus*. Am J Bot 39:167-191

Bange GGJ (1953) On the quantitative explanation of stomatal transpiration. Acta Bot Neerl 2:254-297

Bannan MW (1941a) Variability of wood structure in roots of native Ontario conifers. Bull Torrey Bot Club 68:173-194

Bannan MW (1941b) Vascular rays and adventitious root formation in *Thuja occidentalis* L. Am J Bot 28:457-463

Bannan MW (1942) Notes on the origin of adventitious roots in the native Ontario conifers. Am J Bot 29:593-598

Bannan MW (1951) The reduction of fusiform cambial cells in *Chamaecyparis* and *Thuja*. Can J Bot 29:57-67

Bannan MW (1953) Further observations on the reduction of fusiform cambial cells in *Thuja occidentalis* L. Can J Bot 31:63-74

Bannan MW (1955) The vascular cambium and radial growth in *Thuja occidentalis* L. Can J Bot 33:113-138

Bannan MW (1957) Girth increase in white cedar stems of irregular form. Can J Bot 35:425-434

Bannan MW (1962) The vascular cambium and tree-ring development. In: Kozlowski TT (ed) Tree growth. Ronald Press, New York, pp 3-21

Bannan MW (1965) The rate of elongation of fusiform initials in the cambium of Pinaceae. Can J Bot 43:429-435

Barghoorn ES (1940a) Origin and development of the uniseriate ray in the Coniferae. Bull Torrey Bot Club 67:303-328

Barghoorn ES (1940b) The ontogenetic development and phylogenetic specialization of rays in the xylem of dicotyledons. I. The primitive ray structure. Am J Bot 27:918-928

Barghoorn ES (1964) Evolution of cambium in geologic time. In: Zimmermann MH (ed) The formation of wood in forest trees. Academic Press, New York, pp 3-17

Barlow PW (1976) Towards an understanding of the behaviour of root meristems. J Theor Biol 57:433-451

Barlow PW (1982) 'The plant forms cells, not cells the plant': The origin of de Bary's aphorism. Ann Bot 49:269-271

Barlow PW (1984) Positional controls in root development. In: Barlow PW, Carr DJ (eds) Positional controls in plant development. Cambridge Univ Press, Cambridge, pp 281-318

Barlow PW (1987) Cellular packets, cell division, and morphogenesis in the primary root meristem of *Zea mays* L. New Phytol 105:27-56

Barlow PW, Grundwag M (1974) The development of amyloplasts in cells of the quiescent centre of *Zea* roots in response to removal of the root cap. Z Pflanzenphysiol 73:56-64

Barnett JR (1974) Secondary phloem in *Pinus radiata* D. Don. I. Structure of differentiating sieve cells. NZ J Bot 12:245-260

Barnett JR (1981) Secondary xylem cell development. In: Barnett JR (ed) Xylem cell development. Castle House, Tunbridge Wells (Kent), pp 47-95

Barnett JR (1987) Changes in the distribution of plasmodesmata in developing fibre-tracheid pit membranes of *Sorbus aucuparia* L. Ann Bot 59:269-279

Bary A de (1877) Vergleichende Anatomie der Vegetationsorgane der Phanerogamen und Farne. In: Hofmeister W (ed) Handbuch der Physiologischen Botanik, Bd 3. Engelmann, Leipzig

Bauch J, Liese W, Scholz F (1968) Über die Entwicklung und stoffliche Zusammensetzung der Höftupfelmembranen von Längstracheiden in Coniferen. Holzforschung 22:144-153

Bauch J, Liese W, Schultze R (1972) The morphological variability of the bordered pit membranes in gymnosperms. Wood Sci Tech 6:165-184

Beakbane AB (1961) Structure of the plant stem in relation to adventitious rooting. Nature 192:954-955

Becking JH (1975) Root nodules in non-legumes. In: Torrey JG, Clarkson DT (eds) The development and function of roots. Academic Press, London, pp 507-566

Beeson RC Jr, Montano JM, Proebsting WM (1986) A method for determining the apoplastic water volume of conifer needles. Physiol Plant 66:129-133

Behnke H-D (1971) Über den Feinbau verdickter (nacré) Wände und der Plastiden in den Siebröhren von *Annona* und *Myristica*. Protoplasma 72:69-78

Behnke H-D (1974) Companion cells and transfer cells. In: Aronoff S et al (eds). Phloem transport. NATO Advanced Study Inst Ser A-4. Plenum Press, New York, pp 153-175

Behnke H-D (1983) Special cytology: Cytology and morphogenesis of higher plant cells — Phloem. Progr in Bot 45:18-35

Behnke H-D (1984) Plant trichomes — structure and ultrastructure: General terminology, taxonomic applications, and aspects of trichome-bacteria interaction in leaf tips of *Dioscorea*. In: Rodriguez E, Healey PL, Mehta I (eds) Biology and chemistry of plant trichomes. Plenum Press, New York, pp 1-21

Behnke H-D, Barthlott W (1983) New evidence from the ultrastructural and micromorphological fields in angiosperm classification. Nord J Bot 3:43-66

Behnke H-D, Sjolund RD (eds) (1990) Sieve elements: Comparative structure, induction and development. Springer, Berlin Heidelberg New York

Behrens HM, Gradmann D, Sievers A (1985) Membrane-potential responses following gravistimulation in roots of *Lepidium sativum* L. Planta 163:463-472

Bel AJE van (1990) Xylem-phloem exchange via rays: The undervalued route of transport. J Exp Bot 41:631-644

Bel AJE van, Schoot C van der (1988) Primary function of the protective layer of contact cells: Buffer against oscillations in hydrostatic pressure in the vessels? IAWA Bull 9:285-288

Bell JK, McCully ME (1970) A histological study of lateral root initiation and development in *Zea mays*. Protoplasma 70:179-205

Benayoun J (1983) A cytochemical study of cell wall hydrolysis in the secondary xylem of poplar (*Populus italica* Moench). Ann Bot 52:189-200

Bentley BL (1977a) Extrafloral nectaries and protection by pugnacious bodyguards. Annu Rev Ecol Syst 8:407-427

Bentley BL (1977b) The protective function of ants visiting the extrafloral nectaries of *Bixa orellana* (Bixaceae). J Ecol 65:27-38

Bentrup F-W (1980) Electrogenic membrane transport in plants. A review. Biophys Struct Mechan 6:175-189

Bentrup F-W (1982) Cell electrophysiology and membrane transport. Prog Bot 44:57-63

Bentwood BJ, Cronshaw J (1978) Cytochemical localization of adenosine triphosphate in the phloem of *Pisum sativum* and its relation to the function of transfer cells. Planta 140: 111-120

Benzing DH, Ott DW, Friedman WE (1982) Roots of *Sobralia macrantha* (Orchidaceae): Structure and function of the velamen-exodermis complex. Am J Bot 69:608-614

Bergann F, Bergann L (1962) Über Umschichtungen (Translokationen) an den Sprossscheiteln periklinaler Chimären. Züchter 32:110-119

Bergersen FJ, Kennedy GS, Wittmann W (1965) Nitrogen fixation in the coralloid roots of *Macrozamia communis* L. Johnson. Aust J Biol Sci 18:1135-1142

Bergfeld R, Speth V, Schopfer P (1988) Reorientation of microfibrils and microtubules at the outer epidermal wall of maize coleoptiles during auxin-mediated growth. Bot Acta 101: 57-67

Berlyn GP (1972) Seed germination and morphogenesis. In: Kozlowski TT (ed) Seed biology, vol 1. Academic Press, New York pp 223-312

Berry AM, Torrey JG (1983) Root hair deformation in the infection process of *Alnus rubra*. Can J Bot 61:2863-2876

Bertaud DS, Gandar PW (1986) A simulation model for cell proliferation in root apices. II. Patterns of cell proliferation. Ann Bot 58:303-32

Bertaud DS, Gandar PW, Erickson RO, Ollivier AM (1986) A simulation model for cell growth and proliferation in root apices. I. Structure of model and comparisons with observed data. Ann Bot 58:285-301

Bhojwani SS, Bhatnagar SP (1978) The embryology of angiosperms. 3rd edn. Vikas Publishing House, New Delhi

Biddington NL (1986) The effects of mechanically-induced stress in plants — A review. Plant Growth Reg 4:103-123

Biesboer DD, Mahlberg PG (1978) Accumulation of non-utilizable starch in laticifers of *Euphorbia heterophylla* and *E. myrsinites*. Planta 143:5-10

Biggs AR (1985) Detection of impervious tissue in tree bark with selective histochemistry and fluorescence microscopy. Stain Technol 60:299-304

Biggs AR (1986) Phellogen regeneration in injured peach tree bark. Ann Bot 57:463-470

Biggs AR, Stobbs LW (1986) Fine structure of the suberized cell walls in the boundary zone and necrophylactic periderm in wounded peach bark. Can J Bot 64:1606-1610

Biggs AR, Merrill W, Davis DD (1984) Discussion: Response of bark tissues to injury and infection. Can J For Res 14:351-356

Björkman O, Berry J (1973) High-efficiency photosynthesis. Sci Amer 229:80-87, 91-93

Blyth A (1958) Origin of primary extraxylary stem fibers in dicotyledons. Univ of Calif Publ in Bot 30(2):145-231

Bogar GD, Smith FH (1965) Anatomy of seedling roots of *Pseudotsuga menziesii*. Am J Bot 52:720-729

Böhlmann D (1984) Reaktionsgewebe im Holz der Nadel- und Laubbäume. Mitt Dtsch Dendrol Ges 75:189-202

Boke NH (1940) Histogenesis and morphology of the phyllode in certain species of *Acacia*. Am J Bot 27:73-90

Bond G (1983) Taxonomy and distribution of non-legume nitrogen-fixing systems. In: Gordon JC, Wheeler CT (eds) Biological nitrogen fixation in forest ecosystems: Foundations and applications. Nijhoff/Junk, The Hague, pp 55-87

Bonfante-Fasolo P (1984) Anatomy and morphology of VA mycorrhizae. In: Powell CL, Bagyaraj DJ (eds) VA mycorrhiza. CRC Press, Boca Raton, pp 5-33

Bonner J, Galston AW (1947) The physiology and biochemistry of rubber formation in plants. Bot Rev 13:543-596

Borger GA (1973) Development and shedding of bark. In: Kozlowski TT (ed) Shedding of plant parts. Academic Press, New York, pp 205-236

Borger GA, Kozlowski TT (1972a) Effects of cotyledons, leaves and stem apex on early periderm development in *Fraxinus pennnsylvanica* seedlings. New Phytol 71:691-702 + 2 pl

Borger GA, Kozlowski TT (1972b) Effects of photoperiod on early periderm and xylem development in *Fraxinus pennsylvanica*, *Robinia pseudoacacia*, and *Ailanthus altissima* seedlings. New Phytol 71:703-708 + 2 pl

Borger GA, Kozlowski TT (1972c) Early periderm ontogeny in *Fraxinus pennsylvanica*, *Ailanthus altissima*, *Robinia pseudoacacia*, and *Pinus resinosa* seedlings. Can J For Res 2:135-143

Borger GA, Kozlowski TT (1972d) Effects of light intensity on early periderm and xylem development in *Pinus resinosa*, *Fraxinus pennsylvanica*, and *Robinia pseudoacacia*. Can J For Res 2:190-197

Borger GA, Kozlowski TT (1972e) Effects of temperature on first periderm and xylem development in *Fraxinus pennsylvanica*, *Robinia pseudoacacia*, and *Ailanthus altissima*. Can J For Res 2:198-205

Borghetti M, Edwards WRN, Grace J, Jarvis PG, Raschi A (1991) The refilling of embolized xylem in *Pinus sylvestris* L. Plant Cell Environ 14:357-369

Bosshard HH, Hug UE (1980) The anastomoses of the resin canal system in *Picea abies* (L.) Karst., *Larix decidua* Mill. and *Pinus sylvestris* L. Holz Roh- Werkstoff 38:325-328

Botha CEJ, Evert RF (1981) Studies on *Artemisia afra* Jacq.: The phloem in stem and leaf. Protoplasma 109:217-231

Botosso PC, Gomes AV (1982) Radial vessels and series of perforated ray cells in Annonaceae. IAWA Bull 3:39-44

Boughton VH (1981) Extrafloral nectaries of some Australian phyllodineous acacias. Aust J Bot 29:653-664

Boureau E (1939) Recherches anatomiques et expérimentales sur l'ontongénie des plantules des Pinacées et ses rapports avec la phylogénie. Ann Sci Nat Bot XI Sér 1:1-216

Bowen WR (1963) Origin and development of winged cork in *Euonymus alatus*. Bot Gaz 124:256-261

Boyd JD (1972) Tree growth stresses — Part V: Evidence of an origin in differentiation and lignification. Wood Sci Tech 6:251-262

Boyd JD (1978) Significance of laricinan in compression wood tracheids. Wood Sci Tech 12:25-35

Boyd JD (1980) Biophysical controls on cellulose formation and lignification. In: Little CHA (ed) Control of shoot growth in trees. Proc of joint workshop of IUFRO working parties on xylem physiology and shoot growth physiology. Maritimes Forest Research Centre, Fredericton, N B, Canada, July 20-24, 1980 pp 184-236

Braam J, Davis RW (1990) Rain-, wind-, and touch-induced expression of calmodulin and calmodulin-related genes in *Arabidopsis*. Cell 60: 357-364

Brański S (1970) Relations between anticlinal divisions, intrusive growth and loss of fusiform initials in the cambium of *Pinus sylvestris*. Acta Soc Bot Pol 39:593-615 (in Polish, English summary)

Braun HJ (1963) Die Organisation des Stammes von Bäumen und Sträuchern. Wissenschaftliche Verlagsgesellschaft MBH, Stuttgart

Braun HJ (1967) Entwicklung und Bau der Holzstrahlen unter dem Aspekt der Kontakt--Isolations-Differenzierung gegenüber dem Hydrosystem. I. Das Prinzipder Kontakt-Isolations-Differenzierung. Holzforschung 21:33-37

Braun HJ (1970) Funktionelle Histologie der sekundären Sprossachse. I. Das Holz. Encycl Plant Anatomy IX, 1. Borntraeger, Berlin

Braun HJ (1982) Lehrbuch der Forstbotanik. Fischer, Stuttgart

Braun HJ, Outer RW den (1965) Über die tertiären Wachstumsvorgänge im Bast von *Pseudotsuga taxifolia* (Poir.) Britton. Allg Forst Jagdztg 136:101-105

Braun HJ, Wolkinger F (1970) Zur funktionellen Anatomie des axialen Holzparenchyms und Vorschläge zur Reform seiner Terminologie. Holzforschung 24:19-26

Brawley SH, Wetherell DF, Robinson KR (1984) Electrical polarity in embryos of wild carrot precedes cotyledon differentiation. Proc Natl Acad Sci USA 81:6064-6067

Briarty LG, Coult DA, Boulter D (1970) Protein bodies of germinating seeds of *Vicia faba*. J Exp Bot 21:513-524

Brink RA (1962) Phase change in higher plants and somatic cell heredity. Q Rev Biol 37:1-22

Britz SJ (1979) Chloroplast and nuclear migration. Encycl Plant Physiol (2nd Ser) 7:170-205

Brown CL, Sax K (1962) The influence of pressure on the differentiation of secondary tissues. Am J Bot 49:683-691

Brown RC, Mogensen HL (1972) Late ovule and early embryo development in *Quercus gambelii*. Am J Bot 59:311-316

Brown RM Jr (1985) Cellulose microfibril assembly and orientation: Recent developments. In: Roberts K, Johnston AWB, Lloyd CW, Shaw P, Wollhouse HW (eds) The cell surface in plant growth and development. J Cell Sci Suppl 2:13-32

Brown WV (1975) Variations in anatomy, associations, and origins of Kranz tissue. Am J Bot 62:395-402

Brown WV, Johnson C(Sr) (1962) The fine structure of the grass guard cell. Am J Bot 49:110-115

Brutsch MO, Allan P, Wolstenholme BN (1977) The anatomy of adventitious root formation in adult-phase pecan (*Carya illinoensis* (Wang.) K. Koch) stem cuttings. Hortic Res 17:23-31

Buchen B, Sievers A (1981) Sporogenesis and pollen grain formation. In: Kiermayer O (ed) Cytomorphogenesis in plants. Cell Biol Monogr vol 8. Springer, Berlin Heidelberg New York, pp 349-376

Buchholz JT (1918) Suspensor and early embryo in *Pinus*. Bot Gaz 66:185-228 + 5 pl

Buchholz JT (1925) The embryogeny of *Cephalotaxus fortunei*. Bull Torrey Bot Club 52:311-322 + 1 pl

Buchholz JT, Old EM (1933) The anatomy of the embryo of *Cedrus* in the dormant stage. Am J Bot 20:35-44

Bunce JA (1985) Effect of boundary layer conductance on the response of stomata to humidity. Plant Cell Environ 8:55-57

Bünning E (1951) Über die Differenzierunggänge in der Cruciferen Wurzel. Planta 39:126-153

Bünning E (1952) Morphogenesis in plants. Surv Biol Prog 2:105-140

Bünning E (1965) Die Entstehung von Mustern in der Entwicklung von Pflanzen. Encycl Plant Physiol (1st Ser) 15(1):383-408

Burggraaf PD (1972) Some observations on the course of the vessels in the wood of *Fraxinus excelsior* L. Acta Bot Neerl 21:32-47

Burggraaf PD (1973) On the shape of developing vessel elements in *Fraxinus excelsior* L. Acta Bot Neerl 22:271-278

Burström HG, Uhrström I, Wurscher R (1967) Growth, turgor, water potential, and Young's modulus in pea internodes. Physiol Plant 20:213-231

Burton WG (1950) Studies on the dormancy and sprouting of potatoes. I. The oxygen content of the potato tuber. New Phytol 49:121-134

Büsgen M, Münch E (1931) The structure and life of forest trees. 3rd edn (English translation by T. Thomson). Wiley, New York

Butterfield BG, Meylan BA (1982) Cell wall hydrolysis in the tracheary elements of the secondary xylem. In: Baas P (ed) New perspectives in wood anatomy. Nijhoff/Junk, The Hague, pp 71-84

Butts D, Buchholz JT (1940) Cotyledon numbers in conifers. Ill State Acad Sci Trans 33:58-62

Buvat R (1952a) L'Organisation des méristèmes apicaux chez les végétaux vasculaire. Bull L'Union Nat L'Enseignement Public 40(4):54-66

Buvat R (1952b) Structure, évolution et fonctionnement du méristeme apical de quelques dicotylédones. Ann Sci Nat Ser 11 Bot 13:199-300

Byrne JM (1973) The root apex of *Malva sylvestris*. III. Lateral root development and the quiescent center. Am J Bot 60:657-662

Byrne JM, Heimsch C (1970a) The root apex of *Malva sylvestris*. I. Structural development. Am J Bot 57:1170-1178

Byrne JM, Heimsch C (1970b) The root apex of *Malva sylvestris*. II. The quiescent center. Am J Bot 57:1179-1184

Callaham D, Newcomb W, Torrey JG, Peterson RL (1979) Root hair infection in actinomycete-induced root nodule initiation in *Casuarina*, *Myrica*, and *Comptonia*. Bot Gaz 140:51-59 (Suppl)

Calow pp (1976) Biological machines — A cybernetic approach to life. Arnold, London

Camefort H (1956) Étude de la structure du point végétatif et des variations phyllotaxiques chez quelques gymnospermes. Ann Sci Nat Ser 11 Bot Biol Vég 17:1-185 + 9 pl

Camefort H (1969) Fécondation et proembryogénèse chez les Abiétacées (notion de néocytoplasme). Rev Cytol Biol Vég 32:253-271

Campbell R (1972) Electron microscopy of the development of needles of *Pinus nigra* var *maritima*. Ann Bot 36:711-720 + 5 pl

Canny MJ (1960) The rate of translocation. Biol Rev Camb Philos Soc 35:507-532

Canny MJ (1990) Fine veins of dicotyledon leaves as sites for enrichment of solutes of xylem sap. New Phytol 115:511-516

Carde J-P (1978) Ultrastructural studies of *Pinus pinaster* needles: The endodermis. Am J Bot 65:1041-1054

Carlquist S (1975) Ecological strategies of xylem evolution. Univ California Press, Berkeley

Carlson MC (1938) The formation of nodal adventitious roots in *Salix cordata*. Am J Bot 25:721-725

Carlson MC (1950) Nodal adventitious roots in willow stems of different ages. Am J Bot 37:555-561

Carr DJ (1976) Plasmodesmata in growth and development. In: Gunning BES, Robards AW (eds) Intercellular communication in plants: Studies on plasmodesmata. Springer, Berlin Heidelberg New York, pp 243-289

Carr DJ, Carr SGM, Jahnke R (1980a) Intercellular strands associated with stomata: Stomatal pectic strands. Protoplasma 102:177-182

Carr DJ, Oates K, Carr SGM (1980b) Studies on intercellular pectic strands of leaf palisade parenchyma. Ann Bot 45:403-413

Carrodus BB, Triffett ACK (1975) Analysis of composition of respiratory gases in woody stems by mass spectrometry. New Phytol 74:243-246

Catesson A-M (1964) Origine, fonctionnement et variations cytologiques saisonnières du cambium de l'*Acer pseudoplatanus* L. (Acéracées). Ann Sci Nat Ser 12 Bot 5:229-498

Catesson A-M (1974) Cambial cells. In: Robards A W (ed) Dynamic aspects of plant ultra-structure. McGraw-Hill, London, pp 358-390

Catesson A-M (1980) The vascular cambium. In: Little CHA (ed) Control of shoot growth in trees. Proc Joint Workshop of IUFRO Working Parties on Xylem Physiology and Shoot Growth Physiology. Maritimes Forest Research Centre, Fredericton, New Brunswick, Canada pp 12-40

Catesson A-M (1982) Cell wall architecture in the secondary sieve tubes of *Acer* and *Populus*. Ann Bot 49:131-134

Catesson A-M (1990) Cambial cytology and biochemistry. In: Iqbal M (ed) The vascular cambium. Research Studies Press, Taunton (Somerset). Wiley, New York pp 63-112

Catesson A-M, Roland J-C (1981) Sequential changes associated with cell wall formation and fusion in the vascular cambium. IAWA Bull 2:151-162

Chabot JF, Chabot BF (1975) Developmental and seasonal patterns of mesophyll ultrastructure in *Abies balsamea*. Can J Bot 53:295-304

Chaboud A, Rougier M (1986) Ultrastructural study of the maize epidermal root surface. I. Preservation and extent of the mucilage layer. Protoplasma 130:73-79

Chafe SC (1974a) Cell wall formation and "protective layer" development in the xylem parenchyma of trembling aspen. Protoplasma 80:335-354

Chafe SC (1974b) Cell wall formation in the xylem parenchyma of *Cryptomeria*. Protoplasma 81:63-76

Chafe SC, Doohan ME (1972) Observations on the ultrastructure of the thickened sieve cell wall in *Pinus strobus* L. Protoplasma 75:67-78

Chaffey NJ, Harris N (1985) Plasmatubules: Fact or artefact? Planta 165:185-190

Chaguturu R (1981) C_3-C_4 intermediate species. What's New Plant Physiol 12:21-24

Chalain TMB de, Berjak P (1979) Cell death as a functional event in the development of the leaf intercellular spaces in *Avicennia marina* (Forsskål) Vierh. New Phytol 83:147-155

Chamberlain CJ (1919) The Living Cycads. Univ Chicago Press, Chicago

Chang Y-P (1954a) Bark structure of North American conifers. US Dept Agric Tech Bull 1095:1-86

Chang Y-P (1954b) Anatomy of common North American pulpwood barks. TAPPI Monogr Ser 14

Charlton WA (1980) Primary vascular patterns in root meristems of *Pontederia cordata* and their relevance to studies of root development. Can J Bot 58:1351-1369

Chase WW (1934) The composition, quantity, and physiological significance of gases in tree stems. Univ Minn Agric Exp Stn Tech Bull 99:1-51

Chattaway MM (1948) The wood anatomy of the Proteaceae. Aust J Sci Res B 1:279-302 + 7 pl

Chattaway MM (1951) Morphological and functional variations in the rays of pored timbers. Aust J Sci Res B 4:12-27

Chattaway MM (1955a) The anatomy of bark. IV. Radially elongated cells in the phelloderm of species of *Eucalyptus*. Aust J Bot 3:39-47

Chattaway MM (1955b) The anatomy of bark. V. Eucalyptus species with stringy bark. Aust J Bot 3:165-168 + 2 pl

Cheadle VI, Esau K (1958) Secondary phloem of Calycanthaceae. Univ Calif Publ Bot 29(4): 397-509

Cheadle VI, Esau K (1964) Seconday phloem of *Liriodendron tulipifera*. Univ Calif Publ Bot 36(2):143-252

Chesnoy L, Thomas MJ (1971) Electron microscopy studies on gametogenesis and fertilization in gymnosperms. Phytomorphology 21:50-63

Chessin M, Zipf AE (1990) Alarm systems in higher plants. Bot Rev 56:193-235

Chiang Su-HT (1978) The origin of the first periderm in some dicotyledonous stems. Taiwania 23:61-65

Chouinard L (1959) Sur l'existence d'un centre quiescent au niveau de l'apex radiculaire juvénile de *Pinus banksiana* Lamb. Laval Univ For Res Found Contrib 4 (II):27-31

Christianson ML (1986) Fate map of the organizing shoot apex in *Gossypium*. Am J Bot 73:947-958

Christodoulakis NS, Psaras GK (1987) Stomata on the primary root of *Ceratonia siliqua*. Ann Bot 60:295-297

Church AH (1920) On the interpretation of phenomena of phyllotaxis. Bot Mem 6:1-58 + 18 figs

Clarke AE, Anderson MA, Bacic T, Harris PJ, Mau S-L (1985) Molecular basis of cell recognition during fertilization in higher plants. J Cell Sci Suppl 2:261-285

Clarkson DT, Robards AW (1975) The endodermis, its structural development and physiological role. In: Torrey JG, Clarkson DT (eds) The development and function of roots. Academic Press, London, pp 415-436

Clowes FAL (1950) Root apical meristems of *Fagus sylvatica*. New Phytol 49:248-268 + 1 pl

Clowes FAL (1951) The structure of mycorrhizal roots of *Fagus sylvatica*. New Phytol 50:1-16

Clowes FAL (1954a) The promeristem and the minimal constructional centre in grass root apices. New Phytol 53:108-116

Clowes FAL (1954b) The root cap of ectotrophic mycorrhizas. New Phytol 53:525-529 + 1 pl

Clowes FAL (1956) Localization of nucleic acid synthesis in root meristems. J Exp Bot 7:307-312

Clowes FAL (1958) Development of quiescent centres in root meristems. New Phytol 57:85-88

Clowes FAL (1961) Apical meristems. Blackwell, Oxford

Clowes FAL (1970) The immediate response of the quiescent centre to X-rays. New Phytol 69:1-18

Clowes FAL (1971) The proportion of cells that divide in root meristems of *Zea mays* L. Ann Bot 35:249-261

Clowes FAL (1975) The quiescent centre. In: Torrey JG, Clarkson DT (eds) The development and function of roots. Academic Press, London New York San Francisco, pp 3-19

Clowes FAL (1976) The root apex. In: Yeoman MM (ed) Cell division in higher plants. Academic Press, London, pp 253-284

Clowes FAL (1981a) The difference between open and closed meristems. Ann Bot 48:761-767

Clowes FAL (1981b) Cell proliferation in ectotrophic mycorrhizas of *Fagus sylvatica* L. New Phytol 87:547-555

Clowes FAL (1984) Size and activity of quiescent centres of roots. New Phytol 96:13-21

Clowes FAL, Juniper BE (1968) Plant cells. Blackwell, Oxford

Codaccioni M (1963) Le maintien en survie in vitro du méristème terminal chez le *Castanea sativa*. C R Acad Sci (Paris) 257:2319-2321

Cohen C (1979a) Cell architecture and morphogenesis. I. The cytoskeletal proteins. Trends Biochem Sci 4:73-77

Cohen C (1979b) Cell architecture and morphogenesis. II. Examples in embryology. Trends Biochem Sci 4:97-101

Cooke JR, Baerdemaeker JG De, Rand RH, Mang HA (1976) A finite element shell analysis of guard cell deformations. Trans Am Soc Agric Eng. 19:1107-1121

Cormack RGH (1949) The development of root hairs in angiosperms. Bot Rev 15:583-612

Corson GE Jr (1969) Cell division studies of the shoot apex of *Datura stramonium* during transition to flowering. Am J Bot 56:1127-1134

Cosgrove DJ (1985) Cell wall yield properties of growing tissue. Plant Physiol 78: 347-356

Coutts MP, Lewis GJ (1983) When is the structural root system determined in Sitka spruce? Plant Soil 71:155-160

Cox G, Moran KJ, Sanders F, Nockolds C, Tinker PB (1980) Translocation and transfer of nutrients in vesicular-arbuscular mycorrhizas. III. Polyphosphate granules and phosphorus translocation. New Phytol 84:649-659

Creber GT, Chaloner WG (1990) Environmental influences on cambial activity. In: Iqbal M (ed) The vascular cambium. Research Studies Press, Taunton (Somerset). Wiley, New York, pp 159-199

Crick FHC, Lawrence PA (1975) Compartments and polyclones in insect development. Science 189:340-347

Critchfield WB (1960) Leaf dimorphism in *Populus trichocarpa*. Am J Bot 47:699-711

Cromer AH (1977) Physics for the life sciences, 2nd edn. McGraw-Hill, New York

Cronshaw J (1974a) P-proteins. In: Aronoff S et al (eds) Phloem transport. NATO Advanced Study Inst Ser A-4. Plenum Press, New York, pp 79-115

Cronshaw J (1974b) Sieve element cell walls. In: Aronoff S et al (eds) Phloem transport. NATO Advanced Study Inst Ser A-4. Plenum Press, New York, pp 129-147

Crookston RK (1980) The structure and function of C_4 vascular tissue — some unanswered questions. Ber Dtsch Bot Ges 93:71-78

Cross GL (1937) The morphology of the bud and the development of the leaves of *Viburnum rufidulum*. Am J Bot 24:266-276

Cross GL (1940) Development of the foliage leaves of *Taxodium distichum*. Am J Bot 27: 471-482

Cross GL (1942) Structure of the apical meristem and development of the foliage leaves of *Cunninghamia lanceolata*. Am J Bot 29:288-301

Cross GL (1943) A comparison of the shoot apices of the sequoias. Am J Bot 30:130-142

Curtis JD, Lersten NR (1974) Morphology, seasonal variation, and function of resin glands on buds and leaves of *Populus deltoides* (Salicaceae). Am J Bot 61:835-845

Curtis JD, Lersten NR (1986) Hydathode anatomy in *Potentilla palustris* (Rosaceae). Nord J Bot 6:793-796

Cusset G (1986) La morphogenèse du limbe des dicotylédones. Can J Bot 64:2807-2839

Dacey JWH, Klug MJ (1982) Ventilation by floating leaves in *Nuphar*. Am J Bot 69:999-1003

Dadswell HE, Hillis WE (1962) Wood. In: Hillis WE (ed) Wood extractives and their significance to the pulp and paper industries. Academic Press, New York, pp 3-55

Dale JE (1982) The growth of leaves. Studies in Biol No. 137. Arnold, London

Dale JE (1988) The control of leaf expansion. Annu Rev Plant Physiol Plant Mol Biol 39:267-295

Dart PJ (1975) Legume root nodule initiation and development. In: Torrey JG, Clarkson DT (eds) The development and function of roots. Academic Press, London, pp 467-506

Darvill AG, Albersheim pp (1984) Phytoalexins and their elicitors — A defense against microbial infection in plants. Annu Rev Plant Physiol 35:243-275

Darvill AG, Albersheim P, McNeil M, Lau JM, York WS, Stevenson TT, Thomas J, Doares S, Gollin DJ, Chelf P, Davis K (1985) Structure and function of plant cell wall polysaccharides. J Cell Sci Suppl 2:203-217

Daum CR (1967) A method for determining water transport in trees. Ecology 48:425-431

Davey AJ (1946) On the seedling of *Oxalis hirta* L. Ann Bot 10:237-256

Davies E (1987) Action potentials as multifunctional signals in plants: a unifying hypothesis to explain apparently disparate wound responses. Plant Cell Environ 10:623-631

Davis GL (1966) Systematic embryology of the angiosperms. Wiley, New York

Davis JD, Evert RF (1970) Seasonal cycle of phloem development in woody vines. Bot Gaz 131:128-138

Davis TA (1961) High root-pressures in palms. Nature 192:277-278

Dell B, Malajczuk N, Thomson GT (1990) Ectomycorrhiza formation in *Eucalyptus*. V. A tuberculate ectomycorrhiza of *Eucalyptus pilularis*. New Phytol 114:633-640

Delwich MJ, Cooke JR (1977) An analytical model of the hydraulic aspects of stomatal dynamics. J Theor Biol 69:113-141

DeMason DA (1983) The primary thickening meristem: Definition and function in monocotyledons. Am J Bot 70:955-962

Dengler NG, Mackay LB (1975) The leaf anatomy of beech, *Fagus grandifolia*. Can J Bot 53:2202-2211

Dengler NG, Mackay LB, Gregory LM (1975) Cell enlargement and tissue differentiation during leaf expansion in beech, *Fagus grandifolia*. Can J Bot 53:2846-2865

Dermen H (1945) The mechanism of colchicine-induced cytohistological changes in cranberry. Am J Bot 32:387-394

Dermen H (1947) Periclinal cytochimeras and histogenesis in cranberry. Am J Bot 34:32-43

Dermen H (1953) Periclinal cytochimeras and origin of tissues in stem and leaf of peach. Am J Bot 40:154-168

Dermen H (1955) A 2-4-2 chimera of McIntosh apple. J Wash Acad Sci 45:324-327

Dermen H (1969) Graft-chimeras 1953-1969. Wiss Z Päd Hochsch Potsdam Math- Naturwiss Reihe 13:15-21

Dermen H, Darrow GM (1960) Nature of plant sports. Am Hortic Mag 39:123-173

Dermen H, Diller JD (1962) Colchiploidy of chestnuts. For Sci 8:43-50

Desch HE (1932) Anatomical variation in the wood of some dicotyledonous trees. New Phytol 31:73-126

Deshpande BP (1974) Development of the sieve plate in *Saxifraga sarmentosa* L. Ann Bot 38:151-158

Deshpande BP (1975) Differentiation of the sieve plate of *Cucurbita*: A further view. Ann Bot 39:1015-1022

Deshpande BP, Rajendrababu T (1985) Seasonal changes in the structure of the secondary phloem of *Grewia tiliaefolia*, a deciduous tree from India. Ann Bot 56:61-71

Devaux MH (1900) Recherches sur les lenticelles. Ann Sci Nat Ser 8 Bot 12:1-235

Dexheimer J, Pargney J-C (1991) Comparative anatomy of the host-fungus interface in mycorrhizas. Experientia 47:312-321

Diboll AG, Larson DA (1966) An electron microscopic study of the mature megagametophyte in *Zea mays*. Am J Bot 53:391-402

Dickmann DI (1971) Photosynthesis and respiration by developing leaves of cottonwood (*Populus deltoides* Bartr.). Bot Gaz 132:253-259

Diettert RA (1938) The morphology of *Artemisia tridentata* Nutt. Lloydia (Cincinnati) 1:3-74

Digby J, Wareing PF (1966) The relationship between endogenous hormone levels in the plant and seasonal aspects of cambial activity. Ann Bot 30:607-622

Dimond AE (1966) Pressure and flow relations in vascular bundles of the tomato plant. Plant Physiol 41:119-131

Dittmer HJ (1937) A quantitative study of the roots and root hairs of a winter rye plant (*Secale cereale*). Am J Bot 24:417-420

Dittmer HJ (1949) Root hair variations in plant species. Am J Bot 36:152-155

Dixon AFG (1975) Aphids and translocation. In: Transport in plants: Phloem transport. Encycl Plant Physiol 2nd Ser 1:154-170

Dobbins DR, Kuijt J (1973) Studies on the haustorium of *Castilleja* (Scrophulariaceae). II. The endophyte. Can J Bot 923-931

Dogra PD (1967) Seed sterility and disturbances in embryogeny in conifers with particular reference to seed testing and tree breeding in Pinaceae. Stud For Suec 45:1-97

Dolzmann P (1964) Elektronenmikroskopische Untersuchungen an den Saughaaren von *Tillandsia usenoides* (Bromeliaceae). I. Feinstruktur der Kuppelzelle. Planta 60:461-472

Domingo IL (1983) Nitrogen fixation in southeastern Asian forestry: Research and practice. In: Gordon JC, Wheeler CT (eds) Biological nitrogen fixation in forest ecosystems: Foundations and applications. Nijhoff/Junk, The Hague, pp 295-315

Dormer KJ (1965) Correlations in plant development: General and basic aspects. Encycl Plant Physiol 1st Ser 15(1):452-478

Dormer KJ (1972) Shoot organization in vascular plants. Chapman and Hall, London; Syracuse Univ Press, Syracuse

Dormer KJ (1980) Fundamental tissue geometry for biologists. Cambridge Univ Press, Cambridge

Douliot H (1889) Recherches sur le périderme. Ann Sci Nat Ser 7 Bot 10:325-395

Doyle J (1963) Proembryogeny in *Pinus* in relation to that in other conifers — A survey. Proc R Ir Acad 62B(13):181-216

Doyle J, Brennan M (1971) Cleavage polyembryony in conifers and taxads — A survey. I. Podocarps, taxads and taxodioids. R Dubl Soc Sci Proc A4:57-88

Drabble E (1903) On the anatomy of the roots of palms. Trans Linn Soc Lond, 2nd Ser Bot 6:427-510

Duddridge JA, Read DJ (1984a) The development and ultrastructure of ectomycorrhizas. I. Ectomycorrhizal development on pine in the field. New Phytol 96:565-573

Duddridge JA, Read DJ (1984b) The development and ultrastructure of ectomycorrhizas. II. Ectomycorrhizal development of pine in vitro. New Phytol 96:575-582

Dulieu H (1970) Les mutations somatiques induites et l'ontogénie de la pousse feuillée. Ann Amélior Plant 20:27-44

Dute RR, Rushing AE (1987) Pit pairs with tori in the wood of *Osmanthus americanus* (Oleaceae). IAWA Bull 8:237-244

Dute RR, Rushing AE (1990) Torus structure and development in the woods of *Ulmus alata* Michx., *Celtis laevigata* Willd., and *Celtis occidentalis* L. IAWA Bull 11:71-83

Eames AJ (1961) Morphology of angiosperms. McGraw-Hill, New York

Eames AJ, MacDaniels LH (1947) An introduction to plant anatomy. McGraw-Hill

Edgar E (1961) Fluctuations in the mitotic index in the shoot apex of *Lonicera nitida*. Univ Canterbury (NZ) Publ No 1:1-19

Edwards G, Walker D (1983) C_3, C_4: Mechanisms, and cellular environmental regulation of photosynthesis. Blackwell, Oxford

Edwards KL, Pickard BG (1987) Detection and transduction of physical stimuli in plants. In: Wagner E, Greppin H, Millet B (eds) The cell surface in signal transduction. NATO Advanced Study Ser H12. Springer, Berlin Heidelberg New York, pp 41-66

Ehleringer J, Forseth I (1980) Solar tracking by plants. Science 210:1094-1098

Elias TS (1983) Extrafloral nectaries: Their structure and distribution. In: Bentley B, Elias T (eds) The biology of nectaries. Columbia Univ Press, New York, pp 174-203

Elias TS, Gelband H (1976) Morphology and anatomy of floral and extrafloral nectaries in *Campsis* (Bignoniaceae). Am J Bot 63:1349-1353

Elias TS, Rozich WR, Newcombe L (1975) The foliar and floral nectaries of *Turnera ulmifolia* L. Am J Bot 62:570-576

Ende G van den, Linskens HF (1974) Cutinolytic enzymes in relation to pathogenesis. Annu Rev Phytopathol 12:247-258

Epel BL, Bandurski RS (1990) Apoplastic domains and sub-domains in the shoots of etiolated corn seedlings. Physiol Plant 79:599-603

Erickson RO (1986) Symplastic growth and symplasmic transport. Plant Physiol 82:1153

Erickson RO, Goddard DR (1951) An analysis of root growth in cellular and biochemical terms. Growth 15:89-116 (Suppl)

Erickson RO, Michelini FJ (1957) The plastochron index. Am J Bot 44:297-305

Ervin EL, Evert RF (1967) Aspects of sieve element ontogeny and structure in *Smilax rotundifolia*. Bot Gaz 128:138-144

Erwee MG, Goodwin PB, Bel AJE van (1985) Cell-cell communication in the leaves of *Commelina cyanea* and other plants. Plant Cell Environ 8:173- 178

Esau K (1940) Developmental anatomy of the fleshy storage organ of *Daucus carota*. Hilgardia 13:175-226

Esau K (1943) Vascular differentiation in the pear root. Hilgardia 15:299-311

Esau K (1965a) Plant anatomy, 2nd edn. Wiley, New York

Esau K (1965b) Vascular differentiation in plants. Holt Rinehart Winston, New York

Esau K (1969) The phloem. Encycl Plant Anat, Bd V Teil 2. Borntraeger, Berlin

Esau K (1972) Cytology of sieve elements in minor veins of sugar beet leaves. New Phytol 71:161-168 + 6 pl

Esau K (1973) Comparative structure of companion cells and phloem parenchyma cells in *Mimosa pudica* L. Ann Bot 37:625-632 + 8 pl

Esau K (1977) Anatomy of seed plants, 2nd edn. Wiley, New York

Esau K (1979) Phloem. In: Metcalfe C R, Chalk L (eds) Anatomy of the dicotyledons, 2nd edn, vol 1. Clarendon Press, Oxford, pp 181-189

Esau K, Cheadle VI (1955) Significance of cell divisions in differentiating secondary phloem. Acta Bot Neerl 4:348-357

Esau K, Cheadle VI (1958) Wall thickening in sieve elements. Proc Natl Acad Sci USA 44:546-553

Esau K, Cronshaw J (1967) Tubular components in cells of healthy and tobacco mosaic virus-infected *Nicotiana*. Virology 33:26-35

Esau K, Thorsch J (1985) Sieve plate pores and plasmodesmata, the communication channels of the symplast: Ultrastructual aspects and developmental relations. Am J Bot 72:1641-1653

Eschrich W (1975) Sealing systems in phloem. Encycl Plant Physiol 2nd Ser 1:39-56

Eschrich W (1980) Phloembeladung und verwandte Prozesse (Einführung). Ber Dtsch Bot Ges 93:1-10

Eschrich W (1986) Mechanisms of phloem unloading. In: Cronshaw J, Lucas WJ, Giaquinta RT (eds) Phloem transport. Proc Int Conf Phloem Transport, Aug 1985, Asilomar, Calif. Liss, New York, pp 225-230

Eschrich W, Burchardt R, Essiamah S (1989) The induction of sun and shade leaves of the European beech (Fagus sylvatica L.): Anatomical studies. Trees 3:4-10

Etzler ME (1985) Plant lectins: Molecular and biological aspects. Annu Rev Plant Physiol 36:209-234

Evans LS, Van't Hof J (1975) Is polyploidy necessary for tissue differentiation in higher plants? Am J Bot 62:1060-1064

Evert RF (1960) Phloem structure in Pyrus communis L and its seasonal changes. Univ Calif Publ Bot 32(2):127-19

Evert RF (1961) Some aspects of cambial development in Pyrus communis L. Am J Bot 48:479-488

Evert RF (1963a) Ontogeny and structure of the secondary phloem in Pyrus malus. Am J Bot 50:8-37

Evert RF (1963b) Sclerified companion cells in Tilia americana. Bot Gaz 124:262-264

Evert RF (1977) Phloem structure and histochemistry. Annu Rev Plant Physiol 28:199-222

Evert RF (1982) Sieve-tube structure in relation to function. BioScience 32:789-795

Evert RF (1984) Comparative structure of phloem. In: White RA, Dickinson WC (eds) Contemporary problems in plant anatomy. Academic Press, New York, pp 145-234

Evert RF, Alfieri FJ (1965) Ontogeny and structure of coniferous sieve cells. Am J Bot 52:1058-1066

Evert RF, Deshpande BP (1969) Electron microscope investigation of sieve-element ontogeny and structure in Ulmus americana. Protoplasma 68:403-432

Evert RF, Deshpande BP (1970) An ultrastructural study of cell division in the cambium. Am J Bot 57:942-961

Ewers FW (1982a) Developmental and cytological evidence for mode of origin of secondary phloem in needle leaves of Pinus longaeva (bristlecone pine) and P. flexilis. Bot Jahrb Syst 103:59-88

Ewers FW (1982b) Secondary growth in needle leaves of Pinus longaeva (Bristlecone pine) and other conifers: Quantitative data. Am J Bot 69:1552-1559

Fahn A (1974) Plant anatomy, 2nd edn. Pergamon Press, Oxford

Fahn A (1979) Secretory tissues in plants. Academic Press, London

Fahn A (1986) Structural and functional properties of trichomes in xeromorphic leaves. Ann Bot 57:631-637

Fahn A (1988) Secretory tissues in vascular plants (Tansley Rev 14). New Phytol 108:229-257

Fahn A, Arnon N (1963) The living wood fibres of Tamarix aphylla and the changes occurring in them in transition from sapwood to heartwood. New Phytol 62:99-104 + 1 pl

Fahn A, Werker E (1990) Seasonal cambial activity. In: Iqbal M (ed) The vascular cambium. Research Studies Press, Taunton (Somerset). Wiley, New York, pp 139-157

Fahn A, Zamski E (1970) The influence of pressure, wind, wounding and growth substances on the rate of resin duct formation in Pinus halepensis wood. Isr J Bot 19:429-446

Falconer MM, Seagull RW (1985) Xylogenesis in tissue culture: taxol effects on microtubule reorientation and lateral association in differentiating cells. Protoplasma 128:157-166

Falk S, Hertz CH, Virgin HI (1958) On the relation between turgor pressure and tissue rigidity. I. Experiments on resonance frequency and tissue rigidity. Physiol Plant 11: 802-817

Fassi B, Fontana A, Trappe JM (1969) Ectomycorrhizae formed by Endogone lactiflua with species of Pinus and Pseudotsuga. Mycologia 61:412-414

Fay E de, Sanier C, Hebant C (1989) The distribution of plasmodesmata in the phloem of *Hevea brasiliensis* in relation to laticifer loading. Protoplasma 149:155-162

Faye M, Rancillac M, David A (1980) Determinism of the mycorrhizogenic root formation in *Pinus pinaster* Sol. New Phytol 87:557-565

Fayle DCF (1968) Radial growth in tree roots. Faculty of Forestry, Univ Toronto, Tech Rep 9 pp 1-183

Fayle DCF (1975) Distribution of radial growth during the development of red pine root systems. Can J For Res 5:608-625

Fayle DCF, Farrar JL (1965) A note on the polar transport of exogenous auxin in woody root cuttings. Can J Bot 43:1004-1007

Feldman LJ (1984) The development and dynamics of the root apical meristem. Am J Bot 71:1308-1314

Feldman LJ Torrey JG (1975) The quiescent center and primary vascular tissue pattern formation in cultured roots of *Zea*. Can J Bot 53:2796-2803

Feynman RP, Leighton RB, Sands M (1964) The Feynman lectures on physics, vol II. chapter 31. Tensors. Addison-Wesley, Reading (Mass)

Fincher GB, Stone BA (1981) Metabolism of noncellulosic polysaccharides. Encycl Plant Physiol 2nd Ser 13B:68-132

Fink S (1984) Some cases of delayed or induced development of axillary buds from persisting detached meristems in conifers. Am J Bot 71:44-51

Fischer R, Dengler NG (1977) Mesophyll cell walls in hemlock, *Tsuga canadensis*. Can J Bot 55:1510-1515

Fisher DG (1986) Ultrastructure, plasmodesmatal frequency, and solute concentration in green areas of variegated *Coleus blumei* Benth. leaves. Planta 169:141-152

Fisher DG (1990) Leaf structure of *Cananga odorata* (Annonaceae) in relation to collection of photosynthate and phloem loading: Morphology and anatomy. Can J Bot 68:354-363

Fisher DG, Evert RF (1982) Studies on the leaf of *Ameranthus retroflexus* (Ameranthaceae): Morphology and anatomy. Am J Bot 69:1133-1147

Fleurat-Lessard P (1988) Structural and ultrastructural features of cortical cells in motor organs of sensitive plants. Biol Rev 63:1-22

Fleurat-Lessard P, Millet B (1984) Ultrastructural features of cortical parenchyma cells ("motor cells") in stamen filaments of *Berberis canadensis* Mill and tertiary pulvini of *Mimosa pudica* L. J Exp Bot 35:1232-1341

Foard DE (1971) The initial protrusion of a leaf primordium can form without concurrent periclinal cell divisions. Can J Bot 49:1601-1603

Foard DE, Haber AH (1961) Anatomic studies of gamma-irradiated wheat growing without cell division. Am J Bot 48:438-446

Foard DE, Haber AH, Fishman TN (1965) Initiation of lateral root primordia without completion of mitosis and without cytokinesis in uniseriate pericycle. Am J Bot 56:580-590

Fogel R (1980) Mycorrhizae and nutrient cycling in natural forest ecosystems. New Phytol 86:199-212

Fogel R (1983) Root turnover and productivity of coniferous forests. Plant Soil 71:75-85

Fogel R, Hunt G (1979) Fungal and arboreal biomass in a western Oregon Douglas-fir ecosystem: Distribution patterns and turnover. Can J For Res 9:245-256

Fontana A (1985) Vesicular-arbuscular mycorrhizas of *Ginkgo biloba* L in natural and controlled conditions. New Phytol 99:441-447

Foster AS (1938) Structure and growth of the shoot apex of *Ginkgo bioloba*. Bull Torrey Bot Club 65:531-556

Foster AS (1939) Problems of structure, growth and evolution in the shoot apex of seed plants. Bot Rev 5:454-470

Foster AS (1943) Zonal structure and growth of the shoot apex in *Microcycas calocoma* (Miq.) A. DC. Am J Bot 30:56-73

Foster RC, Marks GC (1966) The fine structure of the mycorrhizas of *Pinus radiata* D. Don. Aust J Biol Sci 19:1027-1038 + 9 pl

Franceschi VR, Giaquinta RT (1983) The paraveinal mesophyll of soybean leaves in relation to assimilate transfer and compartmentation. I. Ultrastructure and histochemistry during vegetative development. Planta 157:411-421

Franck DH (1979) Development of vein pattern in leaves of *Ostrya virginiana* (Betulaceae). Bot Gaz 140:77-83

Frank B (1885) Ueber die auf Wurzelsymbiose beruhende Ernährung gewisser Bäume durch unterirdische Pilze. Ber Dtsch Bot Ges 3:128-145

Franke W (1967) Mechanisms of foliar penetration of solutions. Annu Rev Plant Physiol 18:281-300

Franke W (1969) Ectodesmata in relation to binding sites for inorganic ions and urea on isolated cuticular membrane surfaces. Am J Bot 56:432-436

French V, Bryant PJ, Bryant SV (1976) Pattern regulation in epimorphic fields. Science 193:696-981

Frenzel P (1929) Über die Porengrössen einiger pflanzlicher Zellmembranen. Planta 8:642--665

Frey-Wyssling A (1976) The plant cell wall, 3rd rev edn. Encycl Plant Anatomy 3(4). Borntraeger, Berlin

Friis J, Dute RR (1983) Phloem of primitive angiosperms. II. P-protein in selected species of the Ranalean complex. Proc Iowa Acad Sci 90:78-84

Frost FH (1929) Histology of the wood of angiosperms. I. The nature of pitting between tracheary and parenchymatous elements. Bull Torrey Bot Club 56:259-264 + 1 pl

Fryns-Claessens E, Cotthem H van (1973) A new classification of the ontogenetic types of stomata. Bot Rev 39:71-138

Fuente RK de la, Leopold AC (1966) Kinetics of polar auxin transport. Plant Physiol 41:1481-1484

Fukada E (1968) Piezoelectricity as a fundamental property of wood. Wood Sci Tech 2:299-307

Gahan PB, Rana MA (1985) The quiescent centre and cell determination in roots of *Pisum sativum*. Ann Bot 56:437-442

Gamalei Y (1989) Structure and function of leaf minor veins in trees and herbs. Trees 3:96-110

Gamalei Y (1991) Phloem loading and its development related to plant evolution from trees to herbs. Trees 5:50-64

Gambles RL, Dengler NG (1974) The leaf anatomy of hemlock, *Tsuga canadensis*. Can J Bot 52:1049-1056

Gambles RL, Dengler RE (1982a) The anatomy of the leaf of red pine, *Pinus resinosa*. I. Nonvascular tissues. Can J Bot 60:2788-2803

Gambles RL, Dengler RE (1982b) The anatomy of the leaf of red pine, *Pinus resinosa*. II. Vascular tissues. Can J Bot 60:2804-2824

Garcia-Bellido A (1975) Genetic control of wing disc development in *Drosophila*. In: Porter R, Rivers J (eds) Cell patterning. Ciba Found Symp New Ser 29. Elsevier, Amsterdam, pp 161-182

Garrison R, Wetmore RH (1961) Studies in shoot-tip abortion: *Syringa vulgaris*. Am J Bot 48:789-795

Gatlin LL (1972) Information theory and the living system. Columbia Univ Press, New York

Gedalovich E, Fahn A (1985) Ethylene and gum duct formation in *Citrus*. Ann Bot 56:571-577

Géhu J-M, Géhu J (1977) Les forêts à géophytes des plaines et collines de Nord-ouest de la France. Nat Can 104:47-56

Gerdemann JW (1965) Vesicular-arbuscular mycorrhizae formed on maize and tuliptree by *Endogone fasiculata*. Mycologia 57:562-5

Ghouse AKM, Hashmi S (1983) Periodicity of cambium and the formation of xylem and phloem in *Mimusops elengi* L., an evergreen member of tropical India. Flora: Morphol Geobot Oekophysiol (Jena) 173:479-487

Ghouse AKM, Sabir D (1974) Intrusive growth in the phloem fibers of *Erythrina indica* and *Pongamia glabra*. Isr J Bot 23:223-225

Ghouse AKM, Yunus M (1974) Transfusion tissue in the leaves of *Cunninghamia lanceolata* (Lambert) Hooker (Taxodiaceae). Bot J Linn Soc 69:147-151

Ghouse AKM, Yunus M (1976) Cell length variation in the secondary phloem of *Dalbergia* spp. with increasing age of the cambium. Ann Bot 40:13-16

Giaquinta RT (1980) Translocation of sucrose and oligosaccharides. In: Stumpf PF, Conn EE (eds) The biochemistry of plants, vol 3 (Preiss J, ed), Carbohydrates: Structure and Function. Academic Press, New York, pp 271-320

Giaquinta RT (1983) Phloem loading of sucrose. Annu Rev Plant Physiol 34:347-387

Giddings TH Jr, Staehelin LA (1991) Microtubule-mediated control of microfibril deposition: A re-examination of the hypothesis. In: Lloyd CW (ed) The cytoskeletal basis of plant growth and form. Academic Press, London, pp 85-100

Gifford EM Jr (1951) Early ontogeny of the foliage leaf in *Drimys winteri* var *chilensis*. Am J Bot 38:93-105

Gifford EM Jr, Kupila S, Yamaguchi S (1963) Experiments in the application of H^3-thymidine and adenine-8-C^{14} to shoot tips. Phytomorphology 13:14-22

Gillis PP (1969) Effect of hydrogen bonds on the axial stiffness of crystalline native cellulose. J Polym Sci A-2 7:783-794

Ginzburg C (1963) Some anatomic features of splitting of desert shrubs. Phytomorphology 13:92-97

Girouard RM (1967a) Initiation and development of adventitious roots in stem cuttings of *Hedera helix*. Anatomical studies of the juvenile growth phase. Can J Bot 45:1877-1881 + 3 pl

Girouard RM (1967b) Initiation and development of adventitious roots in stem cuttings of *Hedera helix*. Anatomical studies of the mature growth phase. Can J Bot 45:1883-1886

Godkin SE, Grozdits GA, Keith CT (1983) The periderms of three North American conifers. Part 2: Fine structure. Wood Sci Tech 17:13-30

Goebel K (1880) Beitrage zur Morphologie und Physiologie des Blattes. Bot Zt 38:801-815, 817-826, 833-845

Goebel K (1886) Ueber die Luftwurzeln von *Sonneratia*. Ber Dtsch Bot Ges 4:249-255

Goldsmith MHM (1977) The polar transport of auxin. Annu Rev Plant Physiol 28:439-478

Golinowski WO (1971) The anatomical structure of the common fir (*Abies alba* Mill.) bark. I. Development of bark tissues. Acta Soc Bot Pol 40:149-181

Gómez-Vazquez BG, Engleman EM (1984) Bark anatomy of *Bursera longipes* (Rose) Standley and *Bursera copallifera* (Sessé & Moc.) Bullock. IAWA Bull 5:335-340

González-Fernández A, López-Sáez JF, Moreno P, Giménez-Martin G (1968) A model for dynamics of cell division cycle in onion roots. Protoplasma 65:263-276

Goodwin PB (1983) Molecular size limit for movement in the symplast of the *Elodea* leaf. Planta 157:124-130

Górska-Brylass A (1968) Callose in the cell walls of the developing male gametophyte in Gymnospermae. Acta Soc Bot Pol 37:119-124

Green PB (1976) Growth and cell pattern formation on an axis: Critique of concepts, terminology, and modes of study. Bot Gaz 137:187-202

Green PB (1980) Organogenesis — A biophysical view. Annu Rev Plant Physiol 31:51-82

Green PB (1985) Surface of the shoot apex: A reinforcement-field theory of phyllotaxis. J Cell Sci Suppl 2:181-201

Greenidge KNH (1952) An approach to the study of vessel length in hardwood species. Am J Bot 39:570-574

Gregory RA, Romberger JA (1972) The shoot apical ontogeny of the *Picea abies* seedling. I. Anatomy, apical dome diameter, and plastochron duration. Am J Bot 59:587-597

Gregory RA, Romberger JA (1975) Cambial activity and height of uniseriate vascular rays in conifers. Bot Gaz 136:246-253

Gregory RA, Romberger JA (1977) The shoot apical ontogeny of the Picea abies seedling. IV. Protoxylem initiation and age of internodes. Am J Bot 64:631-634

Griffith MM (1957) Foliar ontogeny in *Podocarpus macrophyllus*, with special reference to transfusion tissue. Am J Bot 44:705-715

Groom P, Wilson SE (1925) On the pneumatophores of paludal species of *Amoora, Carapa*, and *Heritiera*. Ann Bot 39:9-24 + 2 pl

Grosse W, Mevi-Schütz J (1987) A beneficial gas transport system in *Nymphoides peltata*. Am J Bot 74:947-952

Grozdits GA (1982) Microstructure of sequent periderms and ultrastructure of periderm cell walls in *Tsuga canadensis* (L.) Carr. Wood Sci 15:110-118

Grozdits GA, Godkin SE, Keith CT (1982) The periderms of three North American conifers. Part 1: Anatomy. Wood Sci Tech 16:305-316

Grusak MA, Lucas WJ (1986) Cold-inhibited phloem translocation in sugar beet. III. The involvement of the phloem pathway in source-sink partitioning. J Exp Bot 37:277-288

Guinel FC, McCully ME (1987) The cells shed by the root cap of *Zea*: their origin and some structural and physiological properties. Plant Cell Environ 10:565-578

Gunning BES (1976) The role of plasmodesmata in short distance transport to and from the phloem. In: Gunning BES, Robards AW (eds) Intercellular communication in plants: Studies on plasmodesmata. Springer, Berlin Heidelberg New York, pp 203-227

Gunning BES, Overall RL (1983) Plasmodesmata and cell-to-cell transport in plants. Bio-Science 33:260-265

Gunning BES, Pate JS (1969) "Transfer cells" — Plant cells with wall ingrowths, specialized in relation to short distance transport of solutes — their occurrence, structure, and development. Protoplasma 68:107-133

Gunning BES, Pate JS (1974) Transfer cells. In: Robards AW (ed) Dynamic aspects of plant ultrastructure. McGraw-Hill, London, pp 441-480

Gunning BES, Pate JS, Green LW (1970) Transfer cells in the vascular system of stems: Taxonomy, association with nodes, and structure. Protoplasma 71:147-171

Gunning BES, Hughes JE, Hardham AR (1978) Formative and proliferative cell divisions, cell differentiation, and developmental changes in the meristem of *Azolla* roots. Planta 143:121-144

Guttenberg H von (1960) Grundzüge der Histogenese höherer Pflanzen. I. Die Angiospermen. Handbuch der Pflanzenanatomie, Bd 8, Teil 3. Borntraeger, Berlin

Guttenberg H von (1961) Grundzüge der Histogenese höherer Pflanzen. II. Die Gymnospermen. Handbuch der Pflanzenanatomie, Bd 8, Teil 4. Borntraeger, Berlin

Guttenberg H von (1968) Der primäre Bau der Angiospermenwurzel. Handbuch der Pflanzenanatomie, Spez Teil, Band VIII, Teil 5. Borntraeger, Berlin

Guttenberg H von (1971) Bewegungsgewebe und Perzeptionsorgane. Encycl Plant Anatomy, 5, Part 5. Borntraeger, Berlin

Haber AH (1972) Ionizing radiation effects on higher plants. In: Whitson L (ed) Concepts in radiation biology. Academic Press, New York, pp 231-243

Haberlandt G (1882) Vergleichende Anatomie des assimilatorischen Gewebsystems der Pflanzen. Jahrb Wiss Bot 13:74-188 + 6 pl

Haberlandt G (1904) Physiologische Pflanzenanatomie, 3rd edn. Engelmann, Leipzig

Haberlandt G (1909) Physiologische Pflanzenanatomie, 4th edn. Engelmann, Leipzig

Haberlandt G (1914) Physiological Plant Anatomy (Transl from the 4th German edn by M Drummond). MacMillan, London

Haberlandt G (1924) Physiologische Pflanzenanatomie, 6th edn. Engelmann, Leipzig

Haccius B, Fischer E (1959) Embryologische und histogenetische Studien an "monokotylen Dikotylen." III. *Anemone apennina* L. Österr Bot Z 106:373-389

Hagemann W (1970) Studien zur Entwicklungsgeschichte der Angiospermenblätter. Bot Jahrb 90:297-413

Hagemann W (1984) Morphological aspects of leaf development in ferns and angiosperms. In: White RA, Dickison WC (eds) Contemporary problems in plant anatomy. Academic Press, Orlando, pp 301-349

Hagerup O (1930) Über die Bedeutung der Schirmform der Krone von *Acacia seyal* Del. Dan Bot Ark 6(4):1-19 + pl

Haissig BE (1974) Origins of adventitious roots. NZ J For Sci 4:299-310

Hallé F, Oldeman, RAA, Tomlinson PB (1978) Tropical trees and forests: An achitectural analysis. Springer, Berlin Heidelberg New York

Hanover JW, Reicosky DA (1971) Surface wax deposits on foliage of *Picea pungens* and other conifers. Am J Bot 58:681-687

Hanstein J (1868) Die Scheitelzellgruppe im Vegetationspunkt der Phanerogamen Niederrhein Ges Natur-Heilk, Festschr z 50-jährigen Jubiläum Univ Bonn (1868), pp 109-143

Hara N (1957a) Study of the variegated leaves, with special reference to those caused by air spaces. Jpn J Bot 16:86-101

Hara N (1957b) On the types of the marginal growth in dicotyledonous foliage leaves. Bot Mag Tokyo 70:108-114

Harley JL, Smith SE (1983) Mycorrhizal symbiosis. Academic Press, London

Harris JM (1989) Spiral grain and wave phenomena in wood formation. Springer, Berlin Heidelberg New York

Harris N, Chaffey NJ (1985) Plasmatubules in transfer cells of pea (*Pisum sativum* L.). Planta 165:191-196

Harris N, Oparka KJ, Walker-Smith DJ (1982) Plasmatubules: An alternative to transfer cells? Planta 156:461-465

Harris WF, Erickson RO (1980) Tubular arrays of spheres: Geometry, continuous and discontinuous contraction, and the role of moving dislocations. J Theor Biol 83:215-246

Harris WM (1971) Ultrastructural observations on the mesophyll cells of pine leaves. Can J Bot 49:1107-1109

Harris WM (1983) On the development of macrosclereids in seed coats of *Pisum sativum* L. Am J Bot 70:1528-1535

Harte C, Lindenmayer A (1983) Mitotic index in growing cell populations: Mathematical models and computer simulations. Biol Zentralbl 102:509-533

Hartt CE (1973) Mechanism of translocation in sugarcane. Univ Hawaii, Lyon Arbor Lect 4:1-40

Harvey EM (1915) Some effects of ethylene on the metabolism of plants. Bot Gaz 60:193-214

Haupt W (1977) Bewegungsphysiologie der Pflanzen. Thieme, Stuttgart

Hayward HE, Long EM (1942) The anatomy of the seedling and roots of the Valencia orange. US Dept Agric Tech Bull 786:1-31

Head GC (1973) Shedding of roots. In: Kozlowski TT (ed) Shedding of plant parts. Academic Press, New York, pp 237-293

Heaman JC, Owens JN (1972) Callus formation and root initiation in stem cuttings of Douglas-fir (*Pseudotsuga menziesii* (Mirb.) Franco). Can J For Res 2:121-134

Heimerdinger G (1951) Zur Mikrotopographie der Saftströme im Transfusionsgewebe der Koniferennadel. II. Mitteilung. Entwicklungsgeschichte und Physiologie. Planta 40:93-111

Heimsch C (1960) A new aspect of cortical development in roots. Am J Bot 47:195-201

Hejnowicz Z (1967a) Some observations on the mechanism of orientation movement of woody stems. Am J Bot 54:684-689

Hejnowicz Z (1967b) Interrelationship between mean length, rate of intrusive elongation, frequency of anticlinal divisions and survival of fusiform initials in cambium. Acta Soc Bot Pol 36:367-378

Hejnowicz Z (1973) Anatomia Rozwojowa Drzew. Pań Wydawn Nauk, Warsaw (in Polish)

Hejnowicz Z (1980) Anatomia i histogeneza roślin naczyniowych. Pań Wydawn Nauk, Warsaw (in Polish)

Hejnowicz Z (1984) Trajectories of principal directions of growth, natural coordinate system in growing plant organ. Acta Soc Bot Pol 53:29-42

Hejnowicz Z (1989) Differential growth resulting in the specification of different types of cellular architecture in root meristems. Environ Exp Bot 29:85-93

Hejnowicz Z, Kurczyńska EU (1987) Occurrence of circular vessels above axillary buds in stems of woody plants. Acta Soc Bot Pol 56:415-119

Hejnowicz Z, Romberger JA (1973) Migrating cambial domains and the origin of wavy grain in xylem of broadleaved trees. Am J Bot 60:209-222

Hejnowicz Z, Romberger JA (1979) The common basis of wood grain figures is the systematically changing orientation of cambial fusiform cells. Wood Sci Tech 13:89-96

Hejnowicz Z, Romberger JA (1984) Growth tensor of plant organs. J Theor Biol 110:93-114

Hejnowicz Z, Zagórska-Marek B (1974) Mechanism of changes in grain inclination in wood produced by storeyed cambium. Acta Soc Bot Pol 43:381-398

Hejnowicz Z, Nakielski J, Hejnowicz K (1984) Modeling of spatial variations of growth within apical domes by means of the growth tensor. I. Growth specified on dome axis. Acta Soc Bot Pol 53:17-28

Hepler PK (1981) Morphogenesis of tracheary elements and guard cells. In: Kiermayer O (ed) Cytomorphogenesis in plants. Cell Biol Monogr, vol 8. Springer, Berlin Heidelberg New York, pp 327-347

Herbst D (1972) Ontogeny of foliar venation in *Euphorbia forbesii*. Am J Bot 59:843-850

Herman GT, Rozenberg G (1975) Developmental systems and languages. Elsevier/North-Holland, Amsterdam

Heslop-Harrison J (1966) Cytoplasmic continuities during spore formation in flowering plants. Endeavour 25:65-72

Hill AE, Hill BS (1976) Mineral ions. Encycl Plant Physiol 2nd Ser 2(B):225-243

Hill JF (1982) Spacing of parenchyma bands in wood of *Carya glabra* (Mill.) Sweet, pignut hickory, as an indicator of growth rate and climatic factors. Am J Bot 69:529-537

Hill JF (1983) Relationship among vessel diameter, vessel frequency, and spacing of parenchyma bands in wood of *Carya tomentosa* Nutt., mockernut hickory. Am J Bot 70:934-939

Hillis WE (1968) Chemical aspects of heartwood formation. Wood Sci Tech 2:241-259

Hillis WE (1975) Ethylene and extraneous material formation in woody tissues. Phytochemistry 14:2559-2562

Hillis WE (1987) Heartwood and tree exudates. Springer, Berlin Heidelberg New York

Hoefert LL, Gifford EM Jr (1967) Trabeculae in the grapevine infected with leafroll virus. Am J Bot 54:257-261

Holaday AS, Chollet R (1984) Photosynthetic/photorespiratory characteristics of C_3-C_4 intermediate species. Photosynth Res 5:307-323

Holdheide W (1951) Anatomie mitteleuropäischer Gehölzrinden. In: Wetzlar HF (ed) Handbuch der Mikroskopie in der Technik, Bd 5 Teil 1. Umschau, Frankfurt aM, pp 193-367

Holdheide W, Huber B (1952) Ähnlichkeiten und Unterschiede im Feinbau von Holz und Rinde. Holz Roh- Werkstoff 10:263-268

Höll W, Meyer H (1977) Enzyme catalyzed CO_2 fixation by wood preparations of *Tilia cordata* Mill and *Fraxinus excelsior* L. Holzforschung 31:184-187

Holloway PJ (1970) Surface factors affecting the wetting of leaves. Pestic Sci 1:156-163

Holloway PJ (1982) Structure and histochemistry of plant cuticular membranes: An overview. In: Cutler DF, Alvin KL, Price CE (eds) The plant cuticle. Academic Press, London, pp 1-32

Holm T (1925) On the development of buds upon roots and leaves. Ann Bot 39:867-881

Holtzer H (1970) Proliferative and quantal cell cycles in the differentiation of muscle, cartilage, and red blood cells. In: Padykula HA (ed) Control mechanisms in the expression of cellular phenotypes. Symp Int Soc Cell Biol 9:69-88

Holtzer H, Rubenstein N, Fellini S, Yeoh G, Chi J, Birnbaum J, Okayama M (1975) Lineages, quantal cell cycles, and the generation of cell diversity. Q Rev Biophys 8:523-557

Hook DD, Brown CL, Wetmore RH (1972) Aeration in trees. Bot Gaz 133:443-454

Horsley SB, Wilson BF (1971) Development of the woody portion of the root system of *Betula papyrifera*. Am J Bot 58:141-147

Howard ET (1971) Bark structure of the southern pines. Wood Sci 3:134-148

Howard ET (1977) Bark structure of southern upland oaks. Wood Fiber 9:172-183

Hu Y-s, Yao B-j (1981) Transfusion tissue in gymnosperm leaves. Bot J Linn Soc 83:263-272

Hu Y-s, Guan L-q, Tang Z-x (1985) Anatomy of the secondary phloem and the crystalliferous phloem fibers in the stem of *Torreya grandis*. Acta Bot Sinica 27:571-575 + 3 pl

Huber B (1939) Das Siebröhrensystem unserer Bäume und seine jahreszeitlichen Veränderungen. Jahrb Wiss Bot 88:176-242

Huber B (1947) Zur Mikrotopographie der Saftströme im Transfusionsgewebe der Konifernnadel. I. Anatomischer Teil. Planta 35:331-351

Huber B (1961) Grundzüge der Pflanzenanatomie. Springer, Berlin Heidelberg New York

Huber B, Schmidt E (1936) Weitere thermoelektrische Untersuchungen über den Transpirationsstrom der Bäume. Tharandt Forstl Jahrb 87:369-412

Iqbal M, Ghouse AKM (1990) Cambial concept and organisation. In: Iqbal M (ed) The vascular cambium. Research Studies Press, Taunton (Somerset). Wiley, New York, pp 1-36

Isebrands JG, Larson PR (1973) Anatomical changes during leaf ontogeny in *Populus deltoides*. Am J Bot 60:199-208

Isebrands JG, Larson PR (1977) Organization and ontogeny of the vascular system in the petiole of eastern cottonwood. Am J Bot 64:65-77

Isebrands JG, Larson PR (1980) Ontogeny of major veins in the lamina of *Populus deltoides* Bartr. Am J Bot 67:23-33

Isenberg IH (1943) The anatomy of redwood bark. Madroño 7:85-91

Ivanov VB (1983) Peculiarities of cellular organization of root growth as compared to other plant organs. In: Böhm W, Kutschera L, Lichtenegger E (eds) Root ecology and its practical application. Int Symp, Sept 1982, Gumpenstein, Austria. Gumpenstein Verlag, pp 57-62

Jackson WT (1982) Actomyosin. In: Lloyd CW (ed) The cytoskeleton in plant growth and development. Academic Press, London, pp 3-29

Jacobs M, Gilbert SF (1983) Basal localization of the presumptive auxin transport carrier in pea stem cells. Science 220:1297-1300

Jacobs MR (1939) A study of the effects of sway on trees. Commonw For Bur Aust Bull 26:1-17

Jacobs MR (1945) The growth stresses of woody stems. Commonw For Bur Aust Bull 28:1-67

Jacobs MR (1954) The effect of wind sway on the form and development of *Pinus radiata* D. Don. Aust J Bot 2:35-51

Jacobs WP, Morrow IB (1957) A quantitative study of xylem development in the vegetative shoot apex of *Coleus*. Am J Bot 44:823-842

Jacobs WP, Morrow IB (1961) A quantitative study of mitotic figures in relation to development in the apical meristem of vegetative shoots of *Coleus*. Dev Biol 3:569-587

Jaffe LF (1981) The role of ionic currents in establishing developmental pattern. Philos Trans R Soc Lond B 295:553-566

Janczewski E von (1874) Das Spitzenwachsthum der Phanerogamenwurzeln. Bot Ztg 32:113-127

Jane FW (1970) The Structure of Wood. Black, London

Janse JM (1897) Les endophytes radicaux de quelques plantes Javanaises. Ann Jardin Bot Buitenzorg 14:53-201 + 11 pl

Janse JM (1921) La polarité des cellules cambiennes. Ann Jardin Bot Buitenzorg 31:167-180 + 1 pl

Jarvis PG, Morison JIL (1981) The control of transpiration and photosynthesis by the stomata. In: Jarvis PG, Mansfield TA (eds) Stomatal physiology. Cambridge Univ Press, Cambridge, pp 247-279

Jean RV (1982) The hierarchical control of phyllotaxy. Ann Bot 49:747-760

Jean RV (1984) Mathematical approach to pattern and form in plant growth. Wiley, New York

Jeffree CE, Dale JE, Fry SC (1986) The genesis of intercellular spaces in developing leaves of *Phaseolus vulgaris* L. Protoplasma 132:90-98

Jeffrey EC, Wetmore RH (1926) On the occurrence of parichnos in certain conifers. Ann Bot 40:799-811 + 2 pl

Jensen WA (1972) The embryo sac and fertilization in angiosperms. Univ Hawaii, Harold Lyon Arbor Lecture 3:1-32

Jeune B (1982) La morphogenèse de fuilles de *Fraxinus excelsior* L. Bull Soc Bot Fr Lett Bot 129:283-29

Joel DM, Fahn A (1980a) Ultrastructure of the resin ducts of *Mangifera indica* L. (Anacardiaceae). 2. Resin secretion in the primary stem ducts. Ann Bot 46:779-783

Joel DM, and Fahn A (1980b) Ultrastructure of the resin ducts of *Mangifera indica* L. (Anacardiaceae). 3. Secretion of the protein-polysaccharide mucilage in the fruit. Ann Bot 46:785-790

Johansen DA (1950) Plant embryology: Embryogeny of the spermatophyta. Chronica Botanica, Waltham (Mass)

Johnson MA (1943) Foliar development in *Zamia*. Am J Bot 30:366-378

Johri BM, Agarwal S (1965) Morphological and embryological studies in the family Santalaceae. VIII. *Quinchamalium chilense* Lam. Phytomorphology 15:360-372

Jones MGK (1976) The origin and development of plasmodesmata. In: Gunning BES, Robards AW (eds) Intercellular communication in plants: Studies on plasmodesmata. Springer, Berlin Heidelberg New York, pp 81-105

Joshi GV, Karekar MD, Gowda CA, Bhosale L (1974) Photosynthetic carbon metabolism and carboxylating enzymes in algae and mangrove under saline conditions. Photosynthetica 8:51-52

Juniper BE (1976) Junctions between plant cells. In: Graham CF, Wareing PF (eds) The developmental biology of plants and animals. Blackwell, Oxford, pp 111-126

Juniper BE, Jeffree CE (1983) Plant surfaces. Arnold, London

Kahl G, Rosenstock G, Lange H (1969) Die Trennung von Zellteilung und Suberinsynthese in dereprimiertem pflanzlichem Speichergewebe durch Tris-(hydroxymethyl-)aminomethan. Planta 87:365-371

Kaldewey H (1984) Transport and other modes of movement of hormones (mainly auxins). Encycl Plant Physiol 2nd Ser 10:80-148

Kang BG (1979) Epinasty. Encycl Plant Physiol 2nd Ser 7:647-667

Kaplan DR (1969) Seed development in *Downingia*. Phytomorphology 19:253-278

Kaplan DR (1984) Alternative modes of organogenesis in higher plants. In: White RA, Dickison WC (eds) Contemporary problems in plant anatomy. Academic Press, pp 261-300

Kaplan DR, Dengler NG, Dengler RE (1982) The mechanism of plication inception in palm leaves: histogenic observations on the palmate leaf of *Raphis excelsa*. Can J Bot 60: 2999-3016

Karege F, Penel C, Greppin H (1982) Rapid correlation between the leaves of spinach and the photocontrol of a peroxidase activity. Plant Physiol 69:437-441

Kaufert F (1937) Factors influencing the formation of periderm in aspen. Am J Bot 24:24-30

Kausik SB (1974) The stomata of *Ginkgo biloba* L., with comments on some noteworthy features. Bot J Linn Soc 69:137-146

Kausik SB (1976) A contribution to foliar anatomy of *Agathis dammara*, with a discussion on the transfusion tissue and stomatal structure. Phytomorphology 26:263-273

Kausik SB, Bhattacharya SS (1977) Comparative foliar anatomy of selected gymnosperms: Leaf structure in relation to leaf form in Coniferales and Taxales. Phytomorphology 27:146-160

Kauss H (1987) Callose-Synthese: Regulation durch induzierten Ca^{2+}-Einstrom in Pflanzenzellen. Naturwissenschaften 74:275-281

Keeler KH (1977) The extrafloral nectaries of *Ipomoea carnea* (Convolvulaceae). Am J Bot 64:1182-1188

Kennell JC, Horner HT (1985) Megasporogenesis and megagametogenesis in soybean, *Glycine max*. Am J Bot 72:1553-1564

Kevekordes KG, McCully ME, Canny MJ (1988) The occurrence of an extended bundle sheath system (paraveinal mesophyll) in the legumes. Can J Bot 66:94-100

Khan R (1940) A note on "double fertilization" in *Ephedra foliata*. Curr Sci 9:323-325

Kikuyama M, Oda K, Shimmen T, Hayama T, Tazawa M (1984) Potassium and chloride effluxes during excitation of Characeae cells. Plant Cell Physiol 25:965-974

Kinden DA, Brown MF (1975) Electron microscopy of vesicular-arbuscular mycorrhizae of yellow poplar. II. Intracellular hyphae and vesicles. Can J Microbiol 21:1768-1780

Kisser J (1925) Elasticität und Festigkeit pflanzlicher Gewebe. Tabulae Biol 1:27-34

Kisser JG (1958) Die Ausscheidung von ätherischen Ölen und Harzen. Encycl Plant Physiol 1st Ser 10:91-131

Klein G (1923) Zur Ätiologie der Thyllen. Z Bot 15:417-439

Klekowski EJ Jr (1988) Mutation, developmental selection, and plant evolution. Columbia Univ Press, New York

Klekowski EJ Jr, Kazarinova-Fukshansky N, Mohr H (1985) Shoot apical meristems and mutation: Stratified meristems and angiosperm evolution. Am J Bot 72:1788-1800

Kolattukudy PE, Kronman K, Poulose AJ (1975) Determination of structure and composition of suberin from the roots of carrot, parsnip, rutabaga, turnip, red beet, and sweet potato by combined gas-liquid chromatography and mass spectrometry. Plant Physiol 55:567-573

Koller D (1990) Light driven leaf movements. Plant Cell Environ 13:615-632

Kollmann R (1965) Zur Lokalisierung der funktionstüchtigen Siebzellen im sekundären Phloem von *Metasequoia glyptostroboides*. Planta 65:173-179

Kollmann R (1967) Autoradiographischer Nachweis der Assimilat-Transportbahn im sekundären Phloem von *Metasequoia glyptostroboides*. Z Pflanzenphysiol 56:401-409

Kollmann R, Dörr I (1966) Lokalisierung funktionstüchtiger Siebzellen bei *Juniperus communis* mit Hilfe von Aphiden. Z Pflanzenphysiol 55:131-141

Kollmann R, Yang S, Glockmann C (1985) Studies on graft unions. II. Continuous and half plasmodesmata in different regions of the graft interface. Protoplasma 126:19-29

Kollöffel C, Linssen PWT (1984) The formation of intercellular spaces in the cotyledons of developing and germinating pea seeds. Protoplasma 120:12-19

Konar RN, Oberoi YP (1969) Studies on the morphology and embryology of *Podocarpus gracilior* Pilger. Beitr Biol Pflanz 45:329-376

Korn RW (1984) Cell shapes and tissue geometries. In: Barlow PW, Carr DJ (eds) Positional controls in plant development. Cambridge Univ Press, Cambridge, pp 33-52

Kotenko JL (1986) Antheridium formation in *Onoclea sensibilis* L.: Cytoplasmic polarity and the determination of wall positions. Bot Gaz 147:28-39

Kottke I, Oberwinkler F (1986) Mycorrhiza of forest trees — structure and function. Trees 1: 1-24

Kottke I, Oberwinkler F (1987) The cellular structure of the Hartig net: Coenocytic and transfer cell-like organization. Nord J Bot 7:85-95

Kottke I, Oberwinkler F (1990) Comparative investigations of the differentiation of the endodermis and the development of the Hartig net in mycorrhizae of *Picea abies* and *Larix decidua*. Trees 4:41-48

Kozlowski TT (1971) Growth and development of trees, vol I. Seed germination, ontogeny, and shoot growth. Academic Press, New York

Kozlowski TT, Winget CH (1963) Pattern of water movement in forest trees. Bot Gaz 124:301-311

Kramer PJ (1969) Plant and soil water relationships: A modern synthesis. McGraw-Hill, New York

Kramer PJ, Bullock HC (1966) Seasonal variations in the proportions of suberized and unsuberized roots of trees in relation to absorption of water. Am J Bot 53:200-204

Kramer PJ, Kozlowski TT (1979) Physiology of woody plants. Academic Press, New York

Krawczyszyn J, Romberger JA (1979) Cyclical cell length changes in wood in relation to storied structure and interlocked grain. Can J Bot 57:787-794

Krawczyszyn J, Romberger JA (1980) Interlocked grain, cambial domains, endogenous rhythms, and time relations, with emphasis on *Nyssa sylvatica*. Am J Bot 67:228-236

Krenzer EG Jr, Moss DN, Crookston RK (1975) Carbon dioxide compensation points of flowering plants. Plant Physiol 56:194-206

Kübler H (1959a) Studien über Wachstumsspannungen des Holzes — I. Die Ursache der Wachstumsspannungen und die Spannungen quer zur Faserrichtung. Holz Roh- Werkstoff 17:1-9

Kübler H (1959b) Studien über Wachstumsspannungen des Holzes — II. Die Spannungen in Faserrichtung. Holz Roh- Werkstoff 17:44-54

Kubler H (1987) Growth stresses in trees and related wood properties. For Abstr 48:131-189

Kubler H (1991) Function of spiral grain in trees. Trees 5:125-135

Kuijt J (1977) Haustoria of phanerogamic parasites. Annu Rev Phytopathol 17:91-118

Kuijt J, Toth R (1976) Ultrastructure of angiosperm haustoria — A review. Ann Bot 40: 1121-1130

Kumon K, Suda S (1984) Ionic fluxes from pulvinar cells during rapid movement of *Mimosa pudica* L. Plant Cell Physiol 25:975-979

Kundu BC, Rao NS (1955) Origin and development of axillary buds of *Hibiscus cannabinus*. Am J Bot 42:830-837

Kuo J, O'Brien TP (1974) Lignified sieve elements in the wheat leaf. Planta 117:349-353

Kurczyńska E (1986) Terminal vessel and early vessel arrangement in internodes of *Fraxinus excelsior*. Acta Soc Bot Pol 55:3-10

Kursanov AL (1976) Assimilate transport in plants. Nauka, Moscow (in Russian)

Kurt J (1929) Über die Hydathoden der Saxifrageae. Bei Bot Centralbl 46:203-246

Kutschera U (1989) Tissue stresses in growing plant organs. Physiol Plant 77:157-163

Lachaud S (1989) Participation of auxin and abscisic acid in the regulation of seasonal variations in cambial activity and xylogenesis. Trees 3:125-137.

Laetsch WM (1968) Chloroplast specialization in dicotyledons possessing the C_4-dicarboxylic acid pathway of photosynthetic CO_2 fixation. Am J Bot 55:875-883

Laetsch WM (1974) The C_4 syndrome: A structural analysis. Annu Rev Plant Physiol 25:27-52

Lakshmanan KK (1972) Monocot embryo. In: Varghese TM, Grover RK (eds) Vistas in Plant Sciences 2. Searchmates, Sangrur, India, pp 61-110

Lakshmanan KK, Ambegaokar KB (1984) Polyembryony. In: Johri BM (ed) Embryology of angiosperms. Springer, Berlin Heidelberg New York, pp 445-474

Lalonde M, Knowles R (1975) Ultrastructure, composition, and biogenesis of the encapsulation material surrounding the endophyte in *Alnus crispa* var *mollis* root nodules. Can J Bot 53:1951-1971

Lam OC III, Brown CL (1974) Shoot growth and histogenesis of *Liquidambar styraciflua* L. under different photoperiods. Bot Gaz 135:149-154

LaMarche VC Jr (1968) Rates of slope degradation as determined from botanical evidence — White Mountains, California. US Geol Surv Prof Pap 352-I:341-377

Lambert A-M, Vantard M, Schmit A-C, Stoeckel H (1991) Mitosis in plants. In: Giddings TH Jr, Staehelin LA (eds) The cytoskeletal basis of plant growth and form. Academic Press, London, pp 199-208

Lamont BB (1980) Tissue longevity of the arborescent monocotyledon, *Kingia australis* (Xanthorrhoeaceae). Am J Bot 67:1262-1264

Lamont B[B] (1982) Mechanisms for enhancing nutrient uptake in plants, with particular reference to mediterranean South Africa and Western Australia. Bot Rev 48:597-689

Lamont BB, Ryan RA (1977) Formation of coralloid roots by cycads under sterile conditions. Phytomorphology 27:426-429

Lamoreaux RJ, Chaney WR, Brown KM (1978) The plastochron index: A review after two decades of use. Am J Bot 65:586-593

Lamoureux CH (1974) Phloem tissue in angiosperms and gymnosperms. In: Aronoff S et al (eds) Phloem transport. NATO Advanced Study Inst Ser A-4. Plenum Press, New York, pp 1-20

Larson PR (1963) Stem form development of forest trees. For Sci Monogr 5

Larson PR (1975) Development and organization of the primary vascular system in *Populus deltoides* according to phyllotaxy. Am J Bot 62:1084-1099

Larson PR (1976) Procambium vs cambium and protoxylem vs metaxylem in *Populus deltoides* seedlings. Am J Bot 63:1332-1348

Larson PR (1977) Phyllotactic transitions in the vascular system of *Populus deltoides* Bartr. as determined by ^{14}C labeling. Planta 134:241-249

Larson PR (1979) Establishment of the vascular system in seedlings of *Populus deltoides* Bartr. Am J Bot 66: 452-462

Larson PR (1980) Interrelations between phyllotaxis, leaf development and the primary-secondary vascular transition in *Populus deltoides*. Ann Bot 46:757-769

Larson PR (1982) The concept of cambium. In: Baas pp (ed) New perspectives in wood anatomy. Nijhoff/Junk, The Hague, pp 85-121

Larson PR (1984) Vascularization of developing leaves of *Gleditsia tricanthos* L. I. The node, rachis, and rachillae. Am J Bot 71:1201-1210

Larson PR (1985) Rachis vascularization and leaflet venation in developing leaves of *Fraxinus pennsylvanica*. Can J Bot 63:2383-2392

Larson PR, Dickson RE (1973) Distribution of imported ^{14}C in developing leaves of eastern cottonwood according to phyllotaxy. Planta 111:95-112

Larson PR, Fisher DG (1983) Xylary union between elongating branches and the main stem in *Populus deltoides*. Can J Bot 61:1040-1051

Larson PR, Pizzolato TD (1977) Axillary bud development in *Populus deltoides*. I. Origin and early ontogeny. Am J Bot 64:835-848

Läuchli A (1976) Apoplasmic transport in tissues. Encycl Plant Physiol 2nd Ser 2(B):3-34

Lawton JR (1972) Seasonal variations in the secondary phloem of some forest trees from Nigeria. II. Structure of the phloem. New Phytol 71:335-348

Lawton JR (1977) An investigation of the functional phloem in willow. New Phytol 78:189-192

Lawton JRS, Canny MJ (1970) The proportion of sieve elements in the phloem of some tropical trees. Planta 95:351-354

Lee CL (1952) The anatomy and ontogeny of the leaf of *Dacrydium taxoides*. Am J Bot 39:393-398

Lee R, Gates DM (1964) Diffusion resistance in leaves as related to their stomatal anatomy and microstructure. Am J Bot 51:963-975

Lehmann C (1925) Studien über den Bau und die Entwicklungs-Geschichte von Ölzellen. Planta 1:343-373

Leigh EG Jr (1972) The golden section and spiral leaf-arrangement. Conn Acad Arts Sci Trans 44:163-176

Lersten NR (1983) Suspensors in Leguminosae. Bot Rev 49:233-257

Lersten NR, Curtis JD (1982) Hydathodes in *Physocarpus* (Rosaceae:Spiraeoideae). Can J Bot 60:850-855

Lersten NR, Peterson WH (1974) Anatomy of hydathodes and pigment disks in leaves of *Ficus diversifolia* (Moraceae). Bot J Linn Soc 68:109-113

Levin DA (1973) The role of trichomes in plant defense. Quart Rev Biol 48:3-15

Lev-Yadun S, Aloni R (1990) Vascular differentiation in branch junctions of trees: Circular patterns and functional significance. Trees 4:49-54

Leyton L, Juniper BE (1963) Cuticle structure and water relations of pine needles. Nature 198:770-771

Libbenga KR, Bogers RJ (1974) Root-nodule morphogenesis. In: Quispel A (ed) The biology of nitrogen fixation. Elsevier/North Holland, Amsterdam, pp 430-472

Libbenga KR, Harkes PAA (1973) Initial proliferation of cortical cells in the formation of root nodules in *Pisum sativum* L. Planta 114:17-28

Liese W, Bauch J (1964) Über die Wegsamkeit der Hoftüpfel von Coniferen. Naturwissenschaften 51:516

Lindenmayer A (1975) Developmental algorithms for multicellular organisms: A survey of L-systems. J Theor Biol 54:3-22

Lindenmayer A (1984) Models for plant tissue development with cell division orientation regulated by preprophase bands of microtubules. Differentiation 26:1-10

Linsbauer K (1930) Die Epidermis. Handbuch der Pflanzenanatomie, Band IV/Abt 1/Teil 2. Borntraeger, Berlin

Lintilhac PM (1974a) Differentiation, organogenesis, and the tectonics of cell wall orientation. II. Separation of stresses in a two-dimensional model. Am J Bot 61:135-140

Lintilhac PM (1974b) Differentiation, organogenesis, and the tectonics of cell wall orientation. III. Theoretical considerations of cell wall mechanics. Am J Bot 61:230-237

Lintilhac PM (1984) Positional controls in meristem development: A caveat and an alternative. In: Barlow PW, Carr DJ (eds) Positional controls in plant development. Cambridge Univ Press, Cambridge, pp 83-105

Lintilhac PM, Vesecky TB (1980) Mechanical stress and cell wall orientation in plants. I. Photoelastic derivation of principal stresses; with a discussion of the concept of axillarity and the significance of the "arcuate shell zone". Am J Bot 67:1477-1483

Lintilhac PM, Vesecky TB (1981) Mechanical stress and cell wall orientation in plants. II. The application of controlled directional stress to growing plants; with a discussion of the nature of the wound reaction. Am J Bot 68:1222-1230

Lipetz J (1970) Wound-healing in higher plants. Int Rev Cytol 27:1-28

Liphschitz N, Waisel Y (1980) Periderm: structure, origin, development and rhythm of activity. In: Little CHA (ed) Control of shoot growth in trees. Proc Joint Workshop of IUFRO Working Parties on Xylem Physiology and Shoot Growth Physiology. Maritimes Forest Research Centre, Fredericton, New Brunswick, Canada. pp 41-75

List A Jr (1963) Some observations on DNA content and cell and nuclear volume growth in the developing xylem cells of certain higher plants. Am J Bot 50:320-329

Lloyd CW (1984) Toward a dynamic helical model for the influence of microtubules on wall patterns in plants. Int Rev Cytol 86:1-51

Lloyd CW, Barlow PW (1982) The co-ordination of cell division and elongation: The role of the cytoskeleton. In: Lloyd CW (ed) The cytoskeleton in plant growth and development. Academic Press, London, pp 203-228

Lowe SB, Mahon JD, Hunt LA (1982) Early development of cassava (Manihot esculenta). Can J Bot 60:3040-3048

Lumsden CJ (1985) Hierarchical behavior in fit dynamical systems. Bull Math Biol 47:591-612

Lüttge U (1971) Structure and function of plant glands. Annu Rev Plant Physiol 22:23-44

Lüttge U, Higinbotham N (1979) Transport in plants. Springer, Berlin Heidelberg New York

Lüttge U, Osmond CB (1970) Ion absorption in Atriplex leaf tissue. III. Site of metabolic control of light-dependent chloride secretion to epidermal bladders. Aust J Biol Sci 23:17-25

Lyford WH (1975) Rhizography of non-woody roots of trees in the forest floor. In: Torrey JG, Clarkson DT (eds) The development and function of roots. Academic Press, London, pp 179-196

Lyndon RF (1976) The shoot apex. In: Yeoman MM (ed) Cell division in Higher Plants. Academic Press, London, pp 285-314

Mader H (1954) Untersuchungen an Korkmembranen. Planta 43:163-181

Madore M, Webb JA (1981) Leaf free space analysis and vein loading in Cucurbita pepo. Can J Bot 59:2550-2557

Mägdefrau K (1951) Botanik. Carl Winter, Heidelberg

Magel EA, Drouet A, Claudot AC, Ziegler H (1991) Formation of hardwood substances in the stem of Robinia pseudoacacia L. I. Distribution of phenylalanine ammonium lyase and chalcone synthase across the trunk. Trees 5:203-207

Magendans JFC (1983) Anatomy of vein endings in Hedera leaves; influence of dry and wet conditions. Meded Landbouwhogesch Wageningen 83:1-34

Magendans JFC (1985) Anatomy of vein endings in Hedera leaves — Aspects of ontogeny. Agric Univ Wageningen Pap 85-5:1-74

Maheshwari P (1950) An Introduction to the embryology of angiosperms. McGraw-Hill, New York

Maheshwari P, Singh H (1967) The female gametophyte of gymnosperms. Biol Rev 42:88-130

Mahlberg PG (1960) Embryogeny and histogenesis in Nerium oleander L. I. Organization of primary meristematic tissues. Phytomorphology 10:118-131

Mahlberg PG (1961) Embryogeny and histogenesis in Nerium oleander. II. Origin and development of the non-articulated laticifer. Am J Bot 48:90-99

Mahlberg PG, Sabharwal PS (1967) Mitosis in the non-articulated laticifer of Euphorbia marginata. Am J Bot 54:465-472

Mahlstede JP, Watson DP (1952) An anatomical study of adventitious root development in stems of Vaccinium corymbosum. Bot Gaz 113:279-285

Mahmood A (1968) Cell grouping and primary wall generations in the cambial zone, xylem, and phloem in Pinus. Aust J Bot 16:177-195

Mahmood A (1990) The parental cell walls. In: Iqbal M (ed) The vascular cambium. Research Studies Press, Taunton (Somerset). Wiley, New York, pp 113-126

Majer JD (1979) The possible protective function of extrafloral nectaries of *Acacia saligna*. Annu Rep (1978) Mulga Res Cent, Western Aust Inst Tech (Bentley, Western Aust), pp 31-39

Maksymowych R (1973) Analysis of leaf development. Cambridge Univ Press, Cambridge

Maksymowych R, Erickson RO (1977) Phyllotactic change induced by gibberellic acid in *Xanthium* shoot apices. Am J Bot 64:33-44

Malajczuk N, Dell B, Bougher NL (1987) Ectomycorrhiza formation in *Eucalyptus*. III. Superficial ectomycorrhizas initiated by *Hysterangium* and *Cortinacius* species. New Phytol 105:421-428 + 6 figs

Mansfield TA, Hetherington AM, Atkinson CJ (1990) Some current aspects of stomatal physiology. Annu Rev Plant Physiol Plant Mol Biol 41:55-75

Marchant HJ (1982) The establishment and maintenance of plant cell shape by microtubules. In: Lloyd CW (ed) The cytoskeleton in plant growth and development. Academic Press, London, pp 295-319

Marco HF (1939) The anatomy of spruce needles. J Agric Res 58:357-368

Marks GC, Foster RC (1973) Structure, morphogenesis, and ulatrastructure of ectomycorrhizae. In: Marks GC, Kozlowski TT (eds) Ectomycorrhizae: Their ecology and physiology. Academic Press, New York, pp 1-41

Martens P (1938) Nouvelles recherches sur l'origine des espaces intercellulaires. Beih Bot Centralbl 58(A):349-364

Martin FW, Ortiz S (1963) Origin and anatomy of tubers of *Dioscorea floribunda* and *D spiculiflora*. Bot Gaz 124:416-421

Martin G, Josserand SA, Bornman JF, Vogelmann TC (1989) Epidermal focusing and the light microenvironment within leaves of *Medicago sativa*. Physiol Plant 76:485-492

Martin JT, Juniper BE (1970) The cuticles of plants. St Martin's Press, New York

Martin MN (1991) The latex of *Hevea brasiliensis* contains high levels of both chitinase and chitinases/lysozymes. Plant Physiol 95:469-476

Mason PA, Wilson J, Last FT, Walker C (1983) The concept of succession in relation to the spread of sheathing mycorrhizal fungi on inoculated tree seedlings growing in unsterile soils. Plant Soil 71:247-256

Massicotte HB, Peterson RL, Ackerley CA, Piché Y (1986) Structure and ontogeny of *Alnus crispa* — *Alpova diplophloeus* ectomycorrhizae. Can J Bot 64:177-192

Massicotte HB, Peterson RL, Ashford AE (1987) Ontogeny of *Eucalyptus pilularis* — *Pisolithus tinctoris* ectomycorrhizae. I. Light microscopy and scanning electron microscopy. Can J Bot 65:1927-1939

Masuda Y, Yamamoto R (1985) Cell-wall changes during auxin-induced cell extension. Mechanical properties and constituent polysaccharides of the cell wall. In: Brett CT, Hillman JR (eds) Biochemistry of plant cell walls. Soc Exp Biol Sem Ser (Cambridge Univ Press, Cambridge), pp 269-300

Mathew CJ (1980) Embryological studies in Hamamelidaceae: Development of female gametophyte and embryogeny in *Hamamelis virginiana*. Phytomorphology 30:172-180

Mauseth JD (1976) Cytokinin- and gibberellic acid-induced effects on the structure and metabolism of shoot apical meristems in *Opuntia polyacantha* (Cactaceae). Am J Bot 63:1295-1301

Mauseth JD (1983) Introduction to cactus anatomy. Part 5. Secretory cells. Cactus Succulent J 55:171-175

Mauseth JD (1988) Plant anatomy. Benjamin/Cummings, Menlo Park, California

McCully ME (1975) The development of lateral roots. In: Torrey JG, Clarkson DT (eds) The development and function of roots. Academic Press, Londono, pp 105-124

McDaniel CN (1984) Shoot meristem development. In: Barlow PW, Carr JD (eds) Positional controls in plant development. Cambridge Univ Press, Cambridge, pp 319-347

McDougall WB (1921) Thick-walled root hairs of *Gleditsia* and related genera. Am J Bot 8:171-17

McMahon TA (1975) The mechanical design of trees. Sci Am 233:92-102

Mehra PN, Dogra PD (1975) Embryogeny of Pinaceae. I. Proembryogeny. Proc Indian Natl Acad Sci 41B:486-497

Meicenheimer RD (1979) Relationships between shoot growth and changing phyllotaxy of *Ranunculus*. Am J Bot 66:557-569

Meicenheimer RD (1981) Changes in *Epilobium* phyllotaxy induced by N-1- naphthylphthalamic acid and α-4-chlorophenoxyisobutyric acid. Am J Bot 68:1139-1154

Meicenheimer RD (1982) Change in *Epilobium* phyllotaxy during reproductive transition. Am J Bot 69:1108-1118

Meicenheimer RD (1986) Role of parenchyma in *Linum usitatissimum* leaf trace patterns. Am J Bot 73:1649-1664

Meinhardt H (1982) Models of biological pattern formation. Academic Press, London

Meinhardt H (1984) Models of pattern formation and their application to plant development. In: Barlow PW, Carr DJ (eds) Positional controls in plant development. Cambridge Univ Press, Cambridge, pp 1-32

Meins F Jr, Binns AN (1979) Cell determination in plant development. BioScience 29:221-225

Mejstřík V, Kelly AP (1979) Mycorrhizae in *Sequoia gigantea* Lindl et Gard and *Sequoia sempervirens* Endl. Ceská Mykol 33:51-54

Merrill EK (1979) Comparison of ontogeny of three types of leaf architecture in *Sorbus* L (Rosaceae). Bot Gaz 140:328-337

Metcalfe CR (1967) Distribution of latex in the plant kingdom. Econ Bot 21:115-127

Metcalfe CR, Chalk L (1979) Anatomy of the dicotyledons, 2nd edn, vol 1. Clarendon Press, Oxford

Mevi-Schütz J, Grosse W (1988) A two-way gas transport system in *Nelumbo nucifera*. Plant Cell Environ 11:27-34

Meyer CF (1958) Cell patterns in early embryogeny of the McIntosh apple. Am J Bot 45:341-349

Meyer FH (1973) Distribution of ectomycorrhizae in native and man-made forests. In: Marks GC, Kozlowski TT (eds) Ectomycorrhizae: Their ecology and physiology. Academic Press, New York, pp 79-105

Meylan BA, Butterfield BG (1981) Perforation plate differentiation in the vessels of hardwoods. In: Barnett JR (ed) Xylem cell development. Castle House, Tunbridge Wells (Kent), pp 96-114

Milburn JA, McLaughlin ME (1974) Studies on cavitation in isolated vascular bundles and whole leaves of *Plantago major* L. New Phytol 73:861-871

Milindasuta B-E (1975) Developmental anatomy of coralloid roots in cycads. Am J Bot 62:468-472

Miller RH (1985) The prevalence of pores and canals in leaf cuticular membranes. Ann Bot 55:459-471

Miller RH (1986) The prevalence of pores and canals in leaf cuticular membranes. 2. Supplemental studies. Ann Bot 57:419-434

Millington WF, Chaney WR (1973) Shedding of shoots and branches. In: Kozlowski TT (ed) Shedding of plant parts. Academic Press, New York London, pp 149-204

Millington WF, Gunckel JE (1950) Structure and development of the vegetative shoot tip of *Liriodendron tulipifera* L. Am J Bot 37:326-335

Minchin PEH, Thorpe MR (1984) Apoplastic phloem unloading in the stem of bean. J Exp Bot 35:538-550

Mogensen HL (1968) Studies on the bark of the cork bark fir: *Abies lasiocarpa* var. *Arizonica* (Merriam) Lemmon. I. Periderm ontogeny. J Ariz Acad Sci 5:36-40

Mogensen HL, David JR (1968) Studies on the bark of the cork bark fir: *Abies lasiocarpa* var. *Arizonica* (Merriam) Lemmon. II. The effect of exposure on the time of initial rhytidome formation. J Ariz Acad Sci 5:108-109

Mohr H, Sitte P (1971) Molekulare Grundlagen der Entwicklung. BLV, Munich

Mollenhauer HH, Whaley WG, Leech JH (1961) A function of the Golgi apparatus in outer rootcap cells. J Ultrastruct Res 5:193-200

Monson RK, Edwards GE, Ku MSB (1984) C_3 - C_4 intermediate photosynthesis in plants. BioScience 34:563-574

Montain CR, Haissig BE, Curtis JD (1983) Differentiation of adventitious root primordia in callus of *Pinus banksiana* seedling cuttings. Can J For Res 13:195-200

Moore R (1985) Movement of calcium across tips of primary and lateral roots of *Phaseolus vulgaris*. Am J Bot 72:785-787

Moss EH, Gorham AL (1953) Interxylary cork and fission of stems and roots Phytomorphology 3:285-294

Mounts BT (1932) The development of foliage leaves. Univ Iowa Stud Nat Hist 14(5):1-20 + 3 pl

Mueller SC, Brown RM Jr (1982a) The control of cellulose microfibril deposition in the cell wall of higher plants. I. Can directed membrane flow orient cellulose microfibrils? Indirect evidence from freeze-fractured plasma membranes of maize and pine seedlings. Planta 154:489-500

Mueller SC, Brown RM Jr (1982b) The control of cellulose microfibril deposition in the cell wall of higher plants. II. Freeze-fracture microfibril patterns in maize seedling tissues following experimental alteration with colchicine and ethylene. Planta 154:501-515

Mullick DB (1971) Natural pigment differences distinguish first and sequent periderms of conifers through a cryofixation and chemical techniques. Can J Bot 49:1703-1711

Mullick DB (1975) A new tissue essential to necrophylactic periderm formation in the bark of four conifers. Can J Bot 53:2443-2457

Mullick DB, Jensen GD (1973) New concepts and terminology of coniferous periderms: Necrophylactic and exophylactic periderms. Can J Bot 51:1459-1470

Münch E (1930) Die Stoffbewegungen in der Pflanze. Fischer, Jena

Münch E (1938) Statik und Dynamic des schraubigen Baues der Zellwand, besonders des Druck- und Zugholzes. Flora Allg Bot Ztg 132:357-424

Napp-Zinn K (1966) Anatomie des Blattes. I. Blattanatomie der Gymnospermen. Handb d Pflanzenanat, Bd VIII/T 1. Borntraeger, Berlin

Napp-Zinn K (1973) Anatomie des Blattes. II. Blattanatomie der Angiospermen. A. Entwicklungsgeschichtliche und topographische Anatomie des Angiospermenblattes (2 vol). Handb der Pflanzenanat, Bd VIII/T IIA1. Borntraeger, Berlin

Nast CG (1941) The embryogeny and seedling morphology of *Juglans regia* L. Lilloa 6:163-206 + 12 pl

Natesh S, Rau MA (1984) The embryo. In: Johri BM (ed) Embryology of Angiosperms. Springer, Berlin Heidelberg New York, pp 377-443

Nathanielsz CP, Staff IA (1975) A mode of entry of blue-green algae into the apogeotropic roots of *Macrozamia communis*. Am J Bot 62:232-235

Navarro L, Roistacher CN, Murashige T (1975) Improvement of shoot-tip grafting in vitro for virus-free citrus. J Am Soc Hortic Sci 100:471-479

Neilson-Jones W (1969) Plant chimeras. Methuen, London

Nelson T, Langdale JA (1989) Patterns of leaf development in C4 plants. Plant Cell 1:3-13

Neville AC (1988) A pipe-cleaner molecular model of morphogenesis of helicoidal plant cell walls based on hemicellulose complexity. J Theor Biol 131:243-254

Neville AC, Levy S (1985) The helicoidal concept in plant cell ultrastructure and morphogenesis. In: Brett CT, Hillman JR (eds) Biochemistry of plant cell walls. Soc Exp Biol Sem Ser 28:99-124

Neville P (1969) Morphogenèse chez *Gleditsia triacanthos* L. III. Étude histologique et expérimentale de la sénescence des burgeons. Ann Sci Nat Ser 12 Bot 10:301-324

Newcomb W (1981) Nodule morphogenesis and differentiation. Int Rev Cytol Suppl 13:247-298

Niklas KJ (1988) Dependency of the tensile modulus on transverse dimensions, water potential, and cell number of pith parenchyma. Am J Bot 75:1286-1292

Nilsson SB, Hertz CH, Falk S (1958) On the relation between turgor pressure and tissue rigidity. II. Theoretical calculations on model systems. Physiol Plant 11:818-837

Norris RF, Bukovac MJ (1968) Structure of the pear leaf cuticle with special reference to cuticular penetration. Am J Bot 55:975-983

Norstog K (1972) Early development of the barley embryo: Fine structure. Am J Bot 59:123-132

Northcote D H (1972) Chemistry of the plant cell wall. Annu Rev Plant Physiol 23:113-132

Nsimba-Lubaki M, Peumans WJ, Allen AK (1986) Isolation and characterization of glycoprotein lectins from the bark of three species of elder, *Sambucus ebulus*, *S. nigra*, and *S. racemosa*. Planta 168:113-118

Nylund J-E (1980) Symplastic continuity during Hartig net formation in Norway spruce ectomycorrhizae. New Phytol 86:373-378 + 4 pl

Nylund J-E, Unestam T (1982) Structure and physiology of ectomycorrhizae. I. The process of mycorrhiza formation in Norway spruce *in vitro*. New Phytol 91:63-79

O'Brien TP (1970) Further observations on hydrolysis of the cell wall in the xylem. Protoplasma 69:1-14

O'Brien TP (1974) Primary vascular tissues. In: Robards AW (ed) Dynamic aspects of plant ultrastructure. McGraw-Hill, London, pp 414-440

Olesen P, Robards AW (1990) The neck region of plasmodesmata: General architecture and some functional aspects. In: Robards AW et al (eds) Parallels in cell to cell junctions in plants and animals. NATO Advanced Study Inst Ser 6:145-170

Osmond CB, Smith FA (1976) Symplastic transport of metabolites during C_4- photosynthesis. In: Gunning BES, Robards AW (eds) Intercellular communication in plants: Studies on plasmodesmata. Springer, Berlin Heidelberg New York, pp 229-241

Osmond CB, Lüttge U, West KR, Pallaghy CK, Shacher-Hill B (1969) Ion absorption in *Atriplex* leaf tissue. II. Secretion of ions to epidermal bladders. Aust J Biol Sci 22:797-814

Outer RW den (1967) Histological investigations of the secondary phloem of gymnosperms. Meded Landbouwhogesch Wageningen 67(7):1-119

Outlaw WH Jr, Lowry OH (1977) Organic acid and potassium accumulation in guard cells during stomatal opening. Proc Natl Acad Sci USA 74:4434-4438

Overall RL, Gunning BES (1982) Intercellular communication in *Azolla* roots: II. Electrical coupling. Protoplasma 111:151-160

Palevitz BA (1982) The stomatal complex as a model of cytoskeletal participation in cell differentiation. In: Lloyd CW (ed) The cytoskeleton in plant growth and development. Academic Press, London pp 345-376

Panshin AJ, Zeeuw C de (1980) Textbook of wood technology, 4th edn. McGraw-Hill, New York

Parameswaran N, Liese W 1968 Beitrag zur Rindenanatomie der Gattung *Entandrophragma*. Flora Abt B 158:22-40 + 2 pl

Parameswaran N, Kruse J, Liese W (1976) Aufbau und Feinstruktur von Periderm und Lenticellen der Fichtenrinde. Z Pflanzenphysiol 77:212-221

Parthasarathy MV (1974) Ultrastructure of phloem in palms. III. Mature phloem. Protoplasma 79:265-315

Parthasarathy MV (1975) Sieve-element structure. Encycl Plant Physiol 2nd Ser 1:3-38

Parthasarathy MV, Perdue TD, Witztum A, Alvernaz J (1985) Actin network as a normal component of the cytoskeleton in many vascular plant cells. Am J Bot 72:1318-1323

Paszewski A, Zawadzki T (1974) Action potentials in *Lupinus angustifolius* L. shoots. II. Determination of the strength-duration relation and the all-or-nothing law. J Exper Bot 25:1097-1103

Pate JS, Dixon KW (1982) Tuberous, cormous, and bulbous plants: Biology of an adaptive strategy in Western Australia. Univ Western Aust Press, Nedlands

Pate JS, Gunning BES (1969) Vascular transfer cells in angiosperm leaves: A taxonomic and morphological survey. Protoplasma 68:135-156

Pate JS, Gunning BES (1972) Transfer cells. Annu Rev Plant Physiol 23:173-196

Pate JS, Gunning BES, Briarty LG (1969) Ultrastructure and functioning of the transport system of the leguminous root nodule. Planta 85:11-34

Patel JD (1978) How should we interpret and distinguish subsidiary cells? Bot J Linn Soc 77:65-72

Patel RN (1965) A comparison of the anatomy of the secondary xylem of roots and stems. Holzforschung 19:72-79

Patel RN (1975) Bark anatomy of radiata pine, Corsican pine, and Douglas fir grown in New Zealand. NZ J Bot 13:149-167

Patel RN, Shand JE (1985) Bark anatomy of *Nothofagus* species indigenous to New Zealand. NZ J Bot 23:511-532

Pearcy RW, Ehleringer J (1984) Comparative ecophysiology of C_3 and C_4 plants. Plant Cell Environ 7:1-13

Pearson LC, Lawrence DB (1958) Photosynthesis in aspen bark. Am J Bot 45:383-387

Penel C, Gaspar Th, Greppin H (1985) Rapid interorgan communications in higher plants with special reference to flowering. Biol Plant 27:334-338

Periasamy K (1977) A new approach to the classification of angiosperm embryos. Proc Indian Acad Sci 86B:1-13

Persson H (1983) The importance of fine roots in boreal forests. In: Böhm W, Kutschera L, Lichtenegger E (eds) Root ecology and its practical application. Int Symp, Gumpenstein, Austria, Sept 1982, Verlag Gumpenstein, pp 595-608

Peters WJ (1974) Tylosis formation in *Pinus* tracheids. Bot Gaz 135:126-131

Peterson CA (1988) Exodermal Casparian bands: Their significance for ion uptake by roots. Physiol Plant 72:204-208

Peterson CA (1989) Significance of the exodermis in root function. In:Loughman BC, Gasparíková O, Kolek J (eds) Structural and functional aspects of transport in roots. 3rd Int Symp on Structure and Function of Roots, Nitra, Aug 1987, pp 35-40

Peterson CA, Perumalla CJ (1984) Development of the hypodermal Casparian band in corn and onion roots. J Exp Bot 35:51-57

Peterson RL (1970) Bud development at the root apex of *Ophioglossum petiolatum*. Phytomorphology 20:183-190

Peterson RL, Vermeer J (1980) Root apex structure in *Ephedra monosperma* and *Ephedra chilensis* (Ephedraceae). Am J Bot 67:815-823

Peterson TA, Swanson ES, Hull RA (1986) Use of lanthanum to trace apoplastic solute transport in intact plants. J Exp Bot 37:807-822

Philipp M (1923) Über die verkorkten Abschlussgewebe der Monokotylen. Bibl Bot 92:1-28 + 1 pl

Philipson WR, Ward JM, Butterfield BG (1971) The vascular cambium: Its development and activity. Chapman and Hall, London

Philpott J (1953) A blade tissue study of forty-seven species of Ficus. Bot Gaz 115:15-35

Picciarelli P, Alpi A, Pistelli L, Scalet M (1984) Gibberellin-like activity in suspensors of Tropaeolum majus L and Cytisus laburnum L. Planta 162:566-568

Piche Y, Fortin JA, Peterson RL, Posluszny U (1982) Ontogeny of dichotomizing apices in mycorrhizal short roots of Pinus strobus. Can J Bot 60:1523-1528

Pilkington M (1929) The regeneration of the stem apex. New Phytol 28:37-53

Pitman MG, Lüttge U, Kramer D, Ball E (1974) Free space characteristics of barley leaf slices. Aust J Plant Physiol 1:65-75

Pizzolato TD, Larson PR (1977) Axillary bud development in Populus deltoides. II Late ontogeny and vascularization. Am J Bot 64:849-860

Plantefol L (1947) Hélices foliaires, point végétatif et stèle chez les dicotylédones. La notion d'anneau initial. Rev Gén Bot 54:49-80

Plantefol L (1948) La théorie des hélices foliaires multiples: Fondements d'une théorie phyllotaxique nouvelle. Masson and Cie, Paris

Plaut M (1910) Über die Veränderung im anatomischen Bau der Wurzel während des Winters. Jahrb Wiss Bot 48:143-154

Plaut M (1918) Über die morphologischen und mikroskopischen Merkmale der Periodizität der Wurzel, sowie über die Verbreitung der Metakutisierung der Wurzelhaube im Pflanzenreich. In: Festschr 100-jähr Best Württemb landw Hochsch Hohenheim. Ulmer, Stuttgart, pp 129-151

Plymale EL, Wylie RB (1944) The major veins of mesomorphic leaves. Am J Bot 31:99-106

Poethig RS (1987) Clonal analysis of cell lineage patterns in plant development. Am J Bot 74:581-594

Pohlheim F (1969) Über Unterschiede in der Beteiligung des "Dermatogens" an der Mesophyllbildung bei Buxus sempervirens argenteo-marginata hort. Wiss Z Päd Hochsch Potsdam, Math- Naturwiss Reihe 13:167-176

Pohlheim F (1971a) Untersuchungen zur Sprossvariation der Cupressaceae. 1. Nachweis immerspaltender Periklinalchimären. Flora 160:264-293

Pohlheim F (1971b) Untersuchungen zur Sprossvariation der Cupressaceae. 3. Quantitative Auswertung des Scheckungsmusters immerspaltender Periklinalchimären. Flora 160:360-372

Popham RA (1955) Zonation of primary and lateral root apices of Pisum sativum. Am J Bot 42:267-273

Postek MT, Tucker SC (1982) Foliar ontogeny and histogenesis in Magnolia grandiflora L. I. Apical organization and early development. Am J Bot 69:556-569

Postgate J (1987) Nitrogen fixation, 2nd edn. Arnold, London

Poulson ME, Vogelmann TC (1990) Epidermal focusing and effects upon photosynthetic light-harvesting in leaves of Oxalis. Plant Cell Environ 13:803-811

Powers HO (1967) A blade tissue study of forty-seven species and varieties of Aceraceae. Am Mid Nat 78:301-323

Prantl K (1883) Studien über Wachstum, Verzweigung und Nervatur der Laubblätter, insbesondere der Dicotylen. Ber Dtsch Bot Ges 1:280-288

Prat H (1926) Étude des mycorhizes du "Taxus baccata". Ann Sci Nat Bot Ser 10, 8:141-163

Pray TR (1954) Foliar venation of angiosperms. I. Mature venation of Liriodendron. Am J Bot 41:663-670

Pray TR (1955) Foliar venation of angiosperms. II. Histogenesis of the venation of *Lirioden-dron*. Am J Bot 42:18-27

Pray TR (1963) Origin of vein endings in angiosperm leaves. Phytomorphology 13:60-81

Preston RD (1974) The physical biology of plant cell walls. Chapman and Hall, London

Preston RD (1979) Polysaccharide conformation and cell wall function. Annu Rev Plant Physiol 30:55-78

Preston RD (1982) The case for multinet growth in growing walls of plant cells. Planta 155:356-363

Pridgeon AM (1987) The velamen and exodermis of orchid roots. In: Arditti J (ed) Orchid biology, reviews and perspectives, IV. Comstock, Cornell Univ Press, Ithaca

Priestley JH (1930) Studies in the physiology of cambial activity. II. The concept of sliding growth. New Phytologist 29:96-140 + 1 pl

Priestley JH, Scott LI (1939) The formation of a new cell wall at cell division. Proc Leeds Philos Lit Soc Sci Sect 3:532-545

Priestley JH, Swingle CF (1929) Vegetative propagation from the standpoint of plant anatomy. US Dept Agric Tech Bull 151:1-98

Prigogine I, Nicolis G, Lefever R (1975) Models for cellular communication. In: Marois M (ed) From theoretical physics to biology. Elsevier/North-Holland, Amsterdam, pp 49-73

Puławska Z (1965) Correlations in the development of the leaves and leaf traces in the shoot of *Actinidia arguta* Planch. Acta Soc Bot Pol 34:697-712

Pyykkö M (1979) Morphology and anatomy of leaves from some woody plants in a humid tropical forest of Venezuelan Guayana. Acta Bot Fenn 112:1-41

Raatz W (1892) Ueber Thyllenbildungen in den Tracheiden der Coniferenhölzer. Ber Dtsch Bot Ges 10:183-192 + 1 pl

Racette S, Torrey JG (1989) Root nodule initiation in *Gymnostoma* (Casuarinaceae) and *Shepherdia* (Elaeagnaceae) induced by *Frankia* strain HFPGpI1. Can J Bot 67:2873-2879

Raghavan V (1976) Experimental embryogenesis in vascular plants. Academic Press, London

Raghavan V (1986) Embryogenesis in angiosperms: A developmental and experimental study. Cambridge Univ Press, Cambridge

Raghavan V (1990) Origin of the quiescent center in the root of *Capsella bursa-pastoris* (L.) Medik. Planta 181:62-70

Raju MVS, Coupland RT, Steeves TA (1966) On the occurrence of root buds on perennial plants in Saskatchewan. Can J Bot 44:33-37 + 1 pl

Raschke K (1979) Movements of stomata. Encycl Plant Physiol 2nd Ser 7:383-441

Raschke K, Hedrich R, Reckmann U, Schroeder JI (1988) Exploring biophysical and bio-chemical components of the osmotic motor that drives stomatal movement. Bot Acta 101:283-294

Rasdorsky W — also transliterated from Russian as Razdorskij VF, q.v. for later publications

Rasdorsky W (1925) Über die Reaktion der Pflanzen auf die mechanische Inanspruchnahme. Ber Dtsch Bot Ges 43:332-352 + 2 pl

Raunkiaer C (1919) Über Homodromie und Antidromie insbesondere bei Gramineen. Kgl Dan Videnskab Selskab Biol Medd 1(12):1-31

Raven JA (1977) The evolution of vascular land plants in relation to supracellular transport processes. Adv Bot Res 5:153-219

Rayner MC (1926-27) Mycorrhiza. New Phytol 25:1-50 + 1 pl, 65-108 + 1 pl, 171-190 + 2 pl, 248-263, 338-372 + 1 pl; 26:22-45, 85-114

Razdorskij VF (1955) Arhitektonika rastenij. Sovetskaja Nauka, Moscow (in Russian)

Rees LW, Shiue C-J (1957-1958) The structure and development of the bark of quaking aspen. Proc Minn Acad Sci 25-26:113-125

Reid RW, Watson JA (1966) Sizes distributions and numbers of vertical resin ducts in lodge-pole pine. Can J Bot 44:679-525

Rendle BJ (1960) Juvenile and adult wood. J Inst Wood Sci 5:58-61

Reyneke WF, Schijff HP van der (1974) The anatomy of contractile roots in *Eucomis* L'Hérit. Ann Bot 38:977-982

Rich PM (1987) Mechanical structure of the stem of arborescent palms. Bot Gaz 148:42-50

Richards FJ (1948) The geometry of phyllotaxis and its origin. Symp Soc Exp Biol 2:217-245

Richards FJ (1951) Phyllotaxis: Its quantitative expression and relation to growth in the apex. Philos Trans R Soc Lond, Ser B 235:509-564

Riederer M (1991) Die Kutikula als Barriere zwischen terrestrischen Pflanzen und der Atmosphäre. Naturwissenschaften 78:201-208

Riedl H (1937) Bau und Leistung des Wurzelholzes. Jahrb Wiss Bot 85:1-72

Robards AW (1976) Plasmodesmata in higher plants. In: Gunning BES, Robards AW (eds) Intercellular communications in plants: Studies on plasmodesmata. Springer, Berlin Heidelberg New York, pp 15-57

Robards AW, Clarkson DT (1976) The role of plasmodesmata in the transport of water and nutrients across roots. In: Gunning BES, Robards AW (eds) Intercellular communication in plants: Studies on plasmodesmata. Springer, Berlin Heidelberg New York, pp 181-201

Robards AW, Lucas WJ (1990) Plasmodesmata. Annu Rev Plant Physiol Plant Mol Biol 41:309-419

Roberts K, Grief C, Hills GJ, Shaw PJ (1985) Cell wall glycoproteins: Structure and function. J Cell Sci Suppl 2:105-127

Roberts LW (1988) Physical factors, hormones, and differentiation. In: Roberts LW, Gahan PB, Aloni R (joint authors) Vascular differentiation and plant growth regulators. Springer, Berlin Heidelberg New York, pp 89-105

Robinson DE, Grigor JK (1963) The origin of periderm in some New Zealand plants. Trans R Soc NZ Bot 2:121-124

Robinson KR (1985) The responses of cells to electrical fields: A review. J Cell Biol 101:2023-2027

Rodkiewicz B (1970) Callose in cell walls during megasporogenesis in angiosperms. Planta 93:39-47

Rodkiewicz B (1973) Embriologia Roślin Kwiatowych. PWN, Warsaw (in Polish)

Roeckl B (1949) Nachweis eines Konzentrationshubs zwischen Palisadenzellen und Siebröhren. Planta 36:530-550

Rogers SO, Bonnett HT (1989) Evidence for apical initial cells in the vegetative shoot apex of *Hedera helix* cv Goldheart. Am J Bot 76:539-545

Roland J-C (1978a) Early differences between radial walls and tangential walls of actively growing cambial zone. IAWA Bull 1978 (1):7-10

Roland J-C (1978b) Cell wall differentiation and stages involved with intercellular gas space opening. J Cell Sci 32:325-336

Roland J-C, Mosiniak M (1983) On the twisting pattern, texture and layering of the secondary walls of lime wood. Proposal of an unifying model. IAWA Bull 4:15-26

Roland J-C, Vian B (1979) The wall of the growing plant cell: Its three-dimensional organization. Int Rev Cytol 61:129-166

Roland J-C, Reis D, Vian B, Satiat-Jeunemaitre B, Mosiniak M (1987) Morphogenesis of plant cell walls at the supramolecular level: Internal geometry and versatility of helicoidal expression. Protoplasma 140:75-91

Rolinson AE (1976) Rates of cell division in the vegetative shoot apex of rice (*Oryza sativa* L.) Ann Bot 40:939-945

Romberger JA (1963) Meristems, growth, and development in woody plants. US Dept Agric Tech Bull 1293:1-214

Romberger JA, Gregory RA (1974) Analytical morphogenesis and the physiology of flowering in trees. In: Reid CPP, Fechner GH (eds) Proc 3rd N Am For Biol Worksh (Colo State Univ, Sept 1974), pp 132-147

Romberger JA, Gregory RA (1977) The shoot apical ontogeny of the *Picea abies* seedling. III. Some age-related aspects of morphogenesis. Am J Bot 64:622-630

Römheld V, Kramer D (1983) Relationship between proton efflux and rhizodermal transfer cells induced by iron deficiency. Z Pflanzenphysiol 113:73-83

Roth I (1981) Structural patterns of tropical barks. Encycl Plant Anat, 2nd edn, Bd 9/Teil 3. Borntraeger, Berlin

Rüdiger H (1984) On the physiological role of plant lectins. BioScience 34:95-99

Russin WA, Evert RF (1984) Studies on the leaf of *Populus deltoides* (Salicaceae): Morphology and anatomy. Am J Bot 71:1398-1415

Russin WA, Evert RF (1985) Studies on the leaf of *Populus deltoides* (Salicaceae):Ultrastructure, plasmodesmatal frequency, and solute concentrations. Am J Bot 72:1232-1247

Ruth J, Klekowski EJ Jr, Stein OL (1985) Impermanent initials of the shoot apex and diplontic selection in a juniper chimera. Am J Bot 72:1127-1135

Ryder VL (1954) On the morphology of leaves. Bot Rev 20:263-276

SaadEddin R, Doddema H (1986) Anatomy of the 'extreme' halophyte *Arthrocnemum fruticosum* (L.) Moq. in relation to physiology. Ann Bot 57:531-544

Sacher JA (1954) Structure and seasonal activity of the shoot apices of *Pinus lambertiana* and *Pinus ponderosa*. Am J Bot 41:749-759

Sacher JA (1955a) Cataphyll ontogeny in *Pinus lambertiana*. Am J Bot 42:82-91

Sacher JA (1955b) Dwarf shoot ontogeny in *Pinus lambertiana*. Am J Bot 42:784-792

Sachs J (1874) Lehrbuch der Botanik nach dem gegenwärtigen Stand der Wissenschaft, 4th edn. Engelmann, Leipzig

Sachs RM (1965) Stem elongation. Annu Rev Plant Physiol. 16:73-96

Sachs T (1969) Polarity and the induction of organized vascular tissues. Ann Bot 33:263-275 + 3 pl

Sachs T (1970) A control of bud growth by vascular tissue differentiation. Isr J Bot 19: 484-498

Sachs T (1972) The induction of fibre differentiation in peas. Ann Bot 36:189-197

Sachs T (1974) The developmental origin of stomata pattern in *Crinum*. Bot Gaz 135:314-318

Sachs T (1975) The control of the differentiation of vascular networks. Ann Bot 39:197-204

Sachs T (1981a) The control of the patterned differentiation of vascular tissue. Adv Bot Res 9:151-261

Sachs T (1981b) Polarity changes and tissue organization in plants. In: Schweiger HG (ed) International Cell Biology 1980-1981. Papers presented at the 2nd Int Cong on Cell Biology, Berlin, Aug 31-Sept 5, 1980. Springer, Berlin Heidelberg New York, pp 489-496

Sachs T (1984) Axiality and polarity in vascular plants. In: Barlow PW, Carr DJ (eds) Positional controls in plant development. Cambridge Univ Press, Cambridge, pp 193-224

Sachs T, Cohen D (1982) Circular vessels and the control of vascular differentiation in plants. Differentiation 21:22-26

Sack FD (1987) The development and structure of stomata. In: Zeiger E, Farquhar GD, Cowan IR (eds) Stomatal function. Stanford Univ Press, Stanford, pp 59-89

Saint-Côme R (1966) Application des techniques histoautoradiographiques et des méthodes statistiques à l'étude du fonctionnement apical chez le *Coleus blumei* Benth. Rev Gén Bot 73(864):241-323

Sakurada I, Nukushina Y, Ito T (1962) Experimental determination of the elastic modulus of crystalline regions in oriented polymers. J Polym Sci 57:651-6

Salleo S, Lo Gullo MA (1983) Water transport pathways in nodes and internodes of 1-year-old twigs of *Olea europaea* L. G Bot Ital 117:63-74

Salleo S, Lo Gullo MA (1986) Xylem cavitation in nodes and internodes of whole *Chorisia insignis* H B et K plants subjected to water stress: Relations between xylem conduit size and cavitation. Ann Bot 58:431-441

Salleo S, Rosso R, Lo Gullo MA (1982) Hydraulic architecture of *Vitis vinifera* L. and *Populus deltoides* Bartr. 1-year-old twigs. II. The nodal regions as "constriction zones" of the xylem system. G Bot Ital 116:29-40

Salthe SN (1985) Evolving hierarchical systems: Their structure and representation. Columbia Univ Press, New York

Sanders FE, Mosse B, Tinker PB (eds) (1975) Endomycorrhizas. Proc of Symp, Univ of Leeds, July 22-25, 1974. Academic Press, London

Sands R (1975) Radiata pine bark — Aspects of morphology, anatomy, and chemistry. NZ J For Sci 5:74-86

Sanford WW, Adanlawo I (1973) Velamen and exodermis characters of West African epiphytic orchids in relation to taxonomic grouping and habitat tolerance. Bot J Linn Soc 66:307-321

Sato K, Ishida S (1983) Resin canals in wood of *Larix leptolepis* Gord. V. Formation of vertical resin canals. Res Bull Coll Exp For Hokkaido Univ 40:723-740 + 6 pl

Satter RL (1979) Leaf movements and tendril curling. Encycl Plant Physiol 2nd Ser 7:442-484

Satter RL, Galston AW (1981) Mechanisms of control of leaf movements. Annu Rev Plant Physiol 32:83-110

Sauter JJ (1980) The Strasburger cells — Equivalents of companion cells. Ber Dtsch Bot Ges 93:29-42

Sauter JJ (1986) Xylem: Structure and function. Prog Bot 48:388-405

Sauter JJ, Braun HJ (1968) Enzymatic polarity in ray parenchyma cells of conifers in spring. Z Pflanzenphysiol 58:378-381

Sauter JJ, Kloth K (1986) Plasmodesmatal frequency and radial translocation rates in ray cells of poplar (*Populus x canadensis* Moench 'robusta'). Planta 168:377-38

Sauter JJ, Cleve BV, Apel K (1988) Protein bodies in ray cells of *Populus x canadensis* Moench 'robusta'. Planta 173:31-34

Savidge RA, Wareing PF (1981) Plant-growth regulators and the differentiation of vascular elements. In: Barnett JR (ed) Xylem cell development. Castle House, Tunbridge Wells (Kent), pp 192-235

Schaffner M (1906) The embryology of the shepherd's purse. Ohio Nat 7:1-8

Schaffstein G (1932) Untersuchungen an ungegliederten Milchröhren. Beih Bot Centralbl Abt 1 49:197-220

Schmid R (1976) The elusive cambium — another terminological contribution. IAWA Bull 1976 (4):51-59

Schmidt A (1924) Histologische Studien an phanerogamen Vegetationspunkten. Bot Arch 8:345-404

Schmidt HW, Schönherr J (1982) Fine structure of isolated and non-isolated potato tuber periderm. Planta 154:76-80

Schneider B, Herth W (1986) Distribution of plasma membrane rosettes and kinetics of cellulose formation in xylem development of higher plants. Protoplasma 131:142-152

Schneider H (1955) Ontogeny of lemon tree bark. Am J Bot 42:893-905

Schneider-Orelli O (1909) Die Miniergänge von *Lyonetia clerkella* und die Stoffwanderung in Apfelblättern. Zentralbl Bakteriol Parasitenk Abt 2, 24:158-181 + 2 pl

Schnepf E (1969) Physiologie des Protoplasmas: Sekretion und Exkretion bei Pflanzen. Protoplasmatologia VIII 8:1-181

Schnepf E (1974) Gland cells. In: Robards AW (ed) Dynamic aspects of plant ultrastructure. McGraw-Hill, London, pp 331-357

Schnepf E (1984) Pre- and postmitotic reorientation of microtubule arrays in young *Sphagnum* leaflets: Transitional stages and initiation sites. Protoplasma 120:100-112

Schnepf E, Witte O, Rudolph U, Deichgräber G, Reiss H-D (1985) Tip cell growth and the frequency and distribution of particle rosettes in the plasmalemma: Experimental studies in *Funaria* protonema cells. Protoplasma 127:222-229

Scholander PF, Hemmingsen E, Garey W (1961) Cohesive lift of sap in the rattan vine. Science 134:1835-1838

Scholz F, Bauch J (1973) Anatomische und physiologische Untersuchungen zur Wasserbewegung in Kiefernnadeln. Planta 109:105-119

Schönherr J (1976) Water permeability of cuticular membranes. In: Lange OL, Kappen L, Schulze E-D (eds) Water and plant life: Problems and modern approaches. Springer, Berlin Heidelberg New York, pp 148-159

Schönherr J, Huber H (1977) Plant cuticles are polyelectrolytes with isoelectric points around three. Plant Physiol 59:145-150

Schoonraad E, Schijff HP van der (1974) Anatomy of leaves of the genus *Podocarpus* in South Africa. Phytomorphology 24:75-85

Schopf JM (1943) The embryology of *Larix*. Ill Biol Monogr 19(4):1-97

Schoute JC (1913) Beiträge zur Blattstellungslehre. I. Die Theorie. Trav Bot Néerl 10:153-324 + 2 pl

Schramm R (1912) Über die anatomischen Jugendformen der Blätter einheimischer Holzpflanzen. Flora Allg Bot Ztg 104:225-295 + 3 pl

Schroeder JI, Hedrich R (1989) Involvement of ion channels and active transport in osmoregulation and signaling of higher plant cells. Trends Biochem Sci 14:187-192

Schüepp O (1926) Meristeme. Handb d Pflanzenanat, Abt 1 Teil 2 Band IV. Borntraeger, Berlin

Schüepp O (1966) Wachstum und Formbildung in den Teilungsgeweben höherer Pflanzen. Birkhäuser, Basel

Schulman E (1945) Root growth-rings and chronology. Tree-Ring Bull 12:2-5

Schulte PJ, Gibson AC, Nobel PS (1989) Waterflow in vessels with simple or compound perforation plates. Ann Bot 64:171-178

Schulz P, Jensen WA (1977) Cotton embryogenesis: The early development of the free nuclear endosperm. Am J Bot 64:384-394

Schulz R, Jensen WA (1968) *Capsella* embryogenesis: The egg, zygote, and young embryo. Am J Bot 55:807-819

Schumacher W (1967) Die Fernleitung der Stoffe im Pflanzenkörper. Enclycl Plant Physiol 1st Ser 13:61-177

Schwabe WW (1971) Chemical modification of phyllotaxis and its implications. Symp Soc Exp Biol (Control mechanisms of growth and differentiation) 25:301-322

Schwabe WW (1984) Phyllotaxis. In: Barlow PB, Carr DJ (eds) Positional controls in plant development. Cambridge Univ Press, Cambridge, pp 403-440

Schwabe WW, Clewer AG (1984) Phyllotaxis — A simple computer model based on the theory of a polarly-translocated inhibitor. J Theor Biol 109:595-619

Schwendener S (1874) Das mechanische Princip im anatomischen Bau der Monocotylen mit vergleichenden Ausblicken auf die übrigen Pflanzenklassen. Engelmann, Leipzig

Scurfield G (1973) Reaction wood: Its structure and function. Science 179:647-655

Seagull RW (1983) Differences in the frequency and disposition of plasmodesmata resulting from root cell elongation. Planta 159:497-504

Seagull RW (1989) The cytoskeleton. Crit Rev Plant Sci 8:131-167

Selker JML, Sievers A (1987) Analysis of extension and curvature during the graviresponse in *Lepidium* roots. Am J Bot 74:1863-1871

Selvendran RR (1985) Developments in the chemistry and biochemistry of pectic and hemicellulosic polymers. J Cell Sci Suppl 2:51-88

Seth MK, Veena, Agrawal HO (1989) Ring width variation around the circumference in blue pine roots. Phytomorphology 39:165-168

Shah JJ, Unnikrishnan K (1969) The shoot apex and the ontogeny of axillary buds in *Cuminum cyminum* L. Aust J Bot 17:241-253

Shapiro S (1958) The role of light in the growth of root primordia in the stem of the Lombardy poplar. In: Thimann KV (ed) The physiology of forest trees. Ronald Press, New York, pp 445-465

Sharma M, Sharma KC (1987) The ontogeny of apical organization and dormancy breaking in the roots of *Polyalthia longifolia* Benth. & Hook. f. Beitr Biol Pflanz 62:405-414

Shimakura M (1936) On the expansion of bast cells in conifers. Bot Mag Tokyo 50:318-323

Shomer-Ilan A, Beer S, Waisel Y (1975) *Suaeda monoica*, a C_4 plant without typical bundle sheaths. Plant Physiol 56:676-679

Sibaoka T (1966) Action potentials in plant organs. Symp Soc Exp Biol 20:49-73

Siegler EA, Bowman JJ (1939) Anatomical studies of root and shoot primordia in 1-year apple roots. J Agric Res 58:795-803

Sievers A, Behrens HH, Buckhout TJ, Gradmann D (1984) Can a Ca^{2+} pump in the endoplasmic reticulum of the *Lepidium* root be the trigger for rapid changes in membrane potential after gravistimulation? Z Pflanzenphysiol 114:195-200

Sievers A, Buchen B, Volkmann D, Hejnowicz Z (1991) Role of the cytoskeleton in gravity perception. In: Lloyd CW (ed) The cytoskeletal basis of plant growth and form. Academic Press, London, pp 169-182

Sifton HB (1945) Air-space tissue in plants. I. Bot Rev 11:108-143

Sifton HB (1957) Air-space tissue in plants. II. Bot Rev 23:303-312

Sigafoos RS (1964) Botanical evidence of floods and flood-plain deposition. US Geol Surv Prof Pap 485-A:1-35

Silk, WK (1984) Quantitative descriptions of development. Annu Rev Plant Physiol 35:479-518

Simoncioli C (1974) Ultrastructural characteristics of *Diplotaxis erucoides* (L.) DC. Suspensor. G Bot Ital 108:175-189

Singh AP (1987) Fine structure of hydrolysed primary walls in tracheary elements of petiolar xylem in *Eucalyptus delegatensis*. Ann Bot 60:315-319

Singh AP, Mogensen HL (1975) Fine structure of the zygote and early embryo in *Quercus gambelii*. Am J Bot 62:105-115

Singh B (1943) The origin and destribution of inter- and intraxylary phloem in *Leptadenia*. Proc Indian Acad Sci 18B:14-19

Singh H (1978) Embryology of gymnosperms. Encycl Plant Anatomy, Band X/T 2. Borntraeger, Berlin

Sinnott EW, Bloch R (1939) Changes in intercellular relationships during the growth and differentiation of living plant tissues. Am J Bot 26:625-634

Sinnott EW, Bloch R (1941) Division in vacuolate plant cells. Am J Bot 28:225-232

Sinyukhin AM, Britikov EA (1967) Action potentials in the reproductive system of plants. Nature 215:1278-1280

Sinyukhin AM, Gorchakov VV (1968) Role of stem conducting bundles in long distance transmission of stimulation by means of bioelectric pulses. Fiziol Rast 15:477-487 (in Russian); or Sov Plant Physiol 19:400-407 (in English)

Skene DS (1965) The development of kino veins in *Eucalyptus obliqua* L'Herit. Aust J Bot 13:367-378 + 3 pl

Skene DS (1969) A three-dimensional reconstruction of the wood of *Eucalyptus maculata* Hook. Holzforschung 23:33-37

Skene DS (1972) The kinetics of tracheid development in *Tsuga canadensis* Carr. and its relation to tree vigour. Ann Bot 36:179-187

Skene DS, Balodis V (1968) A study of vessel length in *Eucalyptus obliqua* L'Hérit. J Exp Bot 19:825-830

Skutch AF (1927) Anatomy of the leaf of banana, *Musa sapientum* L. var. hort. Gros Michel. Bot Gaz 84:337-391

Smith CW, Lew L-F (1970) Cellular arrangement in the node of various angiosperms. Bot Gaz 131:269-272

Smith DR, Thorpe TA (1975) Root initiation in cuttings of *Pinus radiata* seedlings. I. Developmental sequence. J Exp Bot 26:184-192

Smith FH (1958) Anatomical development of the hypocotyl of douglas-fir. For Sci 4:61-70

Smith GH (1934) Anatomy of the embryonic leaf. Am J Bot 21:194-209

Smithson E (1952) Development of winged cork in *Acer campestre* L. Proc Leeds Philos Lit Soc Sci Sect 6(2):97-103

Smithson E (1954) Development of winged cork in *Ulmus x Hollandica* Mill. Proc Leeds Philos Lit Soc Sci Sect 6(4):211-220

Snow M, Snow R (1931) Experiments on phyllotaxis. I. The effect of isolating a primordium. Philos Trans R Soc Lond B 221:1-43

Snow M, Snow R (1947) On the determination of leaves. New Phytol 46:5-19

Snow R (1965) The causes of bud eccentricity and the large divergence angles between leaves in Cucurbitaceae. Phil Trans Roy Soc Lond B 250:53-77

Soh WY (1972) Early ontogeny of vascular cambium. I. *Ginkgo biloba*. Bot Mag Tokyo 85:111-124

Soh WY (1974) Early ontogeny of vascular cambium. III. *Robinia pseudoacacia* and *Syringa oblata*. Bot Mag Tokyo 87:99-112

Soh WY (1990) Origin and development of cambial cells. In: Iqbal M (ed) The vascular cambium. Research Studies Press, Taunton (Somerset). Wiley, New York, pp 37-62

Sokolowa C (1890) Naissance de l'endosperme dans le sac embryonnaire de quelques gymnospermes. Bull Soc Impér Nat Moscov New Ser 4:446-497

Sonntag P (1887) Über Dauer des Scheitelwachsthums und Entwicklungsgeschichte des Blattes. Jahrb Wiss Bot 18:236-262 + 9 pl

Spanner DC (1974) The electro-osmotic theory. In: Aronoff S et al (eds) Phloem transport. NATO Advanced Study Inst Ser A-4. Plenum Press, New York, pp 563-584

Spanswick RM (1976) Symplasmic transport in tissues. Encycl Plant Physiol 2nd Ser 2(B): 35-53

Spatz H-C, Speck Th, Vogellehner D (1990) Contributions to the biomechanics of plants. II. Stability against local buckling in hollow plant stems. Bot Acta 103:123-130

Speck Th, Spatz H-C, Vogellehner D (1990) Contributions to the biomechanics of plants. I. Stabilities of plant stems with strengthening elements of different cross-sections against weight and wind forces. Bot Acta 103:111-122

Sperry JS, Donnelly JR, Tyree MT (1988) Seasonal occurrence of xylem embolism in sugar maple (*Acer saccharum*). Am J Bot 75:1212-1218.

Spilatro SR, Mahlberg PG (1986) Latex and laticifer starch content of developing leaves of *Euphorbia pulcherrima*. Am J Bot 73:1312-1318

Spurr AR (1949) Histogenesis and organization of the embryo in *Pinus strobus* L. Am J Bot 36:629-641

Srivastava LM (1963) Secondary phloem in the Pinaceae. Univ Calif Publ Bot 36(1):1-142 (includes 35 pl)

Srivastava LM (1964) Anatomy, chemistry, and physiology of bark. Int Rev For Res 1:203277

Srivastava LM (1969) On the ultrastructure of cambium and its vascular derivatives. III. The secondary walls of the sieve elements of *Pinus strobus*. Am J Bot 56:354-361

Srivastava LM (1970) The secondary phloem of *Austrobaileya scandens*. Can J Bot 48:341-359

Srivastava LM (1974) Structure and differentiation of sieve elements in angiosperms and gymnosperms. In: Aronoff S et al (eds) Phloem transport. NATO Advanced Study Inst Ser A-4. Plenum Press, New York London, pp 33-62

Stafstrom JP, Staehelin LA (1988) Antibody localization of extensin in cell walls of carrot storage roots. Planta 174:321-332

Stahl E (1873) Entwickelungsgeschichte und Anatomie der Lenticellen. Bot Ztg 31:561-568, 577-585, 593-601, 609-617 + 1 pl

Steeves TA, Hicks MA, Naylor JM, Rennie P (1969) Analytical studies on the shoot apex of *Helianthus annuus*. Can J Bot 47:1367-1375

Steeves TA, Sussex IM (1989) Patterns in plant development, 2nd edn. Cambridge Univ Press, Cambridge

Steingraeber DA, Fisher JB (1986) Indeterminate growth of leaves in *Guarea* (Meliaceae): A twig analogue. Am J Bot 73:852-862

Steudle E, Jeschke WD (1983) Water transport in barley roots. Planta 158:237-248

Stevens RA, Martin ES (1978) A new ontogenetic classification of stomatal types. Bot J Linn Soc 77:53-64

Stevenson DW, Fisher JB (1980) The developmental relationship between primary and secondary thickening growth in *Cordyline* (Agavaceae). Bot Gaz 141:264-268

Stewart RN (1978) Ontogeny of the primary body in chimeral forms of higher plants. In: Subtelny S, Sussex IM (eds) The clonal basis of development. Academic Press, New York, pp 131-160

Stewart RN, Burk LG (1970) Independence of tissues derived from apical layers in ontogeny of the tobacco leaf and ovary. Am J Bot 57:1010-1016

Stewart RN, Dermen H (1970) Determination of number and mitotic activity of shoot apical initial cells by analysis of mericlinal chimeras. Am J Bot 57:816-826

Stewart RN, Dermen H (1975) Flexibility in ontogeny as shown by the contribution of the shoot apical layers to leaves of periclinal chimeras. Am J Bot 62:935-947

Stewart RN, Dermen H (1979) Ontogeny in monocotyledons as revealed by studies of the developmental anatomy of periclinal chloroplast chimeras. Am J Bot 66:47-58

Stewart RN, Meyer FG, Dermen H (1972) *Camellia* + "Daisy Eagleson", a graft chimera of *Camellia sasanqua* and *C. japonica*. Am J Bot 59:515-524

Stewart RN, Semeniuk P, Dermen H (1974) Competition and accommodation between apical layers and their derivatives in the ontogeny of chimeral shoots of *Pelargonium x hortorum*. Am J Bot 61:64-67

Strasburger E (1872) Die Coniferen und die Gnetaceen. Abel, Leipzig (442 pp + separate atlas of 24 pl)

Strasburger E (1891) Histologische Beiträge. III. Ueber den Bau und die Verrichtungen der Leitungsbahnen in den Pflanzen. Fischer, Jena

Sucoff E (1969) Freezing of conifer xylem and the cohesion-tension theory. Physiol Plant 22:424-431

Sugiyama M, Hara N (1988) Comparative study on early ontogeny of compound leaves in Lardizabalaceae. Am J Bot 75:1598-1605

Sunderland N (1960) Cell division and expansion in the growth of the leaf. J Exp Bot 11:68-80

Süss H, Lengert W, Müller-Stoll WR (1973) Vegetationsperiodische Änderungen der Gefäss- und Holzfaserweite sowie der Länge der Markstrahlzellen bei verschiedenen Laub- und Nadel-holzarten. Holztechnologie 19:235-242

Sussex IM (1952) Regeneration of the potato shoot apex. Nature 170:755-757

Sussex IM, Steeves TA (1967) Apical initials and the concept of promeristem. Phyto-morphology 17:387-391

Swain T (1977) Secondary compounds as protective agents. Annu Rev Plant Physiol 28: 479-501

Swamy BGL (1949) Further contributions to the morphology of the Degeneriaceae. J Arnold Arbor Harv Univ 30:10-38 + 4 pl

Swamy GL (1979) Embryogenesis in *Cheirostylis flabellata*. Phytomorphology 29:199-203

Taiz L (1984) Plant cell expansion: Regulation of cell wall mechanical properties. Annu Rev Plant Physiol 35:585-657

Taiz L, Jones RL (1973) Plasmodesmata and an associated cell wall component in barley aleurone tissue. Am J Bot 60:67-75

Tang SH (1948) Observations on the embryogeny of *Juniperus chinensis*. Bot Bull Acad Sinica 2:13-18

Tangl E (1879) Ueber offene Communicationen zwischen den Zellen des Endosperms einiger Samen. Jahrb Wiss Bot 12:170-190 + 3 pl

Taubert H (1956) Über den Infektionsvorgang und die Entwicklung der Knöllchen bei *Alnus glutinosa* Gaertn. Planta 48:135-156

Taylor DW, Hickey LJ (1990) An Aptian plant with attached leaves and flowers: Implications for angiosperm origin. Science 247:702-704

Terry BR, Robards AW (1987) Hydrodynamic radius alone governs the mobility of molecules through plasmodesmata. Planta 171:145-157

Tetley U (1936) Tissue differentiation in some foliage leaves. Ann Bot 50:523-557 + 2 pl

Thoday D, Davey AJ (1932) Contractile roots. II. On the mechanism of root-contraction in *Oxalis incarnata*. Ann Bot 1st Ser 46:993-1005 + 1 pl

Thomas RL, Cannell MGR (1980) The generative spiral in phyllotaxis theory. Ann Bot 45:237-249

Thompson DW (1942) On growth and form. Cambridge Univ Press, Cambridge; MacMillan, New York

Thompson RG, Fensom DS, Anderson RR, Drovin R, Leiper W (1979) Translocation of ^{11}C from leaves of *Helianthus, Heracleum, Nymphoides, Ipomoea, Tropaeolum, Zea, Fraxinus, Ulmus, Picea*, and *Pinus*: Comparative shapes and some fine structure profiles. Can J Bot 57:845-863

Thomson (Sir) W (Lord Kelvin) (1887) On the division of space with minimum partitional area. Philos Mag J Sci 5th Ser 24:503-514

Thomson WW, Berry WL, and Liu LL (1969) Localization and secretion of salt by the salt glands of *Tamarix aphylla*. Proc Natl Acad Sci USA 63:310-317

Thorsch J, Esau K (1982) Microtubules in differentiating sieve elements of *Gossypium hir-sutum*. J Ultrastruct Res 78:73-83

Thurston EL (1974) Morphology, fine structure, and ontogeny of the stinging emergence of *Urtica dioica*. Am J Bot 61:809-817

Thurston EL (1976) Morphology, fine structure and ontogeny of the stinging emergence of *Tragia ramosa* and *T. saxicola* (Euphorbiaceae). Am J Bot 63:710-718

Timell TE (1980) Formation of compression wood in balsam fir (*Abies balsamea*). III. Ultrastructure of the differentiating phloem. Holzforschung 34:5-10

Timell TE (1986) Compression wood in gymnosperms, 3 vols. Springer, Berlin Heidelberg New York

Ting IP (1982) Plant physiology. Addison-Wesley, Reading (Massachusetts)

Ting IP (1985) Crassulacean acid metabolism. Annu Rev Plant Physiol 36:595-622

Ting IP, Lord EM, Sternberg LSL, DeNiro MJ (1985) Crassulacean acid metabolism in the strangler *Clausia rosea* Jacq. Science 229:969-971

Tippett JT (1986) Formation and fate of kino veins in *Eucalyptus* L'Hérit. IAWA Bull 7:137-143

Tjepkema JD (1983) Hemoglobins in the nitrogen-fixing root nodules of actinorhizal plants. Can J Bot 61:2924-2929

Tjepkema JD, Yocum CS (1974) Measurement of oxygen partial pressure within soybean nodules by oxygen microelectrodes. Planta 119:351-360

Tomlinson PB (1961) Palmae. Vol II of: Metcalfe CR (ed) Anatomy of the monocotyledons. Clarendon Press, Oxford

Tomlinson PB (1974) Development of the stomatal complex as a taxonomic character in monocotyledons. Taxon 23:109-128

Tomlinson PB, Esler AE (1973) Establishment growth in woody monocotyledons native to New Zealand. NZ J Bot 11:627-644

Tomlinson PB, Zimmermann MH (1969) Vascular anatomy of monocotyledons with secondary growth — an introduction. J Arnold Arbor Harv Univ 50:159-179

Torrey JG (1955) On the determination of vascular patterns during tissue differentiation in excised pea roots. Am J Bot 42:183-197

Torrey JG (1976) Initiation and development of root nodules of *Casuarina* (Casuarinaceae). Am J Bot 63:335-344

Torrey JG (1978) Nitrogen fixation by actinomycete-nodulated angiosperms BioScience 28:586-592

Torrey JG, Callaham D (1978) Determinate development of nodule roots in actinomycete-induced root nodules of *Myrica gale*. Can J Bot 56:1357-1364

Torrey JG, Callaham D (1979) Early nodule development in *Myrica gale*. Bot Gaz 140: S10-S14 (Suppl)

Torrey JG, Wallace WD (1975) Further studies on primary vascular tissue pattern formation in roots. In: Torrey JG, Clarkson DT (eds) The development and function of roots. Academic Press, London, pp 91-103

Trachtenberg S, Fahn A (1981) The mucilage cells of *Opuntia ficus-indica* (L.) Mill. — Development, ultrastructure, and mucilage secretion. Bot Gaz 142:206-213

Trachtenberg S, Mayer AM (1982) Biophysical properties of *Opuntia ficus-indica* mucilage. Phytochemistry 21:2835-2843

Trinick MJ (1979) Structure of nitrogen-fixing nodules formed by *Rhizobium* on roots of *Parasponia andersonii* Planch. Can J Microbiol 25:565-578

Tucker CM, Evert RF (1969) Seasonal development of the secondary phloem in *Acer negundo*. Am J Bot 56:275-284

Tucker SC (1963) Development and phyllotaxis of the vegetative axillary bud of *Michelia fuscata*. Am J Bot 50:661-668

Tucker SC (1964) The terminal idioblasts in magnoliaceous leaves. Am J Bot 51:1051-1062

Tukey HB Jr (1970) The leaching of substances from plants. Annu Rev Plant Physiol 21:305-324

Tukey HB Jr, Tukey HB, Wittwer SH (1958) Loss of nutrients by foliar leaching as determined by radioisotopes. Proc Am Soc Hort Sci 71:496-506

Turgeon R, Webb JA (1973) Leaf development and phloem transport in *Cucurbita pepo*: Transition from import to export. Planta 113:179-191

Turgeon R, Webb JA, Evert RF (1975) Ultrastructure of minor veins in *Curcurbita pepo* leaves. Protoplasma 83:217-232

Turing AM (1952) The chemical basis of morphogenesis. Philos Trans R Soc Lond B 237: 37-72

Turner GW (1986) Comparative development of secretory cavities in the tribes Amorpheae and Psoraleae (Leguminosae:Papilionoideae). Am J Bot 73:1178-1192

Turrell FM (1936) The area of the internal exposed surface of dicotyledon leaves. Am J Bot 23:255-264

Tyree MT (1970) The symplast concept: A general theory of symplastic transport according to the thermodynamics of irreversible processes. J Theor Biol 26:181-214

Tyree MT, Dixon MA (1986) Water stress induced cavitation and embolism in some woody plants. Physiol Plant 66:397-405

Tyree MT, Ewers FW (1991) The hydraulic architecture of trees and other woody plants. New Phytol 119:345-360

Tyree MT, Sperry JS (1989) Vulnerability of exlem to cavitation and embolism. Annu Rev Plant Physiol Plant Mol Biol 40:19-38

Ugural AC, Fenster S (1975) Advanced strength and applied elasticity. American Elsevier, New York

Uphof JCT, Hummel K, Staesche K (1962) Plant hairs (by Uphof): Die Verbreitung der Haartypen in den Natürlichen Verwandtschaftsgruppen (by Hummel and Staesche). Encyl Plant Anat IV(5). Borntraeger, Berlin

Van Fleet DS (1950) The cell forms, and their common substance reactions, in the paren-chyma-vascular boundary. Bull Torrey Bot Club 77:340-351

Van Fleet DS (1961) Histochemistry and function of the endodermis. Bot Rev 27:165-220

Vasiliyev AE (1969) Submicroscopic morphology of nectary cells and problems of nectar secretion. Bot Zh 54:1015-1031 (in Russian, English summary)

Veen AH, Lindenmayer A (1977) Diffusion mechanism for phyllotaxis. Plant Physiol 60: 127-139

Venugopal N, Krishnamurthy KV (1987) Seasonal production of secondary phloem in the twigs of certain tropical timber trees. Ann Bot 60:61-67

Venverloo CJ, Hovenkamp PH, Weeda AJ, Libbenga KR (1980) Cell division in Nautilocalyx explants. I. Phragmosome, preprophase band and plane of division. Z Pflanzenphysiol 100:161-174

Vian B (1982) Organized microfibril assembly in higher plant cells. In: Brown RM Jr (ed) Cellulose and other natural polymer systems. Plenum Press, New York, pp 23-43

Vian B, Mosiniak M, Reis D, Roland J-C (1982) Dissipative process and experimental retardation of the twisting in the growing plant cell wall. Effect of ethylene-generating agent and colchicine: A morphogenetic revaluation. Biol Cell 46:301-310

Vian B, Reis D, Mosiniak M, Roland J-C (1986) The glucuronoxylans and the helicoidal shift in cellulose microfibrils in linden wood: Cytochemistry in muro and on isolated molecules. Protoplasma 131:185-199

Vietmeyer ND (1984) The lost crops of the Incas. Ceres (FAO) 17(3)(99):37-40

Vietmeyer ND (1986) Lesser-known plants of potential use in agriculture and forestry. Science 232:1379-1384

Vischer W (1923) Über die Konstanz anatomischer und physiologischer Eigenschaften von Hevea brasiliensis Müller Arg. (Euphorbiaceae). Verh Naturforsch Ges Basel 35:174-185

Vité JP (1967) Water conduction and spiral grain: Causes and effects. Proc XIV IUFRO Kongress München, 1967, Vol IX Sect 41 + WG 22/41:338-351

Vliet GJCM (1976) Radial vessels in rays. IAWA Bull 1976 (3):35-37

Volkmann D (1981) Structural differentiation of membranes involved in the secretion of polysaccharide slime by root cap cells of cress (Lepidium sativum L.). Planta 151:180-188

Volkmann D, Sievers A (1979) Graviperception in multicellular organs. Encycl Plant Physiol 2nd Ser 7:573-600

Wainwright CM (1977) Sun-tracking and related leaf movements in a desert lupine (Lupinus arizonicus). Am J Bot 64:1032-1041

Wainwright SA (1970) Design in hydraulic organisms. Naturwissenschaften 57:321-326

Wainwright SA, Biggs WD, Currey JD, Gosline JM (1976) Mechanical design in organisms. Arnold, London

Waisel Y, Liphschitz N (1975) Sites of phellogen initiation. Bot Gaz 136:146-150

Waisel Y, Noah I, Fahn A (1966) Cambial activity in *Eucalyptus camaldulensis* Dehn. II. The production of phloem and xylem elements. New Phytol 65:319-324

Walbot V (1985) On the life strategies of plants and animals. Trends Genet 1:165-169

Walker WS (1960) The effects of mechanical stimulation and etiolation on the collenchyma of *Datura stramonium*. Am J Bot 47:717-724

Walles B, Nyman B, Aldén T (1973) On the ultrastructure of needles of *Pinus silvestris* L. Stud For Suec 106:1-26

Wangermann E (1967) The effect of the leaf on differentiation of primary xylem in the internode of *Coleus blumei* Benth. New Phytol 66:747-754 + 1 pl

Ward M, Wetmore RH (1954) Experimental control of development in the embryo of the fern, *Phlebodium aureum*. Am J Bot 41:428-434

Wardlaw CW (1957) The reactivity of the apical meristem as ascertained by cytological and other techniques. New Phytol 56:221-229

Wardlaw CW (1965) The organization of the shoot apex. Encycl Plant Physiol 1st Ser 15(1): 966-1076

Wardlaw CW (1970) Cellular differentiation in plants and other essays. Barnes and Noble, New York; Manchester Univ Press, Manchester

Wareing PF (1951) Growth studies of woody species. IV. The initiation of cambial activity in ring-porous species. Physiol Plant 4:546-562

Wareing PF (1976) Origin of cell heterogeneity in plants. In: Graham CF, Wareing PF (eds) The developmental biology of plants and animals. Blackwell, Oxford, pp 29-42

Wareing PF, Al-Chalabi T (1985) Determination in plant cells. Biol Plant 27:241-248

Waring RH, Running SW (1978) Sapwood water storage: its contribution to transpiration and effect upon water conductance through the stems of old-growth Douglas-fir. Plant Cell Environ 1:131-140

Warmbrodt RD (1985) Studies on the root of *Hordeum vulgare* L. Ultrastructure of the seminal root with special reference to the phloem. Am J Bot 72 414-432

Warmbrodt RD, Eschrich W (1985a) Studies on the mycorrhizas of *Pinus sylvestris* L. produced in vitro with the basidiomycete *Suillus variegatus* (Sw. Ex Fr.) O. Kuntze. I. Ultrastructure of the mycorrhizal rootlets. New Phytol 100:215-223

Warmbrodt RD, Eschrich W (1985b) Studies on the mycorrhizas of *Pinus sylvestris* L. produced in vitro with the basidiomycete *Suillus variegatus* (Sw. Ex Fr.) O. Kuntze. II. Ultrastructural aspects of the endodermis and vascular cylinder of the mycorrhizal rootlets. New Phytol 100:403-418

Warren Wilson J, Warren Wilson PM (1961) The position of regenerating cambia — A new hypothesis. New Phytol 60:63-73

Warren Wilson J, Warren Wilson PM (1984) Control of tissue patterns in normal development and in regeneration. In: Barlow PW, Carr DJ (eds) Positional controls in plant development. Cambridge Univ Press, Cambridge, pp 225-280

Warrington SJ, Black HD, Coons LB (1981) Entry of *Pisolithus tinctorius* hyphae into *Pinus taeda* roots. Can J Bot 59:2135-2139

Waterkeyn L (1981) Cytochemical localization and function of the 3-linked glucan callose in the developing cotton fibre cell wall. Protoplasma 106:49-67

Wattendorff J (1974) The formation of cork cells in the periderm of *Acacia senegal* Willd. and their ultrastructure during suberin deposition. Z Pflanzenphysiol 72:119-134

Webb D (1982) Effects of light on root growth, nodulation, and apogeotropism of *Zamia pumila* L. seedlings in sterile culture. Am J Bot 69:298-305

504   References

Webb DT (1983) Developmental anatomy of light-induced root nodulation by *Zamia pumila* L. seedlings in sterile culture. Am J Bot 70:1109-1117

Webb DT, Slone JH (1987) Anatomy of *Macrozamia communis* lateral roots and root nodules formed in vitro studied with light and scanning electron microscopy. Am J Bot 74: 1625-1634

Webb J, Jackson MB (1986) A transmission and cryo-scanning electron microscopy study of the formation of aerenchyma (cortical gas-filled space) in adventitious roots of rice (*Oryza sativa*). J Exp Bot 37:832-841

Weber HC (1980) Untersuchungen an australischen und neuseeländischen Loranthaceae/- Viscaceae. 1. Zur Morphologie und Anatomie der unterirdischen Organe von *Nuytsia floribunda* (Labill.) R. Br. Beitr Biol Pflanz 55:77-99

Weber HC (1987) Evolution of the secondary haustoria to a primary haustorium in the parasitic Scrophulariaceae/Orobanchaceae. Plant Syst Evol 156:127-131

Webster PL, Langenauer HD (1973) Experimental control of the activity of the quiescent centre in excised root tips of *Zea mays*. Planta 112:91-100

Wenham MW, Cusick F (1975) The growth of secondary wood fibres. New Phytol 74:247-261

Went JL van, Willemse MTM (1984) Fertilization. In: Johri BM (ed) Embryology of angiosperms. Springer, Berlin Heidelberg New York, pp 273-317

Werker E (1970) The secretory cells of *Pinus halepensis* Mill. Isr J Bot 19:542-557

Werker E, Fahn A (1969) Resin ducts of *Pinus halepensis* Mill. — their structure, development and pattern of arrangement. Bot J Linn Soc 62:379-411 + 11 pl

Westing AH (1965) Formation and function of compression wood in gymnosperms. Bot Rev 31:381-480

Weston GD, Cass DD (1973) Observations on the development of the paraveinal mesophyll of soybean leaves. Bot Gaz 134:332-335

Wetmore RH (1926) Organization and significance of lenticels in dicotyledons. I. Lenticels in relation to aggregate and compound storage rays in woody stems. Lenticels and roots. Bot Gaz 82:71-88 + 2 pl

Wetmore RH, Garrison R (1966) The morphological ontogeny of the leafy shoot. In: Cutter EG (ed) Trends in plant morphogenesis. Wiley, New York, 187-199

Wetzel S, Greenwood JS (1991) A survey of seasonal bark proteins in eight temperate hardwoods. Trees 5:153-157

White J (1908) Formation of red wood in conifers. Proc R Soc Victoria [Australia] 20(2):107-124

Whitehead RA, Chapman GP (1962) Twinning and haploidy in *Cocos nucifera* Linn. Nature 195:1228-1229

Whitmore TC (1962a) Studies in systematic bark morphology. I. Bark morphology in Dipterocarpaceae. New Phytol 61:191-207

Whitmore TC (1962b) Studies in systematic bark morphology. II. General features of bark construction in Dipterocarpaceae. New Phytol 61:208-220

Whitmore TC (1963) Studies in systematic bark morphology. IV. The bark of beech, oak and sweet chestnut. New Phytol 62:161-169

Whyte LL (1969) Structural hierarchies: A challenging class of physical and biological problems. In: Whyte LL, Wilson AG, Wilson D (eds) Hierarchical structures. American Elsevier, New York, pp 3-16

Wick SM (1991) The preprophase band. In: Giddings TH Jr, Staehelin LA (eds) The cytoskeletal basis of plant growth and form. Academic Press, London, pp 231-244

Wilcox H (1954) Primary organization of active and dormant roots of noble fir, *Abies procera*. Am J Bot 41:812-821

Wilcox H (1962) Growth studies of the root of incense cedar, *Libocedrus decurrens*. I. The origin and development of primary tissues. Am J Bot 49:221-236

Wilcox H (1964) Xylem in roots of *Pinus resinosa* Ait. in relation to heterorhizy and growth activity. In: Zimmermann MH (ed) The formation of wood in forest trees. Academic Press, New York and London, pp 459-478

Wilcox HE (1968) Morphological studies of the roots of red pine, *Pinus resinosa*. II. Fungal colonization of roots and the development of mycorrhizae. Am J Bot 55:688-700

Wilhelm K (1880) Beiträge zur Kenntniss des Siebröhrenapparates dicotyler Pflanzen. Engelmann, Leipzig

Willemse MTM (1974) Megagametogenesis and formation of neocytoplasm in *Pinus sylvestris* L. In: Linskens HF (ed) Fertilization in higher plants. North-Holland, Amsterdam, pp 97-102

Willemse MTM, Went JL van (1984) The female gametophyte. In: Johri BM (ed) Embryology of angiosperms, Springer, Berlin Heidelberg New York, pp 159-196

Williams BC (1947) The structure of the meristematic root tip and origin of the primary tissues in the roots of vascular plants. Am J Bot 34:455-462

Williams EG, Knox RB, Kaul V, Rouse JL (1984) Post-pollination callose development in ovules of *Rhododendron* and *Ledum* (Ericaceae): Zygote special wall. J Cell Sci 69:127-135

Williams RF (1975) The shoot apex and leaf growth: A study in quantitative biology. Cambridge Univ Press, London

Willmer CM (1983) Stomata. Longman, London

Willmer CM, Sexton R (1979) Stomata and plasmodesmata. Protoplasma 100:113-124

Wilson BF (1964a) A model for cell production by the cambium of conifers. In: Zimmermann MH (ed) The formation of wood in forest trees. Academic Press, New York, pp 19-36

Wilson BF (1964b) Structure and growth of woody roots of *Acer rubrum* L. Harv For Pap 11:1-13

Wilson BF (1966) Mitotic activity in the cambial zone of *Pinus strobus*. Am J Bot 53:364-372

Wilson BF (1975) Distribution of secondary thickening in tree root systems. In: Torrey JG, Clarkson DT (eds) The development and function of roots. Academic Press, London, pp 197-219

Wilson BF, Archer RR (1977) Reaction wood: Induction and mechanical action. Annu Rev Plant Physiol 28:23-43

Wilson BF, Archer RR (1979) Tree design: Some biological solutions to mechanical problems. BioScience 29:293-298

Wilson BF, Horsley SB (1970) Ontogenetic analysis of tree roots in *Acer rubrum* and *Betula papyrifera*. Am J Bot 57:161-164

Wilson BF, Wodzicki TJ, Zahner R (1966) Differentiation of cambial derivatives: Proposed terminology. For Sci 12:438-440

Wilson KJ, Nessler CL, Mahlberg PG (1976) Pectinase in *Asclepias* latex and its possible role in laticifer growth and development. Am J Bot 63:1140-1144

Winfree AT (1980) The geometry of biological time. Springer, Berlin Heidelberg New York

Wittmann W, Bergersen FJ, Kennedy GS (1965) The coralloid roots of *Macrozamia communis* L. Johnson. Aust J Biol Sci 18:1129-1134 + 3 pl

Włoch W (1975) Longitudinal shrinkage of compression wood in dependence on water content and cell wall structure. Acta Soc Bot Pol 44:217-229

Włoch W (1981) Nonparallelism of cambium cells in neighboring rows. Acta Soc Bot Pol 50:625-636

Włoch W, Bilczewska E (1987) Fibrillation of events in the cambial domains of *Tilia cordata* Mill. Acta Soc Bot Pol 56:19-35

Wodzicki TJ (1971) Mechanism of xylem differentiation in *Pinus sylvestris* L. J Exp Bot 22:670-687

Wodzicki TJ, Brown CL (1973) Cellular differentiation of the cambium in the Pinaceae. Bot Gaz 134:139-143

Wodzicki TJ, Humphreys WJ (1973) Maturing pine tracheids: Organization of intravacuolar cytoplasm. J Cell Biol 56:263-265

Wodzicki TJ, Witkowska L (1961) On the photoperiodic control of extension growth and wood formation in Norway spruce (*Picea abies* (L.) Karst.). Acta Soc Bot Pol 30:755-764

Wodzicki TJ, Wodzicki AB (1981) Modulation of the oscillatory system involved in polar transport of auxin by other phytohormones. Physiol Plant 53:176-180

Wodzicki TJ, Wodzicki AB, Zajączkowski (1979) Hormonal modulation of the oscillatory system involved in polar transport of auxin. Physiol Plant 46:97-100

Wolpert L (1981) Positional information and pattern formation. Philos Trans R Soc Lond B 295:441-450

Wooding FBP (1966) The development of the sieve elements of *Pinus pinea* Planta 69:230-243

Woodward FI (1987) Stomatal numbers are sensitive to increases in CO_2 from pre-industrial levels. Nature 327:617-618

Worsdell WC (1897) On "transfusion-tissue": Its origin and function in the leaves of gymnospermous plants. Trans Linn Soc Lond Bot 2nd Ser 5:301-319 + 4 pl

Wright M, Osborne DJ (1977) Gravity-regulation of cell elongation in nodes of the grass *Echinochloa colonum*. Biochem Physiol Pflanz 171:479-492

Wu H, Sharpe PJH, Spence RD (1985) Stomatal mechanics. III. Geometric interpretation of the mechanical advantage. Plant Cell Environ 8:269-274

Wutz A (1955) Anatomische Untersuchungen über System und periodische Veränderungen der Lenticellen. Bot Stud 4:43-72 + 11 pl

Wyk AE van (1985) The genus *Eugenia* (Myrtaceae) in southern Africa: Structure and taxonomic value of bark. S Afr J Bot 51:157-180

Wyk P van (1974) Trees of the Kruger National Park, vol 2. Purnell, Cape Town

Wylie RB (1939) Relations between tissue organization and vein distribution in dicotyledon leaves. Am J Bot 26:219-225

Wylie RB (1946) Relations between tissue organization and vascularization in leaves of certain tropical and subtropical dicotyledons. Am J Bot 33:721-726

Wylie RB (1947) Conduction in dicotyledon leaves. Proc Iowa Acad Sci 53:195-202

Wylie RB (1951) Principles of foliar organization shown by sun-shade leaves from ten species of deciduous dicotyledonous trees. Am J Bot 38:355-361

Wylie RB (1952) The bundle sheath extension in leaves of dicotyledons. Am J Bot 39:645-651

Yakovlev MS (1969) Embryogenesis and some problems of phylogenesis. Rev Cytol Biol Vég 32:325-330

Yamamoto F, Kozlowski TT (1987) Effect of ethrel on growth and stem anatomy of *Pinus halepensis* seedlings. IAWA Bull 8:11-19

Yarrow G L, Popham R A (1981) The ontogeny of the primary thickening meristem of *Atriplex hortensis* L. (Chenopodiaceae). Am J Bot 68:1042-1049

Yawney WJ, Schultz RC (1990) Anatomy of a vesicular-arbuscular endomycorrhizal symbiosis between sugar maple (*Acer saccharum* Marsh) and *Glomus etunicatum* Becker & Gerdemann. New Phytol 114:47-57

Yeung EC (1980) Embryogeny of phaseolus: The role of the suspensor. Z Pflanzenphysiol 96:17-28

Yeung EC, Sussex IM (1979) Embryogeny of *Phaseolus coccineus*: The suspensor and the growth of the embryo-proper in vitro. Z Pflanzenphysiol 91:423-433

Yin HC (1941) Studies on the nyctinastic movement of the leaves of *Carica papaya*. Am J Bot 28:250-261

Young DA (1978) On the diffusion theory of phyllotaxis. J Theor Biol 71:421-432

Zagórska-Marek B (1984) Pseudotransverse divisions and intrusive elongation of fusiform initials in the storeyed cambium of *Tilia*. Can J Bot 62:20-27

Zagórska-Marek B (1987) Phyllotaxis triangular unit; phyllotactic transitions as the consequences of the apical wedge disclinations in a crystal-like pattern of the units. Acta Soc Bot Pol 56:229-255

Zahur MS (1959) Comparative study of secondary phloem of 423 species of woody dicotyledons belonging to 85 families. Cornell Univ Agric Exp Sta NY State Coll Agric Mem 358:1-160

Zajączkowski S, Wodzicki TJ (1978) Auxin and plant morphogenesis — A model of regulation. Acta Soc Bot Pol 47:233-243

Zajączkowski S, Wodzicki TJ, Romberger JA (1984) Auxin waves and plant morphogenesis. Encycl Plant Physiol 2nd Ser 10:244-262

Zak B (1971) Characterization and classification of mycorrhizae of Douglas fir. II. *Pseudotsuga menziesii* + *Rhizopogon vinicolor*. Can J Bot 49:1079-1084

Zakrzewski J (1983) Hormonal control of cambial activity and vessel differentiation in *Quercus robur*. Physiol Plant 57:537-542

Zawadzki T, Trebacz K (1982) Action potentials in *Lupinus angustifolius* L. shoots. VI. Propagation of action potential in the stem after the application of mechanical block. J Exp Bot 33:100-110

Zeeuw C de (1941) Influence of exposure on the time of deep cork formation in three northeastern trees. New York St Coll For Syracuse Univ Bull, Tech Publ 56:1-10

Ziegenspeck H (1928) Zur Theorie der Bewegungs und Wachstumserscheinungen bei Pflanzen. Bot Arch 21:449-647

Ziegler H (1964) Storage, mobilization and distribution of reserve material in trees. In: Zimmermann MH (ed) The formation of wood in forest trees. Academic Press, New York, pp 303-320

Ziegler H (1975) Nature of transported substances. Encycl Plant Physiol (2nd ser) 1:59-100

Ziegler H (1987) The evolution of stomata. In: Zeiger E, Farquhar GD, Cowan IR (eds) Stomatal function. Stanford Univ Press, Stanford, pp 29-57

Zimmermann MH (1978) Hydraulic architecture of some diffuse-porous trees. Can J Bot 56:2286-2295

Zimmermann MH (1983) Xylem structure and the ascent of sap. Springer, Berlin Heidelberg New York

Zimmermann MH, Brown CL (1971) Trees — structure and function. Springer, Berlin Heidelberg New York

Zimmermann MH, Jeje AA (1981) Vessel-length distribution in stems of some American woody plants. Can J Bot 59:1818-1892

Zimmermann MH, Potter D (1982) Vessel-length distribution in branches, stem, and roots of *Acer rubrum* L. IAWA Bull 3:103-108

Zimmermann MH, Tomlinson PB (1966) Analysis of complex vascular systems in plants: Optical shuttle method. Science 152:72-73

Zimmermann MH, Wardrop AB, Tomlinson PB (1968) Tension wood in aerial roots of *Ficus benjamina* L. Wood Sci Tech 2:95-104

Zimmermann MH, McCue KF, Sperry JS (1982) Anatomy of the palm *Rhapis excelsa*. VIII. Vessel network and vessel-length distribution in the stem. J Arnold Arbor Harv Univ 63:83-95

Zürcher E, Kucera L, Bosshard HH (1985) Bildung und Morphologie der Thyllen: Eine Literaturübersicht. Vierteljahrsschr Naturforsch Ges Zür 130(3):311-333

Subject Index

*Italic page numbers indicate illustrations.